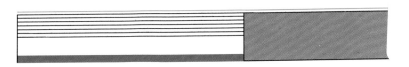

Exploring the Dynamic Universe

An Introduction to Astronomy

The great supernova of 1987 *(right)* in the Large Magellanic Cloud. At left is the Tarantula nebula, site of active star formation. *(National Optical Astronomy Observatories)*

Exploring the Dynamic Universe

An Introduction to Astronomy

Theodore P. Snow
University of Colorado at Boulder

West Publishing Company
St. Paul   New York   San Francisco   Los Angeles

Composition: The Clarinda Company
Copyediting: Pamela S. McMurry
Text and cover design: Janet Bollow
Illustrations: Henry Taly Design; House of Graphics
Cover photo: Supernova 1987A. National Optical Astronomy Observatories.

Copyright © 1988 By WEST PUBLISHING COMPANY
50 W. Kellogg Boulevard
P.O. Box 64526
St. Paul, MN 55164–1003

All rights reserved

Printed in the United States of America

Library of Congress Cataloging-in-Publication Data

Snow, Theodore P. (Theodore Peck)
  Exploring the dynamic universe.

  Bibliography: p.
  Includes index.
  1. Astronomy. I. Title.
QB43.2.S664   1988   520   87-31598
ISBN 0-314-64211-0

To
my teachers
and
my students,
who have taught
me so much.

Contents

Preface	00

Section I

The Nighttime Sky and The Tools of Astronomy — 1

Chapter 1
The Essence of Astronomy — 3

What is Astronomy?	4
A Typical Night Outdoors	5
Astronomical Insight 1.1:	
What is an Astronomer?	6
Astronomical Insight 1.2:	
Astronomy and Astrology	8
A View from Earth	9
From the Earth to the Universe: The Scale of Things	12
Perspective	14
Summary	14
Review Questions	15
Additional Readings	15

Chapter 2
Learning About the Nighttime Sky — 16

The Rhythms of the Cosmos	17
Daily Motions	17
Astronomical Insight 2.1: Celestial Navigation	20
Annual Motions: The Seasons	21
Lunar Motions	23
Planetary Motions	26
Eclipses	28
Historical Developments	29
The Earliest Astronomy	29
Plato and Aristotle	31
Astronomical Insight 2.2:	
The Mythology of the Constellations	32
The Later Greeks: Sophisticated Cosmologies	33
Perspective	36
Summary	36
Review Questions	36
Additional Readings	37

Chapter 3
The Renaissance — 38

Copernicus: The Heliocentric View Revisited	39
Tycho Brahe: Advanced Observations	40

viii   Contents

Johannes Kepler and the Law of Planetary Motion	41
Galileo, Experimental Physics, and the Telescope	44
Astronomical Insight 3.1:	
An Excerpt from *Dialogue on the Two Chief World Systems*	47
Isaac Newton	48
The Laws of Motion	49
Astronomical Insight 3.2,	
The Forces of Nature	50
The Law of Gravitation	52
Energy, Angular Momentum, and Orbits: Kepler's Laws Revisited	54
Astronomical Insight 3.3:	
Relativity	56
Tidal Forces	58
Perspective	59
Summary	59
Review Questions	60
Additional Readings	60

Chapter 4
Messages from the Cosmos: Light and Telescopes   62

The Electromagnetic Spectrum	63
Astronomical Insight 4.1:	
The Perception of Light	64
Continuous Radiation	67
The Atom and Spectral Lines	70
Deriving Information from Spectra	72
Telescopes: Tools for Collecting Light	75
The Need for Telescopes	75
Principles of Telescope Design	77
Major Observatories for All Wavelengths	82
Astronomical Insight 4.2:	
A Night at the Observatory	83
Perspective	135
Summary	136
Review Questions	136
Additional Readings	137

Section II

The Stars	89

Chapter 5
The Sun   90

Basic Properties and Internal Structure	91
Nuclear Reactions	94
Astronomical Insight 5.1: Imitating the Sun	95
Astronomical Insight 5.2:	
WIMPs From the Sun?	97
Structure of the Solar Atmosphere	98
Astronomical Insight 5.3:	
Measuring the Sun's Pulse	99
The Solar Wind	105
Sun Spots, Solar Activity Cycles, and the Magnetic Field	107
Perspective	112
Summary	112
Review Questions	112
Additional Readings	113

Chapter 6
Observations and Basic
Properties of Stars   114

Three Ways of Looking at It: Positions,
    Magnitudes, and Spectra   115
    Positional Astronomy   115
    Stellar Brightnesses   116
    Astronomical Insight 6.1:
        Star Names and Catalogs   117
    Measurements of Stellar Spectra   120
    Astronomical Insight 6.2:
        Stellar Spectroscopy and the Harvard Women 122
Binary Stars   125
Fundamental Stellar Properties   128
    Absolute Magnitudes and Stellar Luminosities   129
    Stellar Temperatures   129
    *The Hertzsprung-Russell Diagram and a New
        Distance Technique*   129
    Stellar Diameters   132
    Binary Stars and Stellar Masses   133
    Other Properties   134
Perspective   135
Summary   136
Review Questions   136
Additional Readings   137

Chapter 7
The Structure and Evolution of Stars   138

Observations of Stellar Evolution   139
    Stellar Structure   141
    The Role of Stellar Mass   142
The Source of Energy: Nuclear Reactions   143
    Astronomical Insight 7.1:
        Stellar Models and Reality   144
Mass Loss and Mass Exchange   147
Star Formation   149
Life Stories of Stars   153
    The Evolution of Stars like the Sun   154
The Middleweights   159
The Heavyweights   162
    Astronomical Insight 7.2:
        The Mysteries of Algol   164
Evolution in Binary Systems   165
Perspective   167

Summary   167
Review Questions   167
Additional Readings   168

Special Insert
The Great Supernova of 1987   169

Chapter 8
Stellar Remnants   175

White Dwarfs, Black Dwarfs   176
    White Dwarfs, Novae, and Supernovae   177
    Astronomical Insight 8.1: The Story of Sirius B   178
Supernova Remnants   180
Neutron Stars   182
    Pulsars: Cosmic Clocks   183
    Neutron Stars in Binary Systems   185
Black Holes: Gravity's Final Victory   188
    Do Black Holes Exist?   190
Perspective   192
Summary   192
Review Questions   192
Additional Readings   193

Section III

The Milky Way — 195

Chapter 9
Structure and Organization of the Milky Way — 196

Variable Stars as Distance Indicators	197
The Structure of the Galaxy and the Location of the Sun	200
Galactic Rotation and Stellar Motions	202
The Mass of the Galaxy	202
The Interstellar Medium	203
Astronomical Insight 9.1:	
The Violent Interstellar Medium	206
Spiral Structure and the 21-Centimeter Line	209
Astronomical Insight 9.2:	
21-Centimeter Emission from Hydrogen	210
The Galactic Center: Where the Action Is	211
Globular Clusters Revisited	213
A Massive Halo?	214
Perspective	215
Summary	216
Review Questions	216
Additional Readings	217

Chapter 10
The Formation and Evolution of the Galaxy — 218

Stellar Populations and Elemental Gradients	219
Stellar Cycles and Chemical Enrichment	221
The Care and Feeding of Spiral Arms	222
Astronomical Insight 10.1:	
A Different Kind of Spiral Arm	224
Galactic History	226
Perspective	227
Astronomical Insight 10.2:	
The Missing Population III	228
Summary	227
Review Questions	229
Additional Readings	229

Section IV

Extragalactic Astronomy — 231

Chapter 11
Galaxies Upon Galaxies — 232

The Hubble Classification System	233
Measuring the Properties of Galaxies	237
The Origins of Spirals and Ellipticals	243
Clusters of Galaxies	243
The Local Group	244
Rich Clusters: Dominant Ellipticals and Galactic Mergers	248
Cluster Masses	250
Superclusters	251
The Origins of Clusters	252
Astronomical Insight 11.1:	
Holes in the Universe	253
Perspective	255
Summary	255
Review Questions	256
Additional Readings	256

Chapter 12
Universal Expansion and the Cosmic Background — 258

Hubble's Great Discovery	259
Hubble's Constant and the Age of the Universe	262

Redshifts as Yardsticks	264	Perspective	307
Astronomical Insight 12.1:		Summary	307
The Distance Pyramid	265	Review Questions	307
A Cosmic Artifact: The Microwave Background	266	Additional Readings	308
The Crucial Question of the Spectrum	267		
Isotropy and Daily Variations	268		
Perspective	269		
Summary	270		
Review Questions	270		
Additional Readings	271		

Section V

The Solar System — 309

Chapter 13
Peculiar Galaxies, Explosive Nuclei, and Quasars — 272

Chapter 15
Origins of the Solar System — 310

Overall Properties of the Planetary System	311
Early Theories of Solar System Formation	312
A Modern Scenario	318
Astronomical Insight 15.1: The Explosive History of Solar System Elements	321
Are There Other Solar Systems?	324
Perspective	325
Summary	326
Review Questions	326
Additional Readings	326

The Radio Galaxies	273
Seyfert Galaxies and Explosive Nuclei	278
The Discovery of Quasars	279
The Origin of the Redshifts	280
Astronomical Insight 13.1: The Redshift Controversy	281
The Properties of Quasars	283
Galaxies in Infancy?	285
Astronomical Insight 13.2: The Double Quasar and Gravitational Lenses	287
Perspective	289
Summary	289
Review Questions	290
Additional Readings	290

Chapter 16
The Earth and Its Companion — 327

The Earth's Atmosphere	328
The Earth's Interior and Magnetic Field	330
A Crust in Action	334

Chapter 14
Cosmology: Past, Present, and Future of the Universe — 292

Underlying Assumptions	293
Einstein's Relativity: Mathematical Description of the Universe	294
Astronomical Insight 14.1: The Mystery of the Nighttime Sky	295
Open or Closed: The Observational Evidence	298
Total Mass Content	298
The Deceleration of the Expansion	300
Astronomical Insight 14.2: The Inflationary Universe	301
The History of Everything	303
What Next?	304
Astronomical Insight 14.3: Particle Physics and Cosmology	306

Astronomical Insight 16.1:
    The Ages of Rock   336
Exploring the Moon   337
    Astronomical Insight 16.2: Lunar Geography   339
    A Battle-Worn Surface and A Dormant Interior   341
The Development of the Earth-Moon System   345
    Formation of the Earth and Its Atmosphere   345
    Origin of the Moon   345
    History of the Moon   347
Perspective   347
Summary   347
Review Questions   348
Additional Readings   348

Chapter 17
The Terrestrial Planets   350

Observations of the Terrestrial Planets   351
General Properties   355
    Astronomical Insight 17.1:
        Visiting Interior Planets   358
Atmospheres of the Terrestrial Planets   358
    Composition of the Terrestrial Atmospheres   359
    Circulation and Seasonal Effects   361
Surfaces and Interiors   364
Evolution of the Terrestrial Planets   369
Perspective   371
Summary   371
Review Questions   371
Additional Readings   372

Chapter 18
The Outer Planets   373

Discovery and Observation   374
    Astronomical Insight 18.1:
        Close Encounters of a Distant Kind   377
General Processes   379
Atmospheres and Interiors of the Gaseous Giants   381
    Atmospheric Composition   382
    Atmospheric Circulation   382
    Interior Conditions   385
Magnetic Fields and Particle Belts   387
Rings and Moons   389
    Satellites of the Giant Planets   389
    Ring Systems   396
    Astronomical Insight 18.2:
        Resolving the Rings   399
Pluto, Planetary Misfit   404
Perspective   405
Summary   406
Review Questions   406
Additional Readings   407

Chapter 19
Space Debris   408

The Minor Planets   409
    Astronomical Insight 19.1:
        Trojans, Apollos, and Target Earth   411
    Kirkwood's Gaps: Orbital Resonances Revisited   412
    The Origin of Asteroids   412
Comets: Messengers from the Past   413
    Halley, Oort, and Cometary Orbits   414
    The Anatomy of a Comet   417
    Astronomical Insight 19.2:
        Encounters with Halley's Comet   418
Meteors and Meteorites   421
    Primordial Leftovers   422
    Dead Comets and Fractured Asteroids   423
Microscopic Particles: Interplanetary Dust and the Interstellar Wind   424
Perspective   426
Summary   426
Review Questions   428
Additional Readings   428

Contents xiii

Section VI

Life in the Universe 429

Chapter 20
The Chances of Companionship 431

Life on Earth 432
  Astronomical Insight 20.1:
    Panspermia Revisited 434
Could Life Develop Elsewhere? 437
The Probability of Detection 438
  Astronomical Insight 20.2:
    The Case for a Small Value of N 441
The Strategy for Searching 442
Perspective 443
Summary 444
Review Questions 445
Additional Readings 445

Appendices

Appendix 1	Symbols Commonly Used in This Text	A1
Appendix 2	Physical and Mathematical Constants	A1
Appendix 3	The Elements and Their Abundances	A2
Appendix 4	Temperature Scales	A3
Appendix 5	Radiation Laws	A3
Appendix 6	Major Telescopes of the World (two meters or larger)	A5
Appendix 7	Planetary and Satellite Data	A6
Appendix 8	Stellar Data	A9
Appendix 9	The Constellations	A11
Appendix 10	Mathematical Treatment of Stellar Magnitudes	A13
Appendix 11	Nuclear Reactions in Stars	A14
Appendix 12	Detected Interstellar Molecules	A16
Appendix 13	Clusters of Galaxies	A17
Appendix 14	The Relativistic Doppler Effect	A18
Appendix 15	The Messier Catalog	A19

Glossary G1

Index I1

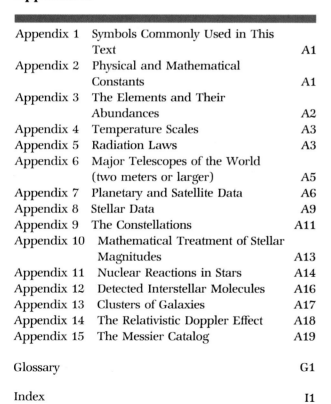

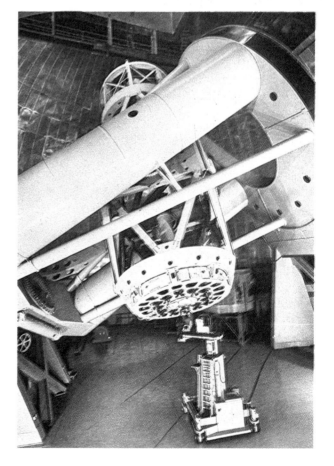

Preface

To study astronomy is, in a sense, the most human thing we can do. What distinguishes us from lower creatures, if not our curiosity, our compulsion to explore and discover? And what exemplifies this compulsion better than the study of the universe?

We probe the heavens (and the Earth) by all possible means, and we do it for no other reason than to learn whatever there is to be known. Astronomy has produced many useful byproducts, or course, and could be (and often is) justified solely on that basis. That is not the real reason for astronomy, however.

This textbook represents an attempt by an astronomer to share both the knowledge and the intellectual gratification of our science. There is considerable beauty in the universe for the eye and mind to behold. Just as it is visually stimulating to gaze at a great glowing nebula or a colorful moon, it is pleasing to the intellect to grasp a new understanding of one of the grand themes of the cosmos. It is hoped that the reader of this book will gain by doing both.

This textbook is intended for the student who has not chosen science as his or her major area of study, but who needs an appreciation of science as a vital aspect of preparation for a career. It is as important for such a student to gain some perspective on the general nature of science as it is to learn a great deal of specific information about a particular discipline in the sciences. For that reason, this text stresses the philosophy and outlook of the scientist as well as the knowledge we have gathered about the physical universe we live in.

It is probably as important for the student to understand how we know what we know as it is to understand what we know. In this era of instantaneous communication and universal access to information, we need more than ever to be able to discriminate among competing hypotheses, to be able to judge the reasonableness of ideas that are advanced. This text in astronomy is written with the underlying theme that to know the workings of science is one of the most important tools we have for meeting the challenges of our technological society.

This book, as its predecessors *The Dynamic Universe* and *Essentials of the Dynamic Universe*, places the emphasis at all times on *how* we learn about the cosmos; on the nature of scientific reason and the methods of scientific progress. The text presents a full overview of our current state of knowledge, while at the same time preparing the student for the changes in our understanding that will surely follow.

This book has been tailored to suit introductory astronomy courses in which the emphasis first is on stellar, galactic, and extragalactic astronomy, with the discussion of the solar system (except for the Sun) coming later. The book is brief enough to be easily covered in a one-term course.

We have made *Exploring the Dynamic Universe* current, by ensuring that the most recent discoveries have been included, such as the influx of new information on the outer planets gained from the *Voyager* flybys, the most recent discoveries on cold matter in the universe gleaned from the infrared maps obtained by the *IRAS* satellite; the news on the nature of comets brought back from the recent apparition of Halley; the many fascinating new results of deep extragalactic surveys and of modern cosmological theory; and, perhaps most spectacular of all, extensive coverage of the great supernova of 1987. We have gone to great lengths to ensure that this most significant event is covered extensively, and that the information is as current as possible. There is a special insert section on the supernova, updated as recently as December 1987, which describes not only what has happened so far, but also the expectations and key questions of astronomers as we wait to see what will be revealed as the infant remnant expands and dissipates.

This book includes many of the successful features of its predecessors. There are extensive tables of data throughout the text and in the appendices; there are Astronomical Insights within the chapters, providing additional anecdotal, historical, or technical information; and there are chapter summaries, review questions, and supplemental reading lists at the end of each chapter. All of these features are designed to enhance student understanding and, perhaps more importantly, student appreciation of the manner and methods of science.

The esthetic appeal of astronomy has been immeasurably enhanced by the full four-color format of the book; this allows the illustrations to be interspersed with the text throughout, rather than being relegated to a few color plate insert sections. This allows the student to appreciate not only the beauty, but also the relevence of astronomical photographs and diagrams, without the need to locate plates in a remote section of the book. Many of the drawings have been updated to use color or to clarify concepts previously less obvious.

The arrangement of the text is largely traditional, with an introductory section on the background of astronomy, both in history and in basic physics; a section, beginning with the Sun, on stars and their lives and deaths; a section on the structure and evolution of our galaxy; a set of chapters on extragalactic astronomy and the universe as a whole; a section on solar system objects and evolution; and a final, brief section on the possibilities that life may exist elsewhere. At the beginning of each of these sections is an introduction that leads the student into the material.

The section on the solar system is much shorter and more concise than is the case in either *The Dynamic Universe* or *Essentials of the Dynamic Universe,* and is written on a comparative planetology basis, instead of an object-by-object sequence. This section opens with a discussion of the overall properties of the solar system and our understanding of its origin, which is followed by a chapter on the Earth and the Moon, which serves not only to describe these two bodies but also to set the stage for comparisons with the other planets and satellite systems. The other three terrestrial planets are then discussed collectively in the next chapter, in the context of what has been learned about the Earth, and this is followed by a similar comparative treatment of the outer planets in the following chapter. The final chapter in this section discusses interplanetary bodies; asteroids, comets, meteoroids, and dust grains.

Supplemental materials for this text include a *Study Guide* authored by Catharine D. Garmany and myself (both University of Colorado), and an *Instructor's Manual* by Stephen J. Shawl (University of Kansas). The *Instructor's Manual* contains helpful discussions of strategies in teaching, provides a large number of exam questions (with answers), and gives complete answers to all the review questions from the main text. The *Study Guide,* intended to help the student get maximum benefit from the text, contains brief chapter summaries, lists of key words and phrases, self-tests, and complete bibliographies of articles on relevent topics, taken from a wide assortment of magazines and journals. In addition to the *Study Guide* and the *Instructor's Manual,* another aid to teaching is offered to large adopters of the text: a set of transparencies for use with overhead projectors, showing a number of useful diagrams and illustrations from the text.

At every step during the preparation of this text, vital assistance was provided by a number of people, whose help is acknowledged with gratitude (with apologies to anyone inadvertantly omitted). The most important guidance and support was provided by my wife, Connie; by the West Publishing Company editor, Denise Simon, and the production editor, Tad Bornhoft. Steve Shawl helped scrutinize the galley proofs, as did M. M. Allen, who also helped gather data for tables and reading lists. K. S. Bjorkman helped to prepare the *Study Guide.*

Among my colleagues at the University of Colorado and elsewhere, many contributed generously to this book, by reviewing portions of the text, by providing new data, and by allowing me the use of illustrations. Particularly generous in this connection were J. M. Shull, L. Esposito (both University of Colorado), and H. Eichhorn (University of Florida). Others in this category include J. Doggett, J. K. Malville, J. C. Brandt (all of the University of Colorado); U. Fink (University of Arizona); L. Frederick (University of Virginia); R. Dreiser (Yerkes Observatory); W. Golisch (NASA Infrared Telescope Facility); M. Phillips (Cerro Tololo Inter-American Observatory); D. Malin (Anglo-Australian Telescope); and C. Covault *(Aviation Week).* Reviewers of the manuscript were:

William Anderson, Phoenix College

Timothy Barker, Wheaton College

Howard Brooks, DePauw University

Neil Comins, University of Maine–Orono

Thomas Johnson, Ferris State College

John Kasher, University of Nebraska
Lawrence Mink, Arkansas State University
John Noble, Western Illinois University
Charles Sawicki, North Dakota State University
Horace Smith, Michigan State University

For all of these people, and to the students whose responses to my teaching philosophies have also helped to shape this book, I am grateful. With their continued input, I trust that this book will continue to evolve, as does our understanding of the dynamic universe.

Exploring
the Dynamic Universe

An Introduction to Astronomy

Section I

The Nighttime Sky and The Tools of Astronomy

We begin our study of astronomy with a discussion of the nature of astronomy and science. Chapter 1 defines astronomy and provides an overview of the nighttime sky and the scale of the universe.

Chapter 2 begins with a description of the motions of celestial objects that can be seen by the unaided eye. It will provide us with an immediate understanding of many of the phenomena that can be seen and appreciated with only our eyes as observing equipment. Thus, armed with all the knowledge our ancestors had, we will see how the ancients fared as they sought to develop a successful picture of the cosmos and their place in it. We will concentrate our historical discussions on the civilizations that arose on the shores of the Mediterranean, for it was here that the foundations of modern astronomy were laid.

Chapter 3 discusses the major developments of the Renaissance, when fresh ideas arose in astronomy, as in all forms of human endeavor. We will learn to appreciate the awesome breakthroughs made by such giants as Copernicus, Brahe, Kepler, Galileo, and Newton, who led the way toward a correct understanding of the universe and the place of our planet in it. We will then move on to the laws of physics that govern such diverse phenomena as planetary orbits, the motions of molecules in a gas, and the tides on the Earth and other celestial bodies.

We would know nothing of the external universe were it not for the light that reaches us from faraway objects, and Chapter 4 describes the nature of light and the way we decipher its messages. We will find that an amazing variety of information can be derived from the spectra of objects like planets and stars. Things once considered forever beyond our grasp are now routinely measured, and in this chapter we will learn how this is done. Chapter 4 includes a description of telescopes and their principles and how they are used to measure light in all portions of the spectrum.

Chapter 1

The Essence of Astronomy

Sunset at Kitt Peak National Observatory. (*National Optical Astronomy Observatories*)

Chapter Preview

What is Astronomy?
A Typical Night Outdoors
The View from Earth
From the Earth to the Universe: The Scale of Things

The oldest of all sciences is perhaps also the most beautiful. No artificial light show can rival the splendor of the heavens on a clear night, and few intellectual concepts can compare with the beauty of our modern understanding of the cosmos.

Today we study astronomy for a variety of technological and practical reasons, but no one loses sight of the underlying majesty, of the human instinct for intellectual satisfaction. To study astronomy is to ask the grandest questions possible, and to find hints at their answers is to satisfy one of humankind's most deeply ingrained yearnings.

In this text we explore astronomy in the modern context, which is highly technical and sophisticated, but we will endeavor to retain the sense of wonder and beauty that has motivated the science since the beginning. Although some may argue that astronomy is not a practical science, we will see that its origins are rooted in very practical requirements for methods of keeping time and maintaining calendars. In this chapter we begin our study by defining astronomy and introducing some simple terminology that will assist us in later chapters.

What is Astronomy?

Astronomy is the science in which we consider the entire universe as our subject. It is the science in which we derive the properties of celestial objects and from these properties deduce the laws by which the universe operates. It is the science of everything.

Technically we might say that astronomy is the science of everything except the earth, or that it is the study of everything beyond the earth's atmosphere, since the earth and its atmosphere fall into the purview of other disciplines such as geophysics or atmospheric science. We will find, however, that the study of astronomy necessarily includes an examination of the properties and evolution of the earth and its atmosphere.

In the modern sense astronomy is probably more aptly called **astrophysics.** Ever since the time of Sir Isaac Newton (the late seventeenth century), the universe has been explored by applying the laws of physics—most of them derived from earthly experiments and observations—to celestial phenomena. Other scientific disciplines enter into our discussions as well: To study the planets, for example, we must know something of geology and geophysics; to analyze molecules

Figure 1.1 An Ancient Astronomical Site. Stonehenge, a stone monument in England, was built according to astronomical alignments in prehistoric times. *(C. D. McLoughlin)*

in space, we must understand the principles of chemistry.

Astronomical observations were made and records kept at least as early as the time of the most ancient recorded history (Fig. 1.1), and we believe that the skies were studied and their cyclical motions pondered long before that. In the earliest times, astronomy had a practical motivation: Knowledge of motions in the heavens made it possible to predict and plan for certain significant events such as the changing of the seasons.

Along with the practical came the whimsical and the spiritual. In ancient times it was believed that events in the heavens exerted some influence over the lives of people on earth, and many early astronomers practiced what we now call astrology. Many a monarch retained the services of an astronomer, not only to foretell the seasons, but also to provide advice on strategies for war, love, politics, and business. In many cultures, religion and astronomy were intimately linked, so astronomers attained the importance that major religious leaders have today.

The rich and diverse Greek civilization that flourished for many centuries before and around the time of Christ developed several sciences, including astronomy, beyond the mystical and spiritual. The Greeks had many preconceived notions about the nature of the universe, but they also made many rational advances in understanding the heavens, setting the stage for the development of modern science many centuries later. Following the era known as the Dark Ages, astronomy (like many other disciplines) experienced the pangs of rebirth during the Renaissance, becoming a rational and methodical science. Battles were yet to be fought between religion and science, but the course of science was set. The work of pioneers such as Copernicus, Galileo, Kepler, and especially Newton placed astronomy on a firm, physical basis.

Modern astronomy still has a practical aspect, but few who pursue the science do so for that reason. Today we study astronomy primarily for the sake of expanding our knowledge; it is research of the purest sort. Even so, we derive many practical and concrete benefits from astronomy; witness, for example, the many technological spin-offs of the space program or the multitude of physical and chemical processes have been discovered through astronomical observation.

No matter how analytical we may be in modern astronomy, however, we never lose sight of the same basic human feelings that inspired our ancestors. Modern astronomers may work with large telescopes and a variety of complex electronic instruments (Fig. 1.2), but we still treasure the moments spent outside, simply watching the skies with the same tools as the ancients.

A Typical Night Outdoors

Imagine that we are sitting outdoors on a fine, clear night, far from city lights or other distractions. This is easy to do and highly recommended; by simply getting out into the countryside we can see many of the beautiful objects that we will be studying in more detail in this text.

Figure 1.2 A Large Modern Telescope. This large reflecting telescope at the Cerro Tololo Inter-American Observatory, located in Chile and operated by the U.S. government, has a diameter of 4 meters, or roughly 13 feet. *National Optical Astronomy Observatories)*

Astronomical Insight 1.1

What Is An Astronomer?

Throughout this text we will be referring to astronomers, scientists, astrophysicists, and physicists. In a book that endeavors to summarize all that we know about the universe and its contents, we can hardly omit a description of the people who devote their time to developing this knowledge.

There are thousands in the United States alone who study astronomy as either a vocation or an avocation. Representative of the latter are the amateur astronomers who engage in a variety of astronomical activities, often on their own but in many cases through local and even nationwide organizations. Telescope making, astrophotography, long-term monitoring of variable stars, public programming, and just plain stargazing are included. If you wish to join such a group, to get advice on buying a telescope, or to learn techniques such as photography of celestial objects, your best bet is to get in touch with an amateur astronomy group. Local clubs exist in most major cities, and regional associations are everywhere. These groups may be difficult to find in the telephone book, but a telescope shop or planetarium is bound to know whom to contact.

The amateur astronomers in the United States outnumber the professionals. There are some 3500 members of the American Astronomical Society, the principal professional astronomy organization in the United States. As a rule these people fall into a limited number of categories: those who do research and teach at colleges and universities; those who do research work at government-sponsored institutions such as the national observatories and federal agencies like NASA; and those who perform research and related engineering functions in private industry, most often with companies involved in aerospace activities.

About eight percent of the members of the American Astronomical Society are women. This is certainly a low percentage, but a recent study indicates that there is little discrimination in the hiring of professional astronomers. The small number of women in astronomy simply reflects the low number who choose to enter the field at the graduate level, so it is

---

Assuming that the Moon is not in one of its brightest phases, the most obvious objects in the sky are the stars. They appear in profusion, scattered across the heavens in a wide range of brightnesses and subtle colors. They twinkle, giving the appearance of vitality. Here and there we may see concentrations of stars in a cluster (Fig. 1.3), or possibly a dimly glowing gas cloud (Fig. 1.4).

If it is a moonless night, we see a broad, diffuse band of light across the sky. This is the Milky Way, our own galaxy of stars, seen edgewise from an interior position (Fig. 1.5). The Milky Way consists of countless stars intermixed with patchy clouds of interstellar gas and dust.

We may also see a few bright, steady objects that do not appear to twinkle. These are the planets, and as many as five of them may be visible on a given night, distributed along a great arc through the sky. Careful observation over several nights will reveal that the planets are all moving gradually along this arc, changing their positions relative to the fixed stars.

The most prominent object in the nighttime sky is usually the Moon, which shines so brightly when near full that it drowns out the light of all but the brightest stars and planets (Fig. 1.6). The Moon, about one-fourth the diameter of the earth but some 250 thousand miles (or about 60 times the radius of the Earth) away, presents various appearances to us, depending on how

Astronomical Insight 1.1

What Is An Astronomer?

at that level that improvement must be made.

Most of the funding for research in astronomy, even for those not working directly for federal labs and observatories, comes from the government. A significant function of an astronomer on a university faculty is to write proposals, usually to the National Science Foundation or to NASA, for support of proposed research programs.

The terms *astronomer* and *astrophysicist* have come to mean pretty much the same thing, although historically there was a difference. An astronomer studied the skies, gathering data but doing relatively little interpretation; an astrophysicist was primarily interested in understanding the physical nature of the universe, and therefore carried out comprehensive analyses of astronomical data or did theoretical work, in both cases applying the laws of physics to phenomena in the heavens. Nearly all modern astronomers do astrophysics to varying degrees, however, ranging from the observational astronomer at one end of the spectrum to the pure theorist at the other. The two terms for people engaged in these pursuits are therefore used interchangeably today. Many modern astronomers call on the fields of engineering (for instrument development), chemistry (in studying planetary and stellar atmospheres and the interstellar medium), geophysics (in probing interior conditions in planets and other solid bodies), and sometimes even biology, but always with an underlying foundation in physics.

If you want to become a professional astronomer, you should be aware from the outset that the field is small and job opportunities are both limited and often subject to the vagaries of federal funding. If you persist, the best course is to study physics, at least through the undergraduate level, and then plan to attend graduate school in astronomy or physics. (The latter option may make you a bit more versatile and is never a handicap when entering astronomy later.) On the bright side, demographic studies have indicated a possible shortage of astronomers for a period beginning in the late 1980s and extending through the rest of this century. Perhaps the timing will be right for you.

Figure 1.3 A Cluster of Stars. This group of stars is held together by mutual gravitational attraction. This cluster, called the Pleiades, is easily visible to the unaided eye in the autumn and winter. *(Palomar Observatory, California Institute of Technology)*

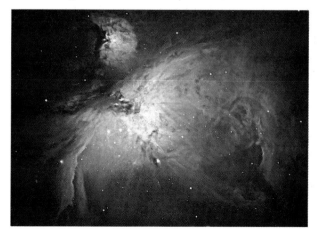

Figure 1.4 A Gaseous Nebula. Clouds of gas and tiny dust particles such as this are the birthplaces of stars. *(© 1981 Anglo-Australian Telescope Board)*

Astronomical Insight 1.2

Astronomy and Astrology

There is an unfortunate tendency in modern society to confuse astrology and astronomy or, worse yet, to consider one a legitimate alternative to the other.

Astrology, the pseudoscience based on the belief that human lives are influenced by the configurations of heavenly bodies, arose at a primitive stage in the development of humankind when the earth was thought to be a flat disk under the dome of the heavens. While ancient Greek astronomers did much to raise the study of the heavens to a scientific level, during the subsequent Dark Ages astrology and the governance of human lives by the stars became once again the primary basis for study of the heavens. This situation changed radically during and after the Renaissance, when scientists began to untangle the true nature of the heavens and the motions of astronomical objects. It is unfortunately true, however, that some people continue to profess a faith in astrology.

One of the basic lessons we have learned about the universe is that it is easy to make mistakes unless we are careful to be objective and accept only those conclusions that can be verified by repeated observations or experiments or by making predictions that can be tested. Astrology fails to meet these criteria. Serious attempts have been made to test astrological lore by statistical analysis of people born under different signs. No trace of a correlation has ever been found.

There are many phenomena that defy the understanding of modern science, and many of them deserve more attention than they are getting. Astrology is not one of them. As a phenomenon, it is worthy of study only by the sociological and psychological sciences, for its effects do not exist in the physical universe. It may be interesting party talk to compare astrological signs, but it would be well to keep in mind the difference between objective reality and subjective impressions. Failure to do so is failure to understand what science is.

---

much of its sunlit portion we see. The Moon is always to be found somewhere along the same east-west strip of the sky, called the **ecliptic,** where the planets travel.

Occasionally we may see brief flashes or trails of light known as **meteors.** These "shooting stars" can be spectacular events, particularly when they arrive with great frequency, as they do during a meteor shower.

Comets are occasional visitors to our sky (Fig. 1.7), as the recent passage of Halley's comet has reminded us. Perhaps once or twice a year a comet is found that is bright enough to be seen with the unaided eye. These largely gaseous bodies orbit the sun, as do the Earth and other planets, but they travel in very elongated paths that bring them close enough to the Sun to heat up and glow visibly only for brief periods of days or weeks. Some of the more spectacular, bright comets were interpreted in ancient times as harbingers of catastrophe.

There is a complex pattern of motions in the sky. Some of it is evident to an alert watcher in an hour or so; other components require careful observations over hours, days, or weeks. The most obvious motion is the steady rotation of the entire sky; objects rise in the east and set in the west, as a reflection of the earth's rotation. (The terms **rotation** and **revolution** are sometimes interchanged, but here *rotation* means the same thing as spin, whereas *revolution* is usually taken to mean orbital motion, such as the motion of the Earth

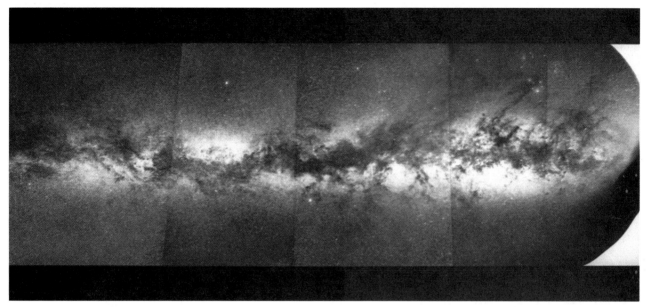

Figure 1.5 The Milky Way This composite photograph shows the hazy band of light across the sky known as the Milky Way. It is a cross-sectional view of the disk of our galaxy, which contains roughly 100 billion stars. *(Mt. Wilson and Las Campanas Observatories, Carnegie Institution of Washington)*

around the Sun or of the Moon around the Earth.) Another motion that can be discerned readily is that of the Moon with respect to the stars. As it orbits the Earth, the Moon moves a distance in the sky equal to its own apparent diameter in just one hour, so it is possible to see its position change with respect to the background stars in a short time. Other cyclical motions, such as those of the planets as they gradually travel along their orbits about the sun, require more patience and care to discover. It is noteworthy, however, that ancient astronomers noticed many of the cycles in the motions of heavenly bodies, some of them quite subtle. That they did so is a testimony to the care and diligence they applied to their studies of the skies.

The View From Earth

When we look at the sky we do not see it in three dimensions, because there are no obvious clues to tell us the distances to the objects we see. This fact led long ago to the concept of the **celestial sphere** (Fig. 1.8), in which the stars and other objects in the sky are said

Figure 1.6 The Moon. Astronauts on one of the *Apollo* missions made this photograph from space. *(NASA)*

Figure 1.7 A Comet. This is a view of Comet Kohoutek during its 1973 passage through the inner solar system. *(NASA)*

to lie on the surface of a sphere centered on the Earth. Although we no longer think of this as literal truth, it is still a convenient device for discussing and visualizing the heavens.

Positions of objects on the celestial sphere are measured in angular units, because lacking knowledge of distances to objects, we have no easy way to determine their actual separations in true distance units such as miles or kilometers. We see only a two-dimesional view. Thus we may specify where a star is by saying how many degrees, minutes, and seconds it is from another star or from a reference direction.

We have noted that the stars rise and set with the daily rotation of the Earth. This gives us a natural basis for timekeeping, and our standard units of time are based on the Earth's rotation. The length of the day is equivalent to the rotational period of the Earth (a more specific definition is given in Chapter 2).

Anyone who has traveled from one hemisphere to the other may have noticed that the stars that are visible are not the same in the north as in the south (Fig. 1.9). The portion of the sky that we can see depends on our latitude, or distance in degrees north or south of the equator. For those of us living in the Northern Hemisphere, a large region of the southern sky is beyond our view. The constellations that we see vary as we travel north or south, a fact that was well known to early astronomers, who deduced from this and other evidence that the Earth is round. One consequence is that astronomers must have telescopes in both hemispheres in order to study the entire sky.

We usually cannot observe celestial objects during daylight, so our view of the heavens at any particular time is limited to the half of the sky that is seen at night. Because of the Earth's motion around the Sun, however, the part of the sky that is visible during the night gradually changes (Fig. 1.10). Therefore, any part of the sky can be observed at night at some time of the year; we need only wait until the appropriate time to observe a given object.

Chapter 1 The Essence of Astronomy   11

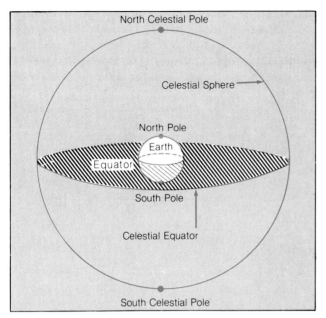

Figure 1.8 The Celestial Sphere. Because we see the sky in only two dimensions, it is useful and convenient to visualize it as a sphere centered on the earth, with the stars and other bodies set on the surface of the sphere. We measure positions of objects on the celestial sphere in angular units, because the actual distances are not directly known. (Throughout this text, we will learn about distance measurements in astronomy.)

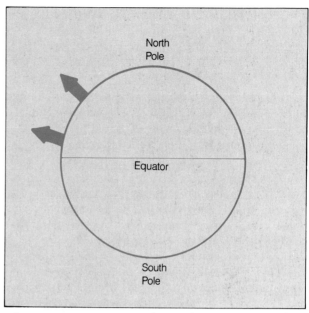

Figure 1.9 Latitude and Our View of the Sky. The portion of the sky that we see depends on where we are with respect to the earth's equator. The arrows indicate the overhead direction from two different latitudes.

Figure 1.10 The Changing View of the Sky With the Seasons. As the Earth orbits the Sun, the portion of the sky that we can see at night changes.

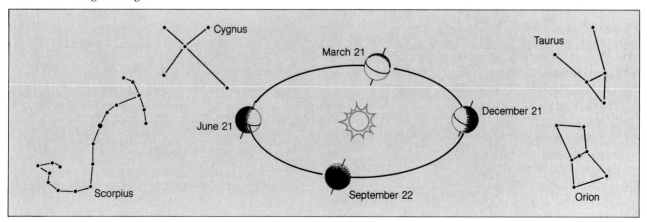

From the Earth to the Universe: The Scale of Things

We have been describing the appearance of the sky to the unaided eye, which has necessarily limited us to nearby objects such as the Sun, the Moon, the planets, and the stars in our part of the local galaxy. It is interesting now to expand our horizons and try to comprehend the scale of the universe beyond this "neighborhood." Some of the distance scales we discuss are listed in Table 1.1.

Even the nearest star is much farther from the Earth than any solar system object. If we take the Sun-Earth distance as a unit of measure (we call this the **astronomical unit** or **AU**), the nearest star is nearly 300 thousand of these units away from us. The most distant known planet, Pluto, is only about 40 AU from the sun, so we see that the stars are much more widely dispersed in space than the objects within our solar system.

Reducing the scale of the solar system may help us visualize the relative distances. For example, if we let the Earth be the size of a basketball (specifically, let us say that its diameter is one foot) and convert the rest of the solar system to the same scale, the Sun would be almost 110 feet in diameter and the distance from the Sun to the Earth would be over two miles. Pluto would be a tennis-ball roughly 90 miles away. The distance from the Sun to alpha Centauri, the nearest star, would be over 600,000 miles, or more than twice the real Earth-Moon distance!

Now consider our galaxy, the vast collection of stars to which the Sun belongs (Fig. 1.11). The Milky Way galaxy contains roughly 100 billion stars, arranged in a huge disklike structure having a diameter of about 100,000 light-years. (A light-year, the distance light travels in a year at its speed of 186,000 miles per second, is equal to about 6 trillion miles.) The distance to alpha Centauri is about 4 light-years, and the most distant stars easily seen with the unaided eye are several hundred light-years away (the majority of the brightest

Figure 1.11 A Galaxy Similar to the Milky Way. We cannot obtain an exterior view of our own galaxy, but it is thought that this one resembles ours. Note that most of the stars lie in a circular disk. *(© 1980 Anglo-Australian Telescope Board)*

Table 1.1   Size Scales in the Universe

Object or Phenomenon	Size* (cm)
Atomic nucleus	10^{-14}
Atom	10^{-8}
Virus	5×10^{-6}
Interstellar dust grain	5×10^{-5}
Bacterium	10^{-4}
Human body cell	5×10^{-3}
Human	2×10^{2}
Earth	1.3×10^{9}
Sun	1.4×10^{12}
Sun-Earth distance	1.5×10^{13}
Distance to nearest star	4.3×10^{18}
Milky Way galaxy	3×10^{23}
Distance to Andromeda galaxy	2×10^{24}
Local group of galaxies	5×10^{24}
Rich cluster of galaxies	10^{25}
Supercluster	10^{26}
The universe	10^{28}

*The sizes listed are meant to be typical, to illustrate the relative scales. For round objects, the diameter is used; for irregular objects, an approximate average dimension is given.

stars in the sky are actually quite nearby, by galactic standards; see Appendix 8). In other words, light from these stars has been traveling for hundreds of years when it reaches our eyes; light from the far side of the galaxy takes about 100,000 years to reach us.

The nearest galaxies beyond the limits of the Milky Way are the Magellanic Clouds, two irregularly shaped, fuzzy patches of light visible only from the Southern Hemisphere (Fig. 1.12). The Magellanic Clouds are roughly 180,000 light-years from the Sun, so we see that they are not very far outside the galaxy (they are considered satellites of the Milky Way, orbiting it in a time of several hundred million years). Light from the Magellanic Clouds takes 180,000 years to reach us. The most distant object visible to the unaided eye is a galaxy similar to the Milky Way, which is called Andromeda after its constellation. Andromeda is about 2.3 million light-years away (Fig. 1.13), so when

Figure 1.12 The Magellanic Clouds. These two irregularly shaped, small galaxies lie just outside the Milky Way galaxy and orbit it, taking hundreds of millions of years for each complete orbit. *(Fr. R. E. Royer)*

Figure 1.13 The Andromeda Galaxy. At a distance of over two million light-years, this galaxy is the most distant object visible to the unaided eye. Without a telescope and a time-exposure photograph, the eye sees only an extended, fuzzy patch of light, rather than the detailed view shown here. *(Palomar Observatory, California Institute of Technology)*

we look at the Andromeda galaxy, we are receiving light that has been traveling more than 2 million years!

Even the distance to the Andromeda galaxy is insignificant compared to the scale of the universe itself. The Milky Way, the Magellanic Clouds, Andromeda, and a number of other galaxies all belong to a concentrated group, or cluster, of galaxies. Most of the other galaxies in the universe also belong to clusters (Fig. 1.14), whose diameters can be as large as tens of millions of light-years. Between clusters of galaxies, space is relatively empty. (Actually, this point is controversial, as we shall see. All we can say for certain is that there are relatively few visible galaxies between clusters.)

Clusters of galaxies are themselves grouped into larger **superclusters,** whose size scales are significant, even relative to the universe itself. A supercluster typically may have a diameter measured in the hundreds of millions of light-years (actually, *diameter* is probably not a good word; superclusters seem to be sheetlike or filamentary structures, not rounded like clusters of galaxies). It is difficult to imagine an organized object (or collection of objects) so large that it takes hundreds of millions of years for light to travel across it.

Beyond the size scale of the superclusters, we approach the scale of the universe itself. It is apparent from a variety of lines of evidence that the universe has an overall size scale measured in billions of light-years. Light reaching us now from the farthest reaches of the universe has been traveling for many billions of years. This has the fascinating implication that we can see the universe as it was at that long-ago time when the light was emitted, a fact used by astronomers who study the origins of the universe.

Considering the sizes and distances of objects in the universe provides us a sobering perspective on ourselves and our tiny planet. It can be quite a revelation to see how much we presume to explain about the universe and how much we think we have learned about the origins, present state, and future evolution of the cosmos and all it contains. It will be wise, however, to keep in mind that there is much that we do not know.

Perspective

The introduction to astronomy provided in this chapter prepares us for a plunge into more detailed discussions. We will begin these in the next chapter by describing the nighttime sky as we see it without telescopes. We will describe in more detail the various bodies and motions observable to the unaided eye, and we will discuss the modern understanding of these phenomena.

Summary

1. Astronomy is the science in which the entire universe is studied. The study of astronomy requires knowledge of several other sciences, such as physics, chemistry, geology, and perhaps biology.
2. On a clear, dark night, the unaided eye can see up to five planets, the Moon, countless stars, and occasional meteors or comets. The planets, Sun, and Moon are always found along a strip of sky called the ecliptic.
3. The Earth's rotation causes all objects in the sky to seem to undergo daily motion, rising in the east and setting in the west.
4. Because we cannot directly see how far away objects are, we can most easily measure their positions in angular units. For convenience, we visualize a celestial sphere, centered on the Earth, on which the astronomical bodies lie.
5. The portion of the sky that can be seen depends on the latitude of the observer, and the portion visible at night depends on the time of year.

Figure 1.14 A Cluster of Galaxies. The faint, fuzzy objects near the center of this photograph are galaxies belonging to a cluster that lies over a billion light-years from the Milky Way. Like our own, each galaxy contains billions of individual stars. *(Palomar Observatory, California Institute of Technology)*

6. The most distant planet in our solar system is about forty times farther from the Sun than the Earth is, whereas the nearest star is some 300,000 times, or four light-years, farther away. Our galaxy is about 100,000 light-years in diameter, the nearest neighbor galaxies are 180,000 to 2.3 million light-years distant, and clusters of galaxies are typically separated by millions to tens of millions of light-years. The size of the universe itself is measured in the billions of light-years.
7. The light we receive from a distant object has been traveling toward us for as long as billions of years, in the case of the most distant galaxies, so we can observe the universe as it was long ago.

Review Questions

1. Discuss and clarify the contrast between astrology and astronomy. Can you think of other popular belief systems that have parallels with astrology?
2. We have said that as many as five planets are visible to the unaided eye. Using information given in Appendix 7, can you decide which five they are?
3. How is it possible to have a "moonless night," as mentioned in the text?
4. What does the fact that the Sun, Moon, and planets all move along the same path through the sky (the ecliptic) say about the structure of the solar system?
5. When we see a familiar object, for example, an airplane, at some distance from us, we can roughly guess how far away it is, by noting how large it appears. Explain why this does not work for stars.
6. We noted in the text that if the Earth were one foot in diameter, the Sun would be about 110 feet in diameter, and would be located about 2 miles away from the Earth. The most distant planet, Pluto, is forty times as far from the sun as the Earth is. If you wanted to build a model solar system, scaled so that Pluto was 10 feet from the sun, how big would the Sun and Earth be, and how far apart?
7. Explain why it is necessary to have astronomical observatories in both the Northern and Southern Hemispheres.
8. Suppose the Sun lies in the same direction as the constellation Aquarius in March. During what month is Aquarius directly overhead at midnight? During what month is Aquarius overhead at sunset? (You may want to consult Fig. 2.6 in order to answer this.)
9. Explain why the galaxy we live in appears to us as a band across the sky.
10. If the universe is 15 billion years old, how far away do we have to look in order to see the origin of the universe?

Additional Readings

We can further appreciate the essence of astronomy, as well as its beauty, by reading a wide range of books and periodicals. Many bookstores contain volumes of astronomical photographs, as well as numerous books on astronomy written for the layperson.

Periodicals that are particularly well suited to students using this text include *Sky and Telescope*, *Mercury*, and *Astronomy*. *Sky and Telescope* is especially recommended for those wishing to carry out projects such as telescope building or astrophotography; it includes monthly charts showing the positions of the stars and planets as they change throughout the year.

Chapter 2

Learning About the Nighttime Sky

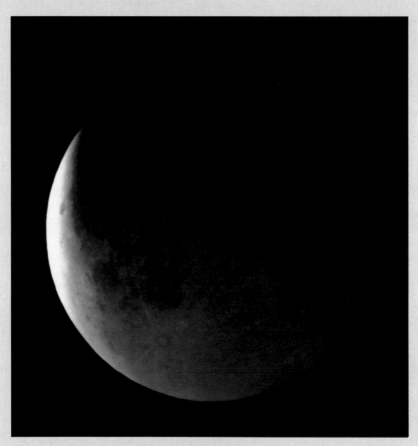

A near-total lunar eclipse. (J. Kloeppel,
Sommers Bausch Observatory,
University of Colorado)

Chapter Preview

The Rhythms of the Cosmos
   Daily Motions
   Annual Motions: The Seasons
   Lunar Motions
   Planetary Motions
Eclipses
Historical Developments
   The Earliest Astronomy
   Plato and Aristotle
   The Later Greeks: Sophisticated Cosmologies

By using their natural senses the ancient astronomers and philosophers learned quite a lot about the cyclical motions in the nighttime sky. Unaided eye observations, made systematically over long periods of time, can reveal the motions of the major solar system objects: the Sun, the Moon, and the brightest planets. It may seem odd to refer to motion of the Sun, because its apparent path through the sky is actually due to the Earth's motion, but this distinction was not known to the ancients. From the viewpoint of the observer on the Earth, it appears that the Sun moves through the sky during the course of a year. Similarly, the Sun, Moon, planets, and stars rise and set daily. We know that this is a reflection of the Earth's rotation, but the early philosophers adopted a perhaps more straightforward interpretation, that the sky rotated about the Earth.

In this chapter we will learn about the patterns and motions in the nighttime sky that were evident to the ancients, and we will briefly describe how these people interpreted what they saw.

Table 2.1 Periods of Significant Motions

Motion	Period*
Sidereal day	$23^h 56^m 4.098^s$
Mean solar day	$24^h 00^m 00^s$
Tropical year (equinox to equinox)	$365^d 5^h 48^m 45^s$
Sidereal year (fixed stars)	$365^d 6^h 9^m 10^s$
Synodic month	$29^d 12^h 44^m 3^s$
Sidereal month	$27^d 7^h 43^m 11^s$
Mercury:	
sidereal period	87.969^d
synodic period	115.88^d
Venus:	
sideral period	224.701^d
synodic period	583.92^d
Mars:	
sidereal period	1.88089^y
synodic period	2.1354^y
Jupiter:	
sidereal period	11.86223^y
synodic period	1.0921^y
Saturn:	
sidereal period	29.4577^y
synodic period	1.0352^y
Uranus:	
sidereal period	84.0139^y
synodic period	1.0121^y
Neptune:	
sidereal period	164.793^y
synodic period	1.00615^y
Pluto:	
sidereal period	247.7^y
synodic period	1.0041^y

*Units for lunar and planetary motions are mean solar days or tropical years.

The Rhythms of the Cosmos

It soon becomes apparent to any observer of the heavens that there are cycles of motions ranging from those that occur daily to those that define the year. These cycles have had a profound influence on mankind and on other life forms on the Earth. Some are due to the Earth's rotation, others to the motions of the Moon or planets, and still others to the Earth's orbital motion about the Sun. Table 2.1 lists the **periods,** or cycle times, for several motions in the solar system. (Unfamiliar terms in the table are explained later in the chapter.)

Daily Motions

The most obvious of the many motions that affect our view of the universe are the daily cycles of all celestial objects due to the rotation of the Earth. The Earth spins on its axis in 24 hours, so we on its surface see a continuously changing view of the heavens. We see the Sun rise and set, along with the Moon, the planets, and most of the stars. Even though we understand that these daily, or **diurnal,** motions are the result of the Earth's spin, we still commonly refer to them as though the objects in the sky were moving around us.

The rotation of the Earth forms the basis for our timekeeping system, since the length of the day is a natural unit of time on which to base our lives, one to which nearly all Earthly species have adapted. The day is divided into 24 hours, each containing 60 minutes, with each minute consisting of 60 seconds. These divisions are based on the numbering system developed several thousand years ago, largely by the Babylonians.

We have to be careful when we define what we mean by a day. What we observe depends on our frame of reference, that is, on whether we look from the Earth or from a point in space that does not move with the Earth. We get one result if we say that the length of a day is the time it takes for the Sun to complete one cycle through the sky (say, from noon one day to noon the next day), and we get a different one if we say that a day is the time it takes for the stars to make one full cycle. We get different results because the Earth moves in its orbit about the Sun at the same time it spins on its axis. Because of this motion, the length of one **solar day** (the time for the Sun to make one full cycle, as seen from the surface of the Earth), is a little longer than one **sidereal day** (the time for the stars to complete a cycle) (Fig. 2.1). Thus, the Sun rises about four minutes later each day, in comparison with the stars.

Similarly, the Moon moves along in its orbit about the Earth during the course of a day, so the Moon rises almost an hour later each day. The planets also move, but so slowly that their diurnal motions are not easily distinguished from those of the stars.

Our "official" day, the one that forms the basis for our timekeeping system, is based on the solar day rather than the sidereal day, which is the true rotation period of the Earth (the term *sidereal* means "with respect to the stars"). Because the Earth's orbital speed is not precisely constant, the length of the solar day varies a little throughout the year. It would be inconvenient to allow our hour, minute, and second to vary along with it, so the average length of the solar day, called the **mean solar day,** has been adopted as our timekeeping standard. The mean solar day is 3 minutes and 56 seconds longer than the sidereal day (see Table 2.1). Special telescopes called **transit telescopes** (Fig. 2.2), which measure precisely the moment when a chosen star passes overhead and therefore can be used to determine the Earth's rotation period, have been used for timekeeping. Official time standards are now based on the vibration frequencies of certain kinds of atoms, with reference to stellar transit observations to make sure that the official time does not get out of synchronization with the stars.

The Earth's rotation causes the stars to travel through the sky in a daily cycle (Fig. 2.3). Our view of these motions depends on how far north or south of the equator we are: if we are north of the equator, stars near the North Pole's projection onto the sky (that is, stars near the **north celestial pole**) appear to circle the pole without setting. If our latitude is 40 degrees north, for example, all the stars within 40 degrees of the north celestial pole stay up all night, while stars farther than 40 degrees from the pole rise and set each day. At the same time, for those of us living in the Northern Hemisphere, a large region of the southern sky is forever beyond our view. The constellations we see vary as we travel north or south, a fact not lost on early astronomers, who concluded from this that the Earth is round.

The system of latitude and longitude used for measuring positions on the Earth's surface has provided the

Figure 2.1 The Contrast Between Solar and Sidereal Days. The arrow indicates the overhead direction from a fixed point on the Earth. From noon one day (left), it takes one sidereal day for the arrow to point again in the same direction, as seen by a distant observer. Because the Earth has moved, however, it will be about four minutes later when the arrow points directly at the Sun again; hence the solar day is nearly four minutes longer than the sidereal day.

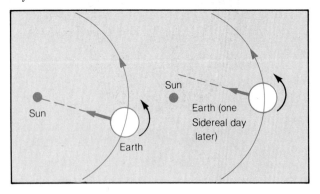

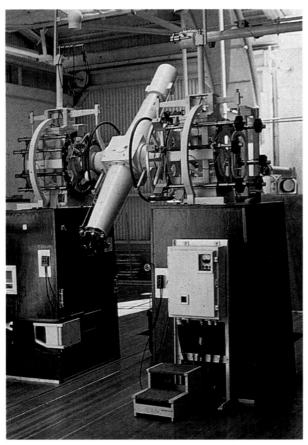

Figure 2.2 A Transit Telescope. Such a device points straight up. It is used to record the times when certain reference stars pass over the meridian and is therefore helpful in measuring the sidereal day. *(U.S. Naval Observatory)*

Figure 2.3 Star Trails Illustrating the Earth's Rotation. The circular trails are created by stars near the south celestial pole, which completed about half a circle during this all-night exposure. *(© 1980 Anglo-Australian Telescope Board)*

basis for the most commonly used system of measuring positions on the sky (Fig. 2.4). In the **equatorial coordinate system,** a star's angular distance north or south of the **celestial equator** (the projection of the Earth's equator onto the sky) is called its **declination,** and its position in the east-west direction is its **right ascension.** Right ascension, measured from a fixed direction on the sky, is expressed in units of hours, minutes, and seconds of time, because it is convenient to specify positions on the sky according to the time they

Figure 2.4 Equatorial Coordinates. Here are illustrated the measurements of star position in declination and right ascension. Declination is the distance (in degrees, minutes, and seconds of angle) of a star north (+) or south (−) of the celestial equator. Right ascension is the distance (in hours, minutes, and seconds of time) of the star to the east of the direction of the vernal equinox, a fixed direction in space.

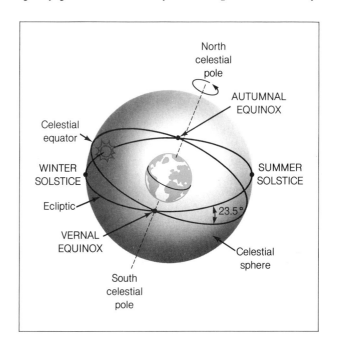

Astronomical Insight 2.1

Celestial Navigation

One of the most time-honored and well-known "practical" applications of astronomy is to find one's position on the Earth by observing the stars. Traditionally this was the only means available to a sailor or someone exploring new territory, whereas today there are many more modern methods (and less unexplored territory).

The essential method is quite simple, although in practice the technique involves mathematical calculations too complex to describe in detail here. The first requirement is to be able to identify one or more objects in the sky whose coordinates are known (usually from a book of tabulated positions, called an **ephemeris**). Thus, familiarity with the constellations and the sky in general is required. Ironically, people who have learned celestial navigation are more likely to know the constellations, as well as a number of individual stars and their names, than is the typical professional astronomer.

Once a known star is recognized and its coordinates found in a catalogue or ephemeris, the navigator has a reference point in the sky whose position is known. The navigator then measures the distance from that position, in angular measure. In particular, the needed measurement is the **zenith distance** of the star, the angle between the navigator's zenith (directly overhead) and the direction to the star. Given this, as well as the sidereal time, it is possible to locate the spot on the Earth's surface where the star is directly overhead and then draw a circle about that spot (on a map) showing all the locations where the star would have the measured zenith distance. This does not pinpoint the navigator's location, so the same procedure is carried out using a second star. Now there are two circles of possible locations, and the two intersect at only two points. Unless badly confused, the navigator normally will know which of the two intersection points is the correct location (since the two points may be hundreds of miles apart). Hence, measurements of only two stars are usually sufficient, but typically a third star is measured as well to be sure there were no errors. In that case there are three circles drawn on the map, and all three should intersect at only one point if the measurements have been made correctly.

The measurement of the zenith distance is done by using a device such as a **sextant** that allows angular separations on the sky to be measured. A sextant usually includes a pointer that can be aimed at the star and a means of giving the user a reference to the vertical direction. Often a bubble in a glass enclosure, much like the one on a carpenter's level, allows the sextant to be held so that it is precisely horizontal.

In practice, the Sun and the Moon can be used for navigation, although they are both difficult to measure precisely because of their large angular extent; also, of course, it is difficult to view the Sun directly because of its brilliance. Furthermore, to use either, it is necessary to have rather complete tables of their positions, which change throughout the year (in the case of the Sun) and throughout the month (in the case of the Moon, whose motion is sufficently rapid to require tables showing the position hourly).

Today navigation of ships and airplanes is done by various means involving technology that has been developed in the past few decades. The well-traveled areas of the Earth are permeated with navigational radio beams that provide directional information, and modern satellite navigation systems provide very accurate data to ships and planes. In addition, devices called **gyroscopes,** which can maintain a given orientation very accurately over long periods, allow navigators to keep track of their positions over lengthy voyages without any reference to external guideposts such as the sky or a satellite navigation system.

Despite all the modern developments cited here, the art of celestial navigation is not completely lost. Military and commercial airline and ship navigators still learn how to do it and presumably could use the method if all else failed.

cross the **meridian,** the north-south line that passes overhead. Thus the coordinates of Sirius, the brightest star in the sky, are right ascension $6^h44^m12^s$ and declination $-16°42'$ (actually, these are the coordinates for the year 1980; one must always specify the year, as explained below). The minus sign indicates a declination south of the equator, which has a declination of 0 degrees; a plus sign would indicate a declination north of the equator.

There is one complication with using the equatorial coordinate system: The Earth slowly wobbles on its axis, in a motion called **precession,** and this causes our coordinates to drift slowly on the sky (Fig. 2.5). It takes some 26,000 years for the Earth to complete one cycle of this motion, which is exactly like the wobbling familiar to anyone who has spun a play top or gyroscope. The motion of our coordinate system is therefore very gradual, causing star positions to shift extremely slowly. Despite the small magnitude of the effect, it was noticed over 2000 years ago, and it must be allowed for by modern astronomers when planning observations.

Annual Motions: The Seasons

We turn our attention now to celestial phenomena caused by the Earth's motion as it orbits the Sun. One aspect of this, the daily eastward motion of the Sun with respect to the stars, has already been mentioned in connection with the difference between solar and sidereal days. It must be emphasized that this is only an apparent motion caused by our changing angle of view as we move with the Earth in its orbit. As the Earth moves about the Sun, it travels in a fixed plane, so that the apparent path of the Sun through the constellations is the same each year. The apparent path of the Sun is called the **ecliptic,** and the sequence of constellations through which it passes is called the **zodiac** (Fig. 2.6). The twelve principal constellations of the zodiac have been identified since antiquity and were once thought by astronomers to have significance for our daily lives. To this day, an astrologer will tell you that your fate is influenced by the sign under which you were born. This is a very old idea; astronomers have refined their understanding of the heavens in the past 2000 years, even if astrologers have not. There is, in fact, no evidence to support the claims of astrologers. Ironically, because of precession, the sign associated with a particular date is no longer the same as the constellation the Sun is passing through on that date.

It happens that the orbital planes of the other planets and of the Moon are closely aligned with that of the Earth, so that the planets and the Moon are always seen near the ecliptic. Hence all of the major objects in the solar system that can be seen by the unaided eye pass through the same sequence of constellations, the zodiac, so it is not surprising that great significance was attached to this sequence by the ancient astronomers.

Besides causing the apparent annual motions of the Sun and planets, the Earth's orbital motion has a second, far more significant, effect on us: It creates our seasons. The Earth's spin axis is tilted with respect to its orbital plane, so that during the course of a year, any point on the Earth's surface is exposed to varying amounts of daylight. Summer in the Northern Hemisphere occurs when the North Pole is tipped toward the Sun; winter occurs during the opposite part of the Earth's orbit, as the pole, which remains fixed in orientation, is tilted away from the Sun (Fig. 2.7). The tremendous seasonal variations in climate at intermediate latitudes are caused by a combination of two effects:

1. The length of the day varies, so that in summer, for example, the Sun has more time to heat the Earth's surface.
2. The angle at which the Sun's rays strike the ground is more nearly perpendicular in the summer, so that the Sun's intensity is much greater, thus heating the surface more efficiently.

Figure 2.5 Precession. The Earth's axis is tilted 23½ degrees with respect to its orbital plane, and it wobbles on its axis, so that an extension of it describes a conical pattern (left) in a time of about 26,000 years. The north celestial pole therefore follows a circular path on the sky (right).

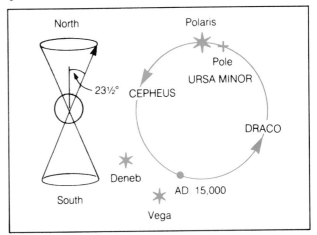

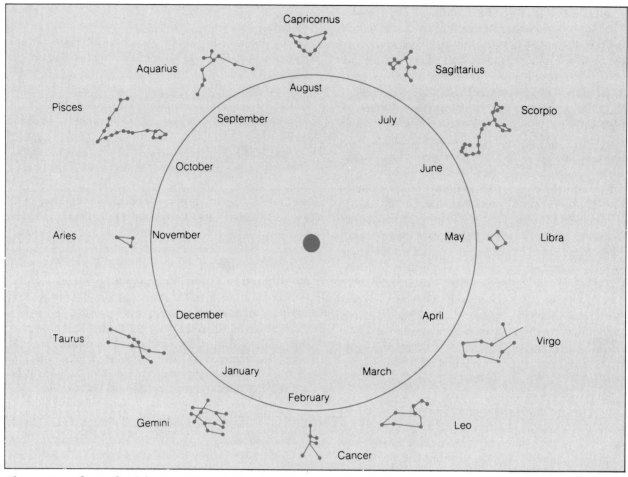

Figure 2.6 The Path of the Sun Through the Constellations of the Zodiac. The dates refer to the position of the Earth each month. To see which constellation the Sun is in during a given month, imagine a line drawn from the Earth's position through the Sun; that line will extend to the Sun's constellation. For example, in March the Sun is in Aquarius.

Figure 2.7. Seasons. The Earth's tilted axis retains its orientation as the Earth orbits the Sun. Thus, at opposite points in the orbit, each hemisphere has winter and summer, depending on whether that hemisphere is tipped toward the Sun or away from it.

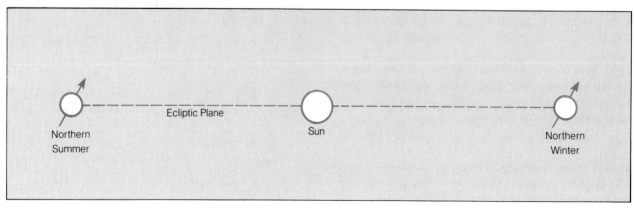

The Earth's axis is tilted 23½° from the perpendicular to the orbital plane. During the year, the Sun, as seen from the Earth's surface, can therefore appear directly overhead as far north and south of the equator as 23½°, defining a region called the **tropical zone** (Fig. 2.8). For those of us who live outside of the tropics, the Sun can never be directly overhead. When the North Pole is tilted toward the Sun, an occasion occurring near June 22 and called the **summer solstice,** the Sun passes directly overhead at 23½° north latitude. On this occasion, sunlight covers the entire north polar region to a latitude as far as 23½° south of the pole. This defines the **Arctic Circle,** and at the time of the solstice the entire circle has daylight for all twenty-four hours of the Earth's rotation. At the pole itself there is constant daylight for six months. At the **winter solstice,** when the South Pole is pointed most closely in the direction of the Sun, the Sun's midday height above the horizon, as viewed from the Northern Hemisphere, is the lowest of the year.

If we follow the Sun's motion north and south of the equator throughout the year, we find that it follows a graceful curve as it traverses its range from +23½° (north) declination to −23½° (south) declination (Fig. 2.9). The Sun crosses the equator twice in its yearly excursion, at the times when the Earth's North Pole is pointed in a direction 90° from the Earth-Sun line. At these times the lengths of day and night in both hemispheres are equal, and these occasions are referred to as the **vernal** (spring) and **autumnal** (fall) **equinoxes,** which take place on about March 21 and September 23, respectively. The direction to the Sun at the time of the vernal equinox coincides exactly with the direction of 0^h right ascension. (The definition of this direction is that it lies along the line of intersection of the Earth's equatorial plane and the orbital plane.)

Lunar Motions

The most prominent object in the nighttime sky is the Moon. The Moon orbits the Earth, its orbital plane nearly coinciding with that of the ecliptic. During each circuit that the Moon makes around the Earth, we see its full cycle of phases. The true orbital period is called the **sidereal period,** because it is the time required for the Moon to go around the Earth, as seen in a fixed reference frame with respect to the stars. The observed cycle of phases, which is the time required for the Moon to return to a given alignment with respect to the

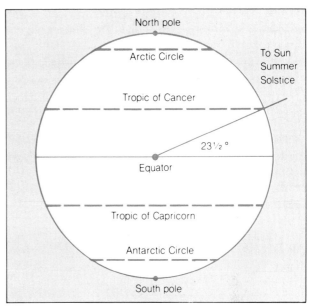

Figure 2.8 The Definition of Latitude Zones on the Earth. At summer solstice, the Sun is overhead at 23½° north latitude, its northernmost point. This defines the Tropic of Cancer, the northern limit of the tropical zone. At the same time the entire area within 23½° of the North Pole is in daylight throughout the Earth's rotation, and the boundary of this region is the Arctic Circle. Similarly, the Antarctic Circle receives no sunlight at all during a complete rotation of the Earth. Six months later, the Sun is overhead at the Tropic of Capricorn.

Figure 2.9 The Path of the Sun Through the Sky. Because of the Earth's orbital motion and the tilt of its axis, the Sun's annual path through the sky has the shape illustrated here.

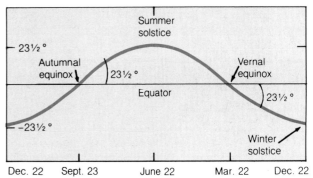

Sun, is called the **synodic period** or the **lunar month.** The difference is akin to the distinction between the solar and sidereal days, discussed in the preceding section. In both cases it is the Earth's motion about the Sun that lengthens the time it takes to complete a full cycle as we see it (Fig. 2.10). The sidereal month is almost 27½ days long, while the lunar month is about 29½ days long (Table 2.1).

The Moon always has the same side facing the Earth, because its rotation period is equal to its orbital period. This is caused by gravitational forces exerted on the Moon by the Earth. (This phenomenon will be discussed fully in Chapters 3 and 5.)

Since the Moon does not emit light of its own but simply reflects sunlight, we easily see only those portions of its surface that are sunlit. As the Moon orbits the Earth and our viewing angle changes, we see varying fractions of the sunlit half of the Moon. Thus the Moon's apparent shape changes drastically during the month; this sequence of shapes is referred to as the **phases** of the Moon (Fig. 2.11). The full cycle of phases is completed during one synodic period, or **lunar month,** of about 29½ days.

The extremes of the cycle are represented by the **full Moon,** which occurs when the moon is directly opposite the Sun, so that we see its entire sunlit hemisphere,

Figure 2.10 Sidereal and Synodic Periods of the Moon. Because the Earth moves in its orbit while the Moon orbits it, the Moon goes through more than one full circle (as seen by an outside observer) to go from one full Moon to the next. Hence the lunar (or synodic) month is about two days longer than the Moon's sidereal period.

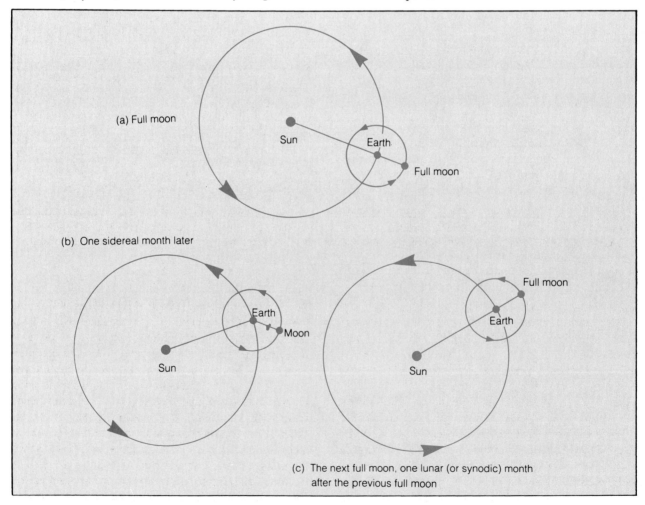

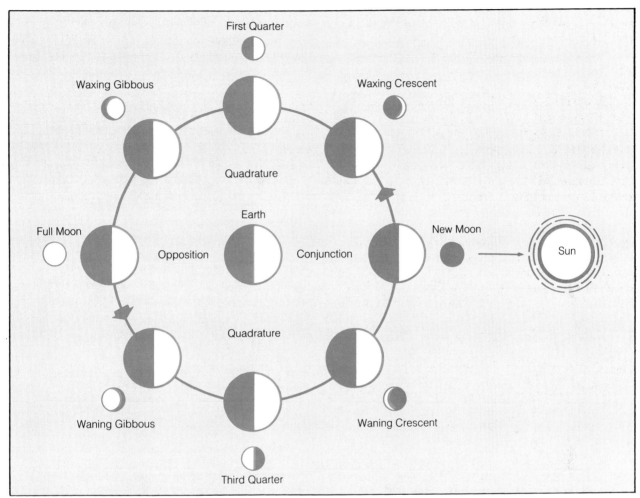

Figure 2.11 Lunar Phases and Configurations. As the Moon orbits the Earth, we see varying portions of its sunlit side. The phases sketched here (outside the circle representing the Moon's orbit) show the Moon as it appears to an observer in the Northern Hemisphere.

and the **new Moon,** when it is between the Earth and the Sun, with its dark side facing us. The full Moon is on the meridian at local midnight, whereas a new Moon is on the meridian at local noon. The new Moon cannot be observed, because only the dark side faces us and because it is so close to the Sun in the sky.

Just as we speak of phases of the Moon, which really refer to its apparent shape as seen from the Earth, we can also speak of its **configurations,** which describe its position with respect to the Earth-Sun direction. For example, when a full Moon occurs, the Moon is at **opposition,** that is, in the direction opposite the Sun; a new Moon occurs at **conjunction,** when the Moon lies in the same direction as the Sun. We can follow the Moon through its phases as we trace its configurations, beginning with the new Moon, which takes place at conjunction. During the first week following conjunction, as the Moon moves toward **quadrature** (the position 90° from the Earth-Sun line), it appears to us to have a crescent shape that grows in thickness each night. This is called the **waxing crescent** phase. When the Moon reaches quadrature, so that we see exactly one-half of the sunlit hemisphere, it has reached the phase called **first quarter.** For the next week, as the Moon goes from quadrature toward opposition and we see more and more of the daylit side, its phase is said to be **waxing gibbous.** After the full Moon, as it again approaches quadrature, the phase is **waning gibbous,** and this time at quadrature, the phase is called **third quarter.** First quarter can be distinguished from third quarter easily by noting the time of night: if the Moon is already up when the Sun sets, it is first quarter, but if it does not rise until midnight, it is third quarter. After third quarter, as the Moon moves closer to the Sun we see a diminishing slice of its sunlit side, and it is in the **waning crescent** phase.

Figure 2.12. Planetary Configurations.

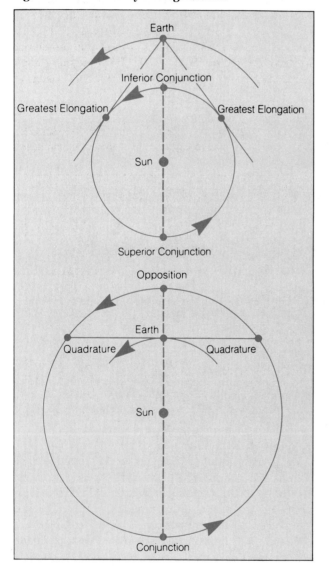

Planetary Motions

Like the Sun and the Moon, the planets move with respect to the background stars. This fact gave the planets their generic name, since *planet* is the Greek word for "wanderer."

The observed motions of the planets are due primarily to their orbital movement about the Sun, although one important aspect of the motion of certain planets is a reflection of the Earth's motion. The planets all orbit the Sun in the same direction as the Earth and in nearly the same plane, so that they appear to move nearly in the ecliptic, through the constellations of the zodiac.

Mercury and Venus, the **inferior planets,** lie within the orbit of the Earth and can never appear far from the Sun in our sky. For Mercury the greatest angular distance from the Sun, called the **greatest elongation,** is about 28°, while Venus can be seen as far as 47° from the Sun. Like the Moon, the planets have specific configurations that refer to their positions with respect

Figure 2.13 The Synodic Period of an Inferior Planet. The inner planets travel faster than the Earth in their orbits, and therefore "lap" the Earth, much as a fast runner laps a slower runner on a track. This illustration shows the approximate situation for Mercury, which has a synodic period of about 116 days, or roughly one third of a year.

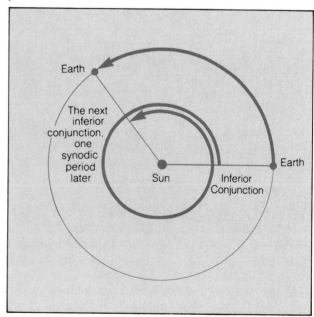

to the Sun-Earth line (Fig. 2.12). An inferior planet is said to be at **inferior conjunction** when it lies directly between the Earth and the Sun and at **superior conjunction** when it is aligned with the Sun, but lies on its far side.

The outer, or **superior**, planets can be seen in any direction with respect to the Sun, including opposition, when they are in the opposite direction from the Sun. Conjunction for a superior planet can occur only when the planet is aligned with the Sun but on the far side of it, in analogy with a superior conjunction for an inferior planet, and **quadrature** occurs when a superior planet is 90° from the direction of the Sun.

Each planet has a sidereal period and a synodic period (Fig. 2.13 and Table 2.1), the former being the true orbital period as seen in the fixed framework of the stars, and the latter being the length of time it takes the planet to pass through the full sequence of configurations, (that is, from one conjunction or one opposition to the next) as seen from the Earth. The situation is much like that of two runners on a track: The time it takes the faster runner to lap the slower one is analogous to the synodic period, while the time it takes simply to circle the track corresponds to the sidereal period.

The motion of the Earth has one very important effect on planetary motions. As we go outward from the Sun, each successive planet has a slower speed in its orbit. (See the discussion of Kepler's laws of planetary motion, Chapter 3.) This means that the Earth, moving faster than the superior planets, periodically passes each of them (this occurs once every synodic period). As the Earth overtakes one of the superior planets, there is an interval of time during which our line of sight to that planet sweeps backward with respect to the background stars, making it appear that the planet is moving backward (Fig. 2.14). The same thing happens when a rapidly moving automobile passes a

Figure 2.14 Retrograde Motion. As the faster-moving Earth overtakes a superior planet in its orbit, the planet temporarily appears to move backward with respect to the fixed stars. This sketch illustrates the modern explanation of something that took ancient astronomers a long time to understand correctly.

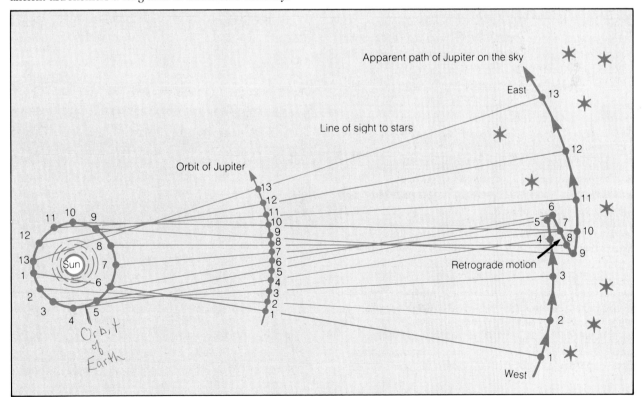

slowly moving vehicle; for a brief moment, it appears to those in the faster vehicle that the slower vehicle moves backward with respect to the fixed background. The apparent backward movement of the planets, called **retrograde motion,** was thought in ancient times to represent a real motion of the superior planets, rather than a reflection of the Earth's motion. This very much complicated many of the early cosmologies, as we shall see later in this chapter.

Eclipses

From the discussion in the preceding sections, it may seem that the Moon should pass directly in front of the Sun on each trip around the Earth and through the Earth's shadow at each opposition, producing alternating solar and lunar eclipses at two-week intervals, but this is obviously not the case. The Moon's orbital plane does not lie exactly in the ecliptic, but is tilted by about 5°. Therefore the Moon usually passes just above or below the Sun as it goes through conjunction and similarly misses the Earth's shadow at opposition. The Moon passes through the ecliptic at only two points on each trip around the Earth, where the planes of the Earth's and the Moon's orbits intersect. Because the Moon's orbital plane wobbles slowly in a precessional motion similar to that of the Earth's spin axis, the line of intersection with the Earth's orbital plane slowly moves around. The combination of this motion, the Moon's orbital motion, and the movement of the Earth around the Sun creates a cycle of eclipses, with the same pattern recurring every eighteen years. (See Table 2.2 for a list of future total solar eclipses.) This cycle of eclipses, called the **Saros,** was recognized in antiquity.

It is purely coincidental that the Moon and the Sun have nearly equal angular sizes, so that the Moon neatly blocks out the disk of the Sun during a solar eclipse (Figs. 2.15, 2.16). The angular diameter of an object is inversely proportional to its distance, meaning that the farther away it is, the smaller it looks. The Sun is much larger than the Moon, but also much more distant. The two objects have the same angular diameter because the ratio of the Sun's diameter to that of the Moon just happens to be compensated by the ratio of the Sun's distance to that of the Moon.

If a total solar eclipse occurs at the time when the Moon is farthest in its slightly noncircular orbit from the Earth, it does not quite block all of the Sun's disk but leaves an outer ring of the Sun visible. This is called an **annular** eclipse. Because a total (or annular) solar eclipse requires a precise alignment of Sun and Moon, an eclipse will appear total only along a well-defined, narrow path on the Earth's surface (Fig. 2.17). There is a wider zone outside of that where the Moon appears to block only a portion of the Sun's disk; people in this zone see a partial solar eclipse.

During a lunar eclipse, when the Moon passes through the Earth's shadow, observers everywhere on

Table 2.2 Total Solar Eclipses of the Future

Date of Eclipse	Duration	Location
March 18, 1988	4.0	Philippines, Indonesia
July 22, 1990	2.6	Finland, arctic regions
July 11, 1991	7.1	Hawaii, Central America, Brazil
June 30, 1992	5.4	South Atlantic
November 3, 1994	4.6	South America
October 24, 1995	2.4	South Asia
March 9, 1997	2.8	Siberia, arctic regions
February 26, 1998	4.4	Central America
August 11, 1999	2.6	Central Europe, central Asia

Figure 2.15 A Total Solar Eclipse. The Moon entirely blocks our view of the Sun. *(Fr. R. E. Royer)*

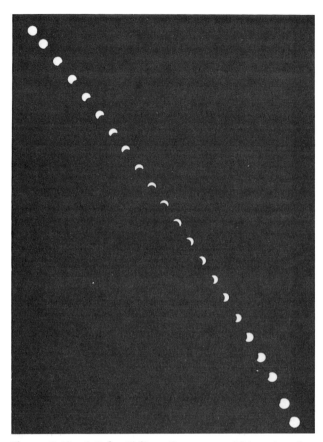

Figure 2.16 A Solar Eclipse Sequence. This series of photographs illustrates the Moon's progression across the disk of the Sun during a partial solar eclipse. *(NASA)*

Figure 2.17 The Moon's Shadow During a Solar Eclipse. This photograph, taken from space, shows the shadow of the Moon on the Earth during a solar eclipse. The eclipse appeared total only to observers on the Earth who were located directly in the center of the shadow's path (that is, in the umbra). *(NASA)*

the Earth see the same portion of the Moon eclipsed. If the entire Moon passes through the **umbra,** the dark inner portion of the Earth's shadow, the eclipse is total, as no part of the Moon's surface is exposed to direct sunlight. If only a portion of the Moon passes through the umbra, then we see a partial eclipse. A penumbral eclipse occurs when all or part of the Moon passes through the **penumbra,** the lighter outer portion of the Earth's shadow; such eclipses are not readily noticeable, however.

Historical Developments

Now that we have discussed most of the phenomena that can be observed without telescopes and thus are aware of nearly all the data available to ancient watchers of the sky, it is appropriate to review the development of human understanding of these phenomena.

It is likely that people were preoccupied with the heavens from the time they first became aware of their environment. The speculative mood that we, in these modern times, can conjure up only by disregarding our daily pressures and escaping into the countryside on a clear night to look at the stars must have dominated the nighttimes of the earliest cultures. In the following discussion, emphasis is placed on developments that occurred around the shores of the Mediterranean, because these developments laid the foundation for the modern understanding of the universe. Parallel developments occurred in many other parts of the world, but in most cases either reached dead ends or were eventually absorbed into the Western scientific culture.

The Earliest Astronomy

The earliest records of astronomical lore have been found in the region east of the Mediterranean now known as Iraq (Fig. 2.18), where the Babylonian culture flourished for many centuries, beginning around 2000 B.C. The Babylonians developed an accurate knowledge of the length of the year, establishing the basis for the modern twelve-month calendar, and bequeathed us our timekeeping and angular measures that are based on the number sixty.

Parallel with the developments in Babylonia, an early Greek civilization arose on the shores of the Mediterranean (Fig. 2.18) and at some unknown time came in contact with the Babylonian culture. The ancient

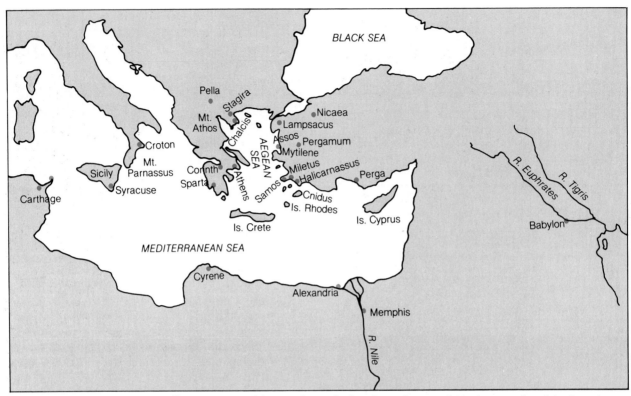

Figure 2.18 The Ancient Mediterranean. This map shows the locations of many of the sites mentioned in the text. (C. Ronan, The Astronomers 1964 [New York: Hill and Wang] with permission.)

Table 2.3 Notable Greek Achievements

Date	Name	Discovery or Achievement
c.900–800 B.C.	Homer	*Iliad* and *Odyssey*; summaries of legends
c.624–547 B.C.	Thales	Rational inquiry leads to knowledge of universe
c.611–546 B.C.	Anaximander	Universal medium; primitive cosmology
c.570–500 B.C.	Pythagoras	Mathematical representation of universe; round Earth
c.500–400 B.C.	Philolaus	Earth orbits central fire
c.500–428 B.C.	Anaxagoras	Moon reflects sunlight; correct explanation of eclipses
c.428–347 B.C.	Plato	Material world imperfect; deduce properties of universe by reason
c.408–356 B.C.	Eudoxus	First mathematical cosmology; nested spheres
c.384–322 B.C.	Aristotle	Concept of physical laws; proof that Earth is round
c.310–230 B.C.	Aristarchus	Relative sizes, distances of Sun and Moon; first heliocentric theory
c.273–? B.C.	Eratosthenes	Accurate size of Earth
c.265–190 B.C.	Apollonius	Introduction of the epicycle
c.200–100 B.C.	Hipparchus	Many astronomical developments; full mathematical epicyclic cosmology
c.100–200 A.D.	Ptolemy	*Almagest*; elaborate epicyclic model

Figure 2.19 Homer's *Odyssey* and *Iliad*. In these two epic poems, Homer described the much more ancient legends of astronomical lore from the civilization that had flourished on the isle of Crete. *(The Granger Collection)*

Figure 2.20 Plato. *(The Granger Collection)*

Greek traditions, known to us largely through the writings of Homer (Fig. 2.19), gave birth to our modern constellation names and other astronomical lore and laid the foundation for the first rational scientific inquiry. (See Table 2.3 for a list of the achievements of the ancient Greeks.)

The first formal scientific thought is associated with the philosopher **Thales** (624–547 B.C.) and his followers, who formed the so-called Ionian school of thought. The principal contribution of Thales was the idea that rational inquiry can go beyond describing the universe to *understanding* it. The Ionians did develop a primitive **cosmology**, or theory of the universe, in which all the basic elements of the universe were formed from water, the primeval substance.

Another major school of thought was developed by **Pythagoras** (c. 570–500 B.C.) and his followers, who believed that natural phenomena could be described mathematically, a belief that is at the heart of all modern science. Pythagoras himself is credited with being the first to assert that the Earth is round and that all heavenly bodies move in circles, ideas that never thereafter lost favor in ancient times.

Plato and Aristotle

One of the most influential characters in the development of Greek philosophy was **Plato** (428–347 B.C.; Fig. 2.20), who established an academy in Athens in about the year 387 B.C., where he taught his ideas of natural philosophy. Plato's fundamental precept was that what we see of the material world is only an im-

Astronomical Insight 2.2

The Mythology of the Constellations

Homer's *Iliad* and *Odyssey*, which appeared around 900–800 B.C., contain the earliest written descriptions of the constellations and their meanings. These descriptions were based on legends from the Minoan civilization, that were already ancient in Homer's time.

Little precise information is available on when and how the mythology of the constellations arose. It has, however, been possible to deduce roughly the era that gave birth to it. Careful examination of the ancient constellations shows them to be distributed symmetrically about a point in the sky that was at the northern celestial pole around 2600 B.C. (but has since shifted away from the pole due to precession), so it is probable that the legends of the constellations were invented at about that time. There are few ancient constellations near the southern celestial pole in regions not visible from the latitude of Crete, giving independent support to the supposition that the legends arose in the Minoan culture.

It is likely that the constellations were not taken as literally as is usually supposed. Rather than being considered faithful depictions of the people and events with which they are associated, the constellations should more properly be viewed as symbolic representations. They were probably not originally designated on the basis of their imagined resemblance to certain characters; instead, areas of the sky were dedicated in honor of prominent figures of mythology, and the familiar pictures of these figures were then fitted to the patterns of bright stars. This helps account for the lack of obvious resemblance between the star patterns and the figures and events they supposedly represent. The constellation names were translated from the Greek of Homer to Latin when the Roman Empire rose to dominance, and the names we are familiar with today are the Latin ones. Interestingly, the names of prominent stars went through another transformation, into Arabic, and our modern star names are Arabic designations, which usually are literal descriptions of the places these stars hold in their constellations. The name Betelgeuse, for example, the name of the prominent red star in the shoulder of Orion, is translated as "the armpit of the giant," or "the armpit of the central one." Thus we use Arabic names for stars in constellations with Latin designations, while both are based on Greek descriptions of the ancient Minoan legends.

Ancient and nonscientific though they are, the constellations have a significant impact on the nomenclature of modern astronomy. The modern constellations refer to very specific regions of the sky with well-defined boundaries that have been agreed to by the international community of astronomers. These "official" constellations are based on those of legend, containing within them the ancient figures of mythology, but their boundaries have been extended so that every part of the sky falls within one constellation or another. A map of the constellations looks a bit like a map of the western United States, where the boundaries are generally straight lines, but the shapes are irregular. As an alternative to the Arabic names for the brightest stars, modern astronomers often use designations based on a star's rank within its constellation, with letters of the Greek alphabet used to indicate the brightness rank. Thus Betelgeuse, the brightest star in Orion, is also called α Orionis.

perfect representation of the ideal creation. This doctrine had the corollary that one can learn more about the universe by reason than by observation, since observation can present us with only an incomplete picture. Hence Plato's ideas of the universe, described in his *Republic*, were based on certain idealized assumptions that he found reasonable. One of the most important of these was that all motions in the universe are perfectly circular and that all astronomical bodies are spherical in shape. Thus he adopted the Pythagorean view that the Sun, the Moon, and all the planets moved in combinations of circular motions about the Earth. Plato evidently thought of these objects as fixed to clear, ethereal spheres that rotated.

The most renowned student of Plato was **Aristotle** (384–322 B.C.), who was the first to adopt physical laws and show why, in the context of those laws, the universe works as it does. He taught that circular motions are the only natural motions and that the center of the Earth is the center of the universe. He also believed that the world is composed of four elements: earth, air, fire, and water. He could demonstrate, in the context of his adopted physical laws, that the universe was spherical in shape and that the Earth was also spherical. He had three proofs of the latter:

1. Only at the surface of a sphere do all falling objects seek the center by falling straight down (it was another premise of his that falling objects were following their natural inclination to reach the center of the universe).
2. The view of the constellations changes as one travels north or south.
3. During lunar eclipses it can be seen that the shadow of the Earth is curved (Fig. 2.21).

By relating his theories to observation in this manner, Aristotle broke with the tradition of Plato to some extent, although he approached the problem in the same manner, letting reason rather than observation guide the way.

Other tenets of Aristotle included the conclusions that the universe is finite in size (this led directly to his belief that the heavenly bodies can follow only circular motions, because otherwise they might encounter the edge of the universe) and that the heavenly bodies were made of a fifth fundamental substance, which he called the aether.

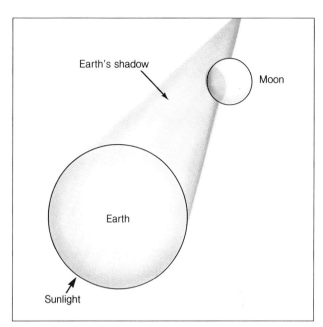

Figure 2.21 Curvature of the Earth's Shadow on the Moon. Only a spherical body can cast a circular shadow for all alignments of the Sun, Moon, and Earth.

The Later Greeks: Sophisticated Cosmologies

After Aristotle's period, the center of Greek scientific thought moved across the Mediterranean Sea to Alexandria, the capital city established in 332 B.C. by Alexander the Great near the site of the present city of Cairo. The first prominent astronomer of this era was **Aristarchus** (c. 310–230 B.C.), the first scientist to adopt the idea that the Sun, not the Earth, is at the center of the universe. This conclusion was based on geometrical arguments showing that the Sun is much larger than the Earth, and therefore more naturally ought to be at the center of the universe. The heliocentric hypothesis of Aristarchus failed to attract many followers at the time, due largely to a lack of any concrete evidence that the Earth was in motion and partly to a general satisfaction with the Aristotelian viewpoint, which had no recognized flaws.

By the third century B.C. the need for more precise mathematical models of the universe became apparent as better observing techniques were developed. A mathematical concept that provided the needed precision while preserving the precepts of Aristotle was the **epicycle,** a small circle on which a planet moves, whose center in turn orbits the Earth following a larger circle called a **deferent** (Fig. 2.22). An important advantage of the epicycle was that it could explain the retrograde motions of the superior planets.

The epicyclic motions of the celestial bodies were refined further by **Hipparchus** (Fig. 2.23), who was active during the middle of the second century B.C. (Very little is known about his life, not even the dates of his birth and death.) Hipparchus, most of whose work was done at his observatory on the island of Rhodes, was one of the greatest astronomers of antiquity. Among his major contributions are:

1. The first use of trigonometry in astronomical work (in fact, he is largely credited with its invention, although many of the concepts were developed earlier).
2. The refinement of instruments for measuring star positions, along with the first known use of a celestial coordinate system akin to our modern equatorial coordinates, which enabled him to compile a catalogue of some 850 stars.
3. Refinement of the methods of Aristarchus for measuring the relative sizes of the Earth, Moon, and Sun.
4. The invention of the stellar magnitude system for estimating star brightness, a system still in use today (with minor modifications).
5. Perhaps most impressive, the discovery of the precession of star positions, which he accomplished by comparing his observations with some that were made 160 years earlier.

It is worthwhile to consider why Hipparchus, with access to the work of Aristarchus (including the knowledge that the Sun is much larger than the Earth), did not adopt the heliocentric view. Apparently he was motivated to reject this idea partly because of one very significant observational reason: He could not detect any apparent shifting of star positions during the course of the year, shifting that he realized should appear as a result of the changing point of view if the Earth moved about the Sun (Fig. 2.24). The lack of any detectable **stellar parallax,** as such an apparent motion is called, helped create resistance to the heliocentric viewpoint

Figure 2.22 The Epicycle. It was realized that planetary motions could be represented by a combination of motions involving an epicycle, which carries a planet as it spins while orbiting the Earth.

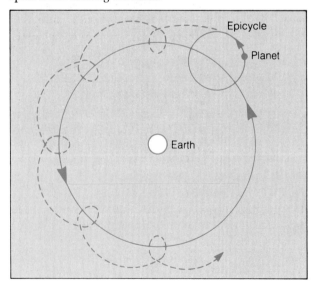

Figure 2.23 A Fanciful Rendition of Hipparchus at his Observatory. *(The Bettmann Archive)*

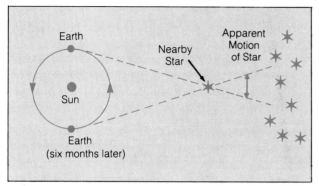

Figure 2.24 Stellar Parallax. As the Earth orbits the Sun, our line of sight toward a nearby star varies, causing the star's position (with respect to more distant stars) to change. The change of position is greatly exaggerated in this sketch; in reality, even the largest stellar parallax displacements are too small to have been measured by ancient astronomers.

for many centuries to come. Other reasons for the resistance to the Sun-centered idea were that there were no obvious flaws in the geocentric model, and no evidence that the Earth moves.

Following the great work of Hipparchus, almost 300 years passed before any significant new astronomical developments occurred. Claudius Ptolemaeus, or simply **Ptolemy** (Fig. 2.25), lived in the middle of the second century A.D. and undertook to summarize all the world's knowledge of astronomy. He did this with the publication of a treatise called the *Almagest*, which was based in part on the work of Hipparchus but also contained some new developments created by Ptolemy. The topics of the thirteen books of the *Almagest* ranged from a summary of the observed motions of the planets to a detailed study of the motions of the Sun and Moon, and from a description of the workings of all the astronomical instruments of the day to a reproduction of the star catalogue of Hipparchus; most importantly, they included the construction of detailed models of the planetary motions (Fig. 2.26). Here Ptolemy made his

Figure 2.25 Claudius Ptolemy. *(The Granger Collection)*

Figure 2.26 The Cosmology of Ptolemy. To account for nonuniformities in the planetary motions, Ptolemy devised a complex epicyclic scheme. The deferent, the large circle on which the epicycle moves, is not centered on the Earth and has a rotation rate that is constant as seen from another displaced point called the equant. This meant that from the Earth, the planet appeared to move faster through the sky when on one side of its deferent than when on the other. Each planet had its own deferent and epicycle, with motion and offsets adjusted to reproduce the observations.

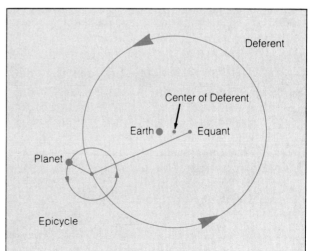

greatest personal contribution, for these models were so accurate in their ability to predict the planetary positions that they were used for the next 1000 years.

The history of Greek astronomy came to an end with the work of Ptolemy. There followed a period known as the Dark Ages, when Western civilization was largely dormant. During this time Arab astronomers preserved many of the Greek traditions, so that the ancient teachings were still firmly entrenched centuries later, when the first stirrings of a new spirit of inquiry began to be felt in Europe.

Perspective

We have now developed a modern picture explaining all of the astronomical phenomena visible to the unaided eye. There is much left for us to learn about the universe and the laws that govern it, but we already have available to us all of the information on which the first astronomers based their crude theories of cosmology. We have seen how the ancient philosophers interpreted these phenomena in developing the earliest concepts of the universe. In the next chapter we shall see how a modern understanding of physics and astronomy was built on these foundations.

Summary

1. The concept of the celestial sphere, along with the celestial equator and the celestial poles, provides a convenient mechanism for discussing the appearance of the sky.
2. The Earth's rotation causes diurnal motions: the daily rising and setting of the Sun, the Moon, the stars, and the planets.
3. The orbital motion of the Earth about the Sun causes annual motions, such as the apparent motion of the Sun through the constellations of the ecliptic, and is responsible, along with the tilt of the Earth's axis, for our seasons.
4. The planetary orbits lie nearly in the same plane, so they are always seen along the ecliptic, in various configurations with respect to the Sun-Earth line. The planets go through temporary retrograde motion due to the relative speed with which they pass or are passed by the Earth.
5. The Moon orbits the Earth, while the Earth orbits the Sun; from the Earth we see the various phases of the Moon as it passes through different configurations.
6. Solar and lunar eclipses occur when the Moon passes directly in front of the Sun and through the Earth's shadow, respectively. The occurrence of these alignments is affected by the tilt and precession of the Moon's orbit.
7. The earliest recorded astronomical data were found in the region known as Babylonia, now Iraq.
8. The first rational inquiry and the earliest cosmological theories arose in the ancient Greek city-states on the shores of the Mediterranean.
9. Pythagoras and his followers developed the notion that the universe can be described by numbers and adopted the belief that the Earth is spherical.
10. Plato and his pupil Aristotle stated underlying principles that they believed to govern the universe and that could not be tested by observation. Their principles included the belief that the Earth is the center of the universe and that all heavenly bodies are spherical.
11. In the third century B.C., Aristarchus used geometrical arguments to conclude that the Sun, not the Earth, is at the center of the universe.
12. Hipparchus, in the second century B.C., made precise observations, applied new mathematical techniques to astronomy, compiled a large star catalog, developed a system for measuring star brightness, discovered precession, and developed the epicyclic theory of planetary motion. Hipparchus did not adopt the Sun-centered cosmology, however, because he could not detect stellar parallax.
13. Ptolemy, in the second century A.D., summarized all astronomical knowledge in the *Almagest* and used the epicyclic theory to construct tables of planetary motion that were used throughout the Dark Ages.

Review Questions

1. Suppose the Earth's speed in its orbit was increased. Would this change the length of the solar day if the sidereal day remained the same as it is now? If so, would it increase or decrease the length of the solar day?

2. Imagine that you live at 30° north latitude. At the time of the winter solstice (i.e., on or about December 21), how far above the southern horizon would the Sun rise at midday? How would your answer differ if you lived at 30° south latitude?
3. Imagine that the Earth's rotation axis is perpendicular to the ecliptic, instead of being tilted 23½° away from perpendicular. What would be the length of the day at the time of the summer solstice for a person living at 40° north latitude?
4. The Moon's synchronous rotation means that its spin and orbital periods are equal. Is it the synodic or the sidereal orbital period that is equal to the rotation period?
5. Would retrograde motion of the planets occur if each planet moved more rapidly than the next one closer to the Sun, rather than more slowly? If it would occur, would it be easily observed?
6. Lunar eclipses always occur at the same phase of the Moon. Which phase is it? At which lunar phase do solar eclipses always take place?
7. The Babylonians adopted a 360-day year, divided evenly into twelve months. Since the year is actually 5¼ days longer than this, how often did they have to add an extra month in order to make things come out even? How do you suppose they allowed for the remaining error that accumulated over many years?
8. Even the best measuring instruments at the time of Hipparchus could only measure angular separations or positions with an accuracy of several arc-minutes. If you assume that the error or uncertainty of a single measurement was 40′, how many years of observation would be required in order for precession, which occurs at a rate of around 50″ per year, to be noticed? How much precession had occurred over the 160-year period covered by the observational data available to Hipparchus?
9. Explain how the epicyclic theories of Hipparchus and Ptolemy satisfied the principles of Plato and Aristotle but also violated some of them in certain ways.
10. Why did Hipparchus, who was aware of the work of Aristarchus, not accept the idea that the Earth orbits the Sun?

Additional Readings

A number of magazines, including *Mercury, Sky and Telescope, Astronomy,* and the *Griffith Observer,* contain practical information for the sky watcher, such as planetary positions and the seasonal appearance of the constellations. There are also annually published handbooks with similar data. One of the most widely used of these is the *Observer's Handbook*, by Roy L. Bishop (Toronto: Royal Astronomical Society of Canada). Some practical exercises in astronomy can be found in such books as *Astronomy: A Self-Teaching Guide*, by Dinah L. Moché (New York: Wiley, 1981).

It is also possible to find readings on the history of astronomy. Below are listed some books relevant to this chapter, including material on ancient astronomical developments that are not covered here, for example, the astronomies of Asia and the Americas. In addition to locating books such as these, it is also useful to browse through professional journals that cover the history of science or of astronomy. These include *Vistas in Astronomy* and the *Journal for the History of Astronomy,* both of which can be found in most science-oriented libraries.

Heath, T. L. 1969. *Greek astronomy.* New York: AMS Press.
Krupp, E. C. 1978. *In search of ancient astronomies.* Garden City, N.J.: Doubleday.
──────. 1983. *Echoes of the ancient skies.* New York: Harper and Row.
Neugebauer, O. 1969. *The exact sciences in antiquity.* New York: Dover.

Chapter 3

The Renaissance

A Renaissance model of the epicyclic
cosmology. (© 1981 O. Gingerich,
with permission of the Houghton
Library, Harvard University)

Chapter 3 The Renaissance 39

Chapter Preview

Copernicus: The Heliocentric View Revisited
Tycho Brahe: Advanced Observations
Johannes Kepler and the Laws of Planetary Motion
Galileo, Experimental Physics, and the Telescope
Isaac Newton
    The Laws of Motion
    The Laws of Gravitation
Energy, Angular Momentum, and Orbits: Kepler's
    Laws Revisited
Tidal Forces

Figure 3.1 Nicolaus Copernicus. *(The Bettmann Archive)*

The fifteenth century saw the beginnings of a reawakening of intellectual spirit in Europe. Some scientific studies began at the major universities, increased maritime explorations brought demands for better means of celestial navigation, and the art of printing was developed, opening the way for widespread dissemination of information.

Major advances in all the sciences accompanied the new developments in other fields of human endeavor. In this chapter we discuss the principal achievements in Renaissance science that led to the development of modern astronomy. In doing so we discuss the accomplishments of five major figures: Copernicus, Brahe, Kepler, Galileo, and Newton.

Copernicus: The Heliocentric View Revisited

Some nineteen years before the epic voyage of Columbus, Niklas Koppernigk (Fig. 3.1) was born in Torun, in the northern part of Poland. As a young man he attended the university in Cracow, where his fondness for Latin, the universal language of scholars, led him to change his surname to Copernicus. At Cracow he developed an avid interest in astronomy, becoming fully acquainted with the Aristotelian view as well as the Ptolemaic model of planetary motions. He persisted in his study of astronomy, and by 1514 he had developed some doubts about the validity of the accepted system.

His reasons for doing so have been the subject of some uncertainty and misconception. It was long assumed that Copernicus was encouraged to adopt the Sun-centered view of the universe because he recognized shortcomings in the geocentric model of Ptolemy. There is, however, no evidence of widespread dissatisfaction with the Ptolemaic system or any records indicating that Copernicus himself found serious inaccuracies in it. His reasons for adopting the heliocentric viewpoint were more subtle.

The basis for his conversion was primarily philosophical. In the mind of Copernicus, his new system presented a pleasing and unifying model of the universe and its motions. While he was no doubt encouraged by the climate of change and cultural revolution that was sweeping Europe with the advent of the Renaissance, he did not adopt the heliocentric view just to be different or to improve the accuracy or reduce the complexity of the accepted view. The Copernican model was, in fact, no more accurate in predicting planetary motions than the Ptolemaic system, and it was just as complex. Copernicus adhered to the notion of perfect circular motions and was obliged to include small epicycles in order to match the observed planetary orbits. Apparently one of the most pleasing aspects of the Sun-centered system to Copernicus was the fact that the relative distances of the planets could be deduced (Fig. 3.2) and were found to have a certain regularity, the spacings between planets growing systematically with distance from the Sun. Copernicus was also able to determine the relative speeds of the planets in their orbits, finding that each planet moves more slowly than the next one closer to the Sun.

Copernicus first circulated his ideas informally sometime before 1514 in a manuscript called *Commentariolis*, and it drew increasing attention over the next several years. The Church voiced no opposition, despite the fact that the ideas expressed in the work were in strong contradiction to commonly accepted Church doctrine. Copernicus was apparently reluctant to publish his findings in a more formal way for fear of raising controversy and was continually rechecking his calculations. Publication of his findings in a book called *De Revolutionibus Orbium Coelestium* (*On the Revolution of the Celestial Sphere*), or simply *De Revolutionibus,* finally took place in 1543, when Copernicus was near death. He did not live to see the profound impact of his work.

Figure 3.2 The Method of Copernicus for Finding Relative Planetary Distances From The Sun. For an inferior planet such as Mercury or Venus, Copernicus knew the angle of greatest elongation, and was therefore able to reconstruct the triangle shown, which provided the Sun-Mercury (or Sun-Venus) distance relative to the Sun-Earth distance (that is, the astronomical unit). For superior planets, similar but slightly more complicated considerations provided the same information.

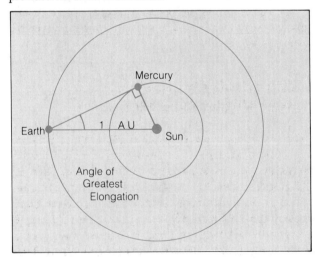

Tycho Brahe: Advanced Observations

Whereas Copernicus had contributed little observationally and had not demanded a close match between theory and observation as long as general agreement was found, there was in fact a strong need for improved precision in astronomical measurements. The next influential character in the historical sequence met this need; if he had not, further progress would have been seriously delayed, as would the eventual acceptance of the Copernican doctrine.

Tycho Brahe (Fig. 3.3) was born in 1546, some three years after the death of Copernicus, in the extreme southern portion of modern Sweden (the region was part of Denmark at the time). Of noble descent, Tycho spent his youth in comfortable surroundings and was well educated, first at Copenhagen University, then at Leipzig, where he insisted on studying mathematics and astronomy despite his family's wish that he pursue a law career. As a result of some notable observations and their interpretation, Tycho eventually developed a strong reputation as an astronomer (and astrologer; the

Figure 3.3 Tycho Brahe. *(The Bettmann Archive)*

distinction was still scarcely recognized) and attracted the attention of the Danish king, Frederick II. In 1575 the king ceded to Tycho the island of Hveen, about fourteen miles north of Copenhagen, along with enough servant support and financial assistance to allow him to build and maintain his own observatory.

Even before this time Tycho had shown an acute interest in astronomical instruments, and with the grant to build his observatory this interest bore fruit. He devised a variety of instruments (Fig. 3.4), which, although they really did not encompass any new principles, were capable of more accurate readings than any before his time.

His observational contribution consisted in part of the unprecedented accuracy of his data and in part of the completeness of his records. Until his time, it had been the general practice of astronomers to record the positions of the planets only at notable points in their travels, such as when a superior planet comes to a halt just before beginning retrograde motion. Tycho made much more systematic observations, recording planetary positions at times other than just the significant turning points in their motions. He also made multiple observations in many instances, allowing the results to be averaged to improve their accuracy. Tycho himself did not attempt any extensive analysis of his data, but the vast collection of measurements that he gathered over the years contained the information needed to reveal the basis of the planetary motions.

Tycho was unable to accept the heliocentric view, primarily because he could find no evidence that the Earth was moving. He tried and failed to detect stellar parallax, which he supposed he should see with his accurate observations if the Earth really moved. Furthermore, as a strict Protestant he found it philosophically difficult to accept a moving Earth when the Scriptures stated that the Earth is fixed at the center of the universe. On the other hand, he realized that the Copernican system had advantages of mathematical simplicity over the Ptolemaic model, and in the end he was ingenious enough to devise a model that satisfied all of his criteria. He imagined that the Earth was fixed with the Sun orbiting it, but that all the other planets orbited the Sun (Fig. 3.5). Mathematically, this is equivalent to the Copernican system in terms of accounting for the motions of the planets as seen from the Earth. The idea never received much acceptance, however, and Tycho is remembered primarily for his fine observations.

Figure 3.4 The Great Mural Quadrant at Hveen. The quadrant was used to measure the angular positions of stars and planets with respect to the horizon. *(The Granger Collection)*

Tycho Brahe died in 1599. The task of seeking the secrets contained in Tycho's data was left to those who followed him, particularly a young astronomer named Johannes Kepler.

Johannes Kepler and the Laws of Planetary Motion

Until now it has been possible to follow developments in a straight sequence, but here we must begin to cover events that occurred at nearly the same time, but in different locations. To complete the thread begun with our discussion of Tycho Brahe, this section is concerned with Johannes Kepler (Fig. 3.6), who worked briefly with Tycho himself and then spent many years on the analysis of the great wealth of observational data

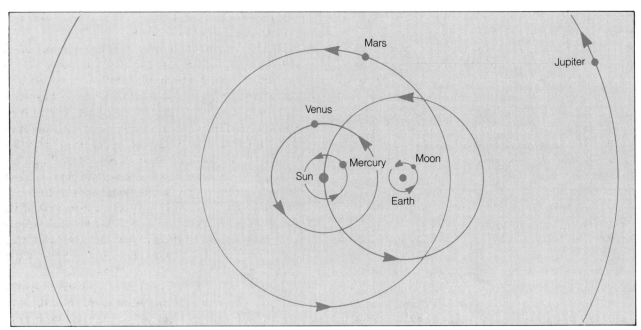

Figure 3.5 Tycho's Model of the Universe. Tycho held that the Earth was fixed, and the Sun orbited the Earth. The planets in turn orbited the Sun. This system was never worked out in mathematical detail, but it successfully preserved both the advantages of the Copernican system and the spirit of the ancient teachings. It also accounted for the lack of observed stellar parallax, since the Earth was fixed.

Figure 3.6 Johannes Kepler. *(The Bettmann Archive)*

that Tycho had accumulated. We must keep in mind, however, that during the same period Galileo Galilei was at work in Italy, and Galileo and Kepler were aware of each other's accomplishments.

Born in 1571, Kepler was a sickly youngster seemingly headed for a career in theology. While attending university, however, he encountered a professor of astronomy who inspired in him a strong interest in the Copernican system, which he adopted wholeheartedly. Kepler thereafter devoted his life to seeking the underlying harmony of the cosmos. He devoted great effort to a search for simple numerical relationships among the planets, and in doing so embarked on several false turnings and erroneous paths. He was a true scientist, however, in that he was always willing to discard his ideas if the data did not support them.

Kepler went to work as Tycho's assistant in 1598, and a year later found himself the beneficiary of the massive collection of data left behind at Tycho's death. Kepler's main mission, due both to the wishes of Tycho and to his own interests, was to develop a refined understanding of the planetary motions and to upgrade the tables used to predict their positions. He set to work first on the planet Mars; the data were particularly extensive, and its motions were among the most difficult to explain in the established Ptolemaic system (or in the Copernican system with its requirement of circular motions only). By a very complex process, Kepler was able to separate the effects of the Earth's motion from those of Mars itself, so that he could map out the path that Mars followed with respect to the Sun.

By 1604 Kepler had determined that the orbit of Mars was some kind of oval, and further experimentation revealed that it was fitted precisely by a simple geometric figure called an **ellipse** (Fig. 3.7). An ellipse is a closed curve defined by a fixed total distance from two points called **foci,** and indeed Kepler found that the Sun was at the precise location of one focus of the ellipse. This discovery was later generalized by Kepler to apply to all of the planets, and it became known as Kepler's first law of planetary motion: The orbit of each planet is an ellipse, with the Sun at one focus.

Further analysis of the motion of Mars revealed a second characteristic, related to the fact that the planet moves fastest in its orbit when it is nearest the Sun and slowest when it is on the opposite side of its orbit, farthest from the Sun. Mathematically, Kepler's second discovery was that a line connecting Mars to the Sun sweeps out equal areas of space in equal intervals of time (Fig. 3.7). Again, Kepler later stated that this law, too, applies to all of the planets.

Kepler's results on the orbit of Mars were published in 1609 in a book entitled *The New Astronomy: Commentaries on the Motions of Mars*. The book received a great deal of attention. In 1619 Kepler published a book entitled *The Harmony of the World*, in which he reported his discovery of a simple relationship between the orbital periods of the planets and their average distances from the Sun. Now known as Kepler's third law, or simply the harmonic law, it states that the square of the period of a planet is proportional to the cube of the semimajor axis (which is half of the long axis of an ellipse). That is, $P^2 = a^3$, where P is the sidereal period of a planet in years, and a is the semimajor axis in terms of the Sun-Earth distance (i.e., in terms of the **astronomical unit** or **AU**). (See Table 3.1.)

In another major work called the *Epitome of the Copernican Astronomy*, published in parts in 1618, 1620, and 1621, Kepler presented a summary of the state of astronomy at that time, including Galileo's discoveries. In this book Kepler generalized his laws, explicitly stating that all of the planets behaved similarly to Mars, something that had clearly been his belief all along.

By the time the *Epitome* was published, the Roman Catholic Church was in a very intolerant frame of

Figure 3.7 The Ellipse. At left is an exaggerated ellipse representing a planetary orbit with the Sun at one focus; at right is a similarly exaggerated ellipse with lines drawn to illustrate Kepler's second law. If the numbers represent the planet's position at equal time intervals, the areas of the triangular segments are equal.

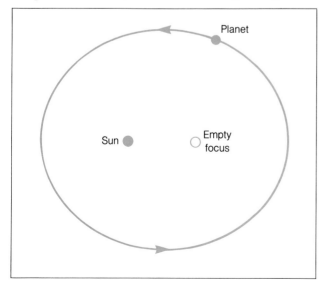 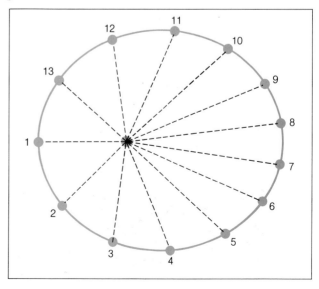

Table 3.1 Testing Kepler's Third Law

Planet	a(AU)	P(years)	a^3	P^2
Mercury	0.387	0.241	0.058	0.058
Venus	0.723	0.615	0.378	0.378
Earth	1.000	1.000	1.000	1.000
Mars	1.523	1.881	3.533	3.538
Jupiter	5.203	11.86	140.85	140.66
Saturn	9.539	29.46	867.98	867.89
Uranus	19.18	84.01	7,055.79	7,057.68
Neptune	30.06	164.8	27,162.32	27,159.04
Pluto	39.44	248.4	61,349.46	61,762.56

Figure 3.8 Galileo Galilei. *(The Granger Collection)*

mind, in contrast with the situation at the time of Copernicus nearly one hundred years before, and Kepler's treatise soon found itself on the *Index of Prohibited Books*, along with *De Revolutionibus*.

In 1627, Kepler published his last significant astronomical work, a table of planetary positions based on his laws of motion, which could be used to predict planetary motions accurately. These tables, which he called the *Rudolphine Tables* in honor of a former benefactor, were used for the next several years. The *Rudolphine Tables* represented an improvement in accuracy over any previous tables by nearly a factor of 100, a resounding and remarkable confirmation of the validity of Kepler's laws. In a very real sense the *Rudolphine Tables* represented Kepler's life's work, since with their publication he completed the task set before him when he first went to work for Tycho. Kepler died in 1630 at the age of 59.

Galileo, Experimental Physics, and the Telescope

Very strong contrasts can be drawn between Kepler and his great contemporary, Galileo Galilei (Fig. 3.8), born in Pisa in northern Italy in 1564. Where Kepler was fascinated with universal harmony and therefore with the underlying principles on which the universe operates, Galileo was primarily concerned with the nature of physical phenomena and was less devoted to finding fundamental causes. Galileo wanted to know how the laws of nature operated, whereas Kepler sought the reason for their existence.

Galileo's approach was level-headed and rational in the extreme. He used simple experiment and deduction in advancing his perception of the universe and has frequently been cited as the first truly modern scientist, although others of his time probably deserve a share of that recognition. A follower of Plato and Aristotle, whose works still dominated in Galileo's time, would proceed by rational thought from standard unproven assumptions; Galileo found it much more sensible to begin with experiment or observation and work toward a recognition of the underlying principles. In doing this Galileo founded an entirely new basis for scientific inquiry, an achievement in many ways more profound than his contributions to astronomy, which were considerable.

Galileo's discoveries in physics concerned the motions of objects and were published in his later years, after his astronomical career had been forcibly ended by Church decree. It was in fact an early interest in *mechanics*, the science of the laws of motion, that lured Galileo away from a career in medicine, the subject of his first studies. Galileo's contributions to physics will be described briefly later in this chapter.

Galileo's astronomical discoveries were quite sufficient to earn him a major place in history, and his flair for debate and habit of ridiculing those whose arguments he disproved made him famous in his own time though not universally loved.

In 1609 Galileo learned of the invention of the telescope and devised one of his own, which he soon put to use in systematic observations of the heavens. Despite the poor quality of the instrument, Galileo made a number of important discoveries almost at once and reported them in 1610 in a publication called the *Starry Messenger*. Here Galileo showed that the Moon was not a perfect sphere but was covered with craters and mountains (Fig. 3.9). He also reported the existence of many more stars than could be seen with the unaided eye and, most significant of all, the fact that Jupiter was attended by four satellites, whose motions he observed long enough to establish that they orbited the parent planet (Fig. 3.10). All of these discoveries, but most especially the latter, violated the ancient philosophies of an idealized universe centered on the Earth (see Table 3.2). The satellites of Jupiter showed beyond reasonable question that there were other centers of motion than the Earth.

Mostly as a result of the reputation Galileo earned with the publication of the *Starry Messenger*, he was

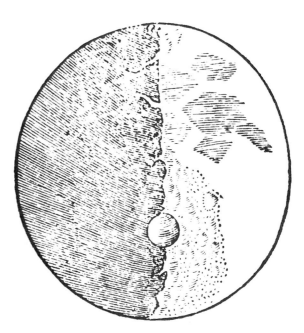

Figure 3.9 Galileo's Sketch of the Moon. *(The Granger Collection)*

Table 3.2 Some of Galileo's Arguments for the Heliocentric Theory

Discovery	Argument
Many faint stars	Difficult to reconcile with the idea of stars as points attached to a crystalline sphere
Craters on the Moon	Moon is not a perfect, immutable heavenly body
Moons of Jupiter	A body other than Earth as a center of motion
Phases of Venus	Explained only if Venus orbits the Sun and shines by reflected sunlight
Sunspots	Spots are on the solar surface, showing that the Sun is not perfect
Variable planetary sizes	Angular size variations explained by motion of planets around the Sun

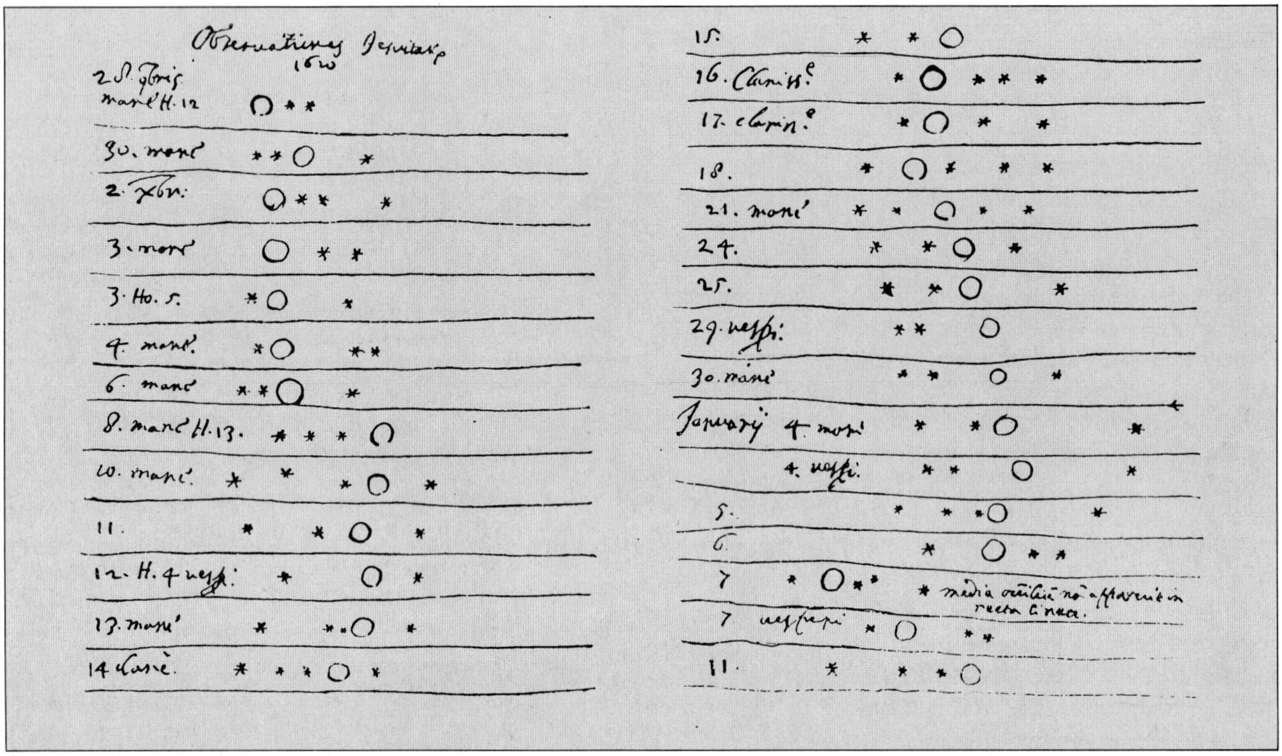

Figure 3.10 Galileo's Sketches of the Moons of Jupiter. This series of drawings by Galileo is often attributed to the *Starry Messenger*, but in fact was made some years later. *(Yerkes Observatory)*

able to negotiate successfully for the position of court mathematician to the Grand Duke of Tuscany, and he moved in 1610 to Florence, where he was to spend the remainder of his long career. Once established there, Galileo continued his observations and soon added new discoveries to his list. He found that Venus changes its appearance much as the Moon does, and he showed that this proves that Venus orbits the Sun, undergoing phases as varying portions of the sunlit side are visible from Earth (Fig. 3.11). He analyzed sunspots, dark blemishes seen crossing the face of the Sun, and showed that they really were *on* the Sun and were not small planets orbiting close to it, as some had suggested. Both of these discoveries refuted ancient teachings.

Galileo began to draw increasingly heavy criticism from the Church, and he made efforts to develop good relations with high-ranking officials in Rome. Nevertheless, in 1616 he was pressured into refuting the Copernican doctrine and for several years thereafter was relatively quiet on the subject. Except for a well-publi-

cized debate on the nature of comets, Galileo spent most of his time preparing his greatest astronomical treatise, which was finally published, after some difficulties with Church censors, in 1632. In order to avoid direct violation of his oath not to support the Copernican heliocentric view, Galileo wrote his book in the form of a dialogue among three characters: One, named Simplicio, represented the official position of the Church; another, Salviati, represented Galileo (although this was, of course, not stated explicitly); and a third, Sagredo, was always quick to see and agree with Salviati's arguments. In this treatise, called the *Dialogue on the Two Chief World Systems*, Galileo, through the character Salviati and at the expense of Simplicio, systematically destroyed many of the traditional astronomical teachings of the Church. The book was published in Italian, rather than the scholarly Latin, and its contents were therefore accessible to the general populace.

Despite a lengthy preface in which Galileo disavowed any personal belief in the heliocentric doctrine,

Astronomical Insight 3.1

An Excerpt from *Dialogue on the Two Chief World Systems*

During the course of three days of discussion, the three characters in Galileo's *Dialogue* thoroughly air all the available evidence and arguments bearing on the understanding of mechanics and the structure and nature of the universe. About midway through, a discussion develops that is central to Galileo's principal point: The Sun, not the Earth, is at the center of the universe. In the following excerpt, we see an example of Salviati's persuasive style, Simplicio's dogged reluctance to give up the ideas of Aristotle, and Sagredo's ready comprehension of Salviati's arguments:

Salviati: Now if it is true that the center of the world is the same about which the circles of the mundane bodies, that is to say, of the Planets, move, it is most certain that it is not the Earth but the Sun, rather, that is fixed in the center of the World. So that as to this first simple and general apprehension, the middle place belongs to the Sun, and the Earth is as far remote from the center as it is from that same Sun.

Simplicio: But from whence do you argue that not the Earth but the Sun is in the center of planetary revolutions?

Salviati: I infer the same from the most evident and therefore necessarily conclusive observations, of which the most potent to exclude the Earth from the said center, and to place the Sun therein, are that we see all the planets sometimes nearer and sometimes farther off from the Earth, with so great differences, that, for example, Venus when it is at the farthest is six times more remote than when it is nearest, and Mars rises almost eight times as high at one time as at another. See therefore whether Aristotle was somewhat mistaken in thinking that it was at all times equidistant from us.

Simplicio: What in the next place are the tokens that their motions are about the Sun?

Salviati: It is shown in the three superior planets, Mars, Jupiter, and Saturn, in that we find them always nearest to the Earth when they are in opposition to the Sun and farthest off when they are towards the conjunction; and this approximation and recession imports thus much, that Mars near at hand appears sixty times greater than when it is remote. As to Venus, in the next place, and to Mercury, we are certain that they revolve about the Sun in that they never move far from it, and in that we see them sometimes above and sometimes below it, as the mutations in the figure of Venus necessarily prove. Touching the Moon, it is certain that it cannot in any way separate itself from the Earth, for the reasons that shall be more distinctly alleged hereafter.

Sagredo: I expect that I shall hear more admirable things that depend upon this annual motion of the Earth than were those dependent upon the diurnal revolution.

In this exchange, Galileo, in the guise of Salviati, advances several arguments based on observations, including the well-known phases of Venus and the less widely quoted arguments having to do with the varying distances of the planets from the Earth. In this and other passages, Galileo went out of his way to mock the followers of Aristotle, who, according to Salviati in an earlier paragraph, "would deny all the experiences and all the observations in the world, nay, would refuse to see them, that they might not be forced to acknowledge them, and would say that the world stands as Aristotle writes and not as Nature will have it."

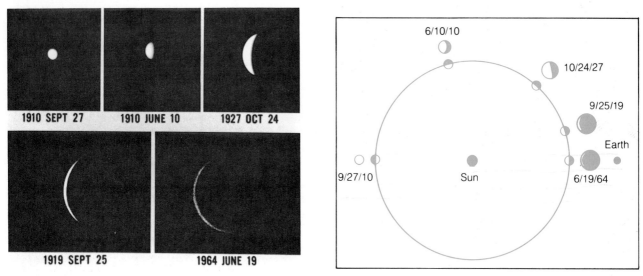

Figure 3.11 The Phases of Venus. As Venus orbits the Sun and its position changes relative to the Sun-Earth line, its phase varies as we see differing portions of its sunlit side. In addition, its apparent size varies because of its varying distance from Earth.

the Church reacted strongly, and within a few months Galileo was summoned before the Roman Inquisition, while further publication of the book was banned. It joined the works of Copernicus and Kepler on the *Index* (where it was to remain until the 1830s), and Galileo himself was sentenced to house arrest, eventually serving this punishment at his country home near Florence. This is where he spent the remainder of his days. Galileo died in 1642, having suffered blindness in the last four years of his life.

Isaac Newton

In the year 1643, a few months after the death of Galileo, Isaac Newton (Fig. 3.12) was born in Woolsthorpe, England. Newton's childhood was unremarkable, except that he showed a growing interest in mathematics and science. At age eighteen he entered college at Cambridge and received a bachelor's degree in early 1665. He then spent two years at his home in Woolsthorpe, largely because the plague made city living rather dangerous. During this time he made a remarkable series of discoveries in the fields of physics, astronomy, optics, and mathematics, in what surely must have been one of the most intense and productive periods of individual intellectual effort in human history.

Newton had a tendency to exhaust a subject, get bored with it, and go on to new fields, so that nothing of his work was published until after some of his discoveries were repeated independently by others. Finally, after persistent urging by his friend (and fellow astronomer) Edmund Halley, in 1687 Newton published a massive work called *Philosophiae Naturalis Principia Mathematica* (Fig. 3.13), now usually referred to simply as the *Principia*. In this three-volume book, Newton established the science of mechanics (which he viewed as merely background material and relegated to an introductory section) and applied it to the motions of the Moon and the planets, developing the law of gravitation as well. His work in optics was published separately in 1704, although it was probably written much earlier than that.

The *Principia* received great notice, particularly in England. As a result, Newton's later life was a public one, with various government positions and less and less time for scientific discovery. With the help of younger associates, he did revise the *Principia* on two occasions, in 1713 and in 1726, making some improvements each time. Newton died in 1727 at the age of 84. His influence lives on in our modern understanding of physics and mathematics. Newton's conclusions on the nature of motions and gravity are still viewed as correct, although it is now realized that there are circumstances in which more complex theories (such as Einstein's theory of relativity) must be used.

Figure 3.12 Isaac Newton. *(The Granger Collection)*

The Laws of Motion

Newton put forth three laws of motion, principles he considered so self-evident that he relegated them to an introductory section of the *Principia*. The first of Newton's laws states the principle of **inertia** (Fig. 3.14), a concept first recognized by Galileo, who realized that an object in motion tends to stay in motion unless something acts to slow or stop it. This was completely contrary to the teachings of Aristotle, who held that the natural tendency of any moving object was to stop, and that it would only continue moving if a force were applied. Aristotle was misled by his failure to recognize friction, a force that tends to stop motion in most everyday situations. Newton expanded the concept of inertia, recognizing it as just one in a series of physical principles that govern the motions of objects and adding the all-important notion of **mass.**

The mass of an object reflects the amount of matter it contains, which in turn determines other properties such as weight and momentum. It is the other properties that are easily observed, not the mass, so mass can be a difficult concept. We can make useful illustrations by considering situations where we alter these other properties but not the mass. For example, the weight of an object varies according to the magnitude of the

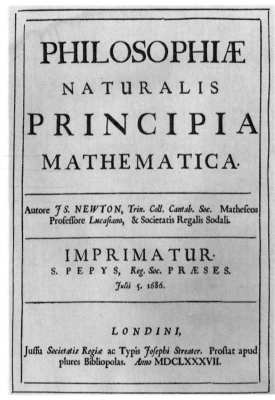

Figure 3.13 The Title Page From an Early Edition of The *Principia*. *(The Granger Collection)*

Figure 3.14 Inertia. An object in motion tends to remain in motion. *(H. Roger Viollet)*

Astronomical Insight 3.2

The Forces of Nature

One of the most difficult concepts to grasp is that of a force, particularly when there is no concrete, visible object to exert it. We think of a force as a push or a pull, and we can visualize it quite easily in certain circumstances, such as when a gardener pushes a wheelbarrow and it moves. It is perhaps less obvious in the case where no movement results, and even less so when no tangible agent exerts the force. In the case of gravity, the force is exerted invisibly and over great distances.

In general terms, a force exerted by a concrete object is referred to as a **mechanical force**, and those exerted without any such agent are **field forces**.

Gravity is the most familiar force created by a field.

There are four basic kinds of field forces in nature that form the basis for *all* forces, mechanical or field. Gravity was the first of the four to be discovered, although in "discovering" gravity and mathematically describing its behavior, Newton did not develop a real fundamental understanding of how it works or why. The reason that gravity was the first to be discovered is simple; no special conditions are required for gravitational forces to be exerted (*all* masses attract each other), and gravity operates over very great distances. After all, Newton deduced its properties by noting how it controls the motions of planets that are separated from the Sun by as much as a billion kilometers.

The second most obvious type of field force, and the second to be discovered and described mathematically, is the **electromagnetic force.** There is an intimate relationship between electric and magnetic fields. The interaction of these fields with charged particles, first described mathematically by the Scottish physicist James Clerk Maxwell, creates forces. The electromagnetic force is actually much stronger than the force of gravity; the electromagnetic force binding an electron to a proton in the nucleus of an atom is 10^{39} times stronger than the gravitational force between the two particles. Like gravity, the electromagnetic force is inversely proportional to the square of the distance between charged particles. Therefore we might expect this force always to dominate gravity, as it does on a subatomic scale, but

---

gravitational acceleration to which it is subjected, but its mass does not change. Astronauts in space may be weightless, but they contain just as much mass as they do on the ground. Mass is usually measured in units of **grams** or **kilograms.** One gram is the mass of a cubic centimeter of water, and a kilogram, which weighs about 2.2 pounds at sea level, is the mass of 1000 cubic centimeters (or one **liter**) of water.

Inertia is very closely related to mass. The more massive an object is, the more inertia it has, and the more difficult it is to start it moving or to stop or alter its motion once it is moving. Newton summarized the concept of inertia in what has become known as his first law of motion:

> A body at rest or in a state of uniform motion tends to stay at rest or in uniform motion unless an outside force acts upon it.

Having stated that a force is required to change an object's state of rest or uniform motion, Newton went on to determine the relationship between force and the change in motion that it produces. To understand this, we must discuss the idea of acceleration.

Acceleration is a general word for *any* change in the motion of an object. It is acceleration when a moving object is speeded up or slowed down or when its direction of motion is altered. It is also acceleration when an object at rest is put into motion. A planet or-

> 1st INERTIA
> 2nd F = MA
> 3rd For every action there is an equal an opp reaction

Astronomical Insight 3.2

The Forces of Nature

this is not so. Most objects in the universe—although they are composed of vast numbers of atoms containing both electrons and protons—have little or no electrical charge, but they always have mass and are therefore subject to gravitational forces. If the planets and the Sun had electrical charges in the same proportion to their masses as the electron and proton do, electromagnetic forces, not gravity, would control their motions.

Besides being immensely stronger, the electromagnetic force also differs from gravity in that it can be either repulsive or attractive, depending on whether the electrical charges are the same or opposite.

The remaining two forces were not discovered until the 1930s (Maxwell's work on electromagnetics was carried out in the late 1800s). Both of these new forces involve interactions at the subatomic level, and they were not discovered until the science of quantum mechanics was developed. One of these forces is the **strong nuclear force**, which is responsible for holding together the protons and neutrons in the nucleus of an atom, and the other is the **weak nuclear force**, some 10^{-5} times weaker than the strong nuclear force. The strong nuclear force is about 100 times stronger than the electromagnetic force, making it the strongest of all, but it operates only over very small distances. Within an atomic nucleus, where protons with their like electrical charges are held together despite their electromagnetic repulsion for each other, it is the strong nuclear force that keeps the nucleus from flying apart. The weak nuclear force plays a more subtle role, showing its effects primarily in certain modifications of atomic nuclei during radioactive decay.

From the smallest to the largest scales, these four forces appear to be responsible for all interactions of matter. It is ironic that at the most fundamental level, the mechanism that makes the forces work is not understood. To do so is one of the principal goals of modern physics. One hope is to develop a mathematical framework that encompasses all of the forces, a framework first sought over sixty years ago by Albert Einstein, and referred to as a **unified field theory**. Progress has been made since Einstein's time, particularly toward unifying the electromagnetic and nuclear forces, but the ultimate goal still eludes us.

---

biting the Sun is undergoing constant acceleration; otherwise it would fly off in a straight line.

Another way of stating Newton's first law would have been to say that in order to accelerate an object a force must be applied to it. Note that this force must be an unbalanced one; that is, there will be acceleration only when a force is applied to an object with no other force to counteract it. Thus, a crate sitting on the floor is subject to the downward force of gravity, but this is balanced by the upward force due to the floor, and there is no acceleration. Newton's second law spells out the relationship among an unbalanced force, the resultant acceleration, and the mass of the object:

> The acceleration of an object is equal to the force applied to it divided by its mass.

$$A = \frac{F}{m}$$

This may be written mathematically as $a = F/m$, where a is the acceleration, F is the unbalanced force, and m is the mass. More commonly, it is written in the equivalent form $F = ma$.

We can visualize simple examples to help illustrate the second law. If one object has twice the mass of another, for example, and equal forces are applied to the two, the more massive one will only be accelerated half as much. Conversely, if unequal forces are applied to objects of equal mass, the one to which the greater force is applied will be accelerated to a greater speed.

Newton's third law of motion is probably more subtle than the first two, although in some circumstances it is quite obvious. It states:

> For every action there is an equal and opposite reaction.

Put in other words, when a force is applied to an object, it pushes back with an equal force (Fig. 3.15). This may sound confusing, because the acceleration is not necessarily equal and because in most common situations, other forces such as friction complicate the picture. Furthermore, there are many static situations in which forces are balanced and no motion occurs.

The third law can be most easily visualized by considering situations where friction is not important. As an example, imagine standing in a small boat and throwing overboard a heavy object, such as an anchor. The boat will move in the opposite direction to the anchor, because the anchor exerts a force on you as you throw it. The "kick" of a gun when it is fired is another example of action and reaction. Technically speaking, when a person jumps off the ground by pushing against the Earth, both he and the Earth are accelerated by the mutual force, but of course the immensely greater mass of the Earth prevents it from being accelerated noticeably.

The third law of motion states the principle on which a rocket works. In this case hot, expanding gas is allowed to escape through a nozzle, creating a force on the rocket. The gas is accelerated in one direction, and the rocket is accelerated in the opposite direction (Fig. 3.16). Anyone who has inflated a balloon and then let go of it, allowing it to zoom through the air, is familiar with the operating principle of a rocket.

The Law of Gravitation

Newton realized, as Galileo had before him, that inertia would cause the planets to fly off in straight lines if no force were acting on them. Newton realized that the planets must be undergoing constant acceleration toward the center of their orbits, that is, toward the Sun. He set out to understand the nature of the force that creates this acceleration.

If you feel confused about the direction of this force, remember what you have just learned about acceleration and inertia: A planet needs no force to keep it moving, but it does require a force to keep its path curving as it travels around the Sun. What is needed is something to push the planets inward toward the Sun (Fig. 3.17). This force can be compared to the tension in a string tied to a rock that you whirl about your head; if you suddenly cut the string, the rock would fly off in whatever direction it happened to be going at the time.

Figure 3.15 Action and Reaction. Some manifestations of Newton's third law are rather subtle, while others are not. Here the books exert a force on a table, and the table exerts an equal and opposite force on the books. There is a static situation; there is no motion. When a cannon is fired, however, the shell is accelerated one way and the cannon the other. The force applied to each is the same, but the cannon has more mass than the shell and is therefore accelerated less, as Newton's second law states. (bottom photo: *U.S. Army*)

What, then, is the string that keeps the planets whirling about the Sun? Newton realized that the Sun itself must be the source of this force, and he made use of Kepler's third law, as well as observations of the Moon's orbit and falling objects at the earth's surface, to discover its properties. He then formulated his law of universal gravitation:

Any two bodies in the universe are attracted to each other with a force that is proportional to the masses of the two

Figure 3.16 Newton's Third Law Applied to the Launch of a Rocket. Hot gases are forced out of a nozzle (or several, as in this case), and in return they exert a force on the rocket that accelerates it. This is the first launch of the space shuttle *Columbia*. (NASA)

bodies and inversely proportional to the square of the distance between them.

The law of gravitation is one of the fundamental rules by which the universe runs. As we shall see in later chapters, it describes the motions of stars about each other or about the center of the galaxy as well as the motions of the Moon and the planets, and it is applicable to the motions of galaxies about one another. Gravity, in fact, is the dominant factor that will determine the ultimate fate of the universe. Table 3.3 shows the relative gravitational forces on a person on earth due to various bodies in the universe.

We are all familiar with the concept of weight; now we see that a person's weight is simply the force of gravity between the person and the Earth. Using Newton's law of gravitation, the two masses are the mass of the person and the mass of the Earth, and the distance between them is the Earth's radius (the Earth acts as though all of its mass were concentrated in a single point at its center).

Many of us are familiar with the notion of gravitational acceleration. The acceleration of gravity at the surface of a planet is proportional to the mass of the planet divided by the square of its radius. (It is found

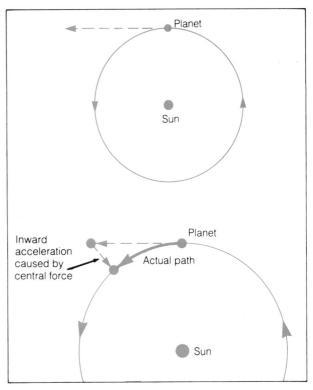

Figure 3.17 Planetary Orbits and Central Force. A planet would fly off in a straight line if no force attracted it toward the center of its orbit. From diagrams like the one below, Newton was able to determine the amount of acceleration required by an orbiting body to keep it in its orbit.

Table 3.3 Gravitational Forces Acting on a Human Standing on Earth

Source of Force	Relative Strength of Force
Earth	1.0
Moon	3.4×10^{-6}
Sun	8.6×10^{-4}
Venus (at closest approach)	1.9×10^{-8}
Jupiter (at closest approach)	3.3×10^{-8}
Nearest star	1.4×10^{-14}
Milky Way galaxy	2.1×10^{-11}
Virgo cluster of galaxies	10^{-15}

Table 3.4 Surface Gravities on Solar System Bodies

Body	Surface Gravity
Earth	1.0 g (= 980 cm/sec^2)
Sun	25.4
Moon	0.17
Mercury	0.38
Venus	0.90
Mars	0.38
Ceres (largest asteroid)	0.000,167
Jupiter	2.70
Saturn	1.20
Uranus	0.90
Neptune	1.10
Pluto	0.08:

by dividing the weight of an object, as expressed by the law of gravitation, by the mass of the object, or $a = F/m = GmM/R^2m = GM/R^2$, where a is acceleration, F is force, and m is the mass of an object on the surface of a planet whose mass is M and radius is R. The mass of the object cancels out, showing that the acceleration of a falling body does not depend on its mass, something Galileo had shown experimentally.)

Knowing that gravitational acceleration is proportional to the mass of a planet and inversely proportional to the square of its radius, we can find the surface gravity on any planet or moon by comparison with the Earth (see Table 3.4). The Moon, for example, has 0.273 times the radius of the Earth, and 0.0123 times the mass. Thus, the acceleration of gravity at the Moon's surface is $0.0123/0.273^2 = 0.165$ that at the surface of the Earth. Therefore astronauts on the Moon weigh approximately one sixth as much as they do on the Earth.

Energy, Angular Momentum, and Orbits: Kepler's Laws Revisited

Even though Newton made use of Kepler's third law in deriving the law of gravitation, the latter is in fact more fundamental. It was soon possible for Newton to show that all three of Kepler's laws follow directly from Newton's laws of motion and gravitation. Kepler's studies of planetary motions had revealed the *result* of the laws of motion and gravitation, while Newton found the *cause* of the motions. To appreciate how this was done, we must further discuss some basic physical ideas.

The concept of **energy** is important for understanding not only orbital motions but also many other aspects of the universe. In an intuitive sense, energy may be defined as the ability to do work. Energy can take on many possible forms, such as electrical energy, chemical energy, heat, and others. All forms of energy can be classified as either **kinetic energy,** the energy of motion, or **potential energy,** stored energy that must be released (i.e., converted to kinetic energy) if it is to do work. A speeding car has kinetic energy because of its motion; a tank of gasoline has potential energy because of its chemical reactivity, that is, tendency to release large amounts of kinetic energy if ignited. Thus a car operates when this potential energy is converted to kinetic energy in its cylinders.

The units used for measuring energy can be expressed in terms of the kinetic energy of specified masses moving at specified speeds. The kinetic energy of a moving object is $\frac{1}{2}mv^2$, where m is its mass and v its speed. In astronomy the most commonly used unit of energy is the **erg,** a small amount of energy that is equivalent to the kinetic energy of a mass of two grams moving at a speed of one centimeter per second.

We often speak in terms of **power,** which is simply energy expended per second. Astronomers tend to use units of ergs per second, which have no special name (the familiar watt is equal to 10^7 ergs per second). Thus, we will speak of the power (or equivalently, the **luminosity**) of a star in terms of its energy output in ergs per second.

Using our understanding of energy, we can now discuss orbital motions in a much more general way than we have previously done. Two objects subject to each other's gravitational attraction have kinetic energy due to their motions and potential energy because they each feel a gravitational force. Just as a book on a table has potential energy that can be converted to kinetic energy if it is allowed to fall, an orbiting body also has potential energy by virtue of the gravitational force acting on it.

Newton's laws and the concepts of kinetic and potential energy can be used to show that there are many types of possible orbits when two bodies interact gravitationally (Fig. 3.18). Not all are ellipses, because if one of the objects has too much kinetic energy (exceeding the potential energy due to the gravity of the other), it will not stay in a closed orbit but will instead follow

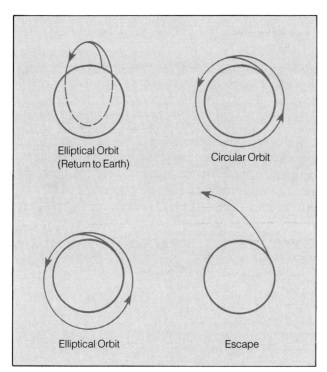

Figure 3.18 Orbital and Escape Velocities. A rocket launched with insufficient speed for circular orbit would tend to orbit the Earth's center in an ellipse, but would intersect the Earth's surface. Given the correct velocity, it will follow a circular orbit. A somewhat larger velocity will place it in an elliptical orbit that does not intersect the Earth. If given enough velocity so that its kinetic energy is greater than its gravitational potential energy, however, it will escape entirely.

an arcing path known as a hyperbola and will escape after one brief encounter. Some comets have so much kinetic energy that after one trip close to the Sun, they escape forever into space, following hyperbolic paths.

If the kinetic energy is less than the potential energy, as it is for all of the planets, then the orbit is an ellipse, as Kepler found. It is technically correct to say that a planet and the Sun orbit a common **center of mass** (Fig. 3.19), rather than saying that the planet orbits the Sun. The center of mass is a point in space between the two bodies where their masses are essentially balanced; more specifically, it is the point where the product of mass times distance from this point for the two objects is equal. Since the Sun is so much more massive than any of the planets, the center of mass for any Sun-planet pair is always very close to the center of the Sun,

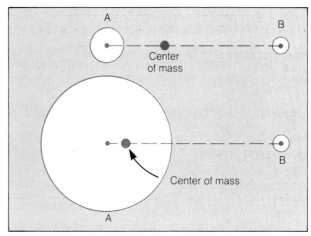

Figure 3.19 Center of Mass. The upper sketch depicts a double star where star A has twice the mass of star B, so the center of mass, about which the two stars orbit, is one-third of the way between the centers of the two stars. The lower sketch shows a case where star A has ten times the mass of star B, so the center of mass is very close to star A. The Sun is so much more massive than any of the planets that the center of mass for any Sun-planet pair is very near the center of the Sun; thus the Sun's orbital motion is very slight.

so the Sun moves very little, and we do not easily see its orbital motion. It is true, however, that the Sun's position wiggles a little as it orbits the centers of mass established by its interaction with the planets, especially the most massive ones. In a double star system, where the two masses are more nearly equal, it is easier to see that both stars orbit a point in space between them.

Thus Kepler's first law as he stated it requires a slight modification: Each planet has an elliptical orbit with the center of mass between it and the Sun at one focus.

The second law can also be restated in terms of Newton's mechanics. Any object that rotates or moves around some center has **angular momentum,** which depends on its mass, its speed, and its distance from the center of motion. In the simple case of an object in circular orbit, the angular momentum is the product mvr, where m is its mass, v its speed, and r its distance from the center of mass.

The total amount of angular momentum in a system is always constant. Because of this, a planet in an elliptical orbit must move faster when it is close to the

Astronomical Insight 3.3

Relativity

In the text, Newton's work is characterized as being nearly modern in the sense that his laws of motion are still thought to be adequate representations of the manner in which forces and bodies interact. Although it is true that Newtonian mechanics is sufficient for most practical applications, there are circumstances where this is not so. By the late 1800s some difficulties with Newton's laws were becoming recognized, and in the first decades of this century these problems led Albert Einstein to develop his theories of special and general relativity.

The circumstances in which Newton's laws fail are those in which we deal with extremely high velocities (that is, velocities that are a significant fraction of the speed of light), in which case special relativity theory must be used; or when very strong gravitational fields are involved, general relativity theory is required. Actually, relativity is always a more accurate representation than Newton's laws, but the two theories give virtually identical results except under the circumstances just cited.

Mathematically, both special and general relativity are too complex to be presented here in any detail. It is possible, however, to provide some intuitive description of what is involved. The dilemma that led Einstein to develop special relativity concerned the speed of light. First let us consider a case involving lower speeds. Suppose a jet aircraft moving at 1000 miles per hour fires a missile forward at a speed relative to the plane of 2000 miles per hour. The speed of the missile with respect to the ground would then be the sum of the plane's speed and the speed with which the missile left the plane, that is, 3000 miles per hour. This is the result expected from Newtonian mechanics.

Now imagine a train moving along a track, and suppose that it turns on its headlight. The light leaves the train at the normal velocity of light relative to the train, and by analogy with the previous example, we might expect that the velocity of the light with respect to a bystander would be equal to the normal velocity of light plus the velocity of the train. For firm theoretical and experimental reasons, however, it is thought to be impossible for the normal speed of light to be exceeded. In other words, an observer on the train would actually measure the same velocity for the light as would an observer standing by the track, in contradiction to the example of the plane and its missile. In Einstein's time an actual measurement of this sort had not been made, but to him there was another reason for postulating that the velocity of light should be the same for all observers: if this were not so, it would be implied that there was some fundamental difference between one frame of reference and others. There would be a "fundamental" frame in which light would have the velocity required by theory, and all other frames would be somehow less fundamental. This would imply that there was somehow a preferred location in the universe, an idea contrary to the lessons learned long before, when it was realized that the Earth was not the center of the universe.

Einstein's solution to the problem involved a surprising postulate: that the rate of passage of time itself depends on the relative velocity between observers. Let us think about the train example again. If we believe that the speed of light must be the same for someone on the train as for someone standing by the track, even though the train is moving, then we are forced to accept Einstein's postulate if both observers are to measure the same velocity of light. For the two observers to measure the same velocity for the light despite their own velocity

Astronomical Insight 3.3

Relativity

difference, it is necessary for the time interval to be different for the two observers.

If the rate of time depends on relative velocity between observers, there are a number of interesting consequences, some of which can be (and have been) tested experimentally. For example, people who travel at high velocity through space will return to earth younger than they would have been if they had stayed home, because time will not pass as rapidly for them. Even the sizes and masses of objects are affected: the length of an object that is moving at high velocity will appear smaller to an observer at rest than to someone moving with the object, and its mass will be greater when measured by the person at rest.

Whereas special relativity states that the laws of physics are the same for observers moving at different *velocities*, general relativity addresses the situation for observers undergoing different *accelerations*. We have seen that a gravitational field such as the Earth's creates acceleration: A falling object increases its speed with every second as it falls. Even without falling, an object is subject to a force because of the acceleration of gravity. We are all familiar with the fact that an object at the surface of the Earth has weight. Now suppose we are somewhere out in space, inside an enclosed spacecraft, and that the spacecraft is accelerating at a rate equivalent to the gravitational acceleration at the Earth's surface. Inside the spacecraft we would "feel" this acceleration, which would create a force (in the direction opposite the spacecraft's motion) exactly identical to the weight we would have on the surface of the Earth. There is no experimental way to tell the difference between the two situations.

Now suppose that the person in the accelerating spaceship fires a gun. The bullet, being "left behind" as the ship continues to increase its velocity, will follow a curving path with respect to the spacecraft, falling toward the rear. This is identical to the curving path of a bullet fired at the surface of the Earth; as the bullet travels toward its target, it drops toward the ground because of gravity. Now suppose that the "bullet" is a beam of light. In the spaceship, the beam would curve away from the direction of the ship's acceleration, just as a material bullet would. For the two situations to be fully equivalent, this implies that a beam of light on Earth also would curve downward because of gravity. Thus, Einstein predicted that a gravitational field should bend light, and this was later confirmed experimentally. Since light always follows the shortest possible path between two points, Einstein characterized the effect of a gravitational field as a "bending" of space itself. Hence, the light could travel the shortest path in the gravitational field, but this path would look like a curve to an observer not subjected to the same gravitational field.

Today we speak of **space-time,** indicating that time itself is thought of as a dimension because of the role it plays in governing the measured properties of an object; we also speak of **curved space-time** because of the curvature that occurs in the presence of a gravitational field. These concepts of special and general relativity will come into play throughout this text, but especially in the later chapters where we discuss situations involving very high velocities or extreme gravitational fields.

center of mass than when it is farther away, so that its velocity compensates for the changes in distance (Fig. 3.20). Thus a planet moves faster in its orbit near **perihelion** (its point of closest approach to the Sun) than at **aphelion** (its point of greatest distance from the Sun). Kepler's second law is really a statement that angular momentum is constant for a pair of orbiting objects.

Kepler's third law was also revised by Newton, and in this case the revision is especially important for our studies. Newton discovered that the relationship between period and semimajor axis depends on the masses of the two objects. Kepler had not realized this, primarily because the Sun is so much more massive than any of the planets that the difference in mass has only a very small effect. Kepler's form of the third law was written $P^2 = a^3$, where P is the planet's period in years and a is the semimajor axis in astronomical units. Newton revised this to:

$$(m_1 + m_2)P^2 = a^3$$

where m_1 and m_2 represent the masses of the two bodies, for example, the Sun and one of the planets. The masses must be expressed in terms of the Sun's mass in this equation. If we use units such as grams for the masses, seconds for the period, and centimeters for the semimajor axis, the equation is complicated by the addition of a numerical factor and is written

$$(m_1 + m_2)P^2 = \frac{4\pi^2}{G}a^3$$

where G is the gravitational constant. Either of these forms of Kepler's third law can be solved for the sum of the masses, and we will see that this is a very important tool for astronomers in deducing the masses of distant objects.

The consideration of orbital motions in terms of kinetic and potential energy leads to the concept of an **escape speed.** If an object in a gravitational field has greater kinetic than potential energy, it will entirely escape the gravitational field. To launch a rocket into space (that is, completely free of the Earth) therefore requires giving it enough upward speed at launch to make its kinetic energy greater than its potential energy due to the Earth's gravitational attraction (Fig. 3.18). It so happens that the speed required to accomplish this is the same for any mass of object, and in equation form it is $v_e = \sqrt{2GM/R}$, where v_e is the escape velocity, G is the gravitational constant, and M and R are the Earth's mass and radius. For the Earth, this speed is 11.2 kilometers per second, or just over 40,000 kilometers per hour. As we will see in later chapters, a planet's escape speed plays a major role in determining the nature of its atmosphere.

Tidal Forces

We have seen that the gravitational force due to a distant body decreases with distance. This means that an object subjected to the gravitational pull of a distant body feels a stronger pull on the side nearest that body and a weaker pull elsewhere. For example, the part of the Earth on the side facing the Moon feels a stronger attraction toward the Moon than do other points on the Earth, and the point on the opposite side from the Moon feels a weaker force. The Earth is therefore subjected to a **differential gravitational force,** which tends to stretch it along the line toward the Moon (Fig. 3.21). Of course, the Sun exerts a similar stretching force on the Earth, but it is too distant to have as strong an effect as the Moon (but its *total* gravitational force on the Earth is much greater than that of the Moon). A **tidal force,** as differential gravitational forces are called, depends on how close one body is to the other,

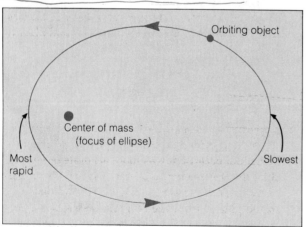

Figure 3.20 Conservation of Angular Momentum. An object moves in an elliptical orbit with varying speed, and the product of its mass times its velocity times its distance from the center of mass (that is, its angular momentum) is constant. Kepler's second law of planetary motion is a rough statement of this fact.

because the key is how rapidly the gravitational force drops off over the diameter of the body subject to the tidal force, and it drops off most rapidly at close distances.

The Earth is a more or less rigid body, so it does not stretch very much due to the differential gravitational force of the Moon. Nevertheless, the tidal forces exerted on the Earth by the Moon create net forces that tend to make the liquid oceans flow toward the points facing directly toward the Moon and directly away from it. These forces are due to the fact that the near side of the Earth feels a stronger attraction toward the Moon than do other points on the Earth's surface, and the far side of the Earth feels a weaker attractive force toward the Moon. As the Earth rotates, the water in the oceans tends to follow the tidal forces created by the Moon, so that in effect the oceans have two huge ridges of water that flow around the Earth as it rotates. Since there are two ridges of water on opposite sides, and the Earth rotates in twenty-four hours, at any given point on its surface one of these ridges passes every twelve hours. Thus we have the ocean's tides, with high tides at any given location separated by about twelve hours.

It is interesting to consider what is happening to the Moon at the same time. It is subjected to a more intense differential gravitational force than the Earth since the Earth is more massive than the Moon. Even though the Moon is a solid body, its shape is deformed by this force, and it has tidal bulges. The Moon is slightly elongated along the line toward the Earth. As we shall see, this has had drastic effects on the Moon's rotation.

There are many other examples of tidal forces, both in the solar system and outside it. The satellites of the massive outer planets are subjected to severe tidal forces, and there are double star systems where the two stars are so close together that they are stretched out of round. We will discuss these in more detail later, along with star clusters and galaxies that are affected by tidal forces and sometimes even tear each other apart.

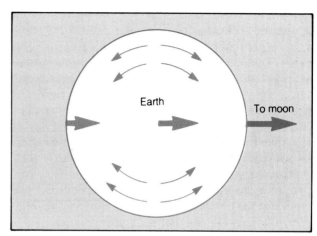

Figure 3.21 The Earth's Tides. The differential gravitational force caused by the Moon tends to stretch the Earth (large arrows). Seawater at any given point on the Earth is subjected to a combination of vertical and horizontal forces, causing it to flow toward either the side of the Earth facing the Moon or the side opposite it (curved arrows).

more than that for us, however; it also provides a basis for understanding the motions of more distant astronomical objects. The law of universal gravitation will be invoked again and again in our study of the solar system and of the rest of the universe because so many important phenomena are explained by it.

Newton's laws of motion and gravitation have stood the test of time rather well. Einstein's theory of general relativity may be viewed as a more complete description of gravity and its interaction with matter than Newton's laws. For most situations, however, Newton's laws are perfectly adequate.

We are ready now to discuss other laws of physics, particularly those that govern the emission and absorption of light.

Perspective

We have now traced the development of astronomy from the introduction of the heliocentric concept to the point where the motions of the bodies in the solar system can be fully described in terms of a few simple laws of physics. The framework of these laws does

Summary

1. Copernicus developed the heliocentric theory because it provided a unifying picture of the universe, not because it was more accurate or simpler than the Earth-centered theory. His new theory enabled Copernicus to calculate the relative distances of the planets from the Sun and arrive at the correct explanations of the seasons and precession.

2. Tycho Brahe made vast improvements in the quantity and quality of astronomical observations of stars and of the planets, accomplishing this by making more precise and complete measurements than his predecessors. Tycho was unable to detect stellar parallax and therefore rejected the heliocentric hypothesis.
3. Kepler sought the underlying harmony among the planets and in the process discovered that each planet orbits the Sun in an ellipse, that a line connecting a planet with the Sun sweeps out equal areas in space in equal intervals of time, and that the square of the period of a planetary orbit is proportional to the cube of its semimajor axis.
4. Galileo used telescopic observations and deductive reasoning to argue for the heliocentric concept of the universe. Despite Church opposition, Galileo brought his ideas before the public in the form of a fictional dialogue between characters of opposing points of view.
5. In the late 1600s, Isaac Newton developed the laws of motion, the law of gravitation, calculus, and made many important contributions to our knowledge of the nature of light and telescopes.
6. Newton's three laws of motion describe the concept of inertia, state that acceleration is proportional to the force exerted on a body and inversely proportional to its mass, and state that for every action there is an equal and opposite reaction.
7. The law of universal gravitation states that any two objects in the universe attract each other with a force that is proportional to the product of their masses and inversely proportional to the square of the distance between them.
8. Newton's laws, along with the concepts of kinetic and potential energy and angular momentum, can be used to explain orbital motions.
9. Kepler's third law was modified by Newton to show that the relationship between the period and semimajor axis of an orbit depends on the sum of the masses of the two objects; this is an important tool for measuring the masses of distant objects.
10. Every body such as a planet or a star has an escape speed, the upward speed at which a moving object has more kinetic energy than potential energy and will therefore escape into space.
11. Differential gravitational forces are responsible for tides on the earth and in the interiors of other planets and satellites.

Review Questions

1. In what way did the heliocentric hypothesis of Copernicus provide a more unified overall concept of the universe than the Earth-centered theory?
2. Kepler's outlook was similar to that of Plato and Aristotle in some ways and rather different in others. Briefly discuss the similarities and differences.
3. Suppose there were a planet at a distance of 2 AU from the Sun. According to Kepler's harmonic law, what would be its sidereal period?
4. How did Galileo's observational discoveries about the Moon and the Sun violate the teachings of Plato and Aristotle?
5. Explain in your own words why the law of inertia (Newton's first law) implies that the planets must experience a force attracting them toward the Sun and not in some other direction.
6. Assume that Saturn is 10 times as far from the Sun as the Earth is and that it has 100 times the mass of the Earth. Compare the gravitational force between Saturn and the Sun with that between the Earth and the Sun.
7. If the diameter of the Earth were suddenly reduced to one fourth its present size (while the Earth's mass remained constant), how would your weight be altered?
8. Imagine a double star in which star A has twice the mass of star B. If the two stars are 3 AU apart, how far from star A is the center of mass between the two stars?
9. When an ice skater in a spin pulls her arms in close to her body, her spin rate speeds up. Explain why, in terms of the conservation of angular momentum.
10. How many high tides per day would there be if the Earth rotated once every twelve hours instead of once every twenty-four hours?

Additional Readings

The references listed here are primarily sources of biographical and historical data on the people discussed in this chapter; many of them contain additional lists of references. Readings on the principles of physics described in this chapter can most easily be found in elementary physics texts, which are available at all levels ranging from the completely nonmathematical to any degree of mathematical sophistication desired.

Beer, A., and P. Beer, eds. 1975. *Kepler. Vistas in astronomy*, Vol. 18. New York: Pergamon Press.

———, and K. A. Strand, eds. 1975. *Copernicus. Vistas in astronomy*, Vol. 17. New York: Pergamon Press.

Cohen, I. B. 1960. Newton in light of recent scholarship. *Isis* 51:489.

———. 1974. Newton. In *Dictionary of scientific biography*, Vol. 10, ed. C. G. Gillispie. New York: Scribner's.

Drake, S. 1980. Newton's apple and Galileo's dialogue. *Scientific American* 243(2):150.

Draper, J. L. E. 1963. *Tycho Brahe: A picture of scientific inquiry in the sixteenth century*. New York: Dover.

Galileo, G. 1632. *Dialogue on the two chief world systems*. Translated by G. de Santillana. Chicago: University of Chicago Press.

Gingerich, O. 1973. Copernicus and Tycho. *Scientific American* 229(6):86.

———. 1973. Kepler. In *Dictionary of scientific biography*, Vol. 7, ed. C. G. Gillispie. New York: Scribner's.

———. 1983. The Galileo affair. *Scientific American* 247(2):132.

Gingerich, O., ed. 1975. *The nature of scientific discovery*. Washington, D.C. Smithsonian Institution Press.

Kuhn, T. 1957. *The Copernican revolution*. Cambridge, Mass.: Harvard University Press.

Whiteside, D. T. 1962. The expanding world of Newtonian research. *History of Science* 1:15.

Chapter 4

Messages from the Cosmos: Light and Telescopes

Refraction of water droplets.
(Photo by Douglas Johnson)

Chapter Preview

The Electromagnetic Spectrum
  Continuous Radiation
  The Atom and Spectral Lines
  Deriving Information from Spectra
Telescopes: Tools for Collecting Light
  The Need for Telescopes
  Principles of Telescope Design
  Major Observatories for All Wavelengths

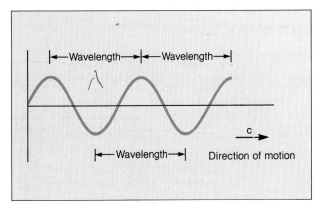

Figure 4.1 Properties of a Wave. Light can be envisioned as a wave moving through space at a constant speed, usually designated c. The distance from one wavecrest to the next is the wavelength, often denoted by the Greek letter lambda, λ. The frequency ν is the number of wavecrests to pass a fixed point per second and is related to the wavelength and the speed of light by $\nu = c/\lambda$.

Some of the tools for unlocking the secrets of the universe became available with the publication of Newton's *Principia* in the 1680s, but others had to wait 200 years or more to be discovered. The laws of motion allowed astronomers to understand how the heavenly bodies move and were of fundamental importance in unraveling the clockwork mechanism of the solar system. To understand the essential nature of a distant object, however—to learn what it is made of and what its physical state is—requires an understanding of what light is and how it is emitted and absorbed, and requires tools to capture the light and analyze it. The only information we can obtain on the nature of a distant object is conveyed by the light that reaches us from it. Fortunately, an enormous amount of information is there, and astronomers have learned much about how to dig this information out.

The Electromagnetic Spectrum

One characteristic of light is that it acts like a wave (Fig. 4.1). It is possible to think of light as passing through space like ripples on a pond (although, as we will discuss shortly, the picture is actually somewhat more complicated than that). The distance from one wavecrest to the next, called the **wavelength**, distinguishes one color from another. Red light, for example, has a longer wavelength than blue light. It is possible to spread out the colors in order of wavelength, using a prism to obtain the traditional rainbow. Newton was the first to discover that sunlight contains all the colors, and he did so by carrying out experiments with a prism. Whenever light is spread out by wavelength, the result is called a **spectrum;** more technically, a spectrum is the arrangement of light from an object according to wavelength. The science of analyzing spectra is called **spectroscopy,** and it will be discussed later in this chapter.

The concept of **frequency** is often used as an alternative to wavelength in characterizing light waves. The frequency is the number of waves per second that pass a fixed point. It is determined by the wavelength and the speed with which the waves move. The speed of light, usually designated c, is constant, and the frequency of light with wavelength λ is $\nu = c/\lambda$. The standard unit for measuring frequency is the **hertz (Hz),** one hertz being equal to one wave per second. The frequency of visible light is typically about 10^{14} Hz.

There are situations in which light acts more like a stream of particles than a wave. Newton developed a "corpuscular" theory of radiation in which he assumed that light consisted of particles, but at the same time others, including most notably Christiaan Huygens, carried out experiments showing light to have definite wave characteristics. There are good arguments for either point of view; for example, the manner in which light waves seem to bend as they pass obstacles the way in which they interfere with each other wave characteristics. On the other hand, light can carry only discrete, fixed quantities can travel in a complete vacuum rather a medium; these are properties of pa

---

Astronomical Insight 4.1

The Perception of Light

Newton's experiments in optics are commonly thought to have marked the beginning of scientific inquiry into the nature of light, but in fact Newton was preceded in this area by many scientists and philosophers. Much of the early research, and even to some extent Newton's work, did not deal with the physical properties of light, but was aimed at understanding how the human eye senses light. The ancient Greeks, the earliest philosophers who studied the nature of light, did not recognize the distinction between light and the act of seeing it.

The Greeks studied the principles of vision in much the same way they studied astronomy: They carried out very few experiments or even systematic observations, choosing instead to unravel the secrets of the universe by reason and logic. The prevalent concept, described by such notable figures as Hippocrates and Aristotle, was that the eye somehow emitted rays or beams with which it sensed things. This was analogous to the sense of touch, where a hand is extended to feel objects. It was known that objects appeared distorted or bent when viewed under water, but this was attributed to bending of the rays that came from the eye, not to effects on something coming from the object being viewed. It would have been possible for the Greeks to experiment with optics by using natural crystals or eyes removed from dead animals, but no such experimentation was done.

In a definitive summary of all that was known about vision, the Greek scientist Galen, in the second century A.D., wrote of an "animal spirit" that flowed from the brain along the optic nerve to the eye, where it was converted into a "visual spirit" in the retina, a membrane covering the rear interior portion of the eyeball. The lens in the front of the eyeball was thought to be responsible for sending out the beam with which the external world was perceived.

Like the astronomical work of Ptolemy, the concepts of the nature of light and vision that were summarized by Galen would be accepted as doctrine for some fifteen centuries until the time of the Renaissance. Some of the ancient Greek ideas, such as the notion that the optic nerve was hollow so that the animal spirit could flow along it from the

---

Out of a variety of seemingly contradictory evidence has developed the concept of the **photon.** A photon is thought of as a particle of light that has a wavelength associated with it. The wavelength and the amount of energy contained in the photon are intimately linked; in general terms, the longer the wavelength, the lower the energy. Thus a red photon carries less energy than a blue one. Mathematically the energy can be expressed as $E = hc/\lambda$, where h is the Planck constant, c the speed of light, and λ is the wavelength. It is important to understand the fact that a photon carries a set amount of energy, not some arbitary or random amount, and that when light strikes a surface, this energy arrives in discrete bundles like bullets, rather than a steady stream. When a photon is absorbed, this energy can be converted into other forms such as heat.

Let us consider for a moment what lies beyond red at one end of the spectrum or violet at the other end. By the mid-1800s, experiments had demonstrated that there is invisible radiation from the Sun at both ends of the spectrum. At long wavelengths, beyond red, is **infrared** radiation, and at short wavelengths is **ultraviolet** radiation. Later it was shown that the spectrum continues in both directions, virtually without limit. Going toward long wavelengths, after infrared light come **microwaves** and **radio** waves; going toward

Astronomical Insight 4.1

The Perception of Light

brain, were especially persistent. Even Leonardo da Vinci, who in the late fifteenth and early sixteenth centuries carried out pioneering work in anatomy, accepted much of the old thinking. Da Vinci took a more modern view in that he believed that light rays passing from an object to the eye play an important role, but he was unwilling entirely to give up the idea that other rays emanate outward from the eye.

In the early seventeenth century Johannes Kepler and the great French philosopher René Descartes developed an understanding of refraction and the formation of an image by a lens, and both correctly viewed vision as the process of image formation on the retina by the lens in the front of the eyeball. Thus when first Christiaan Huygens and then Newton carried out their experiments later in the seventeenth century, it was already well established that seeing was accomplished by sensing something that passes from the object to the eye.

Both men were more concerned with the nature of this radiation than with the properties of the eye, although Newton did discuss the perception of color, which he realized to be entirely a function of the eye. This is an important concept: Neither the rays of light nor the object being viewed is colored; it is the eye that senses wavelength differences and the brain that interprets them as colors. Light rays consist simply of alternating electric and magnetic fields.

Other advances in understanding the physical properties of light were made by scientists who followed Newton and are described in the text. Research has also been aimed at learning how the human eye perceives light. We now know that the eye has a logarithmic response, meaning that it does not perceive the true range of brightnesses that may be displayed by a series of objects, but instead perceives a lesser range. This point is explained more fully in Chapter 6, in the discussion of the stellar magnitude system.

Entire books have been written on the complex interplay between what the brain perceives and physical reality, and of course, this interplay is important in astronomy. To a large extent, modern observational astronomy is the science of recognizing and bypassing the limitations or distortions created by our natural light-gathering system.

---

short wavelengths, after ultraviolet come **X rays and then γ-rays (gamma rays).** All of these different kinds of radiation are just different forms of light, distinguished only by their wavelengths, and together they form the **electromagnetic spectrum** (Fig. 4.2). Electromagnetic radiation is a general term for all forms of light, whether it is visible, X rays, radio waves, or anything else.

The name *electromagnetic radiation* reflects the fact that the wave motion associated with this radiation consists of alternating electric and magnetic fields (Fig. 4.3). These fields propagate through space (or a medium) as they alternate, with the electric and magnetic fields always lying in planes that are perpendicular to each other. The speed of propagation in a vacuum is always the same (almost exactly 300,000 kilometers per second, or about 186,000 miles per second), but it is slightly slower in a medium such as air or glass. The electric and magnetic fields are said to be *in phase* with each other, meaning that they reach their maximum and minimum values simultaneously as the wave propagates, with the relative orientation shown in Figure 4.3.

The light from a typical source such as a light bulb or a distant star consists of vast numbers of photons, each consisting of traveling electric and magnetic

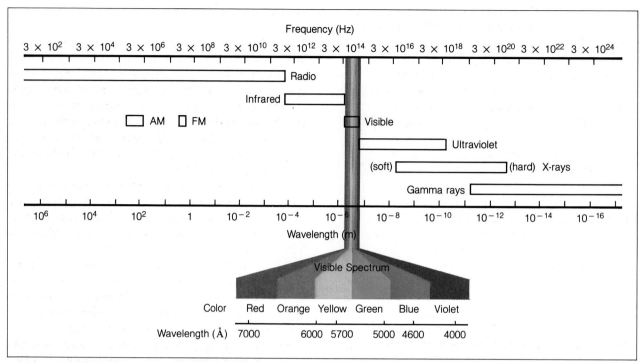

Figure 4.2 The Electromagnetic Spectrum. All of the indicated forms of radiation are identical except for wavelength and frequency.

fields. Each photon has a characteristic orientation as it travels through space, that is, the planes in which its electric and magnetic fields oscillate remain fixed as the photon travels. Normally the orientations in a collection of photons from a source are random, but in some circumstances, they are not. When light is **polarized**, all the photons tend to have the same alignment of their electric and magnetic planes. It is possible to determine from observations whether the light from a source is polarized. As we will learn in later chapters,

Table 4.1 Wavelengths and Frequencies of Electromagnetic Radiation.

Type of Radiation	Wavelength Range*	Frequency Range
Gamma rays	$< 10^{-8}$	$> 3 \times 10^{18}$/Hz
X rays	$1–200 \times 10^{-8}$	$1.5 \times 10^{16} – 3 \times 10^{18}$
Extreme ultraviolet	$200–900 \times 10^{-8}$	$3.3 \times 10^{15} – 1.5 \times 10^{16}$
Ultraviolet	$900–4000 \times 10^{-8}$	$7.5 \times 10^{14} – 3.3 \times 10^{15}$
Visible	$4000–7000 \times 10^{-8}$	$4.3 \times 10^{14} – 7.5 \times 10^{14}$
Near infrared	$0.7–20 \times 10^{-4}$	$1.5 \times 10^{13} – 4.3 \times 10^{14}$
Far infrared	$20–100 \times 10^{-4}$	$3.0 \times 10^{12} – 1.5 \times 10^{13}$
Radio	> 0.01	$< 3 \times 10^{12}$
(Radar)	(2–20)	$(1.5–15 \times 10^{9})$
(FM radio)	(250–350)	$(86–120 \times 10^{6})$
(AM radio)	(18,000–38,000)	$(800–1600 \times 10^{3})$

*All wavelengths are given in centimeters; recall that 1×10^{-8} cm = 1 Angstrom; and 1×10^{-4} cm = 1 micron (10^{-6} meter).

there are astronomical objects that produce polarized light, and this fact provides useful information on the nature of these objects.

Astronomers tend to identify different portions of the electromagnetic spectrum by specialized names such as *infrared* or *ultraviolet*, but we must keep in mind that all are the same, consisting of alternating electric and magnetic fields and differing only in wavelength (and frequency).

The range of wavelengths from one end of the electromagnetic spectrum to the other is immense (see Table 4.1). Visible light has wavelengths ranging from 0.000,04 to 0.000,07 centimeter. In the case of light, a special unit called the **Angstrom** is used, defined such that 1 cm = 100,000,000 Å, or 1 Å = 0.000,000,01 cm = 10^{-8} cm. Thus, visible light lies between 4000 Å and 7000 Å in wavelength. Recently physicists and astronomers have begun to use the *nanometer* (nm), which is equal to 10^{-9} m or 10^{-7} cm. In terms of this unit, visible light lies between 400 and 700 nm.

Infrared light has wavelengths between 7000 Å and a few million Å; that is, between 7×10^{-5} cm and $2-3 \times 10^{-2}$ cm. Microwave radiation (which includes radar wavelengths) lies roughly between 0.1 and 50 cm, with no well-defined boundary separating this region from that of radio waves, which simply includes all longer wavelengths, up to many meters or even kilometers. At the other end of the spectrum, ultraviolet light is usually considered to lie between 100 Å and 4000 Å, while X rays are in the range 1–100 Å, anything shorter than that being considered γ-rays.

Figure 4.3 An Electromagnetic Wave. This illustrates how the electric and magnetic fields oscillate in an electromagnetic wave. The directions of the fields and the direction of the wave travel always have the relative orientations shown here.

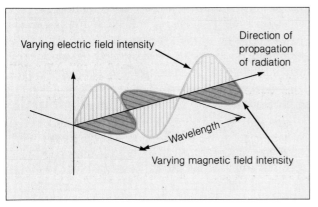

Continuous Radiation

Researchers who followed Newton discovered that the Sun's spectrum contains a number of dark lines, each one corresponding to a particular wavelength. These **spectral lines** provide a great deal of information about a source of light such as a star, and so does the **continuous radiation** (Figs. 4.4 and 4.5), the smooth distribution of light as a function of wavelength. Here we discuss continuous radiation, and in the following sections, the spectral lines.

Any object with a temperature above absolute zero emits radiation over a broad range of wavelengths, simply by virtue of the fact that it has a temperature. Thus, not only stars, but also such commonplace objects as the walls of a room or a human body, emit radiation. The continuous radiation produced because of an object's temperature is called **thermal radiation,** and in this chapter we will restrict ourselves to this type of continuous radiation. In other chapters we will discuss nonthermal sources of radiation.

For glowing objects such as stars, thermal radiation is emitted over a broad range of wavelengths (technically, in fact, at least some radiation is emitted at *all*

Figure 4.4 Continuous Spectrum and Spectral Lines. When sunlight is dispersed by a prism, the light forms a smooth rainbow of continuous radiation, gradually merging from one color into the next, with a maximum intensity in the yellow portion of the spectrum. Superimposed on this continuous spectrum are numerous spectral lines, wavelengths where little or no light is emitted by the Sun.

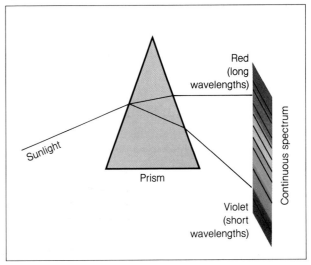

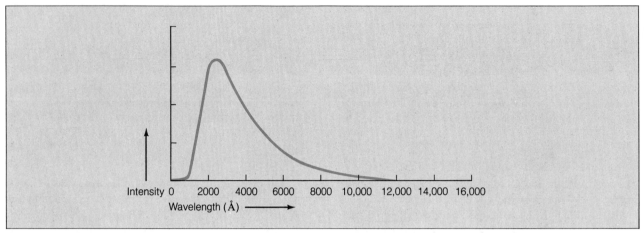

Figure 4.5 An Intensity Plot of a Continuous Spectrum. This kind of diagram shows graphically how the brightness of a glowing object varies with wavelength. The curve shown roughly represents a star of surface temperature 10,000 K, whose continuous radiation peaks near 3000 Å, in the ultraviolet.

wavelengths), with a peak in intensity at some particular wavelength. A simple relationship between the wavelength of maximum emission and the temperature of an object was discovered in 1893 by W. Wien and is now known as Wien's law:

> The wavelength of maximum emission is inversely proportional to the absolute temperature.

The hotter an object is, the shorter the wavelength of peak emission (Fig. 4.6). This immediately explains the variety of stellar colors as being due to a range in stellar temperatures. A hot star emits most of its radiation at relatively short wavelengths and thus appears bluish in color, whereas a cool star emits most strongly at longer wavelengths and appears red in color. The Sun is intermediate in temperature and in color.

When speaking of the colors of stars, we must keep in mind that a star emits light over a broad range of wavelengths, so we do not have pure red or pure blue stars. Our eyes receive light of all colors, and stars therefore are all essentially white. Our impression of color arises from the fact that there is a wavelength (given by Wien's law) at which a star emits more strongly than at other wavelengths, but the star does not emit *only* at that wavelength.

A second property of glowing objects, known as either Stefan's law or the Stefan-Boltzmann law, concerns the total amount of energy emitted over all wavelengths and how this total energy is related to the temperature of an object:

> The total energy radiated per square centimeter of surface area is proportional to the fourth power of the temperature.

This shows that the total energy emitted is very sensitive to the temperature; if you change the temperature by a little, you change the energy by a lot. If, for example, you double the temperature of an object (such as the electric burner on your stove), you increase the total energy it radiates by $2^4 = 2 \times 2 \times 2 \times 2 = 16$. If one star is three times hotter than another, it emits $3^4 = 3 \times 3 \times 3 \times 3 = 81$ times more energy per square centimeter of surface area.

Notice that we have been careful to express this law in terms of energy emitted per square centimeter of surface area. The *total* energy emitted by an object therefore depends also on how much surface area it has. If we talk of stars or other spherical objects, which have a surface area of $4\pi R^2$, where R is the radius, then we can say that the total energy emitted is proportional to the fourth power of the temperature and to the square of the radius. This can be illustrated by considering two stars, one of which is twice as hot but has only half the radius of the other. The hotter star emits $2^4 = 16$ times more energy per square centimeter of surface but has only $½^2 = ¼$ as much surface area, hence it is $16 \times ¼$ or 4 times brighter overall. If, on the other hand, this star were twice as hot and three times as large in radius as the other, it would be $2^4 \times 3^2 = 144$ times brighter.

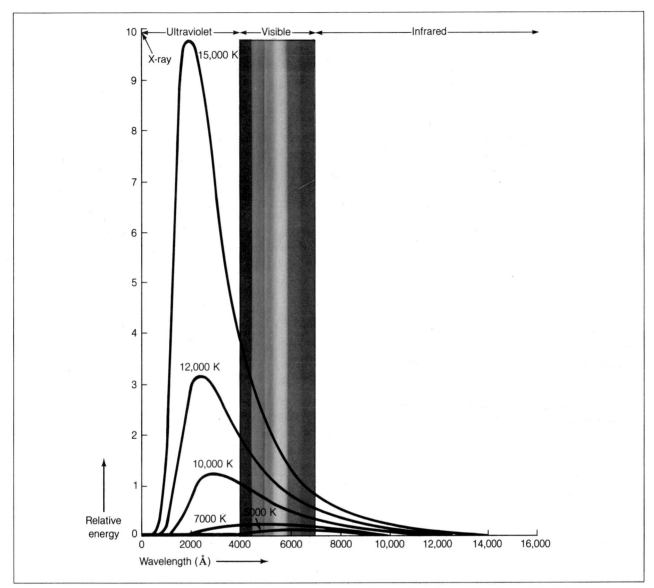

Figure 4.6 Continuous Spectra for Objects of Different Temperatures. This diagram illustrates Wien's law, which says that the wavelength of maximum emission is inversely proportional to the temperature (on the absolute scale).

Both Wien's law and the Stefan-Boltzmann law were first derived experimentally, in much the same manner as Kepler's discovery of the laws of planetary motion. In the case of planetary motions, it remained for Newton to find the underlying reasons for the laws, and he was able to derive them strictly on a theoretical basis. Analogously, Max Planck, the great German physicist who was active early in this century, found a theoretical understanding of thermal emission and was able to derive Wien's Law and the Stefan-Boltzmann law from purely theoretical considerations. The basis of Planck's new understanding was the **quantum** nature of light: the fact that light has a particle nature and carries only discrete, fixed amounts of energy.

Finally, in addition to taking into account the temperature and size of a glowing object, we must consider the effect of its distance. So far we have discussed the energy as it is emitted at the surface, but not how

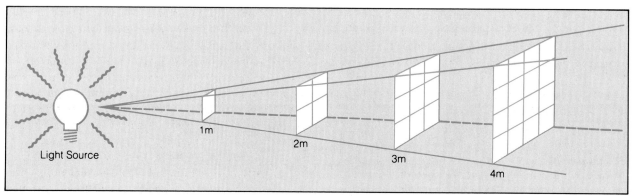

Figure 4.7 The Inverse Square Law of Light Propagation. This shows how the same total amount of radiation energy must illuminate an ever-increasing area with increasing distance from the light source. The area to be covered increases as the square of the distance; hence, the intensity of light per unit of area decreases as the square of the distance.

bright the object looks from afar. What we actually observe, of course, is affected by our distance from the object. For a spherical object that emits in all directions, the **inverse square law** tells us that the brightness decreases as the square of the distance (Fig. 4.7). Thus, if we double our distance from a source of radiation it will appear $½^2 = ¼$ as bright. If we approach, reducing the distance by a factor of two, it will appear $2^2 = 4$ times brighter. As we will see in later discussions of stellar properties, we must know something about the distances to stars before we can compare other properties connected with their brightness.

The Atom and Spectral Lines

The first detailed cataloging of the Sun's spectral lines was carried out in the early 1800s by Joseph Fraunhofer, and the lines are known to this day as **Fraunhofer lines** (Fig. 4.8). In the late 1850s the German scientists Bunsen and Kirchhoff performed experiments and developed theories that made clear the importance of the Fraunhofer lines. Bunsen observed the spectra of flames created by burning various substances and found that each chemical element produced light only at specific places in the spectrum (Fig. 4.9). The spec-

Figure 4.8 Fraunhofer Lines. This is a portion of the Sun's spectrum, showing some of the dark lines first noticed by Wollaston and catalogued by Fraunhofer. *(Deutches Museum, Munich)*

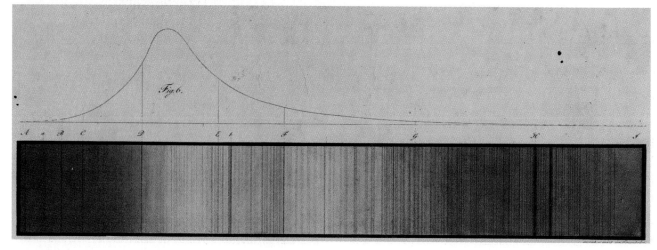

trum of such a flame is dark everywhere except at these specific places, as though the flame were emitting light only at certain wavelengths. For that reason, the bright lines seen in this situation are called **emission lines**.

It was soon noticed that some of the dark lines in the Sun's spectrum coincide exactly in position with some of the bright lines seen by Bunsen in his laboratory experiments. Kirchhoff studied this in detail and was able to show that a number of common elements such as hydrogen, iron, sodium, and magnesium must be present in the Sun because of the coincidence in the wavelengths of the lines. This was the first hint that the chemical composition of a distant object could be determined.

Further studies of the spectral lines revealed various regularities in the arrangement of the lines from a given element. While it was suspected that these regularities must reflect some feature of the structure of atoms, it wasn't until 1913 that the true relationship was discovered by Niels Bohr. By that time it had been established that an atom consists of a small, dense nucleus surrounded by a cloud of negatively charged particles called electrons. The nucleus contains positively charged particles called protons and neutral particles called neutrons. In a normal atom, the number of protons in the nucleus is equal to the number of electrons orbiting it, and the overall electrical charge is zero. The electrons are held in orbit by electromagnetic forces, with laws of the science called **quantum mechanics** governing their motions. Bohr found that the electrons were responsible for the absorption and emission of light, and that they did so by gaining energy (absorption) or by losing it (emission) in the form of photons.

The key to the fixed pattern of spectral lines for each element lay in the fixed pattern of energy levels the electrons could occupy (Fig. 4.10). We can visualize an atom as a miniature solar system, with electrons orbiting the nucleus. Each kind of atom (each element) has its own characteristic number of electrons, and the electrons have in each case a certain set of electron orbits. It may be helpful to visualize a ladder, with each rung representing an orbit or energy level. The ladder for one element, say, hydrogen, has different spacings

Figure 4.9 Emission Lines. The spectrum of a flame is devoid of light except at specific wavelengths where bright emission lines appear.

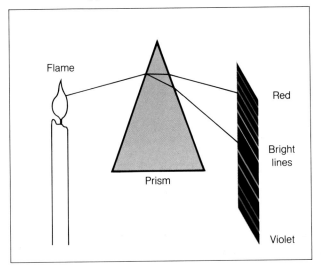

Figure 4.10 The Formation of Spectral Lines. An electron must be in one of several possible orbits, each representing a different electron energy. If an electron absorbs exactly the amount of energy needed to jump to a higher (more outlying) orbit, it may do so. This is how absorption lines are formed, because the wavelength of the photon absorbed corresponds to the energy difference between the two electron orbits, which is fixed for any given kind of atom. Conversely, when an electron drops from one orbit to a lower one, it emits a photon whose wavelength corresponds to the energy difference between orbits, forming an emission line.

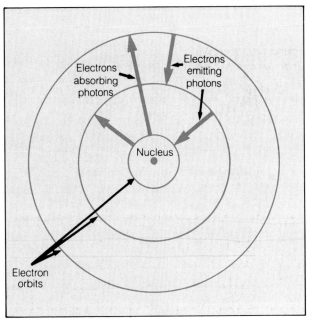

Table 4.2 Principal Lines in the Spectrum of Hydrogen

Series	Lower Level	Line	Wavelength, (Å)
Lyman (ultraviolet)	1	alpha	1216
		beta	1025
		gamma	972
		delta	949
		epsilon	937
		(limit)	(912)
Balmer (visible)	2	alpha	6563
		beta	4861
		gamma	4340
		delta	4101
		epsilon	3970
		(limit)	(3646)
Paschen (infrared)	3	alpha	18,751
		beta	12,818
		gamma	10,938
		delta	10,049
		epsilon	9545
		(limit)	(8204)
Brackett (infrared)	4	alpha	40,512
		beta	26,252
		gamma	21,656
		delta	19,445
		epsilon	18,175
		(limit)	14,585

between the rungs than the ladder for some other element (see Table 4.2). The energy associated with each level increases the higher up the ladder, or the farther from the nucleus, the electron goes.

An electron can absorb a photon of light *only* if the photon carries precisely the amount of energy needed to move the electron to some higher level than the one it is in. In our analogy with a ladder, the electron can jump up only if it will land precisely on a higher rung; that is, it can absorb only a photon whose energy will boost it to another fixed energy level. Since the wavelength of a photon is determined by its energy, this means that an electron can absorb only photons with certain wavelengths. Because each kind of atom has its own unique set of energy levels, each has a unique set of wavelengths at which it can absorb photons.

Thus each kind of atom has its own pattern of spectral lines. A spectral line is an absorption line when the electron receives energy from a photon and jumps to a higher level, and an emission line when an electron drops from a high level to a lower one, releasing a photon. For a given atom the absorption and emission lines occur at the same wavelengths because the spacing of the energy levels determines the wavelengths. Whether the lines are emission or absorption depends simply on whether the electrons are dropping in energy level, giving off photons, or climbing in energy level, absorbing photons.

We can now state three rules of spectroscopy first discovered experimentally by Kirchhoff (Fig. 4.11) and later put in the context of atomic structure:

1. In a hot, dense gas or a hot solid, the atoms are crowded together so that their energy levels overlap and their lines are all blended together, and we see a continuous spectrum.*
2. In a hot, rarefied gas, the electrons tend to be in high energy states and create emission lines as they drop to lower levels.
3. In a relatively cool gas in front of a hot, continuous source of light, the electrons tend to be in low energy levels and absorb radiation from the background continuous source, creating absorption lines.

Deriving Information from Spectra

A great wealth of information is stored in the spectrum of an object. From the analysis of the spectral lines, we can learn such things as the temperature and density of a gas and the velocity of a glowing object with respect to the earth.

One important property of a gas is its degree of **ionization**. Ionization refers to the loss of one or more electrons from an atom when the electrons receive so much energy that they jump entirely free of the atom (Fig. 4.12). The gain in energy that frees an electron can come either from the absorption of a photon with energy exceeding that of the highest electron energy level associated with the atom or from a collision between atoms. In either case, the remaining atom has a net positive electrical charge, having lost one or more negative charges, and is called an **ion**. The likelihood

---

*The production of a continuous spectrum is actually a bit more complex than this. When free electrons combine with ions, they emit at any wavelength, not just in spectral lines. See the discussion of ionization in the following section.

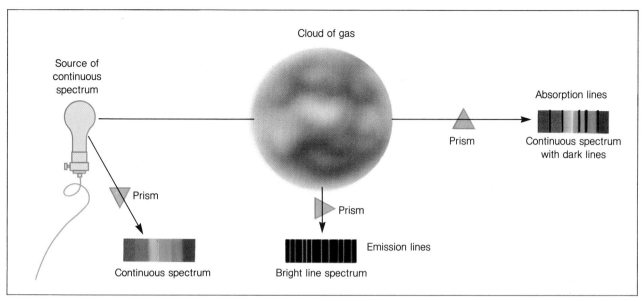

Figure 4.11 Continuous, Emission Line, and Absorption Line Spectra. The positions of the emission and absorption lines match because the same element emits or absorbs at the same wavelengths. Whether it emits or absorbs depends on the physical conditions, as described in Kirchhoff's laws.

of collisional ionizations depends on the temperature of the gas, since the speed of collision depends on temperature. The hotter the gas, the more violent the collisions between atoms, and the greater the degree of ionization.

The spectrum of an atom changes drastically when it has been ionized because the arrangement of energy levels is altered and different electrons are now available to do the absorbing and emitting of photons. The spectrum of atomic helium, for example, is quite different from that of ionized helium, so the astronomer not only can see that helium is present in the spectrum of a star, but also can tell whether it is ionized or not. This provides information on the temperature in the outer layers of the star where the absorption lines are formed. By analyzing the degree of ionization of all the elements seen in the spectrum of a star, it is possible to determine the gas temperature quite precisely.

When an electron moves to an energy state above the lowest possible one, it is said to be in an **excited state** (Fig. 4.12). This can happen as the result of the absorption of a photon of light with appropriate energy or as the result of a collision that is not energetic enough to cause ionization. The degree of excitation of a gas affects its spectrum because the spectral lines created in the gas depend on which energy levels the elec-

Figure 4.12 Excitation and Ionization. In the energy-level diagram on the left, electrons are shown jumping up from the lowest level to higher ones. This process is called excitation, and it can be caused either by the absorption of photons or by collisions between atoms. On the right, electrons are shown gaining sufficient energy to escape altogether, a process called ionization.

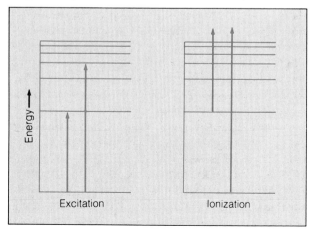

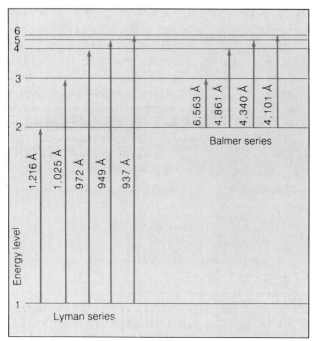

Figure 4.13 The Effect of Excitation on the Spectrum of Hydrogen. As shown here, the absorption lines that can be formed by an atom depend on its degree of excitation. At left are the wavelengths of absorption lines originating in the lowest energy level of hydrogen, while at right are the wavelengths of absorption lines arising from an electron in the first excited level. Note that the Lyman lines are in ultraviolet wavelengths, while the Balmer lines are in the visible portion of the spectrum. This sketch shows only a few of the many energy levels of hydrogen.

trons are in. A hydrogen atom with its electron in the first excited level can produce absorption lines corresponding to the energy separation between this excited level and higher levels, whereas a hydrogen atom with its electron in the lowest state can only produce absorption lines corresponding to transitions of the electron out of that state (Fig. 4.13 and Table 4.2). Therefore, analysis of the spectrum of a gas can tell us how highly excited the gas is, which in turn provides information on other properties such as density (the density affects the frequency of collisions between atoms, which in turn determines how many electrons are in excited levels).

From careful analysis of a star's spectrum, then, it is possible to determine both the temperature of the outer layers of the star (from the ionization) and the density (from the excitation). While most of the examples used

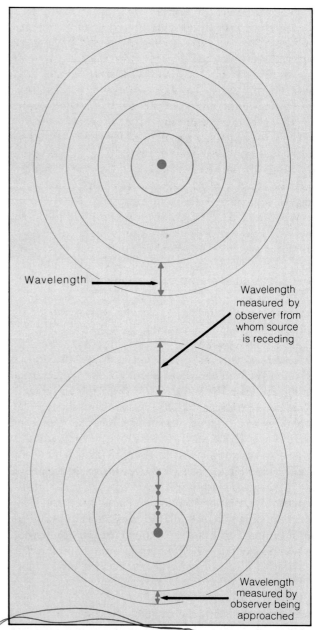

Figure 4.14 The Doppler Effect. The light waves from a stationary source (upper) remain at constant separation (that is, constant wavelength) in all directions, whereas those from a moving source get "bunched up" in the forward direction and "stretched out" in the trailing direction (lower). This causes a blueshift or a redshift for an observer who approaches or recedes from a source of light. Note that it does not matter which is moving, source or observer.

in this discussion have referred to the Sun or the stars, the same considerations can be applied to the spectra of the planets. Because a planet glows only by reflected sunlight, however, its spectrum is the same as the Sun's spectrum, except that some additional features are added by the gas in the planet's atmosphere. These extra spectral lines must be identified and analyzed if the properties of this gas are to be derived.

There is something else that spectral lines can tell us about a distant object, in addition to all the physical data that we have been discussing. We can also learn how rapidly an object such as a star or a planet is moving toward or away from us.

When a source of light such as a star is approaching, the lines in its spectrum are all shifted toward shorter wavelengths than if the light source were at rest, and if the source is moving away from the observer, the lines are all shifted toward longer wavelengths (Fig. 4.14). These two cases are called **blueshifts** (approach) and **redshifts** (recession), respectively, because the spectral lines are shifted toward either the blue or the red end of the spectrum.

A general term for any wavelength shift due to relative motion between source and observer is **Doppler shift,** in honor of the German physicist who first explored the properties of such shifts. The effect applies to waves other than light. Most of us, for example, have noticed the Doppler shift in sound waves when a source of sound passes by. The whistle on an approaching train suddenly changes to a lower pitch at the moment the train passes by, because the wavelength we receive suddenly shifts to a longer one.

In the case of the Doppler shift of light, it is possible to determine the speed with which the source of light is approaching or receding from the simple formula:

$$v = \left(\frac{\Delta\lambda}{\lambda}\right)c,$$

where v is the relative velocity between source and observer, $\Delta\lambda$ is the shift in wavelength (the observed wavelength minus the rest, or laboratory, wavelength of the same line), λ is the laboratory wavelength of the line, and c is the speed of light. If the observed wavelength is greater than the rest wavelength, we have a positive velocity, corresponding to a redshift; if the observed wavelength is less than the rest wavelength, the result is a negative velocity, indicating a blueshift.

It is important to notice that the Doppler shift tells us only about relative motion between the source and the observer: It is not possible to distinguish whether it is the star or the Earth that is moving or whether it is a combination of the two (which is most likely the case). It is also important to keep in mind that the Doppler shift tells us only about motion directly toward or away from the Earth. There is no Doppler shift due to motion perpendicular to our line of sight. If a star is moving with respect to the Earth at some intermediate angle, as is usually the case, then we can determine the part of its velocity that is directed straight toward or away from us, but we cannot determine its true direction of motion or its speed transverse to our line of sight.

Telescopes: Tools for Collecting Light

Now that we know how to analyze light from distant objects, we can turn our attention to the methods for collecting light. The universe is filled with radiation of all wavelengths, so our discussion will necessarily include telescopes designed for all forms of radiation, not just visible light.

The Need for Telescopes

There are several basic benefits of using telescopes. One is to collect light from a large area and bring it to a focus, so that fainter objects can be observed than the eye could see unaided. The human eye has a collecting area only a fraction of a centimeter in diameter, whereas the largest telescopes are several meters in diameter. The light-gathering power of a telescope depends on the area of its collecting surface; since the area of a circle is equal to πr^2, a 2-inch diameter telescope collects four times as much light as a 1-inch telescope, for example. The largest visible-light telescopes are 4 to 6 meters in diameter.

Another basic advantage of telescopes over the unaided eye is that the telescope can be equipped to record light over a long period of time, by using photographic film or a modern electronic detector, while the eye has no capability to store light. A long-exposure photograph taken through a telescope reveals objects too faint to be seen with the eye, even by looking through the same telescope.

By combining both large size and the capability of long exposures, the largest telescopes can detect objects some 100 million times fainter than can be seen by the unaided eye, even under the best conditions.

A third major advantage of large telescopes is that they have superior **resolution,** the ability to discern fine detail. For visible-wavelength telescopes, the Earth's atmosphere creates practical limitations on how fine the resolution can be, but for radio telescopes and optical telescopes in space, the atmosphere is not a hindrance. The resolution of a telescope is normally expressed in terms of the smallest separation between a pair of objects that can be discerned by the telescope. Thus, small resolution is good. The resolution expressed in this way is proportional to the wavelength being observed and inversely proportional to the diameter of the telescope. For visible light (wavelength 5500 Å, for example), a telescope 10 centimeters in diameter (about 4 inches) has a resolution of about 1 arcsecond. (This is really quite good; an arcsecond is about the angular diameter of a dime seen at a distance of two miles!). The human eye, with a pupil diameter of about 1 millimeter, has a resolution of roughly 100 arcseconds, or a little over 1.5 arcminutes. In principle, a large telescope such as the 5-meter Palomar telescope has a resolution far smaller than 1 arcsecond, but in practice, the Earth's atmosphere limits the resolution that is actually achieved. Normally it is impossible to separate objects closer than about 1 arcsecond when looking up through the turbulent atmosphere, and often the blurring effect called "seeing" makes resolution much worse.

We can vastly improve resolution by using telescopes in pairs or groups. This technique, called **interferometry,** makes it possible to measure the precise direction toward a source of radiation by timing the arrivals of the photons at different telescopes. Sources can be mapped with very high precision by using telescopes arranged in series along different alignments, as is done with the *Very Large Array* (Fig. 4.15), which can achieve resolutions of better than 0.1 arcsecond. Interferometry is obviously a very complex undertaking, but it has been successfully applied to radio astronomy, and there are now preliminary plans to do the same for infrared and perhaps visible light. (The shorter the wavelength being observed, the more pre-

Figure 4.15 The *Very Large Array*. This arrangement of twenty-seven radio dishes spread over a 15-mile wide section of New Mexico desert provides high resolution radio maps through the use of interferometry. The individual telescopes can move along railroad tracks, allowing the separations to be adjusted, depending on the type of data needed. *(National Radio Astronomy Observatory, operated by Associated Universities, Inc., under contract with the National Science Foundation)*

cision is required in timing and analyzing the arrivals of photons at separate telescopes. The planned *National New Technology Telescope*, a U.S. project using four large mirrors to achieve the equivalent collecting area of a single 16-meter mirror, is designed to make possible interferometry in infrared and perhaps visible light.) The more widely separated the telescopes are, the better the resolution, and radio astronomers are today engaged in *Very Long Baseline Interferometry* (or simply *VLBI*), using telescopes separated by intercontinental distances.

One technical problem must be overcome in telescope construction: The Earth rotates, and if nothing is done to compensate for this, a star quickly moves out of the field of view. To avoid this problem, telescopes are mounted so that they can be moved by a motor in the direction opposite the Earth's rotation, keeping a target object centered in the field of view. The telescope mounting usually allows the telescope to move independently in declination (north-south motion) and right ascension (east-west motion), so the drive mechanism that compensates for the Earth's rotation needs to operate only in the east-west direction. There is no need to do this, of course, for telescopes that are not attached to the Earth, such as the various observatories in space.

Principles of Telescope Design

To illustrate the general principles of telescope design, we begin by describing telescopes built for visible light. We will then see how those designed for other wavelengths have essentially the same features but some differences of detail.

It is possible to construct a telescope called a **refractor,** which focuses light with lenses (Fig. 4.16), and many of the earliest telescopes were of this type. Today nearly all telescopes are **reflectors,** which use mirrors to focus light. Reflectors can be built much larger and are not hampered by certain technical difficulties associated with lenses, such as the fact that different wavelengths of light are refracted at different angles by a lens.

In a reflector, light is focused by a concave mirror called the **primary mirror** (Figs. 4.17 and 4.18). Because it is usually inconvenient to look at or record an image formed inside the telescope tube, there is most often a secondary mirror that deflects the image outside

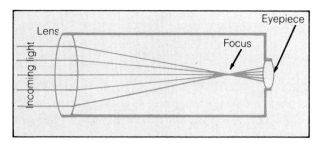

Figure 4.16 The Refracting Telescope. The path of light is bent when it passes through a surface such as that of glass. A properly shaped lens thus can bring parallel rays of light to a focus, where the image of a distant object may be examined. A second lens is used to magnify the image.

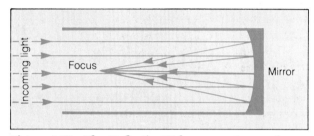

Figure 4.17 The Reflecting Telescope. A properly shaped concave mirror can be used instead of a lens to bring light to a focus. Usually an additional mirror is used to reflect the image outside the telescope tube.

Figure 4.18 A Large Primary Mirror. This photo shows that the proper shape for a telescope mirror is not very highly concave. Only the distorted reflections in this case reveal that the mirror is not flat. This is the 2.4-meter primary mirror for the *Space Telescope*. *(Perkin Elmer Corporation)*

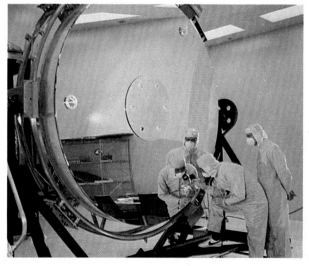

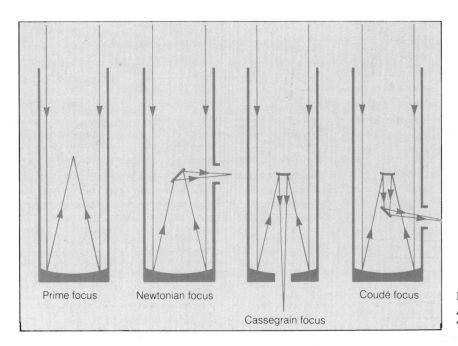

Figure 4.19 Various Focal Arrangements for Reflecting Telescopes.

the tube (Fig. 4.19). This mirror can be a flat mirror, which sends the image out the side (the so-called **Newtonian focus** arrangement), or a convex mirror that reflects the image straight back out of a hole in the bottom of the telescope (the **cassegrain focus**) or to a series of flat mirrors that send the image to a remote location where heavy or large equipment can be used to analyze the light (the **coudé focus**). In very large telescopes, it is possible to observe at the **prime focus,** the point where the light is focused inside the telescope structure by the primary mirror. This arrangement has the advantage of reducing the number of reflections, so that light loss is minimized, but the disadvantage that the observer must spend the night in a tiny, cramped cage mounted inside the telescope. All of these focus arrangements have either a mirror or an observer's cage mounted inside the telescope, blocking some of the incoming light, but the percentage of the light that is lost this way is usually very small.

The biggest telescope for many years was the 200-inch (5-meter) reflector at Mt. Palomar, in southern California (Fig. 4.20), but recently a 6-meter telescope was completed in the Soviet Union (Fig. 4.21).

A very promising new kind of telescope was completed in 1979 at Mt. Hopkins, Arizona. This is the *Multiple Mirror Telescope* (Figs. 4.22 and 4.23), which consists of six separate 72-inch (1.8-meter) primary mirrors with secondary mirrors arranged so that they each focus at the same point. The total collecting area of these six mirrors is equivalent to a single mirror 176 inches (4.5 meters) in diameter. This system is very complicated, and the difficulties of perfectly aligning the six mirrors are enormous, but the savings in cost and the relative ease with which the smaller mirrors can be constructed made the effort worthwhile.

At the focus of any telescope (except for small ones used only for simple viewing of the sky) is an instrument that records light. This could be a simple camera, for making images of the sky, or a **photometer,** an instrument that measures intensities of light using a photocell (a device that converts light into an electrical current). A commonly used instrument is the **spectrograph,** which disperses light according to wavelength so that the spectrum can be recorded and analyzed. Film has traditionally been, and still is, used widely to record the light, but today many kinds of electronic detectors are being developed. These have many advantages over film, both in their accuracy and in such practical matters as storing the data electronically so that it can be transmitted directly into computers for analysis. Auxilliary instruments used in astronomy range from rather compact, lightweight devices that can easily be mounted directly on the telescope for use at the cassegrain focus to large, heavy instruments

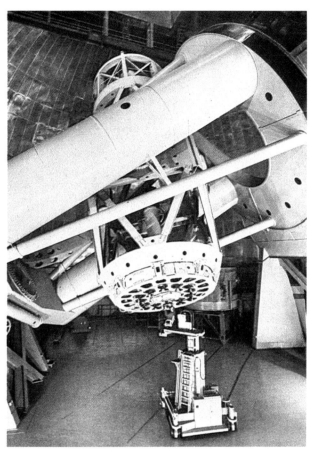

Figure 4.20 The 5-Meter Telescope. This is the largest telescope in the United States, the 200-inch instrument on Mt. Palomar. *(Palomar Observatory, California Institute of Technology)*

Figure 4.21 The 6-Meter Telescope. This is the dome building containing the world's largest telescope, located in the Caucasus Mountains of the southern central Soviet Union. *(A.G.D. Phillip)*

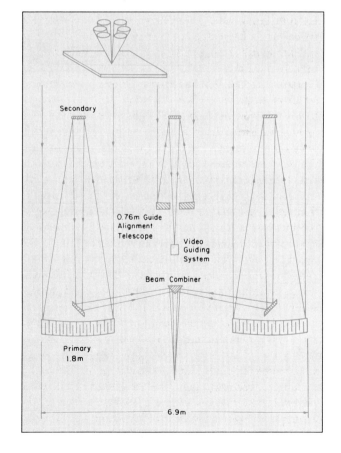

(usually spectrographs) that must be used at the coudé focus.

Telescopes for wavelengths other than visible light have the same basic ingredients as the telescopes just described. There is an element such as a mirror for collecting and focusing light and an instrument at the focus for analyzing and recording the radiation. Telescopes for ultraviolet and infrared wavelengths are virtually identical in design to those for visible light, with

Figure 4.22 A Schematic of the Multiple-Mirror Telescope Design. This drawing shows how the images from two of the six mirrors are brought to a common focus. *(Multiple-Mirror Telescope Observatory, University of Arizona and the Harvard-Smithsonian Center for Astrophysics)*

Figure 4.23 The Multiple-Mirror Telescope, Mt. Hopkins, Arizona. This observatory is operated by the University of Arizona and the Smithsonian Institution. *(Multiple-Mirror Telescope Observatory, University of Arizona and the Harvard-Smithsonian Center for Astrophysics)*

a couple of exceptions: Ordinary mirrors do not reflect well in the ultraviolet, so special surfaces and chemical coatings have to be used; and infrared telescopes glow in the wavelengths they are built to observe, so the instrument must be cooled to reduce the glow (usually by circulating liquid nitrogen or liquid helium through the part of the instrument that contains the detector). Of course, ultraviolet light and large portions of the infrared spectrum do not penetrate the earth's atmosphere, so these telescopes have to be launched into space; therefore, the fields of ultraviolet and infrared astronomy are relatively new ones, with most research occurring only in the past twenty years or so. Some existing and planned space observatories are listed in Table 4.3.

For wavelengths beyond infrared, that is, the radio portion of the spectrum, the radiation reaches the ground, and large earth-based telescopes are used. Radio telescopes usually consist of very large dishes made of metal plating or a wire mesh, and they bring radiation to a focus at a point above the center of the dish (Fig. 4.24). At the focus is suspended a detector. Because the resolution of a telescope is proportional to the wavelength being observed, radio telescopes tend to have very poor resolution. They are built in gigantic sizes partly to compensate for this, since resolution is also inversely proportional to telescope diameter. Even the largest radio dish, however, has relatively poor resolution compared to visible-light telescopes; the giant 1000-foot Arecibo telescope can separate objects in the sky no closer together than about 3 arcminutes, whereas a 10-centimeter (4-inch) visible wavelength telescope has a resolution of about 1 arcsecond. As we discussed earlier, however, interferometry has been used very successfully with radio telescopes in pairs or groups, with the result that radio observatories have achieved much better resolution than visible-light telescopes.

For the shortest wavelengths, below ultraviolet, it is no longer possible to build a simple concave reflector to focus light. Extreme ultraviolet and X-ray radiation simply do not reflect well from any kind of mirror.

Table 4.3 Major Telescopes in Space

Wavelength Region	Telescope*	Agency†	Diameter (m)	Dates
Gamma-ray	(Gamma Ray Observatory: GRO)	NASA		(1990)
X-ray	Einstein Observatory	NASA	0.56	1978–1981
	Exosat	ESA		1982–1986
	(Roentgen Satellite: ROSAT)	Germany-NASA		(1990)
	(Advanced X-ray Astrophyics Telescope Facility: AXAF)	NASA		(1995)
Extreme ultraviolet	(Extreme Ultraviolet Explorer: EUVE)	NASA		(1991)
	(Lyman)	NASA-ESA	0.6	(1992)
Ultraviolet	Copernicus	NASA	0.9	1972–1980
	International Ultraviolet Explorer (IUE)	NASA-ESA-UK	0.41	1978-
	(Hubble Space Telescope: HST)	NASA-ESA	2.4	(1989)
	(Lyman)	NASA-ESA	0.6	(1992)
Visible	(Hubble Space Telescope: HST)	NASA-ESA	2.4	(1989)
	(Hipparcos)	ESA	0.29	(1988)
Infrared	Infrared Astronomical Satellite (IRAS)	Holland-UK-NASA	0.57	1983–1984
	(Shuttle Infrared Telescope Facility: SIRTF)	NASA		(1998)
	(Infrared Space Observatory: ISO)	ESA		?
	(Large Deployable Reflector: LDR)	NASA	10	?
Radio	(Cosmic Background Explorer: COBE)	NASA		(1989)
	(Orbital Very Long Baseline Interferometry: Space VLBI)	NASA		(1995)
	(QUASAT)	ESA		(1996)

*Instruments listed in parentheses are planned or under construction, but not yet launched.
†NASA is the U.S. National Aeronautics and Space Administration; ESA is the European Space Agency, a consortium of several countries.

Figure 4.24 A Large Radio Telescope. At left, the 300-foot dish at the *U.S. National Radio Astronomy Observatory,* in Green Bank, West Virginia. Despite its size, this telescope can be pointed in any desired direction, just as a visible-light telescope can. *(National Radio Astronomy Observatory, operated by Associated Universities, Inc., under contract with the National Science Foundation)*

These wavelengths can be reflected, however, if they strike a surface at a very oblique angle; this is called **grazing incidence,** and an X-ray telescope can be constructed using grazing-incidence reflections. Such a telescope has one or more ringlike primary mirrors at the front and can even have a secondary mirror to bring the radiation to a focus where a detector is located (Fig. 4.25). Like ultraviolet light, X rays do not penetrate the Earth's atmosphere, and observations must be made from space.

The shortest wavelength radiation, the gamma rays, are very difficult to focus. Even grazing-incidence reflections do not work for these extremely high energy photons, and the few crude gamma-ray telescopes built so far have consisted simply of detectors, with no primary light-gathering element to bring the radiation to a focus from a large collecting area. A device called a **collimator** is placed in front of the detector, so that only gamma rays from a certain direction are recorded; without this, there would be no way of knowing where the observed gamma rays came from. Again, gamma rays must be observed from space, because they cannot pass through the Earth's atmosphere.

Figure 4.25 Design of an X-Ray Telescope. A typical X-ray telescope consists of a set of nested rings, each able to reflect X rays to a focal point by means of grazing-incidence reflections. Several rings are used because each one individually has little effective collecting area.

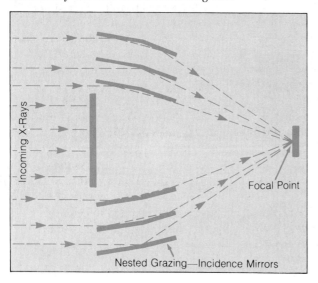

Major Observatories for All Wavelengths

Observatories for visible light and the portions of the infrared spectrum that reach the ground are generally located at remote, high-altitude sites; it is important to locate observatories far from city lights, and being at high altitude decreases the amount of atmosphere above the telescope (this is especially important for infrared telescopes, because water vapor in the atmosphere absorbs at several wavelengths in the infrared). Of course, it is also important to place telescopes at sites where the weather is clear much of the time. Another consideration is the latitude of the site, which determines how far north or south one can observe. Historically, most major observatories were in the Northern Hemisphere, but today there are several large ones in operation in the Southern Hemisphere.

The largest observatories in the U.S. are located in Arizona (Kitt Peak National Observatory (Fig. 4.26), as well as the *Multiple-Mirror Telescope*), California (the 5-meter *Hale Telescope* on Palomar Mountain, and Lick Observatory, near San Jose), and on the island of Hawaii, now the location of the world's largest observatory (Fig. 4.27). The summit of Mauna Kea, nearly 14,000 feet above sea level, has very stable, clear weather, a minimum of atmospheric turbulence overhead, and very dry air, making infrared observations especially suitable. The latitude, about 20° north of the equator, provides access to much of the southern sky.

Figure 4.26 A Panorama of Telescopes at Kitt Peak. This sunset view shows the buildings housing several of the telescopes at Kitt Peak National Observatory near Tucson, Arizona. *(National Optical Astronomy Observatories)*

Astronomical Insight 4.2

A Night at the Observatory

A typical astronomer's night at the observatory will vary quite a bit from one individual to another and especially from one observatory to another. There is a big difference between sitting in a heated room viewing an array of electronics panels and standing in a cold dome peering into an eyepiece while recording data on a photographic plate, yet both extremes (and various possibilities in between) are part of modern observational astronomy.

Despite apparent differences in practice, however, the principles are much the same, regardless of where the observations take place or with what equipment. Here is an outline of a typical night at a large observatory, where a modern electronic detector is used to record spectra of stars.

By midafternoon the astronomer and his or her assistants (an engineer who knows the detector and its workings and a graduate student) are in the dome, making sure the detector is working properly. To do this requires various tests of the electronics and the recording of test spectra from special calibration lamps. Everything is in order, and they eat supper in the observatory dining hall a couple of hours before sunset, returning to the dome to complete preparations for the night's work. A night assistant (an observatory employee) arrives shortly before dark and is given a summary of the plans for the night. It is the night assistant's job to control the telescope and the dome, and he or she is responsible for the safety of the observatory equipment. At large observatories the telescope is considered so complex and valuable that a specially trained person is always on hand, and the astronomer (who may use a particular telescope only a few nights a year) does not handle the controls, except for fine motions required to maintain accurate pointing during the observations.

Before it is completely dark, a number of calibration exposures are taken, using lamps and even the moon, if it is up. These measurements will help the astronomer analyze the data later, by providing information on the characteristics of the spectrograph and the detector.

Finally everything is in order, and if things are going smoothly the telescope is pointed at the first target star just as the sky gets dark enough to begin work. Observing time on large telescopes is difficult to obtain and very valuable; not a moment is wasted.

A pattern is quickly established and repeated throughout the night. After the night assistant has pointed the telescope at the requested coordinates, the astronomer looks at the field of view to confirm which star is the correct one (often by consulting a chart prepared ahead of time). When the star is properly positioned so that its light enters the spectrograph, the observation begins. Depending on the brightness of the star and the efficiency of the spectrograph and the detector, the exposure time may be anywhere from seconds to hours. During this time, the astronomer or the graduate student assistant continually checks to see that the star is still properly positioned, making corrections as needed by pushing buttons on a remote control device.

If the observations are made with an instrument at the coudé focus, the astronomer and assistants spend the entire night in an interior room in the dome building, emerging only occasionally to view the skies, give instructions to the night assistant, or go to the dining hall for lunch (usually served around midnight).

As dawn approaches and the sky begins to brighten, a last-minute strategy is devised to get the most out of the remaining time. Finally the last exposure is completed, and the night assistant is given the OK to close the dome and park the telescope in

(continued on next page)

Astronomical Insight 4.2

A Night at the Observatory

(continued from previous page)

its rest position (usually pointed straight overhead to minimize stress on the gears and support system). The astronomer and the assistants make final calibration exposures before shutting down the equipment and trudging off to get some sleep. The new day for them begins at noon, or shortly thereafter.

The scenario just described depicts a typical night at a major modern observatory such as Kitt Peak, which has extensive resources and the largest telescopes. At smaller facilities, such as a typical university observatory, it is more common to find the astronomer working alone, most often using a cassegrain-focus instrument. In such circumstances the observer must spend the night out in the dome with the telescope (where it can be very cold in the winter) and be content with a homemade lunch. The work pattern is similar, however, as is the desire not to waste any telescope time.

Figure 4.27   The Mauna Kea Observatory.   These are several of the large telescopes on the summit of Mauna Kea, including (from left to right on top of the ridge) the 3.6-meter Canada-France-Hawaii telescope, the University of Hawaii 2.2-meter telescope, and the 3.8-meter United Kingdom infrared telescope. In the foreground are the buildings for two new radio telescopes: a joint Dutch-United Kingdom 15-meter facility (in the large white building), and a 10-meter telescope being built by Cal Tech. The planned 10-meter visible-light *Keck Telescope*, to be the largest in the world for several years, will be built just to the left of the 3-meter NASA *Infrared Telescope Facility* at the far left. *(T. P. Snow)*

Not only does Mauna Kea already have more light-collecting power in its four major telescopes than any other observatory, but the next very large telescopes that are being planned by U.S. astronomers will also be built there. The 10-meter Keck Telescope (Fig. 4.28) is already under construction by the University of California and Caltech, and the U.S. National Observatories are planning for a 16-meter telescope (consisting of four 8-meter mirrors operating together as in the *Multiple Mirror Telescope*). Other major observatory sites are listed in Appendix 6.

Radio observatories generally do not suffer from the same problems of light interference and atmospheric absorption that visible light telescopes have, and they can be located in regions less remote from civilization (the very shortest radio wavelengths are the exception, since they are subject to interference, and telescopes for these wavelengths tend to be built at the same sites as major visible-light observatories). There are major radio observatories in regions where the weather is not always clear, because clouds do not block radio waves. Radio observatories can operate in the daytime as well, because the sky is "dark" in radio wavelengths even if the Sun is up.

The major U.S. radio observatories are located in Green Bank, West Virginia (see Fig. 4.24), near Socorro, New Mexico (the *Very Large Array*, already mentioned), and in Arecibo, Puerto Rico, site of the world's largest single radio antenna, 1000 feet in diameter. Others are in Holland, England, and Australia. It has proven possible to make interferometric observations using radio dishes at observatories separated by hundreds or even thousands of miles, and the U.S. is currently building an array of ten large receivers that will span the continent. This array, called the *Very Long Baseline Interferometer*, will be complete by the early 1990s and will provide unprecedented resolution.

Telescopes for the other wavelengths (parts of the infrared spectrum, ultraviolet light, X rays, and gamma rays) all must be launched into space in order to avoid absorption of these wavelengths by the Earth's atmosphere. In the past 20 years major orbiting observatories for ultraviolet, X-ray, and infrared observations have been operated successfully (Table 4.3). Currently the only facility in operation is the *International Ultraviolet Explorer* (Fig. 4.29), which entered its tenth year

Figure 4.28 The 10-Meter *Keck Telescope*. This illustrates the plans for the large telescope being built jointly by Cal Tech and the University of California, to be placed on Mauna Kea. *(Keck Observatory, University of California and the California Institute of Technology)*

Figure 4.29 The International Ultraviolet Explorer. This is an artist's depiction of the *IUE*. *(NASA)*

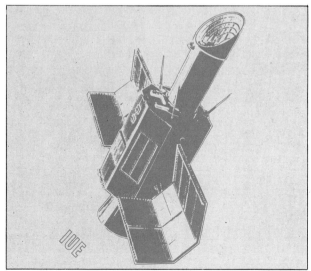

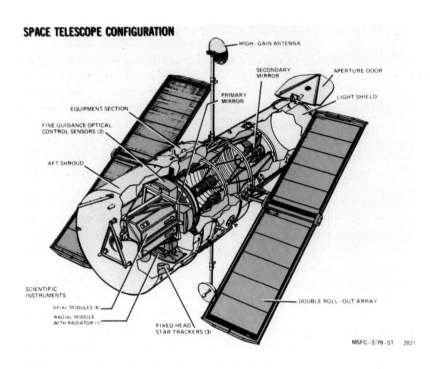

Figure 4.30 The *Hubble Space Telescope.* This is a schematic drawing of the 2.4-meter telescope planned for launch into Earth orbit during 1989. The open end of the telescope is pointed away from us in this view; the near end houses control systems, cameras, and other scientific instruments. The large "wings" are solar panels, which provide electrical power for the spacecraft.

in 1987. Major new infrared and X-ray facilities are being planned, but the next orbiting telescope to be launched will be an ultraviolet and visible-light instrument, the *Hubble Space Telescope* (Fig. 4.30). This joint U.S.-European facility will be the largest telescope yet put into space, with a diameter of 2.4 meters. It will have unprecedented resolution in its visible-light and ultraviolet images, and it will be capable of detecting much dimmer objects than any telescope on the ground. It is expected and hoped that the *Space Telescope* will revolutionize our knowledge of many facets of the universe, from our neighboring planets to the most remote galaxies. The *Space Telescope* was due for launch in late 1986 but has been postponed due to delays in the space shuttle program caused by the *Challenger* disaster. It is now expected that the launch of the *Space Telescope* will occur in late 1989 or early 1990.

Perspective

We have now learned how light is absorbed and emitted by natural processes and how it is gathered and analyzed by various devices. We have seen how the continuous spectrum and overall intensity of light from a star depend on its temperature and size, and we have seen how the spectral lines provide information on the temperature, density, motion, and chemical composition of a star. Telescope technology allows astronomers to derive a remarkable amount of information about distant objects, something we will come to appreciate more and more in the chapters that follow. We are ready now to explore the universe, to see what has been learned with the techniques we have been discussing.

Summary

1. Visible light is just one part of the electromagnetic spectrum, which extends from the gamma rays to radio wavelengths.
2. Light has properties associated with both waves and particles, and these properties are combined in the concept of the photon.
3. The continuous radiation from a star provides information on its temperature, luminosity, and radius through the use of Wien's law, the Stefan-

Boltzmann law, and the more general Planck's law. (quantum mechanics)

4. The observed brightness of a glowing object is inversely proportional to the square of the distance between the object and the observer.

5. Spectral lines are produced by transitions of electrons between energy levels in atoms and ions; absorption occurs when an electron gains energy, and emission occurs when it loses energy, the wavelength in both cases corresponds to the energy gained or lost as the electron changes levels.

6. Each chemical element has its own distinct set of spectral line wavelengths, and therefore the composition of a distant object can be determined from its spectral lines.

7. The ionization and excitation of a gas can be inferred from its spectrum, the former yielding information on the temperature of the gas, and the latter providing data on its density.

8. Any motion along the line of sight between an observer and a source of light produces shifts in the wavelengths of the observed spectral lines, and measurement of this Doppler effect, as it is called, can be used to determine the relative speed of source and observer.

9. The principal reasons for using telescopes are to collect radiation from a large area, to allow light from a source to be collected over longer periods of time than is possible with the unaided eye, and to provide resolution, that is, the ability to discern fine detail.

10. The basic telescope design consists of a large mirror or reflecting surface to bring light to a focus, where an instrument containing film or an electronic detector records it. This general concept works at all wavelengths, from X ray to radio, but the details vary from one wavelength region to the next.

11. The instrument that receives the light at the telescope focus may be used to record images, measure the brightness of objects, or analyze the spectrum of the light.

12. Visible-light and radio telescopes can be located on the earth's surface, but telescopes for X-ray, ultraviolet, and much of the infrared spectrum must observe from above the earth's atmosphere. On the ground, observatory sites must be chosen for clear weather, high atmospheric transmissivity, low turbulence in the air overhead, minimal pollution, remoteness from city lights, and the proper latitude for viewing the desired part of the sky.

13. Many new telescopes are under construction or in the planning stage, for all wavelength regions. One of the next major facilities to go into operation will be the *Space Telescope*, a 2.4-meter telescope for visible and ultraviolet observations, now planned for a 1989 or 1990 launch. The first of several planned very large visible-light telescopes will be the 10-meter *Keck Telescope*, to be in operation on Mauna Kea in the 1990s.

Review Questions

1. Suppose your favorite radio station has a frequency of 1200 kHz (that is, 1200 kilohertz, where one kilohertz equals 1000 hertz). What is the wavelength at which this station transmits?

2. Using Wien's law, calculate the wavelength of maximum emission for the following objects, and comment on whether they would appear to glow in visible light: (a) a star with surface temperature 25,000 K; (b) the walls of a room, with temperature 300 K; (c) the surface of a planet whose temperature is 100 K; (d) gas being compressed as it falls into a black hole, so that its temperature is 10^6 K; and (e) liquid helium, with a temperature of 4 K.

3. Suppose a white dwarf star has a surface temperature of 10,000 K, and a red giant has a temperature of 2000 K. Compare their surface brightnesses (the energy emitted per square centimeter). Now suppose that the white dwarf has a radius of 10^8 cm, and the red giant has a radius of 10^{13} cm. Compare the luminosities (the total energy emitted over the entire surface) of the two stars.

4. Normal stars have *absorption* lines in their spectra, because the relatively cool gas in the outer layers, lying above the hotter interior, absorbs light at specific wavelengths. What would you conclude about a star that has *emission* lines in its spectrum?

5. Suppose the elements iron and beryllium each have a spectral line at the same wavelength. If you found a line at this wavelength in the spectrum of a star, how could you decide which element was responsible?

6. The strong line of hydrogen, whose rest wavelength is 6563 Å, is observed in a number of stars. Calculate the line-of-sight velocity of a star where this line is observed at (a) 6561.8 Å, (b) 6567 Å, and (c) 6593 Å.
7. Compare the light-gathering power of a 4-meter telescope with that of a 1-meter telescope. Which has better resolving power, and by what factor?
8. Suppose you want to make infrared observations of an interstellar dust cloud whose temperature is 500 K. Explain why this would be difficult to do with a telescope at room temperature. How would the situation be improved by cooling the telescope with liquid nitrogen, which has a temperature of 77 K?
9. Explain why ultraviolet telescopes are needed for observations of very hot stars.
10. The planned U.S. *National New Technology Telescope* will consist of four large mirrors using the multiple-mirror telescope concept. The total collecting area will be the equivalent of a single 16-meter mirror. How large must each of the four mirrors be?

Additional Readings

The best sources for further explanation of the properties of light and radiation are elementary physics books, of the sort used as texts in introductory courses. In addition, there are a few nontechnical books on the nature of light and a variety of books and articles on telescopes.

Asimov, I. 1975. *Eyes on the universe.* Boston: Houghton Mifflin.
Bahcall, J. N., and L. Spitzer, Jr. 1982. The space telescope. *Scientific American* 247(1):40.
Beatty, J. K. 1985. HST: Astronomy's greatest gambit. *Sky and Telescope* 69(5):409.
Bragg, W. 1959. *The universe of light.* New York: Dover.
diCicco, D. 1986. The journey of the 200-inch mirror. *Sky and Telescope* 71(4):347.
Evans, David S., and J. Derrall Mulholland. 1986. *Big and bright: A history of the McDonald Observatory.* University of Texas Press.
Field, G. B. 1984. The future of space astronomy. *Mercury* XIII(4):98.
Gordon, M. A. 1985. VLBA—A continent-size radio telescope. *Sky and Telescope* 69(6):487.
Habing, H. J., and G. Neugebauer, 1984. The infrared sky. *Scientific American* 251(5):48.
Harrington, S. 1982. Selecting your first telescope. *Mercury* 11(4):106.
Horodyski, Joseph M. 1986. Reach for the stars: The story of Mount Wilson Observatory. *Astronomy* 14(12):6.
Janesick, J., and M. Blouke. 1987. Sky on a chip: The fabulous CCD. *Sky and Telescope* 74(3):238.
King, H. C. 1979. *The history of the telescope.* New York: Dover.
Kirby-Smith, H. T. 1976. *U.S. observatories: a directory and travel guide.* New York: Van Nostrand.
Labeyrie, A. 1982. Stellar interferometry: A widening frontier. *Sky and Telescope* 63(4):334.
Longair, M. 1985. The scientific challenge of Space Telescope. *Sky and Telescope* 69(4):306.
Meyer-Arendt, J. R. 1972. *Introduction to Classical and Modern Optics.* Englewood Cliffs, N.J.: Prentice Hall.
Miczaika, G. R. 1961. *Tools of the astronomer.* Cambridge, Mass.: Harvard University Press.
Mims, S. S. 1980. Chasing rainbows: The early development of astronomical spectroscopy. *Griffith Observer* 44(8):2.
Overbye, Dennis 1982. Night flight to the stars. *Discover* 3(8):38. (Kuiper Airborne Observatory)
Physics Today 35(11): (several review articles on telescopes for various wavelengths).
Rublowsky, J. 1964. *Light.* New York: Basic Books.
Shore, Lys Ann. 1987. IUE: Nine years of astronomy. *Astronomy* 15(4):14.
Tucker, W., and R. Giacconi, 1985, 1986. The birth of X-ray astronomy. Part 1. *Mercury* XIV(6):178; Part 2. *Mercury* XV(1):13.
Verschuur, Gerrit L. 1987. *The invisible universe revealed: The story of radio astronomy.* New York: Springer-Verlag.

Section II

The Stars

The material in this section is devoted to the stars and their properties. In its four chapters we will learn first about our own star, the Sun, and then about how stars are observed, how their fundamental properties are deduced from observations, and how they work. We will see what processes control the structure of stars, and we will learn how they live, evolve, and die. We will see that stars are dynamic, changing entities and only seem fixed and immutable because the time scales on which they evolve are vastly longer than the human lifetime.

The first chapter of the section summarizes what is known of the Sun, a typical star in many respects, but very atypical in that we can observe it in much more detail than any other star. Information on the Sun will be used as we discuss stars in general in subsequent chapters.

The second chapter of this section discusses the three basic types of stellar observations and how the data obtained from these techniques are used to derive fundamental properties of stars. We systematically consider several stellar properties, discuss how each is determined, and summarize the typical values that are found.

Chapter 7 discusses how stars function and evolve by describing the physical processes that occur in stars, how these processes govern the stellar parameters previously discussed, and how stars change with time as these processes act. The final chapter of the section describes the remnants of stars that have finished their lives, some of the most bizarre and exciting objects in the universe.

Having studied the stars, we will then be ready to consider large groups of stars like our galaxy. A galaxy contains many billions of stars, and its evolution is governed by the evolution of the individual stars within it.

Chapter 5

The Sun

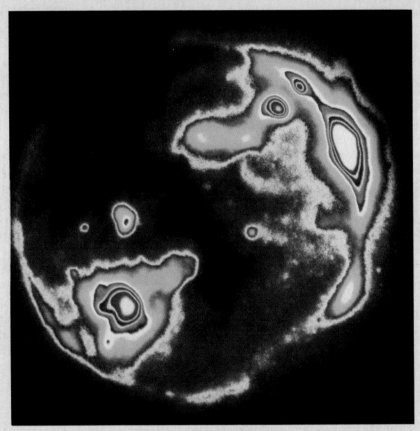

An X-ray image of the Sun,
showing bright active regions
and dark coronal holes.
(NASA/Marshall Flight Center)

Chapter Preview

Basic Properties and Internal Structure
   Nuclear Reactions
Structure of the Solar Atmosphere
The Solar Wind
Sunspots, Solar Activity Cycles, and the Magnetic Field

Our star, the Sun, is rather ordinary by galactic standards. Its mass and size are modest; there are stars as much as a few hundred times larger and a million times more luminous. Its temperature is also moderate, as stars go. In many respects the Sun is entirely a run-of-the-mill entity.

Because the Sun is a typical star, and because we have far more detailed knowledge of it than of other stars, it is useful to begin our discussion of stars with a chapter on the Sun. In subsequent chapters, we will see how the information on the Sun helps us understand other stars. We will also find that studies of other stars help us learn more about the Sun.

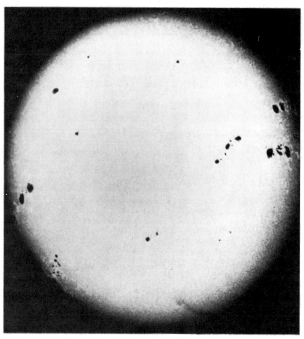

Figure 5.1 The Sun. This is a white-light photograph, showing numerous sunspots. *(Mt. Wilson and Las Campanas Observatories, Carnegie Institution of Washington)*

Basic Properties and Internal Structure

Perhaps the most obvious attribute of the Sun is that it is bright, that it emits vast quantities of light. Its **luminosity**, the amount of energy emitted per second, is about 4×10^{33} ergs per second (for comparison, recall that a 100-watt light bulb emits 10^9 ergs per second). The intensity of sunlight reaching the Earth (above the atmosphere) is about 1.4×10^6 ergs per square centimeter per second. This quantity is known as the **solar constant**, althought it varies slightly due to activity in the Sun's outer layers (discussed later in this chapter).

The Sun is a ball of hot gas (Fig. 5.1). It dwarfs any of the planets in mass and radius (Table 5.1). Its average density is 1.41 grams per cubic centimeter—not much more than that of water—but its center is so highly compressed that the density there is about ten times greater than that of lead. The interior is gaseous rather than solid because the temperature is very high: around 10 million K at the center, diminishing to just under 6000 K at the surface. At these temperatures the gas is partially ionized in the outer layers of the Sun and completely ionized in the core, all electrons having been stripped free of their parent atoms.

A combination of theory and observation tells us quite a bit about the internal structure of the Sun. By solving equations representing known physical laws and accepting only solutions that match the Sun's observed external properties, astronomers have developed theoretical models for the interior. One such model is summarized in Table 5.2.

Table 5.1 The Sun

Diameter: 1,391,980 km ($109.3 D_\oplus$)
Mass: 1.99×10^{33} grams ($332,943\ M_\oplus$)
Density: 1.409 grams/cm^3
Surface gravity: 27.9 Earth gravities
Escape velocity: 618 km/sec
Luminosity: 3.83×10^{33} ergs/sec
Surface temperature: 6500 K (deepest visible layer)
Rotation period: 25.04 days (at equator)

Table 5.2 Structure of the Sun

Zone	Radius (MR$_\odot$* and km)		Temperature (K)	Density (g/cm^3)	Mass Fraction (M$_\odot$)*
Interior	0.00 MR$_\odot$*	0 km	1.6×10^7	160	0.000
	0.04	28,000	1.5×10^7	141	0.008
	0.1	70,000	1.3×10^7	89	0.07
	0.2	139,000	9.5×10^6	41	0.35
	0.3	209,000	6.7×10^6	13	0.64
	0.4	278,000	4.8×10^6	3.6	0.85
	0.5	348,000	3.4×10^6	1.0	0.94
	0.6	418,000	2.2×10^6	0.35	0.982
	0.7	487,000	1.2×10^6	0.08	0.994
	0.8	557,000	7.0×10^5	0.018	0.999
	0.9	627,000	3.1×10^5	0.0020	1.000
	0.95	661,000	1.6×10^5	0.0004	1.000
	0.99	689,000	5.2×10^4	0.00005	1.000
	0.995	692,500	3.1×10^4	0.00002	1.000
	0.999	695,300	1.4×10^4	0.0 000001	1.000
Photosphere	1.000	695,990	6.4×10^3	3.5×10^{-7}	1.000
	1.000	+ 280	4.6×10^3	4.5×10^{-8}	1.000
Chromosphere	1.000	+ 320	4.6×10^3	3.1×10^{-8}	1.000
	1.001	+ 560	4.1×10^3	3.6×10^{-9}	1.000
(Transition)	1.002	+ 1900	8.0×10^3	3.4×10^{-13}	1.000
	1.003	+ 2400	4.7×10^5	4.8×10^{-15}	1.000
Corona	1.003	+ 2400	5.0×10^5	1.7×10^{-15}	1.000
	1.2	+ 140,000	1.2×10^6	8.5×10^{-17}	1.000
	1.5	+ 348,000	1.7×10^6	1.4×10^{-17}	1.000
	2.0	+ 696,000	1.8×10^6	3.4×10^{-18}	1.000

*The radii are expressed in fractions of the Sun's radius R$_\odot$, as well as in kilometers. Above 695,990 km, the kilometer values are heights above the Sun's "surface," or lower boundary of the photosphere. The mass fractions are expressed in units of the Sun's mass M$_\odot$, which has the value 1.989×10^{33} grams. [Data from C. W. Allen, 1973, *Astrophysical Quantities* (London: Athlone Press).]

The Sun is held together by gravity. All of its constituent atoms and ions attract each other, and the net effect is that the solar substance is held in a spherical shape. Gas that is hot exerts pressure on its surroundings, and this pressure, which pushes matter outward, balances the force of gravity, which pulls the matter inward. This balance is called **hydrostatic equilibrium**, with gravity and pressure equaling each other everywhere. The deeper we go into the Sun, the greater the weight of the overlying layers and the more the gas is compressed. The higher the pressure, the greater the temperature required to maintain the pressure, so we find that the pressure and temperature both increase as we approach the center.

The composition of the Sun (Table 5.3) is the same as that of the primordial solar system and of most other stars: About 73 percent of its mass is hydrogen; 25 percent is helium; and there are only traces of everything else. In the outer layers, where no nuclear reactions have taken place, a greater fraction of the mass is in the form of hydrogen (about 79 percent). As we have seen, this is similar to the chemical makeup of the outer planets; it would also represent the terrestrial planets, except that they have lost most of their volatile elements such as hydrogen and helium. It is apparent that all the components of the solar system formed together from the same material.

The ultimate source of all the Sun's energy is in its core, within the innermost 10 percent or so of its radius (Fig. 5.2). Here nuclear reactions create heat and photons of light at γ-ray wavelengths. This light eventually reaches the surface and escapes into space, but it is a

Table 5.3 The Composition of the Sun's Photosphere

Element	Symbol	Number of Atoms*	Fraction of Mass
Hydrogen	H	1.0000	0.735
Helium	He	0.0851	0.248
Lithium	Li	1.55×10^{-9}	7.85×10^{-9}
Beryllium	Be	1.41×10^{-11}	9.27×10^{-11}
Boron	B	2.00×10^{-10}	1.58×10^{-9}
Carbon	C	0.000372	0.00326
Nitrogen	N	0.000115	0.00118
Oxygen	O	0.000676	0.00788
Fluorine	F	3.63×10^{-8}	5.03×10^{-7}
Neon	Ne	3.72×10^{-5}	0.000547
Sodium	Na	1.74×10^{-6}	2.92×10^{-5}
Magnesium	Mg	3.47×10^{-5}	0.000615
Aluminum	Al	2.51×10^{-6}	4.94×10^{-5}
Silicon	Si	3.55×10^{-5}	0.000727
Phosphorus	P	3.16×10^{-7}	7.14×10^{-6}
Sulfur	S	1.62×10^{-5}	0.000379
Chlorine	Cl	2.00×10^{-7}	5.17×10^{-6}
Argon	Ar	4.47×10^{-6}	0.000130
Potassium	K	1.12×10^{-7}	3.19×10^{-6}
Calcium	Ca	2.14×10^{-6}	6.26×10^{-5}
Scandium	Sc	1.17×10^{-9}	3.84×10^{-8}
Titanium	Ti	5.50×10^{-8}	1.92×10^{-6}
Vanadium	V	1.26×10^{-8}	4.68×10^{-7}
Chromium	Cr	5.01×10^{-7}	1.90×10^{-5}
Manganese	Mn	2.51×10^{-7}	1.01×10^{-5}
Iron	Fe	3.98×10^{-5}	0.00162
Cobalt	Co	3.16×10^{-8}	1.36×10^{-6}
Nickel	Ni	1.91×10^{-6}	8.18×10^{-5}
Copper	Cu	2.82×10^{-8}	1.31×10^{-6}
Zinc	Zn	2.63×10^{-8}	1.25×10^{-6}
(all others combined)			(less than 10^{-8} of total)

*The numbers given are relative to the number of hydrogen atoms.

laborious journey. Each photon is absorbed and re-emitted many times along the way (Fig. 5.3), gradually losing energy. In the process the photon becomes a visible light photon, and the energy it loses heats the surroundings. Because a photon travels only a short distance before it is absorbed and because it is re-emitted in a random direction, a photon's progress toward the surface is very slow. It takes an individual photon as long as a million years to migrate from the center of the Sun to the surface, even though the light travel time if it were unimpeded would be only two seconds. If the Sun's energy source were suddenly turned off, we would not be aware of it for a million years!

Throughout most of the solar interior, the gas is thought to be quiescent, without any major large-scale flows or currents. The energy from the core is transported by the radiation wending its slow pace outward, except in the layers near the surface, where convection occurs and heat is transported by the overturning motions of the gas. As we shall see, the bubbling, boiling action in the outer portions of the Sun creates a wide variety of dynamic phenomena on the surface.

The Sun rotates differentially. This is apparent from observations of its surface features, which reveal that like Jupiter and Saturn, the Sun goes around faster at the equator than near the poles. The rotation period is

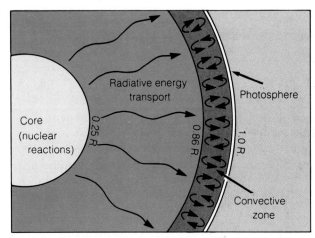

Figure 5.2 The Internal Structure of the Sun. This drawing shows the relative extent of the major zones within the Sun, except that the depth of the photosphere is greatly exaggerated.

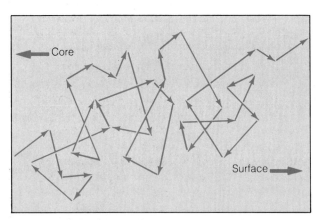

Figure 5.3 Random Walk. A photon is continually being absorbed and re-emitted as it travels through the Sun's interior; each time it is re-emitted, it is in a random direction. Thus its progress from the core, where it is created, to the surface is very slow.

twenty-five days at the equator, twenty-eight days at the middle latitudes, and even longer near the poles. As we will see, the differential rotation probably plays an important role in governing variations in the solar magnetic field, which in turn have a lot to do with the behavior of the most prominent surface feature, the sunspots.

There is evidence that the core of the Sun rotates a little more rapidly than the surface. The only clue for this is found in the presence of very subtle oscillations on the solar surface, which may be wave motions affected by the more rapid spin of the interior. The spin of the solar core is probably a direct result of the collapse and accelerated rotation of the interstellar cloud from which the Sun formed. (The origin of the Sun's rotation will be discussed in Chapter 15.)

Nuclear Reactions

One of the biggest mysteries in astronomy in the early decades of this century was presented by the Sun. The problem was how to account for the tremendous amount of energy it radiates, particularly perplexing in view of geological evidence that the Sun has been able to produce this energy for at least 4 or 5 billion years. Two early ideas—that the Sun is simply still hot from its formation or that it is gradually contracting and releasing stored gravitational energy—were both ruled out since neither mechanism could possibly supply the energy needed to run the Sun for a long enough time.

The first hint at the solution came in the first decade of the 1900s, when Albert Einstein developed his theory of special relativity. He showed that matter and energy are equivalent and that one can be converted into the other according to the famous formula $E = mc^2$, where E is the energy released in the conversion, m is the mass that is converted, and c is the speed of light. This mechanism can produce enormous amounts of energy, and physicists began to contemplate the possibility that somehow this energy was being released inside the Sun and stars.

By the 1920s, following the pioneering work on atomic structure by Max Planck, Niels Bohr, and others, the concept of nuclear reactions began to emerge. Nuclear reactions are transformations like chemical reactions, except that in this case it is the subatomic particles in the nuclei of atoms that react with each other. There are **fusion** reactions, in which nuclei merge together to create a larger nucleus representing a new chemical element; and there are **fission** reactions, in which a single nucleus, usually of a heavy element with a large number of protons and neutrons, splits into two or more smaller nuclei. In either type of reaction, energy is released as some of the matter is converted according to Einstein's formula. In the 1920s, many physicists explored these possibilities, and by the 1930s Hans Bethe had suggested a specific reaction sequence that might operate in the Sun's core.

Astronomical Insight 5.1

Imitating The Sun

Nuclear reactions can produce energy very efficiently and therefore have obvious potential benefits for human society, if methods can be found for producing and controlling nuclear energy on earth. A great deal of effort has gone into developing nuclear power, but the only working reactors today are fission reactors. These have several disadvantages, the primary one being that they produce radioactive waste products that are very difficult to store safely.

Fusion reactions, however, can also produce immense quantities of energy but do not have the danger of runaway reactions that could destroy the reactor. If we could somehow produce and control fusion reactions, we might have a permanent solution to the problem of producing sufficient energy to run our society. Unfortunately the only synthetic fusion reactions to date have been the instantaneous ones that took place in tests of hydrogen bombs.

The difficulty in controlling nuclear fusion is obvious in view of the conditions required for it to occur inside the Sun. The problem is to reproduce in a controlled environment the incredible temperatures and pressures of the Sun's core. Three alternative techniques for doing this are being developed.

The oldest of these ideas, dating back to the 1950s, is to contain the superheated gas in a sort of magnetic bottle. Recall that at sufficiently high temperatures a gas is ionized and consists entirely of charged particles. Such particles are subject to electromagnetic forces and can be trapped within a fixed region by a properly shaped magnetic field. It has been found most feasible to do this by constructing a **torus** (a doughnut-shaped tube), which is twisted into a figure eight to help keep the particles away from the tube wall. The ionized gas can then circulate within the enclosed tube, kept away from the walls by immense magnetic fields created by electromagnets surrounding the tube. The magnetic confinement technique is being studied primarily at Princeton University.

The second type of technique for controlling fusion is being studied primarily at the Lawrence Livermore Laboratories of the University of California. There a **laser** is used to heat material to the required high temperatures, by subjecting it to extremely intense beams of light. The laser, which was invented in the 1960s, produces a very narrow beam of light, in which all the photons have precisely the same wavelength. The power in the beam of light can be immense; lasers have been developed for cutting metal and for performing surgery, for example. The use of lasers to produce fusion reactions involves subjecting small pellets of matter to intense laser beams that instantaneously vaporize the material of the pellet, producing for a brief instant the conditions required for fusion. When this technique has been developed to the point where more energy is produced than is required to power the laser, it will become a useful means of producing energy.

The third and perhaps most promising technique is to place pellets of hydrogen fuel at the impact point of beams of high-velocity atomic particles. The particle accelerators used in many kinds of physics experiments usually involve the acceleration of particles to speeds near that of light. Beams of such particles contain large amounts of energy, and if the energy from several beams can be focused on a hydrogen pellet, it may be possible to heat the pellet to the point where fusion occurs. The largest such experiment, which is being built at the Sandia Laboratories in New Mexico, will bring 36 particle beams together at a single point. This corresponds to an energy of 100 trillion watts of power at that point (lasting for

(continued on next page)

Astronomical Insights 5.1

Imitating The Sun

(continued from previous page) only 50 billionths of a second), enough, it is hoped, to create nuclear fusion and produce even more energy than is put in. This 36-beam accelerator has already been operated, but not yet tested with a fuel pellet in place.

The U.S. government has been supporting fusion research for some time and, we can hope, will continue to do so. It will still take several more years before all the problems and complications have been worked out and fusion becomes a viable source of energy. When that day does come, however, it could have major effects on human society, for in the long run, it could eliminate our dependence for energy on the earth's limited resources.

Bethe envisioned a fusion reaction in which four hydrogen nuclei (each consisting of only a single proton) combine to form a helium nucleus, made up of two protons and two neutrons. The reaction actually occurs in several steps:

1. Two protons combine to form **deuterium,** a type of hydrogen that has a proton and a neutron in its nucleus (one of the two protons undergoing the reaction converts itself into a neutron by emitting a positively charged particle called a **positron**), and another particle called a **neutrino,** which has very unusual properties described in Astronomical Insight 5.2.
2. The deuterium combines with another proton to create an isotope of helium (^{3}He) consisting of two protons and one neutron.
3. Two of these ^{3}He nuclei combine, forming an ordinary helium nucleus (^{4}He, with two protons and two neutrons in the nucleus) and releasing two protons.

At each step in this sequence, heat energy is imparted to the surroundings in the form of the kinetic energy of the particles that are produced, and in Step 2 a photon of γ-ray light is emitted as well.

The net result of this reaction, which is called the **proton-proton chain**, is that four hydrogen nuclei (protons) combine to create one helium nucleus. The end product has slightly less mass than the ingredients, 0.007 of the original amount having been coverted into energy. We will learn in Chapter 7 that other kinds of nuclear reactions take place in some stars, but the vast majority produce their energy by the proton-proton chain.

It is easy to see that the proton-proton chain can produce enough energy to keep the Sun shining for billions of years. The total mass of the Sun is about 2×10^{33} grams. Only the innermost portion of this, perhaps 10 percent, undergoes nuclear reactions, so the Sun started out with about 2×10^{32} grams of mass available for nuclear reactions. If 0.007 of this, or 1.4×10^{30} grams, is converted into energy, then to find the total energy the Sun can produce in its lifetime, we multiply this mass times c^2, finding a total energy of $E = (1.4 \times 10^{30} \text{ g})(3 \times 10^{10} \text{ cm/s})^2 = 1.3 \times 10^{51}$ ergs. The rate at which the Sun is losing energy is its luminosity, which is about 4×10^{33} ergs per second. At this rate, the Sun can last for $(1.3 \times 10^{51} \text{ ergs})/(4 \times 10^{33} \text{ ergs/s}) = 3.2 \times 10^{17}$ seconds, or just about 10 billion years. Geological evidence shows that the solar system is now about 4.5 billion years old, so we can expect the Sun to keep shining for another 5 billion years or more. (In Chapter 8, we shall see what happens to stars like the Sun when their nuclear fuel runs out.)

Nuclear fusion reactions can take place only under conditions of extreme pressure and temperature because of the electrical forces that normally keep atomic nuclei from getting close enough together to react. Nuclei, which have positive charges because all of their electrons have escaped, must collide at extremely high speeds to overcome the repulsion caused by their like

Astronomical Insight 5.2

WIMPs from the Sun?

Nearly all that we know about the reactions in the Sun's core is based on theoretical calculations, and of course astronomers and physicists have been interested in veryifying the results of these calculations. We think that the theory is correct because it neatly accounts for the Sun's observed energy output, but more direct confirmation has been sought.

One way to do this was suggested by the early work of Enrico Fermi, who had postulated the existence of a tiny subatomic particle called the **neutrino**, which, his calculations showed, ought to be released at certain stages in nuclear reactions. The neutrino is a strange little particle having no mass and no electrical charge, whose role is to carry away small amounts of energy, thus balancing the equations describing the reactions. Because of its ephemeral properties, a neutrino hardly interacts with anything at all. (It can interact with other matter only through the *weak nuclear force*, (see Astronomical Insight 3.2).) Thus neutrinos produced in the Sun's core ought to escape directly into space at the speed of light, in sharp contrast with the photons from the core, which, as we have seen, take a million years or more to get out.

A possible test of the reactions going on inside the Sun, therefore, would be to measure the rate at which neutrinos are coming out. Unfortunately, the same properties that allow them to escape the Sun so easily also make them very difficult to catch and count. There is an indirect technique, however, based on the fact that certain nuclear reactions can be triggered by the impact of a neutrino.

A chlorine atom, for example, can be converted into a radioactive form of argon when it encounters a neutrino. Accordingly, an experiment was set up some years ago and is still in operation in which a huge tank containing a chlorine compound is monitored to see how many of its atoms are being converted into radioactive argon atoms, which in turn indicates how many neutrinos are passing through the tank.

Much to everyone's surprise, neutrinos have not been detected coming from the Sun in the expected numbers, and at present no satisfactory explanation has been found. Some of the suggestions are rather startling, such as the idea that the reactions in the Sun have actually stopped, as part of some long-term cyclical behavior, or the possibility that the true source of the Sun's energy is not nuclear reactions at all but a **black hole** in the center (see the discussion of these bizarre objects in Chapter 8).

Very recently, a new ray of hope has been found. Elementary particle theory suggests the possibility of other particles like the neutrino that can interact with matter only through the weak nuclear force. Some of the postulated particles are relatively massive (by subatomic standards) and if they exist could easily account for the loss of energy from the Sun's core that is needed to explain the low flux of neutrinos. The particles under consideration, which have not yet been proven to exist, are lumped under the name *weakly interacting massive particles*, or WIMPs.

An attractive feature of the idea that WIMPs are responsible for carrying away energy from the Sun's core is that this would not only solve the neutrino problem but would also clear up another mystery having to do with the Sun's internal pulsations (see Astronomical Insight 5.3). There has been some difficulty reconciling theoretical calculations of the Sun's internal vibration frequencies with those observed, and the discrepancy would be cleared up nicely if there were some mechanism for energy to escape from the Sun, thus cooling its core and

(continued on next page)

Astronomical Insight 5.2

WIMPs from the Sun?

(continued from previous page) changing the speed of sound in the Sun's interior. This in turn would change the predicted pulsation frequencies just enough to agree with observations.

If WIMPs are created in solar nuclear reactions and carry away energy as they escape directly into space, they appear capable of resolving two major mysteries regarding physical processes inside the Sun. WIMPs also have the potential to solve far-reaching questions about the structure of our galaxy and the distribution of mass in the universe, as we discuss in later chapters. Needless to say, intense efforts are being made by particle physicists to determine whether WIMPs exist, although it may be years before particle accelerators with sufficient power are developed.

---

electrical charges. The speed of particles in a gas is governed by the temperature, and only in the very center of the Sun and other stars is it hot enough (around 10 million degrees) to allow the nuclei to collide fast enough to fuse. This is why only the innermost portion of the Sun can ever undergo reactions. The high pressure in the Sun's core causes nuclei to be crowded together very densely, and this means that collisions will occur very frequently, another requirement for a high reaction rate.

Structure of the Solar Atmosphere

Observations of the Sun's appearance when viewed in different wavelengths of light make it clear that the outer layers are divided into several distinct zones (Fig. 5.4, Table 5.2). The "surface" of the Sun that we see in visible wavelengths is the **photosphere**, which has a temperature ranging between 4000 and 6500 K. When we view the Sun at the wavelength of the strong line of hydrogen at 6563Å, we see the **chromosphere,** a layer above the photosphere where the temperature is 6000 to 10,000 K. Outside that is the very hot, rarefied **corona** (best observed at X-ray wavelengths), whose temperature is 1 to 2 million K. Between the chromosphere and the corona is a thin region called the **transition zone,** where the temperature rapidly rises. Overall, the tenuous gas within and above the photosphere is referred to as the **solar atmosphere.** If we consider the temperature throughout this region, we find that it de-

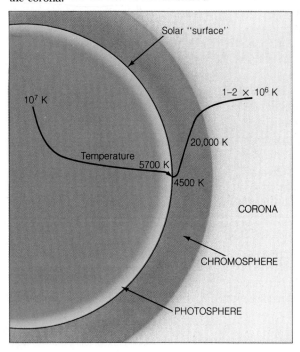

Figure 5.4 The Structure of the Sun's Outer Layers. This diagram shows the relative temperatures of the convective zone, the photosphere, the chromosphere, and the corona.

Astronomical Insight 5.3

Measuring the Sun's Pulse

In the early 1960s, measurements of spectral lines formed in the solar photosphere revealed small Doppler shifts that alternated between blueshifts and redshifts. This seemed to indicate that the Sun's surface was rising and falling, with a period of about five minutes. The reported Doppler shifts were quite small (less than 0.01 Å), near the limit of detectability, and the measurements were correspondingly difficult. Today the existence of this and other periodic oscillations of the Sun's surface is the basis for an entire research field sometimes called *solar seismology*.

Oscillations of the solar surface can be likened to those in a a musical instrument with a resonant cavity. When waves of any kind are created inside a cavity with reflecting walls, certain wavelengths or frequencies will persist, while all others die out. This happens because waves reflected from the walls interfere with incoming waves, the valleys and crests either adding together or canceling each other out. In an enclosed space, the wavelengths that add together constructively are determined by the dimensions of the enclosure; such wavelengths are characteristic of the particular enclosure and correspond to its so-called *resonant frequency*. In a musical wind instrument, this frequency determines the pitch of the tone that is emitted. The musician can control the pitch by altering the dimensions of the internal cavity (by opening and closing valves in most cases).

There are no walls inside the Sun, but physical conditions that vary with depth can create barriers as real as walls for certain types of waves. It has been deduced that sound waves (compressional waves) can persist in the Sun's convective zone and are confined above and below by the manner in which the Sun's temperature varies with depth. Sound waves in a certain range of wavelength are reflected back as they travel farther in or farther out, and this creates a resonant cavity in the outermost portion of the Sun for sound waves whose period of pulsation is about five minutes. These waves create a standing pattern of crests and valleys on the solar surface, and at any particular point on the surface the gas rises and falls with this period.

This phenomenon is quite interesting for its own sake, but it is worthy of careful study for another reason as well: The properties of the waves or oscillations of the solar surface can reveal information on conditions in the solar interior where the waves arise. For example, study of the frequency range of the waves can yield information on exactly how the temperature varies with depth inside the Sun, since it is this variation that determines the location of the lower boundary of the resonant cavity. Perhaps even more interesting, analysis of the surface oscillations can provide information on the internal rotation of the Sun. This is possible because waves traveling over the solar surface in opposite directions are spread out or squeezed together depending on whether they move in the direction of the Sun's rotation or in the opposite direction. Thus waves going in opposite directions have different frequencies, and the difference in frequency depends on the speed of rotation of the Sun in the layer where the waves arise. In effect it is possible to infer the Sun's internal rotational velocity at the lower boundary of the resonant cavity in which the waves oscillate.

The determination of internal rotational speeds using observations of surface oscillations is very complex and requires very sophisticated treatment of extensive quantities of data. At present it has not been possible to carry this effort very far, but it has been determined from the five-minute oscillations that the Sun is rotating more rapidly near the bottom of the convective zone

(continued on next page)

Astronomical Insight 5.3

Measuring the Sun's Pulse

(continued from previous page) than at the surface, which leads to speculation that the rotation near the core may be even more rapid. This could have important implications for the Sun's overall structure and evolution, and a great deal of importance is being attached to further study of this phenomenon.

While the five-minute oscillation of the Sun is now well established and understood, other reported oscillations are not. There is a three-minute oscillation that may be due to a similar resonant cavity at the level of the chromosphere, and there apparently are much longer oscillations that are driven by waves in very deep resonant cavities near the solar core. These controversial oscillations, which have periods of 40 and 160 minutes, may reveal conditions such as the rotation rate deep inside the Sun, and their study is therefore of great importance.

The detection of long-period oscillations is very difficult for a variety of reasons. It is important, for example, to observe the Sun continuously over many cycles of the oscillations, which means to keep observing without interruption for several days. This may sound impossible because of the Sun's daily rising and setting, but in fact it can be done from one of the Earth's poles during local summer. Solar studies have been carried out from the scientific research station at the South Pole. During one study involving six days of continuous observation of the sun, the 160-minute oscillation was convincingly detected.

Another approach to making continuous observations of the oscillations is to establish a worldwide network of observers with similar instruments, so that the Sun can be observed throughout the Earth's twenty-four-hour rotation period. One such network, called the *Global Oscillation Network Group (GONG)*, is being established by the U.S. National Solar Observatories. This network will consist of several identical observing stations located at sites around the world and is expected to provide data on solar oscillations for years to come.

Beyond simply detecting the existence of the long-period oscillations, scientists need to make further refinements in order to use them to probe the Sun's deep interior. To observe them in enough detail to do this requires entirely new technology, and at present several research groups are developing the necessary instruments. The problem is to find a way to observe very small Doppler shifts in spectral lines at many positions on the Sun's surface and to make these measurements repeatedly at very closely spaced intervals of time. The velocities involved are very small, only 1 meter per second or less, so the Doppler shift in a typical visible wavelength line is only a few millionths of an angstrom. To measure such small shifts accurately is a formidable task, but it will probably be possible within a few years. When this has been accomplished, we will mine a wealth of new information on the internal properties of our local star.

---

creases outward through the photosphere, reaching a minimum value of about 4000 K. From there the trend reverses itself, and the temperature begins to rise as we go farther out. The chromosphere, which lies immediately above the temperature minimum, is perhaps 2000 kilometers thick. Above the chromosphere the temperature rises very steeply within a few hundred kilometers to the coronal value of over a million degrees. Clearly something is creating extra heat at these levels; we will shortly consider where this heat arises.

First let us discuss the photosphere, the surface of the Sun as we look at it in visible light. We see this level because it is there that the Sun's density becomes great enough for the gas to be opaque, making it impossible to see any farther into the interior. The Sun's absorption lines (Fig. 5.5) are formed in the

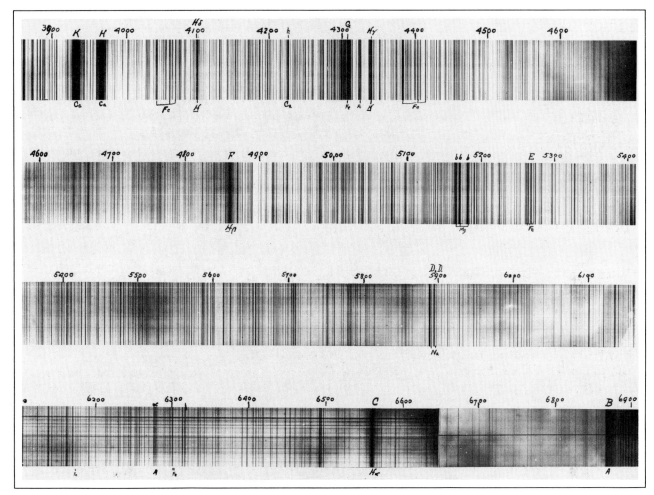

Figure 5.5 Fraunhofer Lines. The major absorption lines (dark lines here) in the Sun's photospheric spectrum are called Fraunhofer lines, in honor of the German scientist who first cataloged many of them in the early nineteenth century. Here we see a photograph of a portion of the solar spectrum. (Also see Fig. 4.8.) *(Mt. Wilson and Las Campanas Observatories, Carnegie Institution of Washington)*

photosphere as the atoms in this relatively cool layer absorb continuous radiation coming from the hot interior.

A photograph of the photosphere reveals a cellular appearance called **granulation** (Fig. 5.6). The traditional interpretation of granulation is that the bright regions are columns of hot, rising gas, and the dark borders are places where cool gas descends in a general convection pattern. Recent data indicates that individual granules travel across the Sun's surface (Fig. 5.6), suggesting instead that granulation represents some kind of surface oscillation or wave phenomenon.

The temperature of the photosphere, roughly 6000 K, is measured by using Wien's Law and the degree of ionization in the gas there; the density, roughly 10^{17} particles per cubic centimeter in the lower photosphere, is found from the degree of excitation, as described in Chapter 4. This is lower than the density of the Earth's atmosphere, which is about 10^{19} particles per cubic centimeter at sea level.

Most of what we know about the Sun's composition is based on the analysis of the solar absorption lines, so strictly speaking, the derived abundances represent only the photosphere. We have no reason to expect strong differences in composition at other levels, however, except for the core, where a significant amount of the original hydrogen has been converted into helium by the proton-proton reaction.

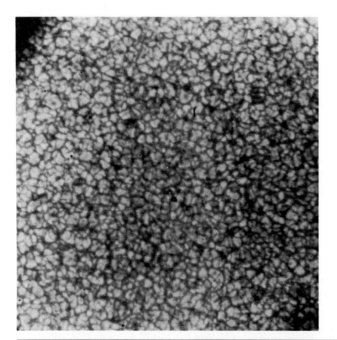

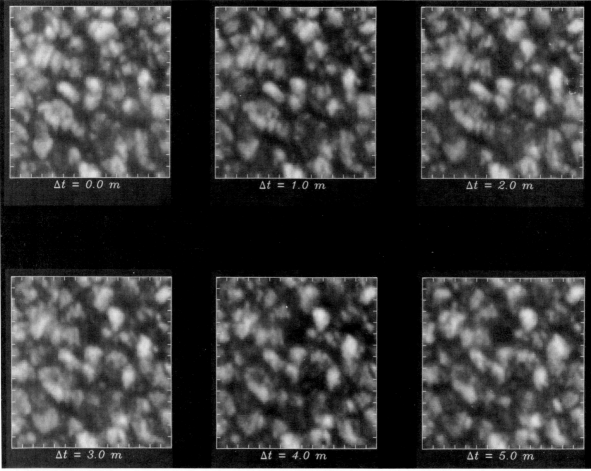

Figure 5.6 Solar Granulation. The upper photograph, obtained by a high-altitude balloon above much of the atmosphere's blurring effect, distinctly shows the granulation of the photosphere, which is due to convective motions in the Sun's outer layers. The six images in the lower figure, obtained from space, show how the granules change from minute to minute. *(top: Project Stratoscope, Princeton University, supported by the National Science Foundation;* bottom: *Courtesy of A. Title, Lockheed Aerospace Corporation)*

The photosphere near the edge of the Sun's disk (the limb) looks darker than in the central portions. This effect, called **limb darkening**, is caused by the fact that we are looking obliquely at the photosphere when we look near the edge of the disk. We do not see as deeply into the Sun there as we do when looking near the center of the disk. The higher-altitude gas we see at the limb is cooler than the gas we see at disk center and thus radiates less.

The chromosphere lies immediately above the temperature minimum. According to Kirchhoff's laws, the fact that this region forms emission lines tells us that the chromosphere is made of rarefied gas that is hotter than the photosphere behind it. When viewed through a special filter that allows light to pass through only at the wavelength of the hydrogen emission line (6563 Å), the chromosphere has a distinctive cellular appearance referred to as **supergranulation**, which is similar to the photospheric granulation but with cells some 30,000 kilometers across instead of about 1000 kilometers (Fig. 5.7). There is also fine-scale structure in the chromosphere in the form of spikes of glowing gas called **spicules** (Fig. 5.8). These come and go, probably due to the magnetic forces that apparently control their motions.

The outermost layer of the Sun's atmosphere is the corona, which extends a considerable distance above the photosphere and chromosphere. The corona is irregular in form: patchy near the sun's surface, but with radial streaks at great heights, suggesting outflow from the Sun (Fig. 5.9). The density of the coronal gas is very low, only about 10^9 particles per cubic centimeter.

As we have already seen, the corona contains highly ionized gas and is very hot. The source of the energy that heats the corona to such extreme temperatures is not well understood, although a general picture has emerged.

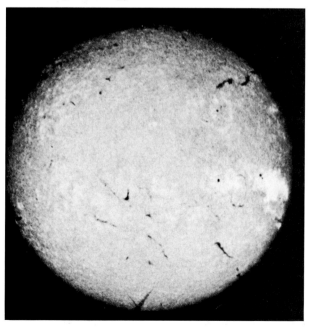

Figure 5.7 The Chromosphere. This photograph, taken through a special filter that allows light to pass through only at the wavelength of the bright hydrogen emission line at 6563 Å, reveals the locations of ionized hydrogen gas on the Sun, primarily in the chromosphere. The light areas (at the right, for example) are active regions associated with sunspots, where the chromosphere glows especially brightly at the observed wavelength. The dark streaks are filaments that do not glow strongly in this emission line. *(NASA)*

Figure 5.8 Spicules. This photograph shows the transient feature of the chromosphere known as spicules. These spikes of glowing gas, which are apparently shaped by the Sun's magnetic field, come and go irregularly. *(Sacramento Peak Observatory, National Optical Astronomy Observatories)*

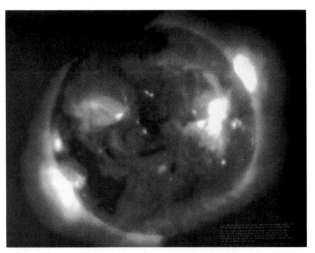

Figure 5.10 An X-ray Portrait of The Corona. This image, obtained by *Skylab* astronauts, shows the Sun in X-ray light, which reveals only very hot regions. The bright regions are places where the corona is especially dense, and the dark regions, known as coronal holes, are places where it is much more rarefied. The structure seen here changes with time. *(NASA)*

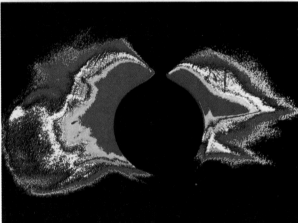

Figure 5.9 The Corona. The upper photograph, obtained during a total solar eclipse, shows the type of structure commonly seen in the corona. There are giant looplike features and an overall appearance of outward streaming. Bits of more intense light from the chromosphere are seen around the edges of the Moon's occulting disk. Visible at left is the planet Venus. The lower image, obtained from space using a disk to block out the Sun itself, illustrates brightness gradations with false colors. *(top: High Altitude Observatory, National Center for Atmospheric Research, sponsored by the National Science Foundation; bottom: NASA)*

X-ray observations reveal that the corona is not uniform but has a patchy structure (Fig. 5.10). In large regions that appear dark in an X-ray photograph of the Sun, the gas density is even lower than in the rest of the corona. These regions are called **coronal holes** and are probably created and maintained by the Sun's magnetic field. The coronal holes, as well as the overall shape of the corona, vary with time (Fig. 5.11), indicating that the corona in general is a dynamic region. Another impressive sign of this is presented by the **prominences** (Figs. 5.12 and 5.13), great geysers of hot gas that arc upward from the surface of the Sun. The prominences are usually associated with sunspots, and both phenomena are linked to the solar activity cycle, which we will discuss shortly.

The hot outer layers of the Sun have provided astronomers with a second major mystery regarding the solar energy budget. In contrast with the mystery of the Sun's internal energy source, the mechanism for heating the chromosphere and corona has not yet been entirely deduced. There is a great deal of energy available in the form of gas motions in the convection zone just beneath the solar surface, and it is generally accepted that the heat of the corona somehow comes from this energy. It is not clear how the energy is transported to such high levels, however, although there are several ideas, each invoking some form of waves. Waves of any kind carry energy, and it has been suggested that either sound waves or magnetic waves of some sort are the agents that transfer energy from the convection zone to the corona.

Recent satellite observations in ultraviolet and X-ray wavelengths have shown that stars similar to the Sun in general type also have chromospheric and coronal zones, so if we can understand how the Sun operates, we will also gain a deeper understanding of how other stars work. Similarly, observations of other stars, particularly the relationship of chromospheric and coronal activity to stellar properties such as age and rotation, can help us learn more about the Sun.

The Solar Wind

The long, streaming tail of a comet always points away from the Sun, regardless of the direction of the comet's motion. The significance of this was fully realized in the late 1950s, when the first U.S. satellites revealed the presence of the Earth's radiation belts and the fact that they are shaped in part by a steady flow of charged particles from the Sun. The solar wind reaches a speed near the Earth of 300 to 400 kilometers per second. It is this flow of charged particles that forces cometary tails always to point away from the Sun.

The solar wind is evidently a natural by-product of the same heating mechanisms that produce the hot corona of the Sun. It was originally thought that particles in this high-temperature region move about with such great velocities that a steady trickle escapes the Sun's gravity, flowing outward into space. Subsequently, however, X-ray observations of the Sun have shown that the situation is not that simple. It is the solar magnetic field that governs the outward flow of charged particles. The coronal holes mentioned earlier are regions where the magnetic field lines open out into

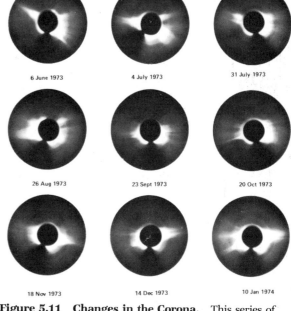

Figure 5.11 Changes in the Corona. This series of photographs was obtained using a special shutter device that blocks out the solar disk and allows the corona to be observed without a total solar eclipse. Here we see that the structure of the corona changes quite markedly over a period of a few months. *(High Altitude Observatory, National Center for Atmospheric Research, sponsored by the National Science Foundation)*

space. Charged particles such as electrons and protons, constrained by electromagnetic forces to follow the magnetic field lines, therefore escape into space only from the coronal holes. The speed of the solar wind is relatively low close to the Sun, but it accelerates as it

Figure 5.12 A Prominence. Here the looplike structures thought to be governed by the Sun's magnetic field are readily seen. This photograph was obtained by *Skylab* astronauts using an ultraviolet filter. *(NASA)*

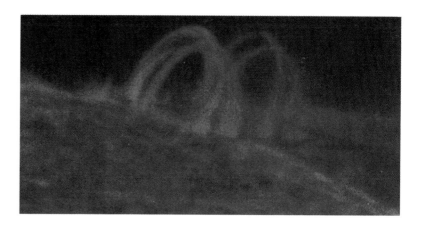

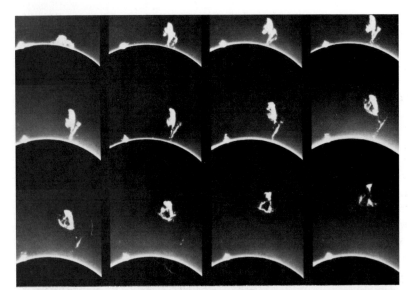
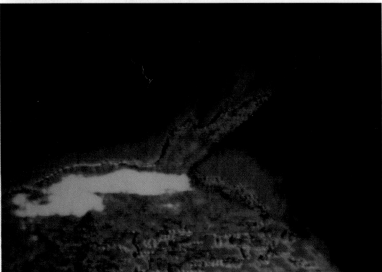

Figure 5.13 An Eruptive Prominence. The striking sequence at top shows an outburst of ionized gas from a prominence seen on the Sun's limb. The photographs were obtained by using a special device to block out the Sun's disk. At bottom is an ultraviolet view of another eruptive prominence, showing the extent reached in less than two hours. (top: *High Altitude Observatory, National Center for Atmospheric Research, sponsored by the National Science Foundation;* bottom: *NASA*)

moves outward, quickly reaching a velocity of 300 to 400 kilometers per second and then remaining nearly constant. The wind has nearly reached its maximum velocity by the time it passes the Earth's orbit, and beyond there it flows steadily outward, apparently persisting at least some distance beyond the orbit of Neptune. (The *Pioneer 10* spacecraft, on its way out of the solar system, still detected solar wind particles as it crossed the orbit of Neptune in mid-1983 and has not yet found a limit to the solar wind.) At some point in the outer solar system, the wind is thought to come to an abrupt halt where it runs into an invisible and tenuous wall of matter swept up from the interstellar medium that surrounds the Sun.

Because the Earth's magnetosphere shields us from the wind particles, most of the direct information we have on the solar wind comes from satellite and space probe observations (Fig. 5.14). Solar wind monitors are placed on board the majority of spacecraft sent to the planets. One striking discovery has been the fact that the wind is not uniform in density but instead seems to flow outward from the Sun in sectors, as if originating only from certain areas on the Sun's surface. This is explained by the X-ray data already mentioned, which indicate that the wind emanates only from the coronal holes. Because the base of the wind is rotating with the Sun, the wind sweeps out through space in a great curve like the trajectory of water from a rotating lawn sprinkler.

Occasional explosive activity on the Sun's surface releases unusual quantities of charged particles, and three or four days later, when this burst of ions reaches the Earth's orbit, we experience disturbances in the ionosphere that can interrupt shortwave radio communications and cause unusually widespread and brilliant auroral displays. These **magnetic storms**, as they are often called, are outward manifestations of a much more complex overall interaction between the Sun and the Earth. At the end of the next section we will discuss the so-called solar-terrestrial relation.

Sunspots, Solar Activity Cycles, and the Magnetic Field

The dark spots on the Sun's disk were observed more than 300 years ago and were cited by Galileo as evidence that the Sun is not a perfect, unchanging celestial object but has occasional flaws. Over the centuries since then, observations of the spots, which individually may last for months, have revealed some very systematic behavior. The number of spots varies, reaching a peak every 11 years, and during the interval between peak numbers of spots their locations on the Sun move steadily from the middle latitudes toward the equator. At the beginning of a cycle, when the sunspot number is maximized, most of them appear in activity bands about 30 degrees north or south of the solar equator. During the next 11 years, the spots tend to move ever closer to the equator, and by the end of the cycle they are nearly on it (Fig. 5.15). By this time the first spots of the next cycle may already be forming at the middle latitudes. A plot of sunspot locations during a cycle clearly shows this effect and is called a "butterfly diagram" because of the shape the pattern of spots makes (Fig. 5.16).

A hint about the origin of the spots was found when their magnetic properties were first measured. This is accomplished by using spectroscopy of the light from the spots and taking advantage of the fact that the energy levels in certain atoms are distorted by the presence of a magnetic field. This distortion causes the

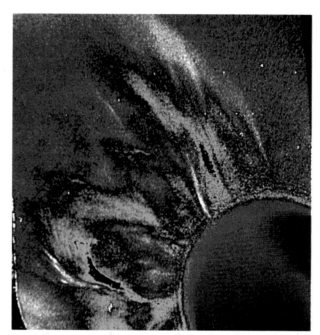

Figure 5.14 The Solar Wind. This far-ultraviolet image was obtained by the *Solar Maximum Mission* in earth orbit. A special color separation technique was used to show the outward flow of gas. *(Data from the Solar Maximum Mission, NASA, and the High Altitude Observatory, National Center for Atmospheric Research, sponsored by the National Science Foundation)*

spectral lines formed by those levels to be split into two or more distinct, closely spaced lines. The degree of line splitting, which is referred to as the **Zeeman effect**, depends on the strength of the magnetic field, so the field can be measured from afar simply by analyzing the spectral lines to see how widely they are split. This technique also works for distant stars.

When the first measurements of the Sun's field were made in the first decade of this century, it was found that the field is especially intense in the sunspots, about a thousand times stronger than in the surrounding gas. It is now thought that the intense magnetic field associated with a sunspot inhibits convection, so that heated gas from below cannot rise up to the surface. Because the spot is cooler than its surroundings, it is not as bright (Fig. 5.17). The typical temperature in a spot is about 4000 K, compared to the roughly 6000 K of the photosphere. Using Stefan's law, we see that the intensity of light emitted in a spot compared to the surroundings is $(4000/6000)^4 = 0.2$; that is, the brightness

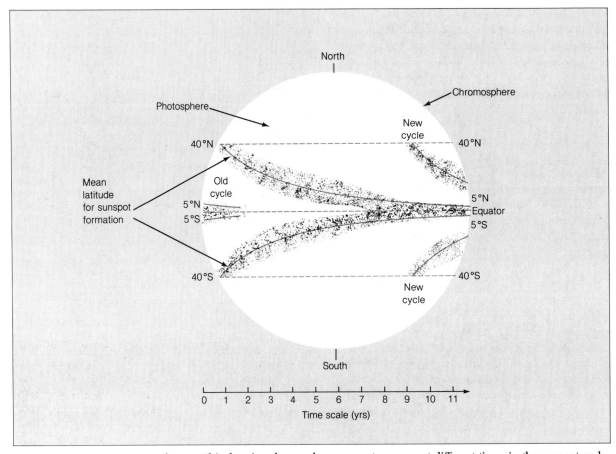

Figure 5.15 Sunspot Locations. This drawing shows where sunspots appear at different times in the sunspot cycle.

Figure 5.16 The Butterfly Diagram. This is a plot of the latitudes of observed sunspots through several cycles of solar activity. During each cycle, the spots gradually shift their favored locations closer and closer to the solar equator. *(Prepared by the Royal Greenwich Observatory and reproduced with the permission of the Science and Engineering Research Council, courtesy of J. A. Eddy)*

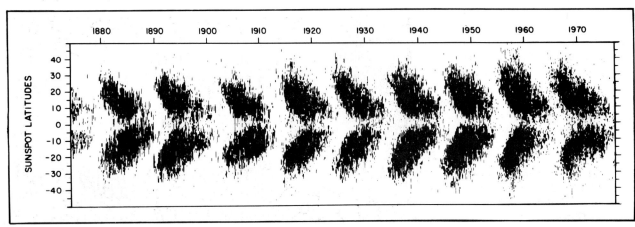

of the solar surface within a sunspot is only about one fifth the brightness in the photosphere outside.

When the sunspot magnetic fields were measured (Fig. 5.18), they were found to act like either north or south magnetic poles; that is, each spot has a specific magnetic direction associated with it. Furthermore, pairs of spots often appear together, the two members of a pair usually having opposite magnetic polarities. The pairs in the Sun's northern and southern hemispheres have their east-west sequences reversed; that is, if a north polar spot leads in one hemisphere, a south polar spot leads each pair in the other hemisphere. During a given 11-year cycle, in every sunspot pair the magnetic polarities always have the same orientation. For example, during one 11-year cycle, the spot to the east in each pair will have a north magnetic polarity, and the one to the west a south magnetic polarity. During the next cycle, all the pairs will be reversed, with the south magnetic spot to the east and the north magnetic spot to the west. Between cycles, when this arrangement is reversing itself, the Sun's overall magnetic field also reverses, with the solar magnetic poles exchanging places. It is actually 22 years before the Sun's magnetic field and sunspot patterns repeat themselves, so the solar magnetic cycle is truly twenty-two years long.

Sunspot groups, often called **solar active regions** (Fig. 5.19), are the scenes of the most violent forms of solar activity, the **solar flares**. These gigantic outbursts of charged particles and visible, ultraviolet, and X-ray emissions are created when extremely hot gas spouts upward from the surface of the Sun (Fig. 5.20). Flares

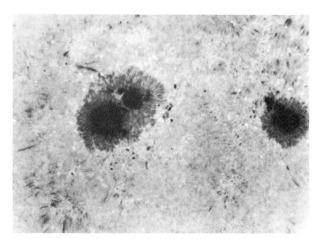

Figure 5.17 Sunspots. This is a telescopic view of a group of spots, showing their detailed structure. They appear dark only in comparison with their much hotter surroundings. *(Mt. Wilson and Las Campanas Observatories, Carnegie Institution of Washington)*

are most common when the greatest density of spots is to be seen on the solar surface. Close examination of flare events shows that the trajectory of the ejected gas is shaped by the magnetic lines of force emanating from the spot where the flare occurred. Charged particles flow outward from a flare, some of them escaping into the solar wind. If the flare occurs where the part of the wind arises that hits the Earth, the Earth is bathed by an extra dose of solar wind particles some three days later, creating the effects on radio communications and aurorae mentioned in the previous sec-

Figure 5.18 Magnetic Fields on the Sun. The *magnetogram*, or map of magnetic field strength (right), was made through the measurement of the splitting of spectral lines by magnetic fields (the Zeeman effect). Here colors (dark blue and yellow) are used to indicate regions of opposing magnetic polarity. Note that the polarities in the southern hemisphere are reversed with respect to those in the north. At left is a visible-light photograph of the Sun taken at the same time, so that the correspondence between the surface magnetic field and the visible sunspots can easily be seen. *(National Optical Astronomy Observatories)*

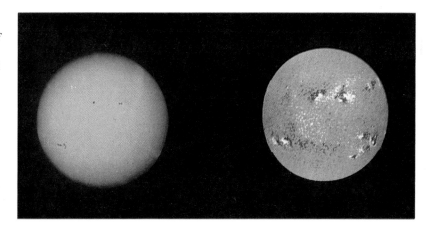

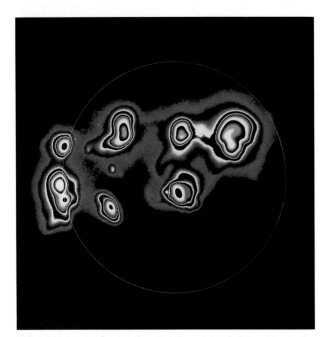 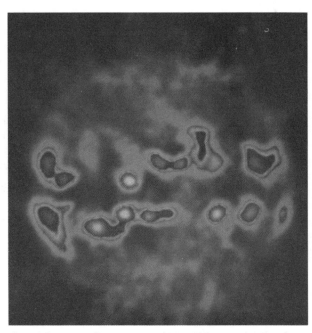

Figure 5.19 Solar Active Regions. At left is an X-ray image and at right is a radio map. In both, bright colors indicate the solar active regions, which emit strongly in X-ray and radio wavelengths.(left: *NASA*; right: *National Radio Astronomy Observatory, operated by Associated Universities, Inc., under contract with the National Science Foundation*)

Figure 5.20 A Major Flare. The gigantic looplike structure in this ultraviolet photograph obtained from *Skylab* is one of the most energetic flares ever observed. Supergranulation in the chromosphere is easily seen over most of the disk. *(NASA)*

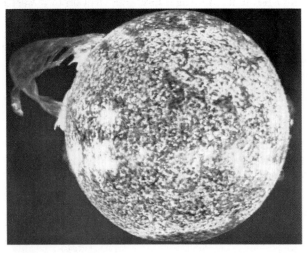

tion. Apparently flares occur when twisted magnetic field lines suddenly reorganize themselves, releasing heat energy and allowing huge bursts of charged particles to escape into space.

The combination of all these bits of data on sunspots, magnetic fields, and solar activity cycles has led to the development of a complex theoretical picture that successfully accounts for many of the observed phenomena. This theory envisions ropes or tubes of magnetic field lines inside the Sun, connecting its north and south magnetic poles. At some places these tubes become kinked and loops break through the surface, creating pairs of sunspots with opposing magnetic polarities where they emerge and reenter. Early in the sunspot cycle the magnetic tubes break through the surface at the middle latitudes, but later, as the Sun's magnetic field is moving toward reversal of the poles, they break through near the equator. This accounts for the latitude-dependence of the spots during a cycle, as shown by the butterfly diagram. When the solar magnetic field reverses itself every 11 years, so do the magnetic ropes; when the new cycle begins, the sunspot

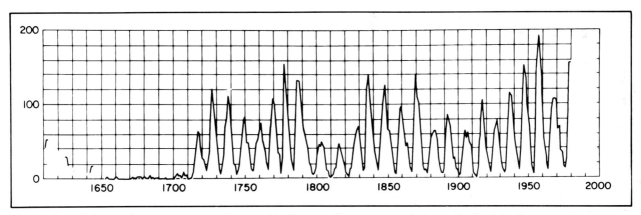

Figure 5.21 The Sun's Long-Term Activity. This diagram illustrates the relative level of activity (in terms of sunspot numbers) over three centuries. It is clear that the level varies; note the span of about fifty years (1650–1700), known as the Maunder minimum, when there was little activity. There is evidence that the Sun has long-term cycles that modulate the well-known 22-year period. (*J. A. Eddy*)

pairs have their polarities reversed compared with the pairs of the previous cycle.

Records of sunspot counts maintained over centuries have allowed astronomers to study the long-term behavior of the Sun. It has been found that the 22-year cycle is not perfectly repeatable and that there have been longer-term variations in solar activity (Fig. 5.21). Most striking was a prolonged period in the late 1600s when the cycle seemed to stop, with no marked periods of sunspot maxima. This epoch of reduced sunspot activity has been called the **Maunder minimum**, after its discoverer, E. W. Maunder (also the originator of the butterfly diagram). Only time will tell whether the Maunder minimum was part of some much longer cycle in the Sun's behavior.

The origins of the Sun's magnetic field and its periodic pole reversals probably lie in a dynamo similar to those thought to be at work in the interiors of the planets that have fields. Unlike the terrestrial planets, however, the Sun is not rigid, and consequently differential rotation may play a role in creating the instability that causes the dynamo to reverse itself every 11 years. From this solar behavior we might speculate that the magnetic fields of Jupiter and Saturn, which also rotate differentially, may also reverse themselves from time to time. (The fact that the Earth's magnetic field also reverses from time to time shows that differential rotation is not absolutely required for reversals, although it may allow them to occur more frequently.)

X-ray and ultraviolet observations of other stars show that many, particularly relatively cool stars like the Sun, not only have chromospheres and coronae but also have activity cycles. We deduce from this that these stars must also have magnetic fields that reverse themselves periodically. Statistical studies of large numbers of stars help reveal how their activity cycles depend on such phenomena as rotation and stellar age. For example, we will learn (in Chapter 15) that the Sun's rotation has slowed since the time of its formation. From studies of other stars we find that younger stars generally tend to rotate more rapidly than older ones. The degree of chromospheric and coronal activity also diminishes with age, suggesting that rotation of a star is responsible for its activity cycle. Thus the Sun probably had more intense activity when it was younger than it does now. This is just one example of the kinds of things we can learn by combining information on the Sun and on other stars like it.

The solar activity cycle may have some indirect effects on the climate of the Earth. Eleven-year patterns have been reported in the occurrence of droughts, and it seems possible that such patterns are related to the solar cycle, although it is not known how. During the time of the Maunder minimum, the Earth's climate was in chaos, with terrible droughts in many areas and particularly severe winters in Europe and in North America.

There are other aspects of the relationship between

solar activity and the Earth's atmosphere. As the solar wind fluctuates in intensity, the Earth's magnetosphere varies in extent. We have already seen that solar flares, which occur most often during a sunspot maximum, create important effects on the ionosphere. The chemistry of the upper atmosphere may also be strongly influenced by variations in the solar ultraviolet emission, which are linked in turn to the solar activity cycle. The entire question of solar-terrestrial relations is an important one that merits and is receiving further attention.

Perspective

Our Sun, the source of nearly all our energy, is a very complex body. As stars go, it is apparently normal in all respects, so we imagine that other stars are just as complex, even though we cannot observe them in such detail.

We have explored the Sun, both in its deep interior and in the outer layers that can be observed directly. We know that nuclear fusion is the source of all the energy, and that the size and shape of the Sun are controlled by the balance between gravity and pressure throughout the interior. Somehow heat is transported above the surface, keeping the chromosphere and the corona hotter than the photosphere. Perhaps most intriguing of all is the solar activity cycle and its relationship to the Sun's complex magnetic field.

We are now prepared to study stars in general, using the Sun as an example to help us understand some of what we learn about other stars.

Summary

1. The Sun is an ordinary star, one of billions in the galaxy.
2. The Sun is a spherical, gaseous object whose temperature and density increase toward the center. The ultimate source of the light the Sun emits from its surface is nuclear fusion in its core.
3. The energy produced in the core by nuclear fusion is slowly transported outward by radiation, except near the surface, where energy is transported by convection.
4. The nuclear reaction that powers the Sun is the proton-proton chain, in which hydrogen is fused into helium.
5. The outer layers of the Sun govern the nature of the light it emits. These layers consist of the photosphere, with a temperature of about 6000 K; the chromosphere, with a temperature ranging between 6000 and 10,000 K; and the corona, with a temperature over 1,000,000 K.
6. The photosphere, which is the visible surface of the Sun, creates absorption lines, while the hotter chromosphere and corona create emission lines.
7. The excess heat in the outer layers is somehow transported there from the convective zone just below the surface, but the mechanism for transporting the heat is not well known.
8. The Sun emits a steady outward flow of ionized gas called the solar wind. The wind originates from coronal holes and is controlled by the solar magnetic field.
9. The sunspots occur in 11-year cycles, which are part of the 22-year cycle of the Sun's magnetic field reversals. The spots are regions of intense magnetic fields where flux tubes from the solar interior break through the surface.
10. Studies of other stars show that those similar to the Sun in temperature also have chromospheres, coronae, and activity cycles.
11. The solar activity cycle may have important effects on the earth's climate.

Review Questions

1. Why is lead a solid material when the Sun, which is many times denser at its core than lead, is gaseous?
2. Why do nuclear reactions occur only in the innermost core of the Sun? Explain in terms of hydrostatic equilibrium.
3. Explain why the use of special filters to isolate the wavelengths of certain spectral lines allows us to examine distinct layers of the Sun separately.
4. Why are the granules in the Sun's photosphere bright and the descending gas that surrounds them relatively dark?
5. Use Kirchhoff's laws to explain why the photo-

sphere produces absorption lines, but the chromosphere and corona produce emission lines.
6. What would we conclude about the temperature just below the outermost layers of the photosphere if the Sun's disk appeared brighter, instead of darker, at the edges?
7. Why does it take about three days for the effects of a solar flare to begin to occur in the Earth's magnetosphere?
8. Explain why we say that the Sun's activity cycle is 22 years long, even though the number of spots peaks every 11 years.
9. Describe how energy produced in the Sun's core by nuclear reactions heats the chromosphere and corona. In other words, describe how the energy gets from the core to the outer layers, step by step.
10. We noted in this chapter that other stars like the Sun apparently slow their spin rates with age, as the Sun has. What does this tell us about the possibility that other stars might have planetary systems?

Additional Readings

Eddy, J. A. 1977. The case of the missing neutrinos. *Scientific American* 236(5):80.

Gibbon, E. G. 1973. *The quiet sun.* NASA Publication NAS 1.21. Washington, D.C.: NASA.

Gough, D. 1976. The shivering sun opens its heart. *New Scientist* 70:590.

Harvey, J. W., J. R. Kennedy, J. W. Leibacher 1987, GONG: To see inside our Sun. *Sky and Telescope* 74(5):470.

Newkirk, G., and K. Frazier. 1982. The solar cycle. *Physics Today* 35(4):25.

Parker, E. N. 1975. The sun. *Scientific American* 233(3):42.

Waldrop, Mitchell. 1985. First sightings: Solar systems elsewhere. *Science 85* 6(5):26.

Walker, A. B. C., Jr. 1982. A golden age for solar physics. *Physics Today* 35(11):61.

Wallenhorst, S. G. 1982. Sunspot numbers and solar cycles. *Sky and Telescope* 64(3):234.

Wilson, O. C., A. H. Vaughan, and D. Mihalas 1981. The activity cycle of stars. *Scientific American* 244(2):104.

Wolfson, R. 1983. The active solar corona. *Scientific American* 248(2):104.

Chapter 6

Observations and Basic Properties of Stars

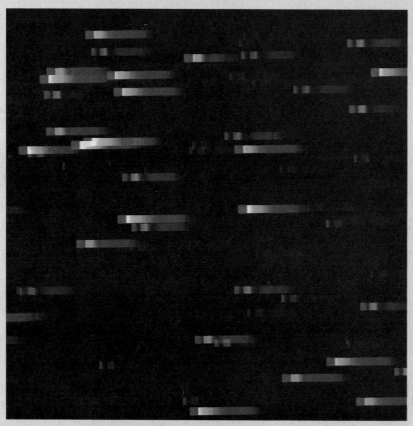

The Hyades star cluster,
observed through a prism,
revealing the spectra of the stars.
*(University of Michigan
Observatories)*

Chapter Preview

Three Ways of Looking at Stars: Positions,
    Magnitudes, and Spectra
  Positional Astronomy
  Stellar Brightness
  Measurements of Stellar Spectra
Binary Stars
Fundamental Stellar Properties
  Absolute Magnitudes and Stellar Luminosities
  Stellar Temperatures
  The Hertzsprung-Russell Diagram and a New
    Distance Technique
  Stellar Diameters
  Binary Stars and Stellar Masses
  Other Properties

Everything that we can learn about a star is contained in the light that we receive from it. Fortunately a lot of information is there, and astronomers have learned how to extract much of it.

In this chapter we will first discuss the three basic types of astronomical observations and then learn how fundamental parameters of stars are measured using these techniques.

Three Ways of Looking at Stars: Positions, Magnitudes, and Spectra

Nearly all observations of stars fall into one of three categories: measurements of position, brightness, or spectra. The first two of these methods have rather long histories. Positional measurements date back to the first human observations of the skies, and brightness measurements (although rather crude ones) were made by Hipparchus in the second century B.C. In the following sections, each method is discussed separately.

Positional Astronomy

The science of measuring star positions is called **astrometry**. Ancient astronomers used simple devices such as quadrants and sextants (Fig. 6.1) for the measurement of angular positions on the sky, but the principal technique used today is to photograph the sky and carefully measure the positions of the star images on the photographic plates. If a number of photographs of a given portion of the sky are taken and measured separately, the results can be averaged to produce a more accurate determination than is possible from a single photograph. Today it is possible to measure a stellar position to a precision of less than 0.01 arc-second. When the *Space Telescope* is in operation in the late 1980s, even more accurate measurements will be possible, because the fuzziness of star images caused by the Earth's atmosphere will be eliminated.

Astrometric measurements are made by modern astronomers for a variety of reasons. For example, such measurements are important for the analysis of stellar motions and what these motions have to tell us about the structure of the galaxy (see Chapter 9). Astrometric data are also very important for cataloging stars with sufficient accuracy that they can be observed by other telescopes. (A major task now underway is the preparation of star lists and charts for *Space Telescope* obser-

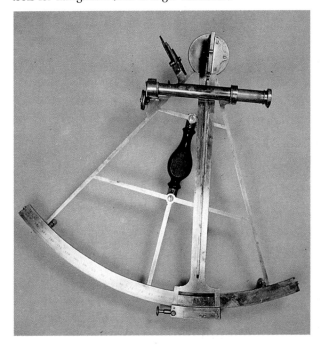

Figure 6.1 A Sextant. Instruments of this type were used for centuries for the measurement of star positions relative to each other. In modern times, sextants are used as tools for navigation. *(The Granger Collection)*

vations, which will involve much fainter stars than are listed in current catalogs.) Finally, and perhaps most importantly, positional measurements can be used to measure distances to stars, making use of their parallax motions.

Recall from Chapter 2 that stellar parallax is the apparent shifting in a nearby star's position due to the orbital motion of the Earth (Fig. 6.2). Ancient and medieval astronomers were unable to detect stellar parallax, leading most of them to reject the idea that the Earth orbits the Sun. Eventually the heliocentric theory won out, but stellar parallax remained undetected until 1838. The reason stellar parallaxes had defied earlier observers now became obvious: Even for the closest stars, the maximum shift in position was just over 1 arcsecond (therefore the parallax angle, one-half the maximum shift, is much less than one arcsecond). The stars are simply much farther away than the ancient astronomers had dreamed possible. The annual parallax motion of the star alpha Centauri, the star nearest to the Sun, is less than the angular diameter of a dime as seen from a distance of about two miles, a very small angle indeed.

The successful detection of stellar parallax led to a direct means of determining distances to stars. The amount of shift in a star's apparent position due to our own motion about the Sun depends on how distant the star is; the closer it is, the bigger the shift. The parallax angle is defined as one half the total angular shift of a star during the course of a year. Once it is measured, this angle tells us the distance to a star in a unit of measure called the **parsec.** A star whose parallax is 1 arcsecond is, by definition, 1 parsec away (the word *parsec*, in fact, is a contraction of *parallax-second*, meaning a star whose parallax is 1 arcsecond). In mathematical terms, the distance to a star whose parallax is π is $d = 1/\pi$. Thus, if a star has a parallax of 0.4 arcsecond, it is $1/0.4 = 2.5$ parsecs away. Recall that the closest star has a parallax of less than 1 arcsecond; we conclude therefore that even this star is more than 1 parsec away.

In more familiar terms, a parsec is equal to 3.26 light-years, or 3.08×10^{18} centimeters, or 206,265 astronomical units (AU). The use of parallax measurements to determine distances is a very powerful technique; it is the only direct means astronomers have for measuring how far away stars are. Unfortunately, parallaxes are large enough to be measured only for stars rather near us in the galaxy. The smallest parallax that can be detected is a bit less than 0.01 arc second, and accurate measurements can be made only for parallax angles of 0.01 arcsecond or larger, corresponding to distances within one hundred parsecs. The galaxy, on the other hand, is more than 30,000 parsecs in diameter! Clearly other distance-determination methods are needed if we are to probe the entire galaxy. One very powerful method will be described later in this chapter.

Stellar Brightness

The second general type of stellar observation is the measurement of the brightness of stars. This was first attempted in a systematic way over two thousand years ago, when Hipparchus established a system of brightness rankings that is still used today. Hipparchus ranked the stars in categories called **magnitudes,** from first magnitude (the brightest stars) to sixth (the faintest visible to the unaided eye). In his catalog of stars, and in most since then, these magnitudes are listed along with the star positions.

The magnitude system has been modernized, and all astronomers use the same technique for measurement rather than relying on subjective impressions. It was discovered in the mid-1800s that what the eye perceives as a fixed *difference* in intensity from one magnitude to the next actually corresponds to a fixed inten-

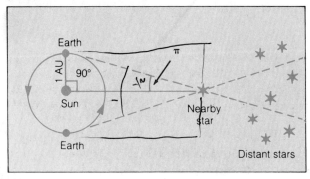

Figure 6.2 Stellar Parallax. As the Earth orbits the Sun, our line of sight toward a nearby star varies in direction enough to make that star appear to move back and forth with respect to more distant stars. The parallax angle π is defined as one half of the total amount of apparent annual motion. If π is 1 arcsecond, the Sun-star distance is 206,265 times the Sun-Earth distance; this defines the parsec—the distance to a star whose parallax angle is 1 arcsecond.

Astronomical Insight 6.1

Star Names and Catalogs

From the earliest times when astronomers systematically measured star positions, they began to compile lists of these positions in catalogs. Hipparchus developed an extensive catalog more than two thousand years ago (see Chapter 2), as did early astronomers in China and other parts of the world.

To list stars in a catalog of positional measurements requires some kind of system for naming or numbering the stars, and a variety of such systems have been employed. The ancient Greeks designated stars by the constellation and the brightness rank within the constellation; for example, the brightest star in Orion is α Orionis, the second brightest is β Orionis, the next brightest is γ Orionis, the next is δ Orionis, and so on (the constellation names are spelled in their Latin versions, since this system was perpetuated by the Romans and the Catholic Church after the time of the Greeks). Today many stars are still referred to by their constellation rankings, using the Greek alphabet to designate the rank.

After the time of Ptolemy, when Western astronomy went into decline for some 1300 years, Arab astronomers occupying northern Africa and southern Europe carried on astronomical traditions. These people assigned proper names to many of the brightest stars, and these names, such as Betelgeuse (α Orionis), Rigel (β Orionis), and Bellatrix (γ Orionis), are also still in use.

With the advent of the telescope, when many new stars were discovered that were too faint to be seen with the naked eye, catalogs rapidly outgrew the old naming systems. Generally each catalog now assigned numbers to the stars, usually in a sequence related to their positions. Thus modern catalogs, such as the *Henry Draper Catalog*, the *Boss General Catalog*, the *Yale Bright Star Catalog*, and the *Smithsonian Astrophysical Observatory Catalog*, list stars by increasing right ascension (from west to east in the sky, in the order in which the stars pass overhead at night). In some cases, separate listings are made for different zones of declination (that is, for different strips of sky, separated in the north-south direction). Because each of these catalogs includes a different particular set of stars, each has its own numbering system, and often no cross-reference to other catalogs is provided. Hence a given star may have a constellation-ranked name, an Arabic name, and numbers in a variety of catalogs, but only by matching up the coordinates is it possible to be sure you are referring to the same star in different catalogs. Actually, things are not quite so bad as that, since some modern catalogs do provide cross-references to others. Nevertheless, one of the principal tasks of a would-be astronomer is to become familiar with catalogs and their use.

---

sity *ratio*. Measurements showed that a first magnitude star is about 2.5 times brighter than a second magnitude star; a second magnitude star is 2.5 times brighter than a third magnitude star; and so on (Fig. 6.3). The ratio between a first magnitude star and a sixth magnitude star was found to be nearly 100. In 1850 the system was formalized by the adoption of this ratio as exactly 100; thus, the ratio corresponding to a one-magnitude difference is the fifth root of 100, or $(100)^{1/5} = 2.512$. Hence a first magnitude star is 2.512 times brighter than a star of second magnitude, a sixth magnitude star is $2.512 \times 2.512 = 6.3$ times fainter than a fourth magnitude star, and so on.

Magnitudes are most commonly measured through the use of **photometers**; these devices produce an electric current when light strikes them and are used in

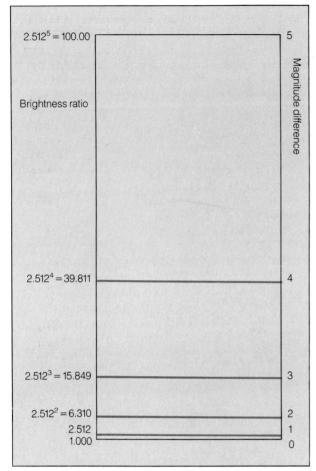

Figure 6.3 Stellar Magnitudes. This diagram shows schematically how brightness ratios and magnitude differences are related. To determine the relationship between ratios and fractional magnitudes requires calculations using logarithms (see Appendix 10).

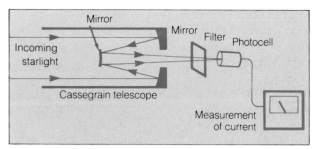

Figure 6.4 Measuring Stellar Magnitudes. This is a schematic illustration of photometry, the measurement of stellar brightnesses. Light from a star is focused (often through a filter that screens out all but a specific range of wavelengths) onto a photocell, a device that produces an electric current in proportion to the intensity of light. The current is measured and, by comparison with standard stars, converted into a magnitude.

many familiar applications, such as door-openers in modern buildings. The amount of electrical current produced is determined by the intensity of light, so an astronomer need only measure the current to determine the brightness of a star (Fig. 6.4) and hence its magnitude.

Once magnitudes could be measured precisely, it was found that stars have a continuous range of brightnesses and do not fall neatly into the various magnitude rankings. Therefore fractional magnitudes must be used; Deneb, for example, has a magnitude of 1.26 in the modern system, although it was formerly classified simply as a first magnitude star. Each of the former categories is found to include a range of stellar brightnesses. This is especially so for the first magnitude stars, some of which turned out to be as much as two magnitudes brighter than others. To measure these especially bright stars in the modern system requires the adoption of magnitudes smaller than 1. Sirius, for example, which is the brightest star in the sky, has a magnitude of -1.42. By using negative magnitudes for very bright objects, astronomers can extend the system to include such objects as the Moon, whose magnitude when full is about -12, and the Sun, whose magnitude is -26 (thus the Sun is about 25 magnitudes—or a factor of $2.512^{25} = 10^{10}$—brighter than Sirius). Table 6.1 lists the magnitudes of a variety of objects.

Of course there are stars fainter than the human eye can see, so the magnitude scale must also extend beyond sixth magnitude. With moderately large telescopes, stars as faint as fifteenth magnitude can be measured, and with long-exposure photographs taken with the largest telescopes, stars as faint as twenty-fifth magnitude are revealed. Such a star is nineteen magnitudes—or a factor of $2.512^{19} = 4 \times 10^{7}$—fainter than the faintest star visible to the unaided eye. The *Space Telescope* is expected to detect stars as faint as twenty-eighth magnitude, which is a factor of $2.512^{3} = 15.9$ fainter yet.

The stellar magnitudes we have discussed so far all refer to visible light. As we learned in Chapter 4, however, stars emit light over a much broader wavelength band than the eye can see. We also learned that the wavelength at which a star emits most strongly depends on its temperature. By measuring a star's brightness at two different wavelengths, it is therefore possible to learn something about its temperature. For this purpose astronomers use filters that allow only certain wavelengths of light to pass through. A star's brightness is typically measured through two such filters, one that passes yellow light and one that passes blue light, resulting in the measurement of V (for visual, or yellow) and B (for blue) magnitudes (Fig. 6.5). Since a hot star emits more light in blue wavelengths than in yellow, its B magnitude is *smaller* than its V magnitude. For a cool star, the situation is reversed, and the B magnitude is larger than the V magnitude (Remember that the magnitude scale is backward in the sense that a lower magnitude corresponds to greater brightness.)

The difference between the B and V magnitudes is called the **color index**. The exact value of this index is a function of the temperature of a star, so that stellar temperatures may be estimated simply by measuring the V and B magnitudes. A very hot star might have a color index $B - V = -0.3$, while a very cool one might typically have $B - V = +1.2$.

One additional type of magnitude should be mentioned, although it is a very difficult one to measure directly: The **bolometric magnitude** includes all the light emitted by a star at all wavelengths. Determining a bolometric magnitude requires ultraviolet and infrared, as well as visible, observations and can best be

Table 6.1 **Apparent Magnitudes of Familiar Objects**

Object	Apparent Magnitude
Sun	−26.74
Moon (full)	−12.73
100-watt bulb at 100 ft	−13.70
Venus (greatest elongation)	−4.22
Jupiter (opposition)	−2.6
Mars (opposition)	−2.02
Sirius (brightest star)	−1.45
Mercury (greatest elongation)	−0.2
Alpha Centauri (nearest star)	−0.1
Large Magellanic Cloud	+0.1
Saturn (opposition)	+0.7
Small Magellanic Cloud	+2.4
Andromeda galaxy (farthest visible object)	+3.5
Brightest globular cluster (47 Tucanae)	+4.0
Orion nebula	+4
Uranus (opposition)	+5.5
Faintest object visible to naked eye	+6.0
Neptune (opposition)	+7.9
Crab nebula	+8.6
3C273 (brightest quasar)	+12.80
Pluto (opposition)	+14.9

Figure 6.5 The Color Index. Here are continuous spectra of two stars, one much hotter than the other. The wavelength ranges over which the blue *(B)* and visual *(V)* magnitudes are measured are indicated. We see that the hot star is brighter in the B region than in the V region of the spectrum; therefore, its B magnitude is *smaller* than its V magnitude, and it has a *negative* $B - V$ color index. The opposite is true for the cool star.

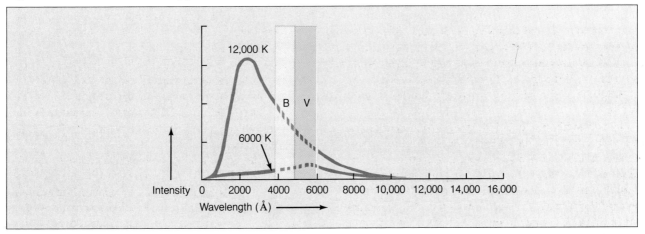

done using telescopes in space. This is particularly true for hot stars, which emit a large fraction of their light in ultraviolet wavelengths. For cool stars, which emit little ultraviolet but a lot of infrared radiation, bolometric magnitudes can be measured from the earth's surface with fair accuracy. The bolometric magnitude of a star is always smaller than the visual magnitude, because more light is included when all wavelengths are considered. Often bolometric magnitudes are estimated by comparison with "standard" stars, stars whose properties are well known and assumed to represent all stars of similar type. The normal means of comparison is to measure the visual magnitude of the target star and then add a **bolometric correction,** a negative correction factor based on standard stars. We will see later in this chapter how bolometric magnitudes are related to other properties of stars, particularly luminosity.

Measurements of Stellar Spectra

The first observations of stellar spectra, made in the mid-1800s before it was understood how spectral lines are formed, were accomplished using **spectroscopes,** simple devices that allow the observer to see the light from a star spread out according to wavelength, but not to record it. In the late 1800s introduction of the technique of photographing spectra allowed a systematic study of stellar spectra to get under way. A photograph of a spectrum is called a **spectrogram** (Fig. 6.6).

Among the first astronomers to systematically examine spectra of a large number of stars was the Roman Catholic priest Angelo Secchi, who in the 1860s cataloged hundreds of spectra using a spectroscope. He found that the appearance of the spectra varied considerably from star to star, although they were consistent in one respect: They all showed continuous spectra with absorption lines. The work of Kirchhoff soon showed that this was due to the relatively cool outer layers of a star, which absorb light from the hotter interior. At first it was thought that the differing appearances of stellar spectra were caused by differences in the chemical composition of the stars, and in one of the early classification schemes stars were assigned to categories based on their compositions. The basis of the modern classification system for stellar spectra was founded by a group of astronomers at Harvard, most notably Annie J. Cannon (Fig. 6.7). Cannon found a smooth sequence of types of spectra, in which the pattern of strong absorption lines changed gradually from one type to the next. Having already assigned letters of the alphabet to the various types, she placed them in the sequence O, B, A, F, G, K, M (Fig. 6.8).

It was later realized that the differing appearances of stellar spectra were due not to differences in chemical composition, but to differences in temperature. The hotter a star is, the more highly ionized the gas is in its outer layers. The degree of ionization in turn governs the pattern of spectral lines that will form. (If this is confusing, review the discussion of ionization in Chapter 4.) Therefore Cannon's sequence of spectral types

Figure 6.6 A Stellar Spectrum. This photograph of a stellar spectrum shows the manner in which they are usually displayed, as negative prints. Hence light features are absorption lines, wavelengths at which little or no light is emitted by the star. *(From Abt, H., A. Meinel, W. W. Morgan, and R. Tapscott. 1968. An atlas of low-dispersion grating stellar spectra [Tucson: Kitt Peak National Observatory])*

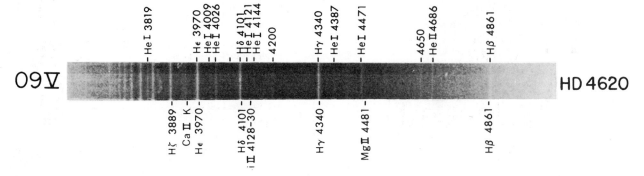

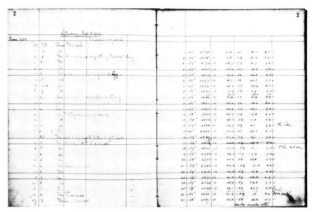

Figure 6.7 Annie J. Cannon. A member of the Harvard College Observatory for almost fifty years, Cannon classified the spectra of several hundred thousand stars. At right is a page from one of her notebooks. Today she is recognized as the founder of modern spectral classification. *(Harvard College Observatory)*

was a *temperature* sequence: the hottest stars are the O stars, and the coolest are the M stars. She developed a fine enough eye for subtle differences between spectra to assign subclasses to each of the major classes. In this system, the Sun is a G2 star, being an intermediate between types G and K.

Figure 6.8 A Comparison of Spectra. Here we see several spectra representing different spectral classes. They are arranged in order of decreasing stellar temperature (top to bottom). A few major absorption lines are identified. *(Mt. Wilson and Las Campanas Observatories, Carnegie Institution of Washington)*

In the modern classification system, a few key spectral lines establish the type of a given star (Fig. 6.9). For the O stars, ionized helium, which requires a very high temperature for its formation, is the principal species that reveals the spectral type. For the slightly cooler B stars, it is atomic helium; for the A stars, atomic hy-

Figure 6.9 Ionization for Stars of Various Spectral Types. This diagram shows which ions appear prominently in the spectra of stars of different classes. Note that the degree of ionization evident for the hot stars (left) is much greater than for the cool ones (right).

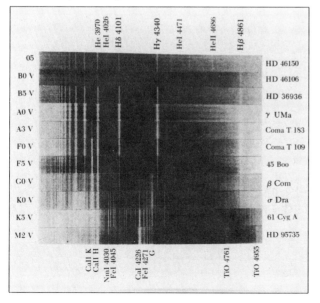
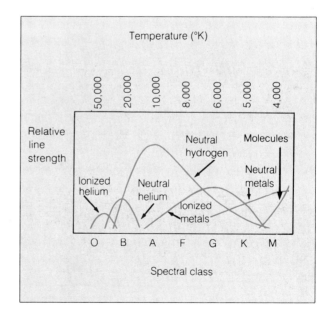

Astronomical Insight 6.2

Stellar Spectroscopy and The Harvard Women

There is some irony in the fact that Harvard University, long a stronghold of the all-male tradition in American colleges, was the institution that nurtured some of the nation's first leading women astronomers. Even today there are few women among professional astronomers, but they were even more underrepresented at the turn of the century, when the foundations of modern stellar spectroscopy were developed at the Harvard College Observatory.

When Edward C. Pickering became director of the observatory in 1877, Secchi's work (described in the text), based on visual inspection of stellar spectra through a spectroscope, was the only attempt to classify stars according to the appearance of their spectra. Henry Draper, an American amateur astronomer, had in 1872 become the first to photograph the spectrum of a star. Upon Draper's death in 1882, his widow endowed a new department of stellar spectroscopy at Harvard. As director, Pickering hired among his assistants a number of women, several of whom went to work on the problem of classifying spectra of stars.

An innovative technique was used to photograph spectra of large numbers of stars. A thin prism was placed in front of the telescope, so that each star image on the photographic plate at the focus was stretched, in one direction, into a spectrum. If color film had been used (it wasn't), each stellar image would have looked like a tiny rainbow. Pickering and his group refined this **objective prism** technique to the point where all stars in a field of view, even those as faint as ninth or tenth magnitude, would appear as spectra sufficiently exposed for classification. Thus it became possible to amass stellar spectra in vast quantities.

The problem of sorting out and classifying the spectra fell to Pickering's associates. One of them, Williamina Fleming, published the first *Draper Catalog of Stellar Spectra* in 1890. This catalog assigned some 10,351 stars in the Northern Hemisphere to spectral classes A through N, in a simple elaboration of a rudimentary classification scheme adopted earlier by Secchi. However, a number of Fleming's classes were later dropped.

While the first catalog was being prepared, a niece of Henry Draper, Antonia Maury, joined the staff and set to work on the analysis of spectra of bright stars. Spectra for these objects could be photographed using thicker prisms (actually, a series of two or three thin prisms) so that the spectra were more widely spread out according to wavelength, allowing greater detail to be seen. Antonia Maury concluded, on the basis of these high-quality spectra, that stars should be grouped into three distinct sequences rather than just one. She had found that some stars had unusually narrow spectral lines, others were rather broad, and a third group was in between. It was later found that her sequence *c*, the thin-lined stars, are giant stars. (The lower atmospheric pressure in these stars causes the lines to be less broadened than in main sequence stars.) Maury's discovery helped Ejnar Hertz-

Astronomical Insight 6.2

Stellar Spectroscopy and The Harvard Women

sprung confirm the distinction he had recently found between giant and main sequence stars, and he thought her work to be of fundamental importance. Unfortunately, Maury's separate sequences were not adopted in the further work on classification at Harvard, which was thereafter based on a single sequence.

The most important of the Harvard workers in spectral classification arrived on the scene in 1896. Her name was Annie J. Cannon, and she gradually modified the classification system to the present sequence, finding that the arrangement O, B, A, F, G, K, M was a logical ordering, with smooth transitions from one type to the next. (At this time, there was still no knowledge of the fact that this was a temperature sequence.) Cannon was able to distinguish the gradations so finely that for each major division she established the ten subclasses that are in use today. In 1901 she published a catalog of classifications for 1122 stars and then embarked on her major task: the classification of over 200,000 stars whose spectra appeared on survey plates covering both hemispheres. By this time she was so skilled that she could reliably classify a star in a few minutes, and most of the job was done in a four-year period between 1911 and 1914. The resulting *Henry Draper Catalog* appeared in nine volumes of the *Annals of the Harvard College Observatory*, the final one being published in 1924. Pickering died in 1919, before the catalog was complete, and was succeeded as director by Harlow Shapley.

Annie Cannon continued her work, later publishing a major *Extension* of the catalog, along with a number of other specialized catalogs. She died in 1941, and the American Astronomical Society subsequently established the Annie J. Cannon prize for outstanding research by women in astronomy.

The successes of the Harvard women were not confined to stellar spectroscopy. Another major area of interest to Pickering and later to Shapley was the study of variable stars, and Henrietta Leavitt, who joined the group as a volunteer in 1894, played a leading role in this area. Following some early work on the establishment of standard stars for magnitude determinations, by 1905 Leavitt was at work identifying variables from comparisons of photographic plates. (She was eventually to discover over 2000 of them.) In the process she examined the Magellanic Clouds for variables and noticed that their periods of pulsation correlated with their average brightnesses. This was a discovery of profound importance; the period-luminosity relationship for variable stars was to become an essential tool for establishing both the galactic distance scale and the intergalactic scale (see discussions in Chapters 9 and 11).

The Harvard women, including those discussed here and several whose names are now obscure, were a remarkable group. They were responsible for a number of major advances in the science of astronomy at a time when its basis in physics was just becoming clear. In this day of awakening recognition of the proper role of women in all areas, it is fitting to consider and appreciate the pioneering work done by this group.

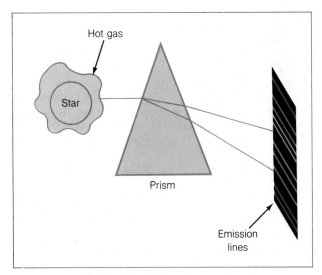

Figure 6.10 Stellar Emission Lines. Very few stars have emission lines in their visible-wavelength spectra (but many do in ultraviolet wavelengths). When such lines are present, it usually signifies the existence of hot gas above the surface. According to Kirchhoff's second law, a rarefied, hot gas produces emission lines.

drogen. The F stars have strong hydrogen lines, but also have lines due to certain metallic elements that are ionized once (that is, these atoms have lost just one electron). The G stars have a mixture of ionized and atomic metals, and in the K and M stars these elements are nearly all in atomic form. The cooler M stars also have strong molecular lines.

Having established this classification system around the turn of the century, Cannon cataloged nearly a quarter of a million stars, a monumental task that took about five years (although the publication of the results, called the *Henry Draper Catalog*, took place over a much longer period). This catalog has been a fundamental reference for generations of astronomers, and the system of classification established by Cannon has, with some modification, been in use since its development.

Some stars do not fit neatly into the standard spectral classes, and these are often referred to as "peculiar" stars. In most cases these are stars whose chemical composition is unusual, at least at the surface where the spectral lines form. Most stars have the same

Figure 6.11 A Pulsating Variable Star. The curved line shows how the brightness (in magnitude units) varies as the star expands and contracts. The sequence of sketches illustrates how the expansion and contraction phases are related in sequence to the variations in brightness. The surface temperature also varies.

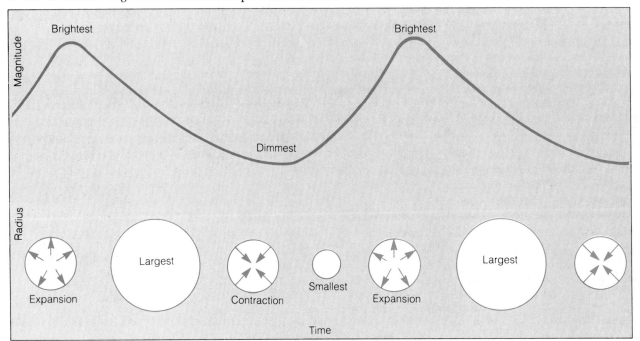

basic composition. (See the list of relative abundances of the elements in the Sun in Table 5.3.) Others are unusual because they have emission lines in their spectra, which, according to Kirchhoff's laws, means that they must be surrounded by hot, rarefied gas (Fig. 6.10). The so-called peculiar stars are probably quite normal but in short-lived stages of evolution, so there are not many of them around at any one time.

The variable stars represent another kind of unusual star. The majority of these are stars whose brightness fluctuates regularly as they alternately expand and contract (Fig. 6.11). As in the case of the peculiar stars, the variables are normal stars in special stages of their lifetimes where particular combinations of atmospheric pressure and ionization conditions produce instabilities that cause the pulsations. The most widely known pulsating variable stars are the δ **Cephei** stars, giants whose spectral type varies between F and G as they pulsate. As we will see in Chapter 9, these stars, as well as the less luminous **RR Lyrae** variables, are very useful tools in measuring distances, because their luminosities can be inferred from their periods of pulsation. Other pulsating stars include the **Mira** stars, or **long-period variables,** M supergiants that take a year or longer to go through a complete cycle; there is also a variety of shorter-period variables that are not quite so regular as the δ Cephei and RR Lyrae stars. Some stars vary erratically, even explosively, and these will be discussed in Chapter 8.

Binary Stars

About half of the stars in the sky are members of double or **binary** systems, where two stars orbit each other regularly. All types of stars can be found in binaries, and their orbits also come in many sizes and shapes. In some systems the two stars are so close together that they are actually touching, and in others they are so far apart that it takes hundreds or thousands of years for them to complete one revolution.

Binary systems can be detected by each of the three different types of observations we have discussed (Table 6.2). Positional measurements, of course, tell us when two stars are very near each other in the sky. Sometimes this occurs by chance, when a nearby star happens to lie nearly in front of one that is in the background (Fig. 6.12); this is called an **optical double** and is not a true binary system, since the two stars do not orbit each other. In a **visual binary,** a pair of stars is seen close together and measurements show that they are in motion about one another. Accurate positional measurements are needed to reveal the orbital motion, because the two stars are very close together in the sky and they appear to move very slowly about each other. The separation between the two stars must be many AUs in order for them to appear separately as seen from the Earth, and therefore their orbital period is many years.

It is possible for binary systems to be recognized

Table 6.2 Types of Binary Systems

Type of System	Observational Characteristic	Type of System	Observational Characteristic
Visual binary	Both stars visible through telescope (usually requires photograph); change of relative position confirms orbital motion.	Spectroscopic binary	Spectral lines undergo periodic Doppler shifts, revealing orbital motion. If two sets of spectral lines are seen shifting back and forth (which occurs only if the two stars are comparable in magnitude), it is a double-lined spectroscopic binary; if only one set of lines is seen (in the more common situation where one star is much brighter than the other), it is a single-lined spectroscopic binary.
Astrometric binary	Positional measurements reveal orbital motion about center of mass.		
Spectrum binary	Spectrum shows lines due to two distinct spectral classes, thus revealing presence of two stars.		
Eclipsing binary	Apparent magnitude varies periodically due to eclipses as stars alternately pass in front of each other.		

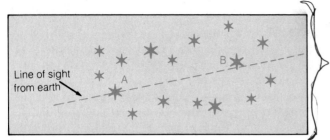

Figure 6.12 An Optical Double. A chance alignment of two stars at different distances may give the appearance of a double star. Here stars A and B appear very close together, as seen from Earth, but are in fact widely separated.

even when only one of the two stars can be seen. If positional measurements over a lengthy period of time reveal a wobbling motion of a star as it moves through space (Fig. 6.13), it may be inferred that the star is orbiting an unseen companion. Such systems, detected because of variations in position, are called **astrometric binaries.**

Simple brightness measurements can also tell us when a star that may appear to be single is actually part of a binary system. If we happen to be aligned with the plane of the orbit, so that the two stars alternately pass in front of each other as we view the system, the observed brightness will decrease each time one star is in front of the other. This system is called an **eclipsing binary** (Fig. 6.14).

Figure 6.13 An Astrometric Binary. Careful observation of the motion of a star across the sky in some cases reveals a curved path such as the one shown here. Such a motion is caused by the presence of an unseen companion star, so that the visible star orbits the center of mass of the binary system as it moves across the sky.

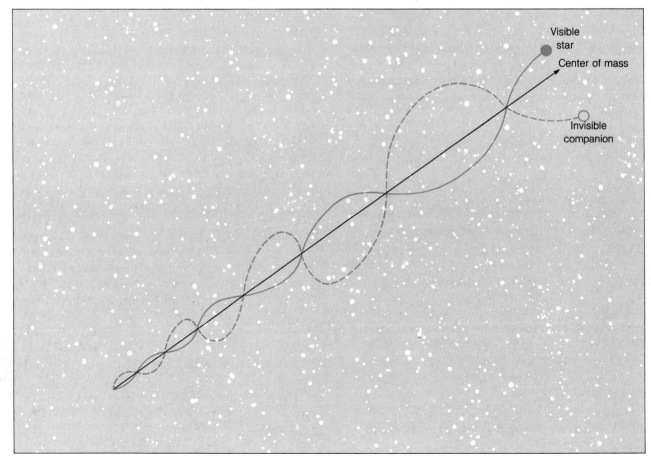

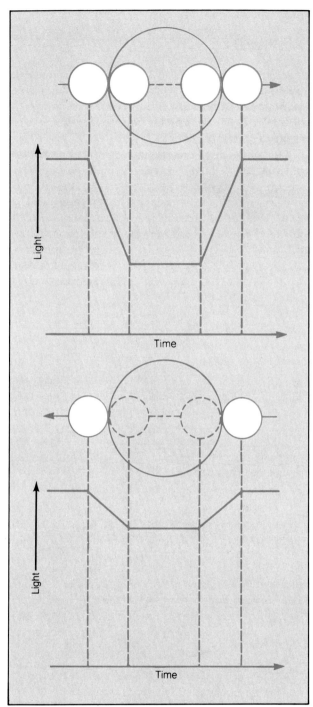

Figure 6.14 An Eclipsing Binary. If our line of sight happens to be aligned with the orbital plane of a double star system, the two stars will alternately eclipse each other, and this causes brightness variations in the total light output from the system, as shown. In the upper figure, the smaller star is passing in front of the larger one; in the lower sketch, the smaller one is passing behind. The segmented graph in each case illustrates how the total brightness of the system varies as these eclipses occur.

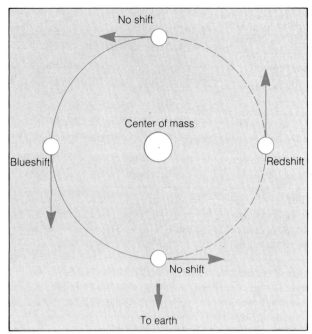

Figure 6.15 A Spectroscopic Binary. As a star orbits the center of mass of a binary system, its velocity relative to the Earth varies. This produces alternating redshifts and blueshifts in its spectrum, as the star recedes from and approaches us. For simplicity, only one of the stars is shown here to be moving, although in reality both do.

Finally, spectroscopic measurements can also be used to recognize binary systems even when a star appears single. If two stars are actually present and have nearly equal brightnesses, spectral lines from both will appear in the spectrum. If we find a star with line patterns representing two different spectral classes, we can be sure that two different stars are present; such a system is called a **spectrum binary.** Even if it is not clear that two different spectra are present (that is, if the two stars have similar spectral classes, or if one is too faint to contribute noticeably to the spectrum), the Doppler effect may still reveal the fact that the star is double. Because the stars are orbiting each other, they move back and forth along our line of sight (as long as we are not viewing the system face-on), and this motion produces alternating blueshifts and redshifts of the spectral lines (Fig. 6.15). Binaries in which these periodic Doppler shifts are seen are called **spectroscopic binaries;** they are more common than spectrum binaries.

A given double star system may fall into more than one of these categories. For example, a relatively nearby system seen edge-on could be a visual binary if both stars can be seen in the telescope; it may also be

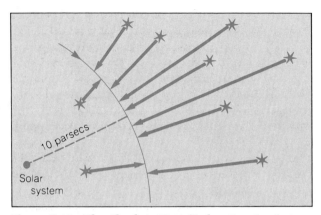

Figure 6.16 The Absolute Magnitude. Imagine that we can move stars from their true positions to a uniform distance from us of ten parsecs. The magnitude a star would have at this distance is called the absolute magnitude; it is directly related to the stellar luminosity, since the distance effect has been accounted for.

Table 6.3 Visual Absolute Magnitudes of Selected Objects

Object	Absolute Magnitude
Typical bright quasar	−28
Brightest galaxies	−25
Andromeda galaxy	−21.1
Milky Way galaxy	−20.5
Type I supernova (maximum brightness)	−18.8
Large Magellanic Cloud	−18.7
Type II supernova (maximum brightness)	−17
Small Magellanic Cloud	−16.7
Typical globular cluster	−8
The most luminous stars	−8
Typical nova outburst	−7.7
Vega (bright star in summer sky)	+0.5
Sirius	+1.41
Alpha Centauri (nearest star)	+4.35
Sun	+4.83
Venus (greatest elongation)	+28.2
Full moon	+31.8
100-watt bulb	+66.3

an eclipsing binary if the two stars alternately pass in front of each other; and it almost surely will be a spectroscopic binary since the motion back and forth along our line of sight is maximized when we view the orbit edge-on.

Binary stars merit our attention partly because there are so many of them, but even more importantly because of what they can tell us about the properties of individual stars. As we will see in the next section, much of our basic information on the nature of stars comes from measurements of binary systems.

Fundamental Stellar Properties

Before being able to understand how stars work, astronomers need to determine their physical properties and how they are related to each other. A number of fundamental quantities characterize a star, and in these sections we will see how the observations described previously can be used to determine quantities such as the luminosities, the temperatures, the radii, and above all, the masses of stars. At several steps along the way it is also important to know the distances to stars, and we will learn about a new, very powerful distance determination method.

Absolute Magnitudes and Stellar Luminosities

The luminosity of a star is the total amount of energy it emits from its surface, in all wavelengths, and is usually measured in units of ergs per second (see Chapter 4). To make such a measurement it is necessary to take into account the distance to a star and the light received from it at all wavelengths, not just those to which the human eye responds. In order to eliminate the effect of distance, astronomers use the **absolute magnitude,** defined as the magnitude a star would have if it were seen at a standard distance of 10 parsecs (Fig. 6.16; Table 6.3). This distance was chosen arbitrarily but is used by all astronomers. Thus a comparison of absolute magnitudes reveals differences in luminosities, because all the effects of distance have been canceled out.

It helps to consider a specific example. Suppose a certain star is 100 parsecs away and has an **apparent magnitude** (observed magnitude) of 7.3. To determine the absolute magnitude, we must find out what this star's magnitude would be if it were only 10 parsecs away. The inverse square law tells us that since the star would be a factor of 10 closer, it would appear a factor of $10^2 = 100$ brighter, or exactly 5 magnitudes brighter. Therefore, its absolute magnitude is 7.3 − 5 = 2.3. It is always possible to derive the absolute magnitude from the measured apparent magnitude and the distance, by following a similar argument. Other examples are given in the questions at the end of this chapter and in Appendix 10.

Let us now return to the question of luminosities. By determining the absolute magnitudes of stars, we can compare their luminosities, since the distance effect has been removed. Since we must allow for light emitted by a star at all wavelengths, the bolometric magnitude is the quantity that is used. Thus, if one star's bolometric magnitude is 5 magnitudes smaller than another's, we know that its luminosity is a factor of 100 greater. In order to express the luminosity in terms of ergs per second, a comparison is made with a standard star for which a direct measurement of the luminosity has been made by actually measuring the intensity of light at all wavelengths and allowing for distance.

What astronomers find when they determine stellar luminosities is that the values from star to star can vary over an incredible range. Stars are known with luminosities from as small as 10^{-4} that of the Sun to as great as 10^6 that of the Sun, a range of 10 billion from the faintest to the most luminous! The luminosity is by far the most highly variable parameter for stars; the others that we shall discuss cover ranges of only a few hundred or so from one extreme to the other.

Stellar Temperatures

Earlier in this chapter, we saw that the color of a star, like its spectral class, depends on its temperature, so that the temperature can be deduced by observing either. Recall that the color index, the difference between the blue (B) and visual (V) magnitudes, is a measured quantity that indicates temperature. A negative value of $B - V$ means that the star is brighter in blue than in visual light and therefore is a hot star. A large positive value indicates a cool star. A specific correlation of color index with temperature has been developed and is used to determine temperatures from observed values of $B - V$.

More refined estimates of temperature can be made from a detailed analysis of the degree of ionization, which is done by measuring the strengths of spectral lines formed by different ions. This is basically the same as simply estimating the temperature from the spectral class, since in either case the point is that the strengths of spectral lines of various ions depend on how abundant those ions are, which in turn depends on the temperature.

The temperature referred to here may be called the surface temperature, although stars do not have solid surfaces. We are really referring to the outermost layers of gas, where the absorption lines form. This region is called the atmosphere of a star, and it actually has some depth, although it is very thin compared to the radius of the star.

Stellar temperatures, as we have already seen, range from about 2000 K for the coolest M stars to 50,000 K or more for the hottest O stars.

The Hertzsprung-Russell Diagram and a New Distance Technique

We have seen that temperature and spectral class are intimately related, and we have learned how astronomers deduce the luminosities of stars. In the first decade of this century, the Danish astronomer Ejnar Hertzsprung and, independently, the American astrophysicist Henry Norris Russell (Fig. 6.17), began to consider how luminosity and spectral class might be related to each other. Each gathered data on stars whose luminosities (or absolute magnitudes) were known, and found a close link between spectral class (temperature) and absolute magnitude (luminosity). This relationship is best seen in the diagram constructed by Russell in 1913 (Fig. 6.18), now called the **Hertzsprung-Russell**, or **H-R diagram**. In this plot of absolute magnitude (on the vertical scale) versus spectral class (on the horizontal axis), stars fall into narrowly defined regions, rather than being randomly distributed (also see Fig. 6.19). A star of a given spectral class cannot have just any absolute magnitude, and vice versa.

The great majority of stars fall into a diagonal strip running from the upper left (high temperature, high luminosity) to the lower right (low temperature, low luminosity) of the H-R diagram. This strip has been given the name the **main sequence.**

Figure 6.17 Henry Norris Russell. One of the leading astrophysicists of the era when a physical understanding of stars was first emerging, Russell made many major contributions in a variety of areas. *(Princeton University)*

Figure 6.18 The First H-R Diagram. This plot showing absolute magnitude versus spectral class was constructed in 1913 by Henry Norris Russell. *(Estate of Henry Norris Russell, reprinted with permission)*

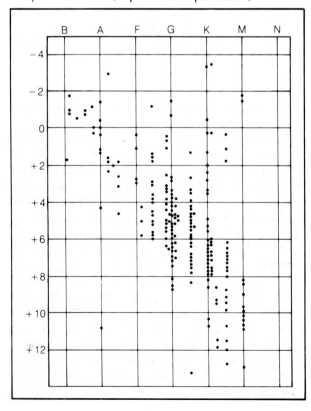

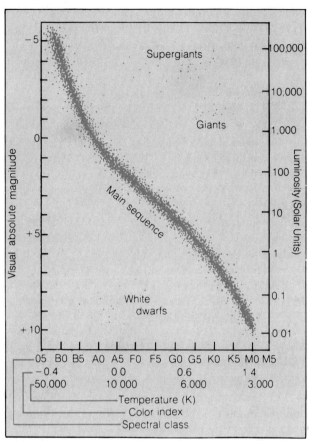

Figure 6.19 A Modern H-R Diagram. This diagram shows a large number of stars and gives alternative units on both axes for the two general parameters: luminosity and temperature.

One group of a few stars fails to fall on the main sequence but appears in the upper right (low temperature, high luminosity) of the diagram. Since the spectra of these stars indicate that they are relatively cool, their high luminosities cannot be due to greater temperatures than the main sequence stars of the same type. The only way one star can be a lot more luminous than another of the same temperature is by having a lot more surface area. (Recall that the two stars will emit the same amount of energy per square centimeter of surface.) Hertzsprung and Russell realized that these extra-luminous stars sitting above the main sequence must be much larger than those on the main sequence, and they called them **giants** and **supergiants**.

The distinction among giants, supergiants, and main sequence stars (commonly known as **dwarfs**) has been incorporated into the spectral classification system used by modern astronomers. A **luminosity class** (Fig. 6.20) has been added to the spectral type with which we are already familiar. The luminosity classes, designated by Roman numerals following the spectral type, are I for supergiants (this group is further subdivided into classes Ia and Ib), II for extreme giants, III for giants, IV for stars just a bit above the main sequence, and V for main sequence stars, or dwarfs (not to be confused with white dwarfs). Thus, a complete spectral classification for the bright summertime star Vega, for example, is A0V, meaning that it is an A0 main sequence star. Betelgeuse, the red supergiant in the shoulder of Orion, has the full classification M2Iab (because it is intermediate between luminosity classes Ia and Ib). It is usually possible to assign a star to the proper luminosity class by examining subtle details of its spectrum. For example, giant and especially supergiant stars have relatively low atmospheric pressures, which affect the spectral line widths and the state of ionization of certain elements.

Another group of stars (which has become known mostly since the time when Russell first plotted the H-R diagram) does not fall into any of the standard luminosity classes but appears in the lower left (high temperature, low luminosity) corner of the diagram. Since these stars are hot but not very luminous, they must be very small, and they have been given the name **white dwarfs.** These objects have some very bizarre properties, which will be discussed in Chapter 8.

The H-R diagram can be used to find distances to stars, even stars that are very far away (Fig. 6.21). The idea is really very simple: If we know how bright a star is intrinsically (how much energy it is actually emitting from its surface), and we measure how bright it appears to be, we can determine how far away it is because we know that the difference between its intrinsic brightness and its observed brightness is due to the distance. The only problem lies in knowing the intrinsic brightness (luminosity) of the star, and this is where the H-R diagram comes in.

Once we determine the spectral class of a star, we can place it in the H-R diagram (as long as we are sure we know the luminosity class, so that we know whether it is on the main sequence or is a giant or supergiant). Once we have placed it on the diagram, we simply read off the vertical axis the absolute magnitude of the star, which is a measure of its luminosity. A comparison of the absolute magnitude with the observed apparent magnitude then amounts to the same thing as a comparison of the intrinsic and apparent

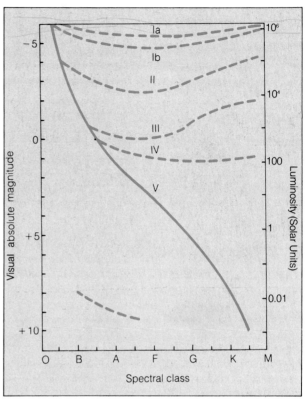

Figure 6.20 Luminosity Classes. This H-R diagram shows the locations of stars of the luminosity classes described in the text. A complete spectral classification for a star usually includes a luminosity class designation, if it has been determined.

brightnesses of the star, and from such a comparison the distance can be found.

The calculation of a star's distance from the difference between the apparent and absolute magnitudes can best be illustrated by considering a few simple cases. For example, if the difference $m - M$ (the apparent minus the absolute magnitude), which is called the **distance modulus,** is 5, the star appears 5 magnitudes (a factor of 100) fainter than it would at the standard distance of 10 parsecs. A factor of 100 in brightness is created by a factor of 10 change in distance, so this star must be ten times farther away than it would be if it were at 10 parsecs distance; therefore, it is $10 \times 10 = 100$ parsecs away. Similarly, a star whose distance modulus $m - M$ is 10 is 1000 parsecs away. If $m - M = 15$, the distance is 10,000 parsecs. It should be obvious that if $m = M$ (so that $m - M = 0$), the distance to the star must be 10 parsecs, because this is the distance that defines the absolute mag-

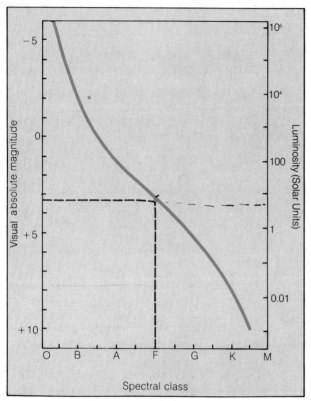

Figure 6.21 Spectroscopic Parallax. Knowledge of a star's spectral class (including the luminosity class) allows the absolute magnitude, and hence the distance, to be determined. This figure shows how the absolute magnitude (+3.3) of an F0 main sequence star is read from the H-R diagram. The distance can then be found by comparing the absolute and apparent magnitudes of the star (as explained in the text and Appendix 10).

nitude. As a rule of thumb, it helps to remember that for every 5 magnitudes of difference between the apparent and absolute magnitudes, the distance increases by a factor of 10.

This method is very powerful, because it can be used for very large distances. All that is needed is to be able to place a star on the H-R diagram so that its absolute magnitude can be determined and to measure its apparent magnitude. Because this distance-determination technique depends on classifying the spectrum of a star in order to place it on the H-R diagram, it is called the **spectroscopic parallax** method. (The word *parallax* is used by astronomers as a general word for distances, even though technically speaking no parallax is measured in this case.)

Stellar Diameters

We have already seen that a star's position in the H-R diagram depends partly on its size, since luminosity is related to total surface area. If two stars have the same surface temperature (and therefore the same spectral class), but one is more luminous than the other, we know that it must also be larger. The Stefan-Boltzmann law (see Chapter 4) specifically relates the three quantities luminosity, temperature, and radius; use of this law allows the radius to be determined if the other two quantities are known.

Eclipsing binaries provide another means of determining stellar radius that is independent of other properties. Recall that these are double star systems in which the two stars alternately pass in front of each other as we view the orbit edge-on. The eclipsing binary is very likely also to be a spectroscopic binary, so that the speeds of the two stars in the orbits can be measured from the Doppler effect. We therefore know how fast the stars are moving, and from the duration of the eclipse we know how long it takes one to pass in front of the other; the simple formula

$$\text{distance} = \text{speed} \times \text{time}$$

then gives us the diameter of the star that is being eclipsed. Even if no information on the orbital velocity is available, the relative diameters of the two stars can be deduced from the relative durations of the alternating eclipses.

Eclipsing binaries provide the most direct means of measuring stellar sizes, but unfortunately there are not many of them. In most cases the radii are estimated from the luminosity and temperature as described above. In a few cases stellar radii have been measured directly by use of speckle interferometry, a sophisticated technique for clarifying the image of a star by removing the blurring effects of the Earth's atmosphere. Only relatively near, large stars can be measured this way, however.

Stars on the main sequence do not vary greatly in radius, ranging from perhaps 0.1 times the Sun's radius for the M stars at the lower right-hand end to 10 or 20 solar radii at the upper left. Of course, large variations in size occur as we go away from the main sequence, either toward the giants and supergiants, which may be 100 times the size of the Sun, or toward the white dwarfs, which are as small as 0.01 times the size of the Sun.

Binary Stars and Stellar Masses

The mass of a star is the most important of all its fundamental properties, for it is the mass that governs most of the others. This point will be discussed at some length in the next chapter.

Unfortunately, there is no direct way to see how much mass a star has. The only way to measure a star's mass is by observing its gravitational effect on other objects, and this is possible only in binary star systems, where the two stars hold each other in orbit by their gravitational fields. The fact that binary systems are common provides many opportunities to determine masses by analyzing binary orbits.

The basic idea is rather simple, although the application may be quite complex, depending on the type of binary system. Kepler's third law, in the form derived by Newton, is used. Remember that if the period P is measured in years, the average separation of the two stars (the semimajor axis a) in astronomical units, and the masses m_1 and m_2 in units of solar masses, then Kepler's third law is

$$(m_1 + m_2)P^2 = a^3$$

We need only observe the period and the semimajor axis in order to solve for the sum of the two masses. Careful observations of the sizes of the individual orbits (actually, of the relative distances of the two stars from the center of mass; see Fig. 6.22) also yield the ratio of the masses; when both the sum and the ratio are known, it is simple to solve for the individual masses.

Complications arise when some of the needed observational data are difficult to obtain. The period is almost always easy to measure with some precision, but not so the semimajor axis a. The main problems are that the apparent size of the orbit is affected by our distance from the binary (so the distance must be well known if a is to be accurately determined) and that the orbital plane is inclined at a random, unknown angle to our line of sight (so the apparent size of the orbit is foreshortened by a unknown amount). In some cases it is possible to unravel these confusing effects by carefully analyzing the observations. Even in cases where this is not possible, some information about the masses of the stars can still be gained, but usually only in terms of broad ranges of possible values rather than precise answers.

The masses of stars vary along the main sequence from the least massive stars in the lower right to the

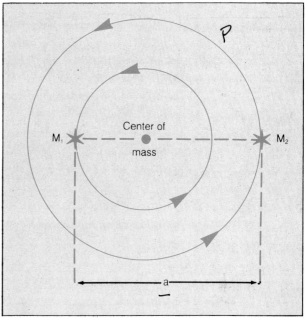

Figure 6.22 Binary Star Orbits. This figure illustrates the terms used in Kepler's third law. Two stars of masses m_1 and m_2 (m_1 larger than m_2 in this case) orbit a common center of mass, each making one full orbit in period P. The semimajor axis a that appears in Kepler's third law is actually the sum of the semimajor axes of the two individual orbits about the center of mass; this sum corresponds to the average distance between the two stars.

most massive at the upper left. The M stars on the main sequence have masses as low as 0.05 solar mass, while the O stars reach values as great as 60 solar masses. It is likely that stars occasionally form with even greater masses (perhaps up to 100 solar masses), but as we will see in the next chapter, such massive stars have very short lifetimes, so it is rare to find one.

The giants, supergiants, and white dwarfs have masses comparable to those of main sequence stars. Hence their obvious differences from main sequence stars in other properties such as luminosity and radius have to be caused by something other than extreme or unusual masses. This topic is discussed in the next chapter.

For main sequence stars, there is a smooth progression of all stellar properties from one end of the H-R diagram to the other (Table 6.4). The mass and the radius vary by similar factors, while the luminosity changes much more rapidly along the sequence. The

Table 6.4 Properties of Main Sequence Stars

Spectral Type	Mass ($M_\odot$*)	Temperature*	B − V*	Luminosity ($L_\odot$*)	M_V*	B.C.*	Radius ($R_\odot$*)
O5V	40	40,000	−0.35	5×10^5	−5.8	−4.0	18
B0V	18	28,000	−0.31	2×10^4	−4.1	−2.8	7.4
B5V	6.5	15,000	−0.16	800	−1.1	−1.5	3.8
A0V	3.2	9900	0.00	80	+0.7	−0.4	2.5
A5V	2.1	8500	+0.13	20	+2.0	−0.12	1.7
F0V	1.7	7400	+0.27	6	+2.6	−0.06	1.4
F5V	1.3	6580	+0.42	2.5	+3.4	0.00	1.2
G0V	1.1	6030	+0.58	1.3	+4.4	−0.03	1.1
G5V	0.9	5520	+0.70	0.8	+5.1	−0.07	0.9
K0V	0.8	4900	+0.89	0.4	+5.9	−0.19	0.8
K5V	0.7	4130	+1.18	0.2	+7.3	−0.60	0.7
M0V	0.5	3480	+1.45	0.03	+9.0	−1.19	0.6
M5V	0.2	2800	+1.63	0.008	+11.8	−2.3	0.3

* Masses are in units of $M_\odot$, the solar mass (1.99×10^{33} g); luminosities are in units of $L_\odot$, the solar luminosity (3.83×10^{33} erg/sec), and radii are in units of $R_\odot$, the solar radius (6.96×10^{10} cm). The temperatures are *effective temperatures*, which are a measure of surface temperature. The heading B − V stands for the color index, M_V stands for visual absolute magnitude, and B.C. stands for bolometric correction.

mass of an O star is perhaps 100 times that of an M star, while the luminosity is greater by a factor of 10^8 or more. There is a so-called **mass-luminosity relation** among main sequence stars, a numerical expression in which the luminosity of a star is proportional to an exponential power of the mass. In its simplest form, this relation states that the luminosity is proportional to the cube of the mass. More precise versions of the mass-luminosity relation reflect some variation in the exponent along the main sequence.

Other Properties

Several other properties of stars can be determined primarily from the analysis of their spectra. Perhaps the most fundamental of these is the composition. We have

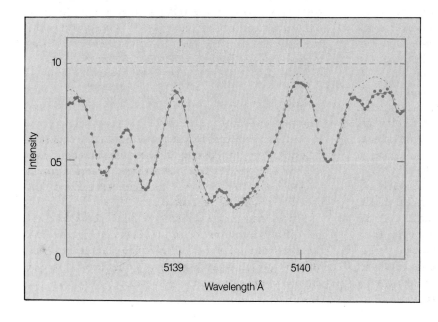

Figure 6.23 Determination of Stellar Composition. This diagram illustrates a modern technique for measuring the chemical composition of a star. A theoretical spectrum (solid line) is compared with the observed spectrum (dotted and dashed line). The abundances of elements assumed present are varied in the computed spectrum until a good match is achieved. *(D. L. Lambert)*

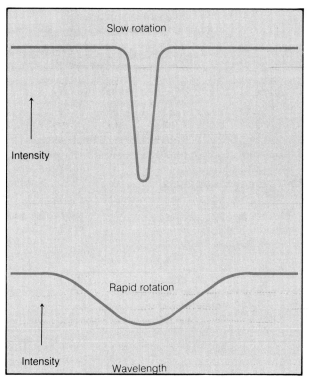

Figure 6.24 The Effect of Rotation on Stellar Absorption Lines. Rotation of a star broadens its spectral lines due to the Doppler effect. Here the same line is shown as it would appear in the spectrum of a star rotating slowly (top) and one that is spinning rapidly (bottom).

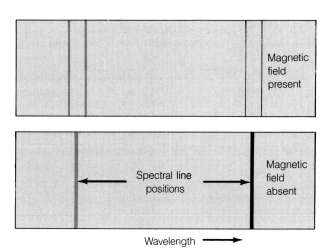

Figure 6.25 The Measurement of a Magnetic Field. The presence of a magnetic field splits certain absorption lines of some elements, in a process called Zeeman splitting. The amount of separation is a measure of the strength of the magnetic field.

already seen that stars are generally made of the same material in the same proportions, but it required a sophisticated analysis to show this. It was necessary first to develop techniques for taking the effects of temperature into account, for these effects, as we have already learned, are dominant factors in controlling the strengths of lines in a stellar spectrum. In modern work on stellar composition, complex computer programs are used to calculate simulated spectra with different assumed compositions until a match with an observed spectrum is found (Fig. 6.23).

Another stellar property that can be learned from the spectrum is the rotational velocity at the surface, because the Doppler effect causes broadening of the spectral lines (Fig. 6.24). A spectral line forms over the entire disk of a star, and in general one edge of the disk is approaching the Earth while the other is receding. Hence some of the gas creates a blueshift, and some a redshift (the gas in the central portions of the disk has little or no shift), so the rate of rotation of the star determines the degree of broadening of the spectral lines.

A final property, whose importance is difficult to assess because of the difficulty of measuring it in many cases, is the magnetic field of a star. In certain stars the Zeeman effect (described in Chapter 5) can be used; this is the splitting of some spectral lines due to the presence of a magnetic field (Fig. 6.25). The Zeeman effect can be applied only to very slowly spinning stars whose spectral lines are very narrow, so that the splitting can be seen. In the many stars where the lines are too broad for this we have no good method of measuring the magnetic field strength.

Perspective

In this chapter we have learned the basics of stellar observations by discussing the three fundamental types of observations and the characteristics of stars revealed by each. Considerable astronomical terminology has been introduced in the process.

We have gone on to learn how all the basic properties of stars are derived from observational data. We can categorize, classify, and describe stars in any way we wish. Now we are ready to see how they work and *why* the various quantities are related the way they are and no other way.

Summary

1. There are three basic types of stellar observation: position, brightness, and spectroscopy.
2. Positional astronomy (astrometry) has developed techniques capable of measuring position to an accuracy approaching 0.001 arcsecond, which is sufficient to detect proper motions, binary motions in some cases, and stellar parallaxes for stars up to 1000 parsecs away.
3. Stellar brightness measurements are commonly carried out using the stellar magnitude system, in which a difference of one magnitude corresponds to a brightness ratio of 2.512.
4. Magnitudes can be measured at different wavelengths, allowing the determination of color indices, or over all wavelengths to determine bolometric magnitudes.
5. Stellar spectra contain patterns of absorption lines that depend on the surface temperatures of stars and therefore can be used to assign stars to spectral classes that represent a sequence of temperatures.
6. Peculiar stars are those that do not conform to the usual spectral classes, most often because of unusual surface compositions but sometimes because they have emission lines.
7. Binary star systems can be detected by positional variations of one or both stars (astrometric binaries), brightness variations (eclipsing binaries), or composite or periodically Doppler shifted spectral lines (spectrum or spectroscopic binaries).
8. Stellar luminosities are determined from knowledge of the distances to stars and their apparent magnitudes. The absolute magnitude, a measure of luminosity, is the magnitude a star would have if it were seen from a distance of 10 parsecs.
9. Stellar temperatures can be inferred from the $B - V$ color index, estimated from the spectral class, or determined from the degree of ionization in the star's outer layers.
10. The Hertzsprung-Russell diagram shows that the luminosities and temperatures of stars are intimately related and that stars that do not fall on the main sequence are either larger (red giants or supergiants) or smaller (white dwarfs) than those on the main sequence.
11. The distance to a star can be measured by determining its spectral class and then using the H-R diagram to infer its absolute magnitude, which is then compared to its apparent magnitude to yield the distance; this technique is called spectroscopic parallax.
12. Stellar diameters can be determined directly in eclipsing binaries from knowledge of the orbital speed and the duration of the eclipses.
13. Masses of stars are derived in binary systems by observing the period and the orbital semimajor axis and using Kepler's third law.

Review Questions

1. Does the Sun have annual parallax motion, as seen from the Earth?
2. How much fainter is a twelfth magnitude star than an eleventh magnitude star? How does a fourth magnitude star compare in brightness with a fifth magnitude star?
3. Briefly discuss the role of stellar parallax in the development of the heliocentric theory.
4. Explain why the bolometric magnitude of a star is always smaller than the visual magnitude.
5. Explain, in terms of what you learned about ionization in Chapter 4, why stars of different temperature have different absorption lines in their spectra.
6. If a certain star has absolute bolometric magnitude $M_{bol} = 1.5$ and another has $M_{bol} = 4.5$, which is more luminous, and by how much?
7. If one star is four times hotter than another and has twice as large a radius, how does its luminosity compare with that of the other star? (Note: This question requires you to review some material in Chapter 4.)
8. Explain why a G2 star that lies well above the position of the Sun on the H-R diagram must have a larger radius than the Sun. If its luminosity is 100 times greater than that of the Sun, how much larger is its radius?
9. Summarize the various reasons why it is difficult to determine the masses of a pair of stars in a binary system, even though in the ideal case this can be done.
10. Explain why the spectral lines of a particular element, such as hydrogen, can have quite different strengths from one star to another, yet the abundance of the element may be the same in the two stars.

Additional Readings

Probably the best sources of additional information on stars and their observational properties are general astronomy textbooks, of which there are a number readily available in nearly any library. A few more specific resources are listed here as well.

Abt, H. A. 1977. The companions of sun-like stars. *Scientific American* 236(4):96.

Aller, L. H. 1971. *Atoms, stars, and nebulae.* Cambridge, Mass.: Harvard University Press.

Beyer, Steven L. 1986. *The star guide.* Boston: Little, Brown, Inc.

De Vorkin, D. 1978. Steps towards the Hertzsprung-Russell diagram. *Physics Today* 31(3):32.

Evans, D. S., Barnes, T. G., and Lacy, C. H. 1979. Measuring diameters of stars. *Sky and Telescope* 58(2):130.

Getts, Judy, 1983, Decoding the Hertzsprung-Russell diagram, *Astronomy* 11(10):16.

Kaler, James B. 1987, The B stars: Beacons of the sky, *Sky and Telescope,* 74(2):174.

———1986, Cousins of our sun: The G stars, *Sky and Telescope,* 72(5):450.

———1986, The K stars: Orange giants and dwarfs, *Sky and Telescope,* 72(2):130.

———1986, M stars: Supergiants to dwarfs, *Sky and Telescope,* 71(5)450.

———1986, The origins of the spectral sequence, *Sky and Telescope,* 71(2):129.

———1987, The spectacular O stars, *Sky and Telescope* 74(5):464.

———1987, The temperate F stars, *Sky and Telescope* 73(2):131.

———1987, White sirian stars: Class A, *Sky and Telescope* 73(5):491.

McAlister, H. A. 1977. Binary star speckle interferometry. *Sky and Telescope* 53(5):346.

Mihalas, D. 1973. Interpreting early type stellar spectra. *Sky and Telescope* 46(2):79.

Philip, A. G. D., and Green, L. C. 1978. The H-R diagram as an astronomical tool. *Sky and Telescope* 55(5):395.

Rubin, Vera, 1986, Women's work: Women in modern astronomy, *Science 86* 7(6):58.

Struve, O., and Zebergs, V. 1962. *Astronomy of the 20th century.* New York: Crowell Collier and Macmillan.

Upgren, A. 1980. New parallaxes for old: A coming improvement in the distance scale of the universe. *Mercury* 9(6):143.

Welther, B. 1984. Annie Jump Cannon: Classifer of the stars. *Mercury* XIII(1):28.

Chapter 7

Stellar Structure: The Structure and Evolution of Stars

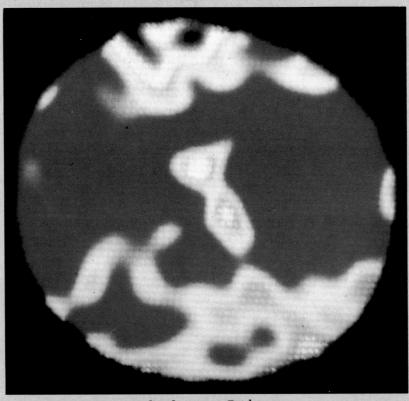

Surface features on Betelgeuse, detected by interferometry. *(National Optical Astronomy Observatories)*

Chapter Preview

Observations of Stellar Evolution: Star Clusters
    Stellar Structure
    The Role of Stellar Mass
    The Source of Energy: Nuclear Reactions
Mass Loss and Mass Exchange
Star Formation
Life Stories of Stars
    The Evolution of Stars like the Sun
    The Middleweights
    The Heavyweights
    Evolution in Binary Systems

In the previous chapter we learned how to measure the properties of stars—how to determine masses, temperatures, luminosities, diameters, and many other characteristics. Now we are ready to discuss our understanding of why stars have the properties that are observed and how these properties change with time.

Stars are dynamic, that is, they do not simply stay the same forever; instead, they change. The changes are usually very slow compared to human lifetimes, so we tend to think of the stars as permanent. The process of deducing the life cycles of stars has been compared to the way in which we can determine the life cycle of a tree: rather than waiting decades to see a single tree grow from a seedling, become mature, gradually age, and finally die, we deduce that it goes through these stages by observing different trees in the different stages. Similarly, we discover how stars age and die by observing many different stars in different stages of evolution.

We will begin this chapter with a discussion of the observational evidence that indeed stars change as they age. We then discuss briefly the physical reasons for these changes, and finally we will describe in some detail the life stories of stars.

Observations of Stellar Evolution: Star Clusters

Normally stars do not change rapidly enough for us to actually observe the changes (although there are circumstances where changes can be sudden, as we shall see). Most of what we know about stellar evolution comes from a combination of theoretical models and observations of stars in various stages of their lives. The detective work required to piece together the life stories of stars is simplified by the fact that stars often form and evolve in groups called **clusters** (Table 7.1),

Table 7.1 Selected Clusters of Stars

Cluster	Type	m_v	Distance (pc)	No. Stars	Diameter (pc)	Age (years)
47 Tuc	Globular	4.0	5,100	10^4	10	1.5×10^{10}
M103	Open	6.9	2,300	30	5	1.6×10^7
h Persei	Open	4.1	2,250	300	16	1.0×10^7
χ34	Open	5.6	440	240	14	1.0×10^8
Perseus	Open	4.1	167	80	12	1.0×10^7
Pleiades	Open	1.3	127	120	4	5.0×10^7
Hyades	Open	0.6	42	100	5	6.3×10^8
S Mon	OB	4.3	740	60	6	6.3×10^6
τ CMa	Open	3.9	1,500	30	3	5.0×10^6
Praesepe	Open	3.7	159	100	4	4.0×10^8
o Velorum	Open	2.6	157	15	2	2.5×10^7
M67	Open	6.5	830	80	4	4.0×10^9
θ Carinae	Open	1.7	155	25	3	1.3×10^7
Coma	Open	2.8	80	40	7	5.0×10^8
ω Cen	Globular	3.6	5,000	10^3	20	1.5×10^{10}
M13	Globular	5.9	7,700	10^5	11	1.5×10^{10}
M22	Globular	5.1	3,000	10^5	9	1.5×10^{10}
M15	Globular	6.4	14,000	10^6	11	1.5×10^{10}

where we can see directly the differences among stars as they age. There are several types of clusters, including **open** or **galactic clusters** such as the well-known Pleiades cluster (Fig 7.1); **OB associations,** loose groupings dominated by stars of types O and B (Fig. 7.2); and the magnificent **globular clusters,** huge spherical conglomerations containing hundreds of thousands of stars (Fig. 7.3).

It is thought that the stars in a given cluster all formed at the same time from the same material. Thus all the stars in a cluster are the same age, and astronomers can deduce the ways in which stars of different types evolve. The fact that the stars in a cluster are located together in space is also very helpful, for it means that the stars are all at the same distance from us, and so we know that the differences we see in such quantities as brightness are real differences, not the result of different distances.

It is most helpful to make an H-R diagram for a cluster of stars (Fig. 7.4), because this shows us graphically the contrasts in the properties of the stars in the cluster. Often such a diagram has apparent magnitude instead of absolute magnitude on its vertical axis, because we can assume that all the stars are at the same distance, and thus their ranking according to apparent magnitude is the same as their ranking according to absolute magnitude. (It is sometimes possible to deduce the absolute magnitudes of stars in a cluster by constructing such a diagram and matching it with the standard H-R diagram where absolute magnitudes are used. By doing this, we can find the distance to the cluster because we then know both the apparent and absolute magnitudes of the stars in it. This technique is called **main sequence fitting.**)

Observations of many different clusters show that the main sequence in the cluster H-R diagrams is often

Figure 7.1 The Pleiades. This relatively young galactic cluster is a prominent object in the Northern Hemisphere during the late fall and winter. *(Palomar Observatory, California Institute of Technology)*

Figure 7.2 An OB Association. The bright blue-white stars in this photo are members of a loose cluster called an OB association. The brightest member (whose image is heavily overexposed) is the star 15 Monocerotis, and the cluster is called NGC2264. Associations such as this are often not gravitationally bound, and the stars become dispersed in time. (© *1981 Anglo-Australian Telescope Board*)

cut off; that is, the main sequence does not always extend all the way up to the upper left-hand corner, where the most luminous, most massive stars are usually found (Fig. 7.4). Instead, these clusters have stars in the upper right-hand region, that is, red giants and supergiants. We deduce that the stars that start on the upper main sequence must change into red giants and supergiants, so that as a cluster ages, stars somehow change and become cooler and redder, at the same time becoming more luminous. The older a cluster is, the farther down is the point where the main sequence ends. The age of a cluster can be estimated from the location of this **main sequence turnoff** (Fig. 7.5). Thus, not only do clusters allow us to see how different stars evolve, but they also allow us to determine the ages at which stars change.

Recall that the main sequence in the H-R diagram is really a sequence of stars with different masses, with the low-mass stars at lower right and the most massive stars at upper left. Now we see that the most massive stars evolve the most rapidly; they are the first to leave the main sequence. In the next section we will see why this is so.

Other direct observational evidence of stellar evolution is sometimes found (such as when stars undergo rapid change during transitions from one phase to another), but most evolutionary processes are too slow for us to see. Hence, the study of stars in clusters is our best observational tool for understanding stellar evolution. In the next section, we describe the inner workings of stars, in preparation for the subsequent discussion of their life stories.

Figure 7.3 The Globular Cluster M13. (*U.S. Naval Observatory*)

Stellar Structure

A star is a spherical ball of hot gas. Even though the star's density may be very high in the deep interior, the high temperature keeps the material in a gaseous state throughout. The central density may be more than 100 grams/cm^3 (for comparison, keep in mind that water has a density of 1 gram/cm^3, and ordinary rock has a density near 3 grams/cm^3), and the temperature in the core is often tens of millions of degrees. Under these conditions the gas in the inner regions of a star is completely ionized, consisting entirely of free electrons and

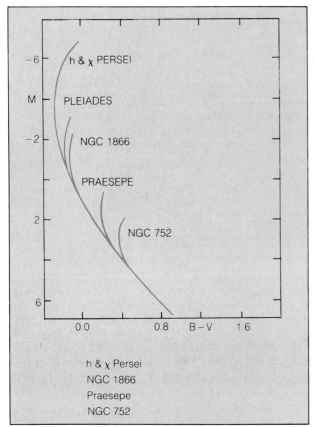

Figure 7.4 H-R Diagrams for Star Clusters. Hear, plotted on the same axes, are H-R diagrams (using color index rather than spectral type) of several clusters. Different symbols are used in the righthand portion to distinguish among the giant stars in different clusters.

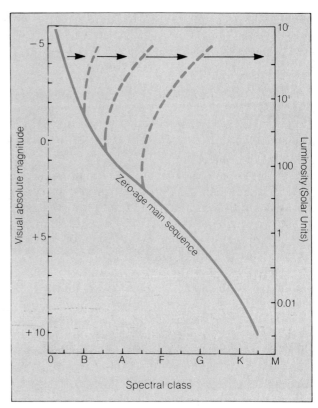

Figure 7.5 The Evolution of the Main Sequence. As a cluster ages, the stars on the upper main sequence move to the right. The farther down the main sequence this has happened, the older the cluster.

atomic nuclei. The outer layers of a star consist of gas that may be only partially ionized, that is, the atoms still retain some of their electrons, but not all of them (recall that varying degrees of ionization result in varying spectral types). The visible surface of a star is called the **photosphere**; the spectral lines form in this layer as light from the interior passes through on its way out into space. Below the photosphere, the temperature and density increase steadily toward the center of the star.

The Role of Stellar Mass

Our knowledge of the interior of a star comes from using equations to represent the relationships among various physical parameters such as temperature, density, and pressure and then solving the equations. One of the most important relationships that we include in such calculations is the balance between the inward force of gravity and the outward force of pressure. We know that these two forces must balance everywhere inside a star, because if they did not, the star would either collapse (if gravity dominated) or expand (if pressure were stronger). The balance between the two is called **hydrostatic equilibrium** (Fig. 7.6).

The assumption of hydrostatic equilibrium alone tells us a great deal about interior conditions. For example, we know that the inward gravity force must become stronger and stronger as we go deeper and deeper into the interior, because the farther in we go, the greater the amount of mass above. In order for pressure to balance gravity at all points in a star, the pressure must therefore increase as we go in. This in turn means that the density and temperature rise as we go in, because high pressure means that the gas is even more compressed, and compression causes a gas to get hot. Even the size of a star is determined by hydrostatic equilibrium, because the volume the star fills is the result of how much compression there is. Since the mass of a star is the factor that governs its inward gravitational force, it is mass that determines the hydrostatic equilibrium and therefore determines all the other properties of a star (Fig. 7.7).

Thus we see that mass is the single most important property a star has. This explains why the main sequence on an H-R diagram is really a sequence of stars of different masses. The stars on the main sequence may be identical in composition, but they have widely different temperatures, diameters, and luminosities because they have different masses. In a cluster of stars, then, the differences among the stars must be due to differences in mass. The fact that the stars on the upper main sequence evolve and change most rapidly tells us that the rate of change of a star depends on its mass. The more massive it is, the more rapidly it evolves.

The Source of Energy: Nuclear Reactions

We already learned (in Chapter 5) that nuclear fusion reactions in which hydrogen is converted into helium are the energy source for the Sun. These reactions can take place only in the inner core of the Sun, where the temperature is high enough for the nuclei to collide with sufficient speed to overcome their repulsive electrical forces. The energy source of other stars is similar,

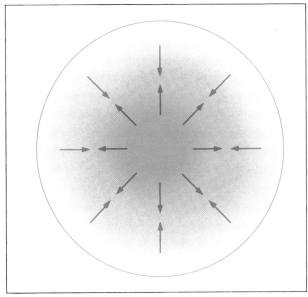

Figure 7.6 Hydrostatic Equilibrium. This cutaway sketch of a star shows its spherical shape. The arrows represent the balanced inward forces of gravity and outward pressure, and the shading indicates that the density increases greatly toward the center. Equilibrium is reached when the core becomes sufficiently hot to attain the pressure necessary to counterbalance gravity.

although in some, the reactions that occur are not the same as the ones that take place in the Sun (see Appendix 11 for details of the reactions discussed here). The reaction that produces the Sun's energy is the proton-proton chain, a simple sequence involving the merging of hydrogen nuclei (protons) as helium nuclei are built up. This reaction produces most of the energy for stars

Figure 7.7 The Importance of Mass. A star's mass is the single quantity that governs all its other properties for a given composition. The sequence shown here is the same in general, but the details vary for different chemical compositions.

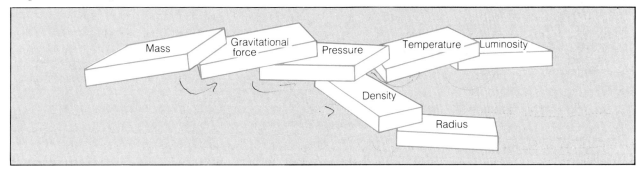

Astronomical Insight 7.1

Stellar Models and Reality

In this chapter we discuss the physical processes that are believed to govern the internal conditions and external properties of stars. We will find that, despite the remoteness of stars and the fact that all of our observational information is derived only through the analysis of light emitted from their surfaces, we know quite a bit about their insides. It may seem surprising that we could have any confidence at all in our theories of stellar interiors and atmospheres, but the fact that we do derives from our faith that the laws of physics apply to distant stars just as they do on earth. Given that, stars are actually rather simple objects compared to a solid planet or a biological system.

Stars are made entirely of gas, and this is the major reason for the simplicity of describing them theoretically. Gases obey certain well-defined laws, which are verified through experiment and theory. Add the fact that stars are basically symmetric and stable (not changing rapidly with time), and we can invoke additional physical laws in the form of equations for the forces acting on gas particles in a star and for the production and transport of energy. Within the context of the adopted physical laws, it can be shown that there is only one solution to the set of equations that describe a star's structure and properties. This is a very powerful statement, for it implies that all we need to do is be sure to include the correct laws of physics in our calculations, and we will get the correct solution.

Aside from the fact that our known laws of physics may be incomplete, the major difficulty in calculating theoretical models of stars lies in the fact that some of the processes occurring in stars are quite complex when examined in sufficient detail. For example, it is easy to treat the transport of energy from the core to the surface of a star when the energy is carried solely in the form of radiation, but it is very difficult when the energy is transported by flowing streams of heated gas in the process we call convection. The calculations are simple when a star is spherically symmetrical, but they become more complex when the star is rotating so rapidly that its shape becomes somewhat flattened. Magnetic fields in stars are probably very complex, if our Sun is any example, and theory has not yet fully treated even the Sun's magnetic field. These are cases where we know the laws of physics well enough, we believe, but their application is made complex by the vast amount of detail that must be incorporated into our calculations.

Despite the complications, stellar model calculations are quite good. We know this because it is possible to achieve very good agreement between observed quantities (such as stellar luminosity, radius, surface temperature, and others) and those calculated from theory. The first test of any theoretical model for a star is to see whether the model successfully reproduces the observed surface conditions. If it does we are confident that the model must also represent the interior fairly well.

Today's stellar structure theorists are busily refining their models, not so much because they are incorrect as because it is becoming increasingly possible to add details to the calculations. Just as new telescope technology increases the ability of astronomers to observe astronomical objects, new advances in computer technology increase the capability of theorists to represent reality in ever greater detail. The first stellar models were calculated with slide rules or slow, mechanical desk calculators, and it took weeks or months to calculate a single model. In the past two or three decades, model complexity has increased in proportion to computer capacity and speed, and it is now possible to calculate many models (for different stages in the evolution of a star, for example) in hours. Today we are entering a new generation of "supercomputers"—machines with vastly greater speed and capacity for massive calculations— and no one will benefit from this more than the stellar structure theorist. As the capabilities of the computers increase, so will the ability of astrophysicists to represent reality in their calculations.

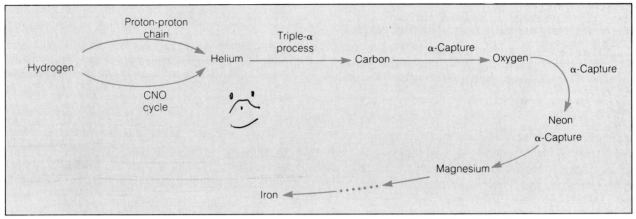

Figure 7.8 The Enrichment of Heavy Elements. This diagram schematically illustrates the sequence of element formation that occurs in a stellar core as one nuclear fuel after another is exhausted. The first step, the conversion of hydrogen into helium, occurs in all stars, but the number of subsequent steps that a star goes through depends on its mass. The heaviest element that can be formed by stable reactions inside stars is iron.

on the lower main sequence, but the higher temperatures inside upper main sequence stars allow another reaction, called the **CNO cycle,** to be more efficient. In this cycle the net result is the same as that of the proton-proton chain, but the steps along the way are different. Carbon is involved as a **catalyst,** that is, it helps the reactions occur but is not used up in the process, and both nitrogen and oxygen appear as intermediate products. The CNO cycle, like the proton-proton chain, converts four hydrogen nuclei into a helium nucleus, releasing energy in the process as some of the initial mass is transformed.

All stars on the main sequence produce energy by converting hydrogen to helium or, as astronomers often say, by "burning" hydrogen. A star departs from the main sequence when the hydrogen fuel is used up (as discussed later in this chapter). There can be subsequent reaction stages in a star's lifetime, however, depending on whether the core of the star gets hot enough (Fig. 7.8). For example, helium nuclei can merge to form carbon nuclei in a sequence called the **triple-alpha reaction** (because helium nuclei are called alpha particles, and three of them merge to form a carbon nucleus). After carbon has been formed, there is a series of **alpha-capture reactions** in which helium nuclei are added to form new elements, each one made heavier than the previous one by the addition of the two protons and two neutrons that make up a helium nucleus. Each reaction stage requires a higher temperature than the one before, and each produces more energy than the one before, so that each uses up

its fuel more rapidly. As a result, the longest-lived phase of a star's life is its main sequence, or hydrogen-burning, phase.

We can estimate any star's main sequence lifetime in the same manner we used for the Sun in Chapter 5, by estimating how long it will take to use up the available hydrogen fuel (Fig. 7.9). Recall that we assumed that 10 percent of the Sun's mass would undergo reactions, and that 0.007 of this would be converted into energy according to the equation $E = mc^2$. Solving this

Figure 7.9 Stellar Lifetimes. The large bucket on the left represents the large quantity of energy produced by a massive star over its lifetime; the water gushing out of the hole at its bottom depicts its high luminosity, or rate of energy loss. The small bucket at right represents a low-mass star, which produces much less energy over its lifetime, but which loses it so slowly (that is, it has such a low luminosity) that is outlives the more massive star by a wide margin.

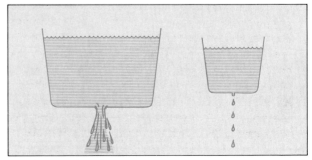

Table 7.2 Main Sequence Lifetimes

Spectral Type	Hydrogen-Burning Lifetime
O5V	5×10^5 yrs
B0V	9×10^5
B5V	8×10^7
A0V	4×10^8
A5V	1×10^9
F0V	3×10^9
F5V	5×10^9
G0V	9×10^9
G5V	1×10^{10}
K0V	2×10^{10}
K5V	4×10^{10}
M0V	2×10^{11}
M5V	3×10^{11}

equation gives an estimate of the total energy available to the Sun from hydrogen burning over its lifetime; to find the length of time the Sun will be able to keep shining, we divide the total available energy by the rate of energy loss, the luminosity. For the Sun we found a lifetime of 10 billion years.

The lifetimes of other stars can be estimated easily by comparison with the Sun. We need only consider the mass relative to that of the Sun, because this tells us how much fuel the star has, and the luminosity of the star, because this tells us how rapidly the available energy is lost. We find that more massive stars have shorter lifetimes, because their luminosities are very much greater than that of the Sun while their masses may be only a few times greater. For example, a star of 20 times the Sun's mass has a luminosity roughly 10,000 times that of the Sun, so its lifetime is (20/10,000) = 0.002 times as great (or about 2 million years). A low-mass star, on the other hand, has such a low luminosity that its lifetime is far greater than that of the Sun (Table 7.2).

The energy produced by nuclear reactions is in the form of heat (that is, motion of gas particles) and electromagnetic radiation (gamma rays). This energy is transported outward from the core by two mechanisms: radiation and convection. Radiation is a slow process, because the photons of light are continually absorbed and re-emitted along the way, each time in a random direction. It has been estimated that a photon of light produced in the core of the Sun takes as long

Figure 7.10 Energy Transport Inside of Stars. In a star on the upper portion of the main sequence, energy is transported by convection in the inner zone and radiation in the outer regions. The opposite is true of lower main sequence stars.

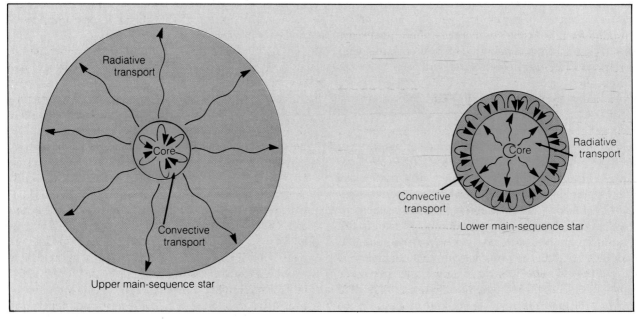

as a million years to make its way to the surface. In the process, it gradually loses energy, starting as a gamma ray but emerging from the photosphere as a photon of visible light (the lost energy goes into heating the gas).

The second form of energy transport, convection, occurs when the temperature drops rapidly going outward toward the star's surface. Where this happens, the gas begins to overturn and hot gas rises. Most stars have both convection and radiative energy transport, but the locations where they occur are different for upper main sequence stars than for those on the lower main sequence (Fig. 7.10). In the upper main sequence stars it is thought that convection occurs in the core but that radiative transport is dominant in the outer layers, whereas in the Sun and other lower main sequence stars, the inner regions are radiative while the outer levels undergo convection.

Now that we know something about how energy is produced and transported inside stars, we are ready to move on to a discussion of other processes that affect the evolution of stars. Among the most important are those that can alter the mass of a star.

Mass Loss and Mass Exchange

We have seen that the mass of a star is the primary factor in determining not only its other properties but also the rate at which the star runs out of nuclear fuel and evolves. We have tacitly assumed that a star's mass does not change during its lifetime. This is a fair assumption for most stars, but there are important exceptions. Some stars can either gain or lose mass, and these processes affect how they evolve.

Mass loss occurs in stars that are very luminous, and is aided by their luminosity. The hot stars in the upper left portion of the H-R diagram are observed to have gas moving outward at speeds of more than a thousand km/sec (Fig. 7.11). The apparent driving mechanism for these **stellar winds** is a pressure force that is exerted by photons of light. When a photon is absorbed by a gas particle, it gives the particle a tiny push. Ordinarily this **radiation pressure** has no noticeable effect, but in extremely luminous stars, it can be sufficient to literally blow away the outer layers. The result can be a loss of mass that is quite significant; a star that begins with a mass of perhaps twenty times the Sun's mass may end up with only half of that. This process influences how the star evolves, because as we have seen, the rate of evolution depends on the mass of the star.

The very luminous cool stars, the red giants and supergiants, also lose mass at significant rates (Fig. 7.11). These stars are so large that their surface gravities are very low and the outermost layers are only loosely attached. Therefore only a weak force is required to cast these layers free, but because the photons of light from these stars are mostly in the red and infrared, where they have relatively little energy, it is not certain that radiation pressure acting on gas particles alone can cause the gas to flow outward. Instead, it is thought that tiny solid particles form by condensation in the outer layers of these stars, and that radiation pressure acts more effectively on these particles, expelling the outer layers. As we will learn in Chapter 9, interstellar space is permeated by tiny interstellar dust grains, many of which form in the outer layers of red giants. Infrared observations reveal the dust that forms in the outer layers of red giants and supergiants, and these observations tell us that large quantities of mass are lost by these stars. Like the stellar winds of the hot stars, the winds from these cool stars have very important effects on their evolution.

Cool stars, like the Sun, have chromospheres and coronae (Fig. 7.12). It is thought that these superheated regions are given their high temperatures as a result of the transfer of energy from the convection zone at the stellar surface to higher levels. The red giants and supergiants do not seem to have sufficiently high temperatures to have coronae, but they have chromospheres, while main sequence stars have both.

Stars can gain, rather than lose, mass under certain circumstances. A star can gain mass if it is in a binary system where the partner star is losing mass (Fig. 7.13). This process is important only in very close binaries, where the two stars are separated by only a few times their diameters at most (some are so close that the two stars actually touch). In such a system, one star can gain mass from another if the other star has a stellar wind or if it becomes so large that it literally spills over onto its companion. There is a point in space between the two stars where the gravitational attraction to each is equal; if any matter from one star flows outward past that point, it will be attracted toward the other star. Usually the orbital motions cause the matter to spiral inward instead of falling directly onto the other star, forming a disk of material called an **accretion disk** around the star that is receiving matter. The

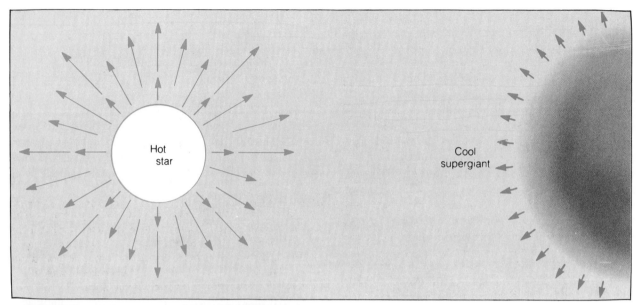

Figure 7.11 Stellar Winds. In the drawing at top a luminous hot star (left) ejects gas at a very high velocity; the lengths of the arrows indicate that the gas accelerates as it moves away from the star. A luminous cool star (especially a K or M supergiant) is so extended in size that the surface gravity is very low, and material drifts away at relatively low speeds. In both types of stars, radiation pressure probably helps accelerate the gas outward. At bottom is a photograph of a hot star with an extensive bubble of gas around it, created by a stellar wind. *(Photo: © 1984 Anglo-Australian Observatory)*

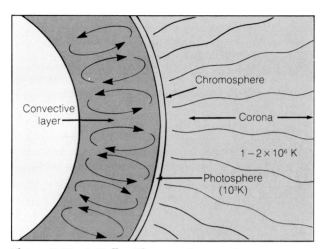

Figure 7.12 A Stellar Chromosphere and Corona. Stars in the lower portion of the main sequence all have chromospheres and coronae, analogous to the sun (see Chapter 13 for more details). The photosphere is the region in which the star's continuous spectrum and absorption lines are formed; the chromosphere is a thin, somewhat hotter region just above; and the corona is a very hot, extended region outside of that. The source of heat for the chromosphere and corona is probably related to convective motions in the star's outer layers.

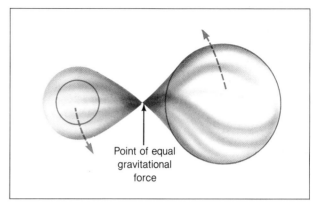

Figure 7.13 Mass Transfer in a Binary System. The more massive star in a binary will be the first to evolve and swell up as a red giant. In the process, its outer layers may come sufficiently close to the companion star to be pulled to it gravitationally. As we will see in later chapters, such a transfer of mass can radically affect both stars.

process of mass exchange in a binary system can completely reverse the relative masses of the two stars: the one that was more massive initially can end up the less massive.

As we describe the life stories of stars later in this chapter, we will see how important mass loss or mass accretion is for some stars. Before that, however, we will discuss observations and theoretical models of the process by which stars form.

Star Formation

Earlier in this chapter we mentioned OB associations, loosely bound groups of O and B stars. We now know that these are very young clusters, probably the youngest in the galaxy, because of the fact that they contain stars on the extreme upper portion of the main sequence. In many cases the stars are moving away from the center and escaping the association, so that within a few million years no concentration of stars will remain.

Stars form from interstellar gas and dust; we even find that young clusters may still be embedded in the remains of the cloud from which they formed (Figs. 7.14–7.16). The volumes between stars in our galaxy are permeated with a very rarefied medium of gas and fine solid particles known as interstellar dust. This material is discussed in Chapter 9; here we note that there are large concentrations known as interstellar clouds, and it is in these concentrations that the density may build up to the point where stars can form. The best way to probe these dark and dusty regions where visible light cannot penetrate is to observe them in infrared wavelengths; infrared radiation penetrates much farther through clouds of gas and dust than does visible light, and the dust around a newborn star is hot and glows at infrared wavelengths. To see infant stars we must look for infrared sources buried in dark clouds (Fig. 7.17).

In our galaxy many regions have been found where star formation is apparently taking place. One of the best observed of these is in the sword of Orion within the Orion nebula, a great, glowing cloud of gas. A very young cluster of stars is located there, and infrared observations reveal a number of infrared sources embedded within the dark cloud associated with the cluster (Fig. 7.18).

In other similar regions we find faint variable stars called **T Tauri** stars. These appear to be newborn stars

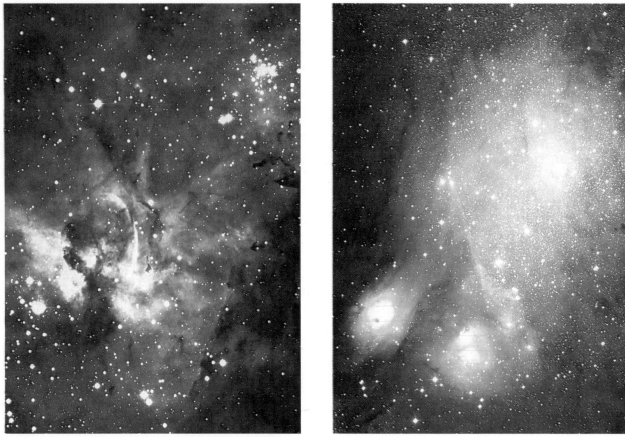

Figure 7.14 Regions of Young Stars and Nebulosity. Here are two regions populated by recently formed stars and their associate interstellar gas and dust. (© 1981, 1984 Anglo-Australian Telescope Board)

Table 7.3 Steps in Star Formation (1 Solar Mass)

1. *Gravitational collapse of interstellar cloud:* Collapse proceeds most rapidly at the center, where the density grows most quickly. The cloud core remains cool because heat escapes in the form of infrared radiation.

2. *Slow contraction of core:* Once the core density is high enough so that the cloud becomes opaque to infrared radiation, heat is trapped, and the temperature and pressure in the core rise. The central portion of the cloud contracts very slowly for a while.

3. *Infall:* Material from the outer portions of the cloud continues to fall in, creating heat and shock waves. When this happens, the hydrogen molecules in the core begin to break up due to the additional heating. The breakup of molecules takes up heat from the gas, reducing the pressure, so that a new collapse phase occurs.

4. *Final contraction:* After the second collapse, the internal pressure again rises to balance gravity, and the central condensation enters a new phase of very slow contraction. At this point, the protostar is still surrounded by gas and dust and is usually observable only as an infrared source.

5. *Nuclear ignition:* The protostar continues to heat gradually as the contraction goes on. Eventually, the core becomes hot enough to start nuclear fusion reactions, and the contraction stops because there is an internal source of energy to provide pressure that balances gravity fully. The star is now on the main sequence. Observationally, it may still be associated with gas and dust and may still undergo a T Tauri phase.

Figure 7.15 A Dark Cloud. This is the well-recognized Horsehead nebula. Many stars in the process of formation have been detected by infrared technique within the dark and dusty regions seen here. (© *1984 Anglo-Australian Telescope Board*)

Figure 7.16 Traces of A Newborn Star. The wispy, glowing tail here is a streak of gas that is being ionized by a beam of high-velocity gas ejected from a young star. The star itself is hidden from view by the dark cloud. (*B. J. Bok*)

in the process of shedding the excess gas and dust from which they formed; in fact, spectroscopic measurements indicate that some material may still be falling into these stars, as though the collapse of the cloud from which they are forming is not yet complete.

A picture has emerged of how stars form from the assorted observations described here (Table 7.3). This picture is based in part on theoretical calculations as well as on observations.

The process begins with the gravitational collapse of an interstellar cloud (Fig. 7.19). It is not certain what causes a cloud to begin to fall in on itself, but once it starts, gravity takes care of the rest.

Calculations show that the innermost portion collapses most quickly, while the outer parts of the cloud are still slowly picking up speed. The temperature in the core builds up as the density increases, but for a long time the heat escapes as the dust grains in the interior radiate infrared light. Eventually, however, the cloud becomes opaque in the center, and radiation can no longer escape directly into space. After this the heat builds up much more rapidly, and the resulting pressure causes the collapse to slow to a very gradual shrinking. Material still falling in crashes into the dense core, creating a violent shock front at its surface. The luminosity of the core, which by this time may have a temperature of a thousand degrees or more, depends on its mass. If the mass is large—that is, several times the mass of the Sun—the luminosity may be sufficient to blow away the remaining material by radia-

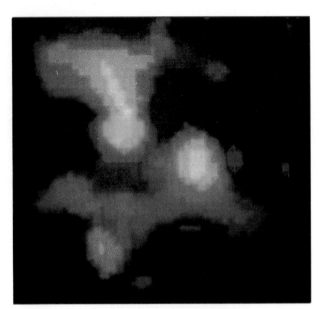

Figure 7.17 An Embedded Infrared Source. In this infrared photograph of a region within the Orion nebula, blue indicates the location of an intense, relatively warm source that is thought to be a newborn star. *(R. D. Gehrz, J. Hackwell, and G. Grasdalen, University of Wyoming)*

tion pressure. For a lower-mass **protostar,** as the central dense object is now called, material continues to fall in. Eventually the density of this material becomes low enough that the protostar can be seen through it. Regardless of mass, a point is reached where a starlike object becomes visible to an outside observer; this stage may be identified with the **T Tauri stars,** unstable young stars that eject material in strong winds (discussed in Chapter 15). If the obscuring matter is swept away violently, the object may appear to brighten up dramatically in a very short time. There have been a few cases where a previously rather faint star brightened suddenly by several magnitudes.

In any case, the protostar continues to shrink slowly, growing hotter in its core. The shrinking stops when a sufficiently high temperature is reached to permit nuclear reactions to begin in the center. This is a major landmark in the process of forming a star; once the reactions have started, the star is on its way to becoming stable (no further shrinking), and it lies on the main sequence where it will spend most of its life converting hydrogen into helium in its core.

It is convenient to trace the star's path on the H-R diagram as it forms (Fig. 7.20). When it is a cold, dark

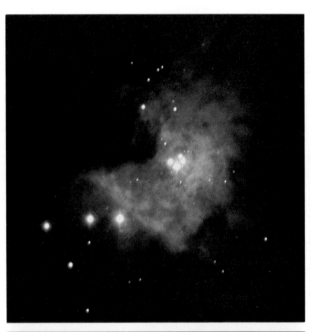

Figure 7.18 The Orion Nebula. At top is a normal, long-exposure photograph showing the familiar nebula. Below is an infrared image of the same region. Here little of the glowing gas can be seen, but the young, hot stars that are embedded within the outer portions are clearly visible. Note that the Trapezium, a close group of four brilliant young stars, is visible in both images. *(top: U.S. Naval Observatory; bottom: © 1984 Anglo-Australian Telescope Board)*

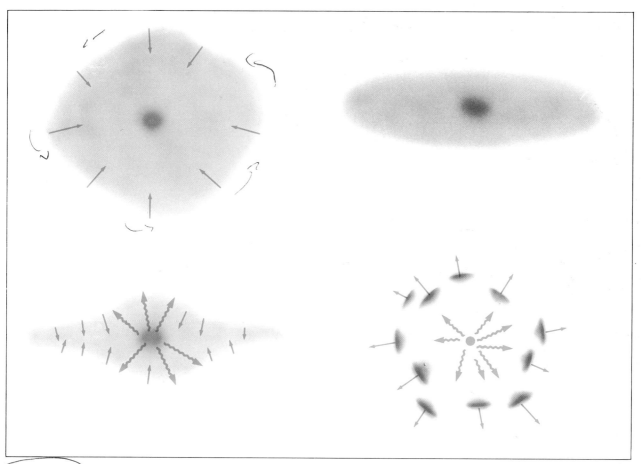

Figure 7.19 Steps in Star Formation. A rotating interstellar cloud collapses, most rapidly at the center. At first infrared radiation escapes easily and carries away heat, but eventually the central condensation becomes sufficiently dense that this radiation is unable to escape and the core becomes hot. After a lengthy period of slow contraction, nuclear reactions begin in the protostar. At some stage, in a process not well understood, the young star develops a strong wind that helps clear away remaining debris.

cloud, its position is far to the right and down, well off the scales of temperature and luminosity appropriate for stars. As the cloud heats up in the core and begins to glow in infrared wavelengths, it moves up and to the left, eventually attaining sufficient luminosity to fall onto the standard H-R diagram, but still off to the right of the main sequence. When it reaches the phase of slow contraction, it moves gradually to the left and down, finally reaching the main sequence when the reactions begin.

The entire process, from the beginning of cloud collapse until the newly formed star is on the main sequence, takes many millions or even billions of years for the least massive stars but only a few hundred thousand years for the most massive stars. In a cluster with stars of various masses, the most massive stars may form, live, and die before the least massive ones even reach the main sequence. The H-R diagram for such a cluster shows stars off to the right of the main sequence at the lower end (Fig. 7.21).

Life Stories of Stars

We are now in a position to combine our knowledge of basic stellar properties, stellar structure, star forma-

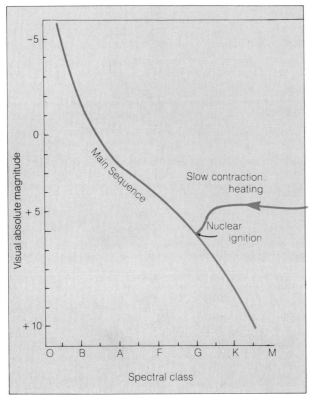

Figure 7.20 The Path of a Newly Forming Star on the H-R Diagram. As a protostar heats up, it eventually becomes hot and luminous enough to appear in the extreme lower right-hand corner of the H-R diagram. When the core becomes opaque, the rapid collapse slows to a gradual shrinking, and the protostar moves across to the left as it heats. Eventually nuclear reactions begin, and the star is on the main sequence.

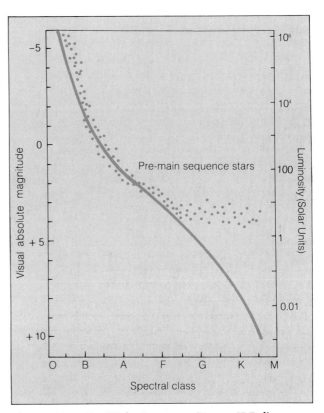

Figure 7.21 Pre-Main Sequence Stars. H-R diagrams of very young clusters often show stars that have not yet reached the main sequence but are still moving toward it. At the same time, some of the massive, short-lived stars at the top of the main sequence may have already evolved away from it, having used up their hydrogen fuel.

tion, and observations and theory of how stars change as they age. The result is a comprehensive picture of the evolution of stars.

The Evolution of Stars Like the Sun

Since so much of a star's evolutionary behavior depends on its mass, we will discuss separately examples of different masses, beginning with a star of the sun's mass (Table 7.4). In each case, it will prove convenient to refer to a star's location on the H-R diagram and how this location changes as the star evolves. (Fig. 7.22 shows the evolutionary path of a star like the Sun.)

Such a star, when it arrives on the main sequence, is similar to the Sun in all its properties. Thus it is a G2 star with a surface temperature around 6000 K and a luminosity of about 10^{33} ergs per second. Its composition initially is nearly 80 percent hydrogen by mass and more than 20 percent helium, with only traces of the other elements. It has convection in the outer layers and a chromosphere and corona.

In the core, the proton-proton chain converts hydrogen into helium. As we saw in Chapter 5, this process lasts some 10 billion years for a star of one solar mass. During this time, the core gradually shrinks as the hydrogen nuclei are replaced by a smaller number of the heavier helium nuclei, and the internal temperature rises. This causes an increase in luminosity, so that the

Table 7.4 Evolution of a 1-$M_\odot$ Star

1. *Hydrogen burning:* During the 10^{10} years the star spends on the main sequence, hydrogen is converted to helium through the proton-proton chain. The luminosity gradually increases as the core becomes denser and hotter.

2. *Development of degenerate core and evolution to red giant:* When hydrogen in the core is used up, the core contracts until it is degenerate. A hydrogen shell source outside the core still undergoes reactions, causing the outer layers to expand and cool. The star becomes a red giant.

3. *Helium flash:* The core eventually gets hot enough to ignite the triple-alpha reaction, in which helium is converted to carbon. Because the core is degenerate and cannot expand and cool to counteract the new source of heat, the reaction rapidly takes place throughout the core. So much heat energy is added so quickly that degeneracy is destroyed, and the reactions become stable.

4. *Stable helium burning:* Once reactions are taking place in the core again, the major energy production is there, and the shell source becomes less important. The outer layers contract, and the surface heats up; the star moves back to the left on the H-R diagram.

5. *Second red giant stage:* When the core helium is used up, the energy again comes from a shell (an inner shell where helium is converted to carbon, and an outer shell where hydrogen is still converted to helium). The outer layers expand again, and the star becomes a red giant for the second time.

6. *Mass loss:* Through a stellar wind and pulsational instabilities, the star loses its outer layers, exposing the hot core. The core continues shrinking but is never again hot enough for nuclear reactions. The star becomes a planetary nebula periodically as the outer layers are shed.

7. *White dwarf:* The core again becomes degenerate as it shrinks. Eventually the outer layers are gone altogether, leaving only the hot, degenerate core, which is a white dwarf.

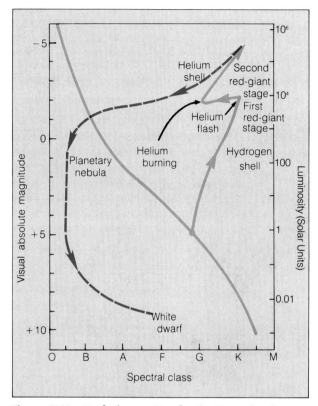

Figure 7.22 Evolutionary Track of a Star Like the Sun. This H-R diagram illustrates the path a star of one solar mass is thought to follow as it completes its evolution. The dashed portion, following the second red giant stage, is less certain than the earlier stages.

star gradually moves up in the H-R diagram. Thus the main sequence is not a perfectly narrow strip but has some breadth because stars on it slowly move upward as their luminosities increase. The starting point, the lower edge of the main sequence, is called the **zero-age main sequence (ZAMS;** Fig. 7.23) because this is where newly formed stars are found.

When the hydrogen in the core is gone, reactions there will cease. By this time, however, the temperature in the zone just outside has nearly reached the point where reactions can take place there, and with a little more shrinking and heating of the core the proton-proton chain begins again in a spherical shell surrounding the core (Fig. 7.24). At this point the star has an inert helium core, and it is producing all of its energy in the hydrogen-burning shell, which steadily moves outward inside the star, eating its way through the available hydrogen.

The core continues to shrink and heat. This heating enhances the nuclear reactions in the shell, causing them to produce more and more energy. Because this source of energy is closer to the surface of the star than the core, the structure of the star is altered. Excess energy in the outer layers of the star, created by the presence of the energy source outside the core, causes the layers to expand. As they do so, the surface cools, and the star moves to the right on the H-R diagram. It also

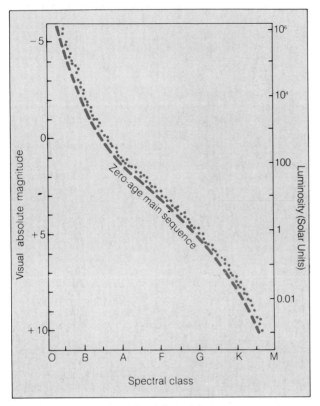

Figure 7.23 The Width of the Main Sequence. As stars on the main sequence gradually convert hydrogen into helium in their interiors, the cores shrink a little, and become hotter. This increases the luminosity, and the stars gradually move upward on the H-R diagram. As a result, the main sequence, which consists of stars of a variety of ages, is not a narrow strip, but has some breadth.

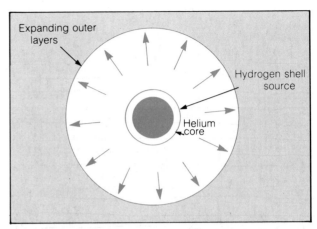

Figure 7.24 The Hydrogen Shell Source. After the hydrogen in the core of a star is completely used up, the core, now composed of helium, continues to shrink and heat. This causes a layer of gas outside the core to reach the temperature required for nuclear reactions. A shell source, in which hydrogen is converted into helium, is ignited. The star's outer layers expand and cool, and the star becomes a red giant.

moves upward (Fig. 7.25) because its luminosity increases (due both to the enhanced energy production rate in the shell and to the fact that convection starts in the outer layers (Fig. 7.26) increasing the rate at which luminous energy reaches the star's surface and is emitted into space). The star becomes a red giant, reaching a size of 10 to 100 times its main-sequence radius.

The helium core becomes extremely dense. At first the temperature is not high enough for any new nuclear reactions to start there (the triple-alpha reaction, in which helium nuclei combine to form carbon, requires a temperature of about 100 million K). As the core continues to shrink, the matter there takes on a very strange form. A new kind of pressure gradually takes over from the ordinary gas pressure that has been supporting the core. The new pressure is created by a property of electrons that prevents them from being squeezed too close together. The gas in the stellar core contains many free electrons, and eventually their resistance to being compressed becomes the dominant pressure that supports the core against further collapse. When this happens, the gas is said to be **degenerate**.

A degenerate gas has many unusual properties. One of them is that the pressure no longer depends on the temperature, and vice versa. If the gas is heated further, it will not expand to compensate as an ordinary gas would. This has important consequences if nuclear reactions should start in the degenerate region of a star.

We now have a red giant star, with highly expanded outer layers. Near the center is a spherical shell in which hydrogen is burning in nuclear reactions, and inside this is a degenerate core containing helium nuclei. The core is small, but it may contain as much as a third of the star's total mass.

During the red giant phase, the helium core continues to be heated by the reactions going on around it. Eventually the temperature becomes sufficiently high for the triple-alpha reaction to begin. When it does, there are spectacular consequences. Ordinarily a gas expands when it is heated, thus limiting how hot it can get (because an expanding gas tends to cool). In a de-

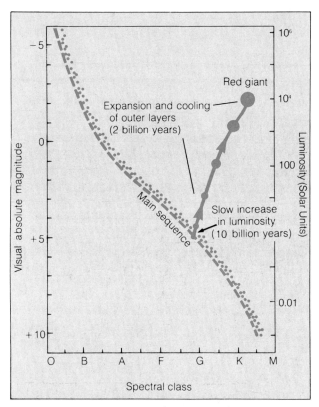

Figure 7.25 The Development of a Red Giant. As a star's outer layers expand because of the shell hydrogen source, they cool. At the same time, the surface area increases, raising the star's luminosity. The star therefore moves up and to the right on the H-R diagram.

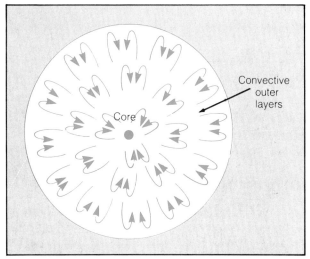

Figure 7.26 The Internal Structure of a Red Giant. The stellar core, which actually contains most of the mass of the star, is very small compared to the huge extent of the outer layers. Most of the volume of the star consists of relatively rarefied gas, constantly overturning because of convection.

generate gas, however, no such expansion occurs, because pressure does not depend on temperature in the usual way. When the reactions begin in the degenerate core of a red giant, the temperature goes up quickly, while the core retains the same density and pressure. The increased temperature speeds up the reactions, producing more heat, which accelerates the reactions even further. There is a rapid runaway effect, and in an instant (literally seconds) reactions occur throughout a large fraction of the core. This spontaneous runaway reaction is called a **helium flash**.

The helium flash has dramatic consequences for the star's interior, but calculations show that there would be little or no immediate effect visible to the outside observer. The overall direction of the evolution is changed, however, and this would be apparent after thousands of years.

The helium flash adds enough heat to the core to destroy its degeneracy. The core then returns to a more normal state where temperature and pressure are linked and the triple-alpha reaction continues in a more stable, steady fashion. The structure of the star returns to a more normal state, with the primary energy source at the core. The outer layers of the star begin to retract and heat, and the star moves back to the left on the H-R diagram.

Old clusters, particularly globular clusters, have a number of stars in this stage of evolution, and they form a sequence called the **horizontal branch** (Fig. 7.27). This is seen only in clusters old enough for stars as small as one solar mass to have evolved this far; as we saw in the previous chapters, this requires an age of 10 billion years or more.

In due course the helium in the core becomes exhausted, leaving an inner core of carbon. Helium still burns in a shell around the core; it is even possible to have a situation where there is an active hydrogen-burning shell farther out in the star at the same time. The star expands and cools, becoming a red giant for the second time. This second red giant stage is short-lived, however, lasting perhaps one million years. A degenerate core again develops, but there will not be another dramatic flare-up in its interior. In fact, no fur-

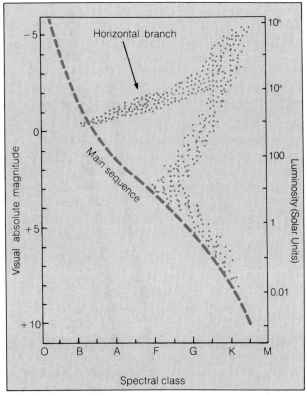

Figure 7.27 The H-R Diagram for a Globular Cluster. These star clusters are very old, and all the upper main sequence stars have evolved to later stages. Here we see a number of stars in or approaching the red giant stage, and a number on the horizontal branch, where they move following the helium flash. Only very old clusters have a horizontal branch. (Besides their great ages, globular clusters also have a relatively low content of heavy elements, and this affects how the stars evolve. This point is discussed more fully in Chapter 10.)

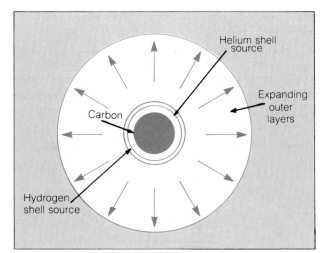

Figure 7.28 A Star Late in its Lifetime. Here the core has completed both the hydrogen-burning and helium-burning stages and is composed of carbon. Outside the core, there may still be an active hydrogen shell source, as well as a helium-burning shell, and the star is becoming a red giant for the second time.

ther nuclear reactions occur in the core of this low-mass star because the temperature never reaches the extremely high levels needed to cause heavy elements like carbon to react.

A star in its second red giant stage may be viewed as having two distinct zones (Fig. 7.28): the relatively dense interior, consisting of a carbon core inside a helium-burning shell surrounded by a hydrogen-burning shell, and the very extended, diffuse outer layers. The inner portion may contain up to 70 percent of the mass but occupies only a small fraction of the star's volume.

During the red giant phases, particularly the second one, the star loses mass by means of a stellar wind, as described in this chapter. The amount of mass lost can be significant, although it probably would not affect the evolution of a star that begins with the mass of the Sun. As we will see in the next section, however, this mass-losing phase can have very important consequences for stars that begin with more mass than the Sun.

As the nuclear fuel in the interior runs out, the core shrinks. The details of what happens to the outer layers at this time are not clear, but we know that some material is lost through the stellar wind. Apparently this mass loss occurs in isolated episodes for some stars, so that shells of gas are ejected periodically. The star gradually gets rid of its outer layers.

The evidence that this occurs is primarily observational, since a successful theory of how it happens has not yet been developed. Astronomers have found a number of objects that seem to consist of a hot, compact star surrounded by a shell of expanding gas (Fig. 7.29). These shells often glow with a bluish green color due to their emission lines, and through a telescope or on a photograph they bear some resemblance to the planets Uranus and Neptune. Because of this resemblance, they are called **planetary nebulae**. Some, such as the Dumbbell nebula or the Ring nebula, are well-known objects visible with a small telescope.

The star that remains in the center of a planetary nebula lies far to the left in the H-R diagram. Its surface temperature may be as high as 100,000 K or more, far hotter than any main sequence star. Evidently, in the process of ejecting the shell, the star's surface temperature increases as hot inner gas becomes exposed on the surface. In essence such a star is the naked stellar core left behind when the outer layers were cast off.

Some nuclear reactions may still be going on in a shell inside the star, but they do not last much longer because the fuel is depleted. The density, already very high, increases as the stellar remnant condenses under the force of gravity. A larger and larger fraction of the star's interior becomes degenerate. As the star shrinks, it stays hot, but its luminosity decreases due to its diminishing surface area, and it moves downward on the H-R diagram. Eventually the shrinking stops and it becomes a stable star again, but now it is a very small object, so dim that it lies well below the main sequence in the lower left-hand corner of the H-R diagram. It is now called a **white dwarf**.

A white dwarf is a bizarre object in many ways. It is made of degenerate matter whose peculiar properties govern its internal structure. It has a mass as great as that of the Sun (some are even slightly more massive), yet it is approximately the size of the Earth. This means that its density is incredibly high, roughly a million times that of water. A cubic centimeter of white dwarf material would weigh a ton at the surface of the Earth! The properties of white dwarfs and the means by which they are observed are described in the next chapter.

The Middleweights

Let us now consider a star of significantly greater mass than the Sun, say five to ten solar masses (Table 7.5). For this star, much of the evolution is similar to that of the one solar-mass star, although there are some major differences. One of the contrasts, of course, is the timescale, because this star will have perhaps 100 times the solar luminosity and accordingly will use up its fuel faster. Let us escort this star through its development, starting again on the main sequence (Fig. 7.30).

Our middleweight star lands on the middle to upper portion of the zero-age main sequence when it has completed its formation and nuclear reactions begin. Its spectral type is B8 if its mass is five solar masses,

Figure 7.29 Planetary Nebulae. At top is the Dumbbell Nebula; at bottom, the Helix Nebula. In both, the blue-green color is produced by emission lines of twice-ionized oxygen. The central stars, which are responsible for ejecting the gas of the nebulae, are visible at the center of each. (top: *Lick Observatory photograph*; bottom: © *1979 Anglo-Australian Telescope Board*)

and B2 if it is ten solar masses. The nuclear reaction in the core is the CNO cycle, producing helium from hydrogen. There is no convection in the outer layers, but there is in the core. The surface temperature is roughly 12,000 K to 20,000 K, depending on the mass.

Like a star of one solar mass, the more massive star becomes gradually more luminous while on the main

Table 7.5 Evolution of a 5-$M_\odot$ Star

1. *Hydrogen burning:* While on the main sequence, the star converts hydrogen to helium in its core by the CNO cycle. The time on the main sequence is much shorter than for a 1-$M_\odot$ star, about 100 million years instead of 10 billion.

2. *First red giant stage:* When the core hydrogen is gone, a shell source powers the star, causing the outer layers to expand and cool as the star becomes a red giant. The core does not become degenerate because of its high temperature, and there is no helium flash.

3. *Subsequent red giant stages:* Each time a nuclear fuel source in the core is used up, the shell source outside the core becomes the dominant source of energy, and the star becomes a red giant. The core then contracts and heats until a new fuel ignites, and the star shrinks again, moving back to the left on the H-R diagram. A star may undergo several red giant loops in this way, progressively burning helium, carbon, oxygen, neon, and other elements in the sequence of alpha-capture reactions.

4. *Mass loss:* During each red giant or supergiant phase, mass is lost through stellar winds. Eventually the loss of mass and the requirement for even higher temperatures to start new reaction phases combine to halt all nuclear reactions.

5. *Final stage:* The form of remnant left depends on how much mass remains after all the red giant stages. If the final mass is under 1.4 $M_\odot$, the star will end as a white dwarf whose composition is determined by the number of nuclear reaction stages that took place. If the final mass is between 1.4 and 2 or 3 $M_\odot$, then the remnant will be a neutron star. If the final mass is over the neutron star limit, a black hole results. It is thought that most stars with initial masses near 5 $M_\odot$ lose enough mass to become either white dwarfs or neutron stars.

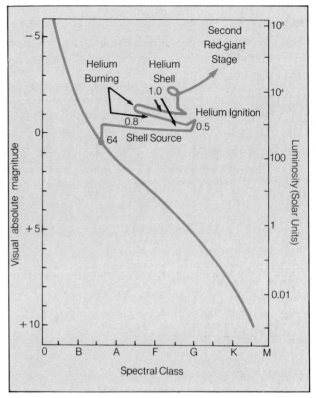

Figure 7.30. The Evolutionary Path of a Five-Solar-Mass Star. This star goes through a more complicated series of stages than a less massive star. There are probably additional red giant stages following those indicated here, but calculations of the evolution have not yet been carried that far. The numbers indicate times in units of 10^6 years.

sequence, as the core becomes more compressed. The star moves upward a little; this occurs on all parts of the main sequence, broadening the entire strip on the H-R diagram. In the case of the middleweight star, the hydrogen in the core is gone (after about 100 million years) before a shell outside the core is ignited, and for a brief period the star simply contracts and heats, with no reactions taking place. During this time, it turns sharply to the left in the H-R diagram. After only 1 or 2 million years, however, hydrogen begins burning again in a shell outside the core, and the star reverses itself and heads for the red giant region. Within a few hundred thousand years, it reaches the upper right-hand portion of the H-R diagram. In this case the core is not degenerate, because the temperature is high enough to prevent a collapse to the necessary density, so no helium flash occurs. When the core is hot enough, the triple-alpha reaction begins quietly, and the star contracts and moves to the left in the H-R diagram.

This star does not move far to the left, however, and never really leaves the red giant region before it develops a new reaction shell outside the core, and expands and cools once again, reversing itself and heading back to the right.

What happens next is not clear, but a sketchy picture is beginning to emerge. The core temperature continues to increase, and it is possible for new nuclear reactions to begin. These reactions, called **alpha-capture reactions,** start with carbon and helium nuclei.

(Remember that a helium nucleus, consisting of two protons and two neutrons, is called an alpha particle.) A carbon nucleus can capture an alpha particle and form an oxygen nucleus, which in turn can capture another alpha particle to form neon, and so on. If the core temperature reaches high enough levels, a complex sequence of reactions can occur, gradually building up a supply of heavy elements in the star's core. Each time one fuel source is exhausted and shell burning begins, the star expands and moves upward and to the right in the H-R diagram; each time a new reaction starts in the core, the star settles down and moves to the left. Thus, a massive star can execute a number of left-to-right loops in the red giant region of the H-R diagram. The number of loops depends on the mass, but the situation is very complicated, and theoretical calculations do not give definitive answers.

One thing that complicates matters is that extremely luminous red giants lose mass through strong, dense winds, as described in this chapter. The rate of mass loss can be sufficient to alter the mass of the star significantly in a few hundred thousand years. Reduction of the mass of a star will curtail its evolution. The central density and temperature will be reduced compared to what they would have been, and this in turn reduces the number of reaction stages that the star will go through. At some difficult-to-determine point all reactions cease. Then the red giant will contract and move down and to the left in the H-R diagram, possibly passing through a phase as a planetary nebula.

Whether these middleweight stars end up as white dwarfs is uncertain. It has been shown theoretically, and confirmed from observations, that a white dwarf cannot exist if the star's mass is greater than a certain limit (about 1.4 times the mass of the Sun). A mass above this limit creates sufficient gravitational force to overcome the pressure of the degenerate electron gas in the core, and the star collapses catastrophically. As we shall see, this violent ending certainly occurs in the most massive stars; in the case of a star of five to ten solar masses, however, a white dwarf may form if enough mass is lost to bring the star below the limit. The stellar wind and planetary nebula ejection processes might reduce the mass far enough, so it is possible that many of the stars that start out in our middleweight division end up as white dwarfs.

Even so, there are contrasts with the white dwarfs formed from the less massive stars. Recall that for a star of one solar mass, the most advanced nuclear reaction that occurs is the triple-alpha process, which converts helium into carbon. The star of five to ten solar masses goes well beyond this point, however, so heavier elements are built up in its core. The resulting white dwarf may have an interior made of oxygen, neon, or even heavier elements, while white dwarfs produced by lower-mass stars consist of carbon or perhaps only helium.

What of the relatively massive stars that do not shed enough material to become white dwarfs? These stars contract beyond the white dwarf stage and therefore exceed even the stupendous densities achieved by the white dwarfs. There seem to be two possible end results: The star may contract to another kind of degenerate state where it can stabilize, or it may collapse indefinitely and never again reach stability. In either case, the collapse may be violent, with the infalling matter creating an outburst that we recognize as a supernova explosion.

In the first of these two possibilities, the collapse of the interior is stopped when a new kind of degenerate pressure is created. As the collapse occurs, the protons and electrons are forced so close together that a certain kind of nuclear reaction takes place, one in which neutrons are formed. Neutrons, like electrons, can reach a state where they cannot be squeezed any further; this creates a degenerate neutron gas pressure that may be sufficient to balance the force of gravity and create a kind of super white dwarf called a **neutron star.** Like a white dwarf, a neutron star has a strict limit on its mass (two or three solar masses). If the stellar core has more mass than this, the second possibility mentioned above occurs: The collapse never stops. We will return to this case in the next section, where very massive stars are discussed.

The neutron star has even more fantastic properties than a white dwarf. Its diameter is about 10 kilometers, yet it contains more than 1.4 solar masses, and its density far exceeds even that of the white dwarf. A cubic centimeter of neutron star material would weigh a billion tons at the earth's surface!

Observations have confirmed that neutron stars can form during supernova explosions, because there have been cases where a neutron star was found in the center of the expanding cloud of debris left over from such an event. One of these apparently formed in historical times, when the supernova that created the present-day Crab Nebula occurred. (This was the "guest star" observed by Chinese astronomers in 1054.)

Neutron stars are exceedingly dim objects; they are very difficult to see directly and were for a long time thought to be unobservable. The ways in which they have been observed will be described in the next chapter. For now, we have finished our life story of the middleweight star, and we are ready to see how the most massive stars evolve.

The Heavyweights

The most massive stars form in basically the same manner as their less massive relatives, but they do so more quickly. All phases of their evolution (Fig. 7.31) are speeded up in comparison to the less massive stars. The massive stars in a cluster can go through the entire evolution from birth to death before the less massive ones even reach the main sequence.

Stars containing twenty or more solar masses start their main sequence lifetimes as O stars. They have surface temperatures of 30,000 K or more and luminosities of 10^5 to 10^6 times that of the Sun. The dominant nuclear reaction during the main sequence phase is the CNO cycle. These stars have some excess surface heating, which forms coronae, but it is not thought likely that convection occurs in the surface layers. There is convection in the cores of these stars, however.

Like all the other stars, these stars undergo gradual core contraction as the hydrogen there is converted into helium. By contrast with the less massive stars, however, the O stars do not move upward in the H-R diagram; they move straight across to the right. This happens because O stars steadily lose matter through high-velocity winds (described in this chapter), so that the tendency to increase their core temperature and luminosity is offset by the decreased mass of the overlying layers.

The details of the later stages of evolution for these massive stars are not at all clear. Most likely several stages of nuclear burning take place as the star goes through numerous loops in the red giant region of the H-R diagram, leading ultimately to the formation of iron in the core. At the same time, the wind may rip away so much of the star's substance that the outer portion of the core is exposed, creating a kind of "peculiar" object called a **Wolf-Rayet star** (Fig. 7.32). These hot stars have very strong winds and excess quantities of carbon (a result of the triple-alpha reaction) or nitrogen and oxygen (products of the CNO cycle) at their surfaces. When the outer layers have been stripped off, we get a direct view of the innards of the star and can see material that has recently undergone nuclear reactions. The convection that is taking place in the core helps bring this processed matter up to the surface.

When iron has been formed in the core of a star, no further nuclear reactions can support the star against the inward force of gravity. Indeed, any further reactions that could occur would help gravity, because reactions involving elements heavier than iron are all **endothermic**, meaning that more energy is required to create the reaction than is produced. Such reactions actually cool the star's interior, thereby reducing the pressure and allowing gravity to dominate further.

Apparently the following sequence of events can occur (Fig. 7.33). When the core is composed of iron, the

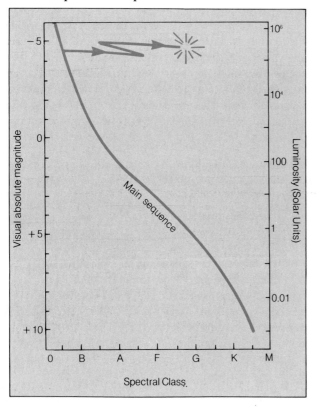

Figure 7.31 Evolution of an O Star. The evolutionary track of a very massive star is quite simple: it moves to the right (not up, because mass loss by a stellar wind counteracts the effects of expansion), briefly moves to the left during helium burning, then goes back to the right. It may go back and forth (rapidly) a few times, but within a total lifetime of about one million years, it exhausts all nuclear fuel and explodes in a supernova.

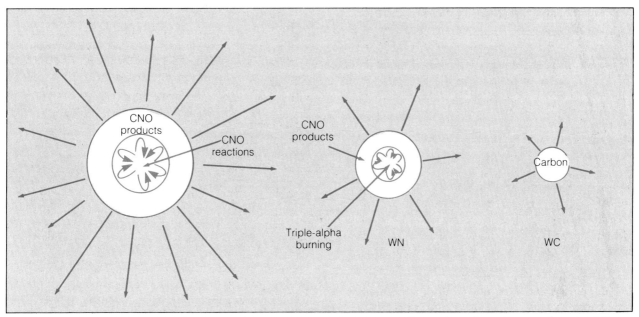

Figure 7.32 Formation of a Wolf-Rayet Star. A massive star loses its outer layers through a stellar wind; meanwhile its core is undergoing CNO cycle reactions, gradually being transformed into helium. When the outer layers have been blown away, the exposed inner region of the star, which contains enhanced quantities of nitrogen and oxygen as by-products of the CNO cycle, is a type of Wolf-Rayet star known as a WN star. If the star has progressed to helium burning by the triple-alpha process, it is a WC star with an overabundance of carbon.

reactions stop, and the star begins to collapse, having no means of support. As the collapse begins, the compression of the core causes endothermic reactions to start, and suddenly there is an effective vacuum at the center, as the heat there goes into the reactions. The star now collapses very quickly, in free-fall, and the star implodes. It is as though the rug were suddenly pulled out from under the outer layers, and they come crashing down into the center of the star. Furthermore, the reactions that occur produce vast quantities of neutrinos, tiny subatomic particles that can escape directly into space (see Chapter 5), carrying away energy. The loss of neutrinos from the core enhances the vacuum effect, making the infall of the outer layers even more

Figure 7.33. A Supernova Explosion. When the last possible reaction stage is completed, the star's outer layers collapse inward because the source of internal pressure is gone. The collapse is so violent that the star "bounces back," and explodes as a supernova. The explosion is actually caused by the outward pressure of neutrinos created in rapid nuclear reactions.

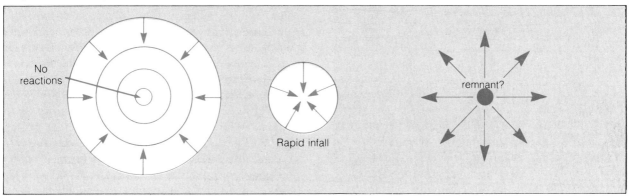

Astronomical Insight 7.2

The Mysteries of Algol*

The bright double star known as Algol (or β Persei) has played a leading role in the discovery of two important astronomical phenomena. Both of these phenomena—the existence of binary stars and the process of mass transfer in close binaries—went unrecognized for some time; they seemed too radical for most astronomers to accept.

The first report that Algol's brightness varies came as long ago as 1667, when the Italian astronomer Geminiano Montanari noted that the star occasionally appeared fainter than normal. At that time a few variable stars were known, but no explanation for them had been found. It was not until 1782 that anyone rediscovered the variability of Algol; in that year the English astronomer John Goodricke found that the light variations were periodic, with a brief decrease in brightness occurring every 69 hours. Goodricke made the then novel suggestion that the dimmings could be due to the presence of a companion star that orbited Algol, eclipsing it each time it passed in front. At that time there was controversy over earlier suggestions that some stars appeared to be double, and the prevailing view (including that of the eminent English astronomer William Herschel) was one of skepticism.

In independent studies of Goodricke's idea, the Swiss astronomer Daniel Huber and an obscure English clergyman named William Sewall soon analyzed the light variations of Algol to deduce crudely the properties of the suspected companion. The writings of both men, published originally in 1787 and 1791, were ignored by their contemporaries and then lost, not to be rediscovered until well into the twentieth century. Meanwhile the notion that stars could be double gradually became accepted, as additional Algol-type stars were found, and observations of visual binaries became common. Interestingly, it was William Herschel who conclusively proved the existence of double stars in 1803 when he was able actually to measure orbital motion in a visual binary system. In 1889 spectroscopic observations of Algol itself finally revealed the orbital motion of the bright star about its companion, establishing that stars can be double even though they present a single point image when viewed in a telescope or photographed.

Much more recently, Algol played a similar leading role in

---

sudden. This causes a massive shock, and the material bounces back violently in a supernova explosion. A good deal of matter from the star is dispersed into space as a result of the explosion.

For a short time during the supernova explosion, densities and temperatures reach incredible levels, sufficient to cause the formation of very heavy elements. Most of the atoms in the universe that are heavier than iron were created in supernova explosions. Of course some or all of the intermediate weight elements that were created in the star's interior during its more placid stages of evolution are also dispersed into space by the explosion, so the process enriches the interstellar gas with a wide spectrum of elements.

What remains of the original star? Apparently in many cases nothing at all is left, except a cloud of cosmic debris. In other cases, a fraction of the star's final mass remains in the form of a highly condensed remnant. This may be a neutron star, if the remnant mass is below the limit of two or three solar masses.

If too much matter is contained in this remnant, however, a neutron star cannot form, because the force of gravity is too strong. The star simply continues to collapse. Nothing can ever stop it. The resulting object, first envisioned mathematically nearly seventy years ago, is called a **black hole**, and today there is observational evidence that such things exist.

Although a black hole literally cannot be seen, there

Astronomical Insight 7.2

The Mysteries of Algol

the discovery of mass transfer in binary systems. During the twentieth century, Algol has been widely observed, using photometric and spectroscopic techniques, and a paradox was revealed as an understanding of stellar evolution arose. Analysis of the data on Algol using Kepler's third law and the shape of the light curve (a graph showing the brightness of the star versus time) showed that the visible star is a B main sequence star having a mass of 3.7 times that of the sun, whereas the fainter star is a G giant star having a mass of only 0.81 times the sun's mass. From observations of cluster H-R diagrams and stellar evolution theory, it was known that more massive stars evolve faster than less massive ones. Thus the fact that the *less* massive member of the Algol system was already a red giant while the *more* massive star was still on the main sequence presented a major mystery, which became known as the Algol paradox.

The solution to this puzzle was found only in the past twenty years or so, as ultraviolet observations have revealed massive stellar winds from some kinds of stars, and other evidence (mostly from telescopes in space) for mass transfer between stars in close binary systems has been found. Evidently the G star in the Algol system originally contained more mass, and was once an upper main sequence star, while its companion started out on the lower main sequence. As the first star, being more massive, became a red giant, it swelled up to such a large size that matter began to flow across to the companion. The first star became less massive and fainter than the companion, which became hotter and brighter as its mass grew. Today the original situation is reversed, with the star that started as the less massive one now being the more massive. In the future another reversal may occur, as the B star becomes a red giant and swells up, spilling mass back onto its companion.

The existence of binary stars and the process of mass transfer in close binaries were both profound discoveries in astronomy. It is ironic that hints of both were provided by Algol, but in both cases the evidence was found long before astronomers were ready to understand and accept the implications.

*Historical material for this article is largely from Ashbrook, J. 1984. *The Astronomical Scrapbook* (Cambridge, England: Cambridge University Press & Sky Publishing), p. 347.

---

are indirect means by which its presence may be detected. These means, and the information about black holes that can be obtained from observations, are described in some detail in the next chapter. For now, we return to the question of the evolution of massive stars.

It seems almost certain that all stars starting out with masses in excess of ten solar masses or so end their lives in supernova explosions. The race between mass loss and the nuclear reactions in the core determines how many nuclear reaction stages the star undergoes, as well as how much mass it will have left when all the possible reactions are finished. The outcome of the race is difficult to call theoretically. Hence it is difficult to predict exactly which stars will end up as neutron stars and which will form black holes. Some may simply explode, leaving no remnant behind.

Evolution in Binary Systems

It was pointed out earlier in this chapter that some special effects may occur in certain binary systems. If the two stars are sufficiently far apart, say several astronomical units or more, then most likely each evolves on its own, as though the companion were not there. If, however, they are very close together, significant interactions can drastically alter the normal course of events (Fig. 7.34).

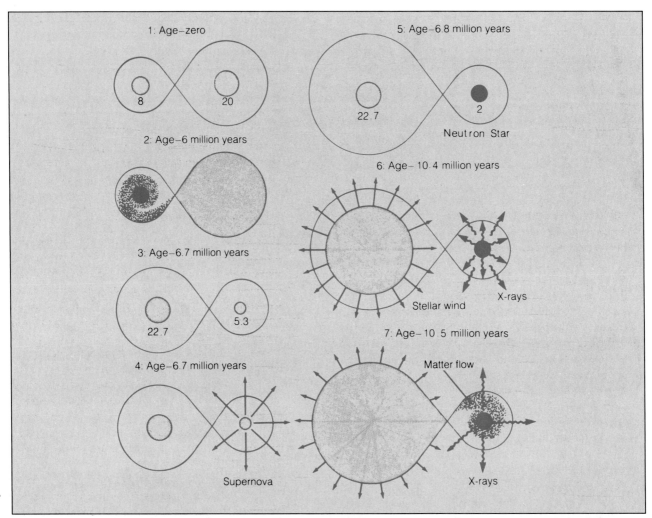

Figure 7.34 Mass Exchange and Evolution in a Close Binary System. The star that is initially more massive evolves faster, and as a red giant it deposits material on the companion. This speeds up the evolution of the second star, so that it expands and returns matter to the first star. If the first one is by this time a compact object, such as a neutron star, the consequences of this stage of mass transfer can be violent (see Chapter 8).

If either star is massive enough to have a strong wind, matter can be transferred to the other star. Even if no strong wind is present initially, as soon as either star reaches the red giant stage, it can swell up sufficiently to dump material onto the other through the point between them where their gravitational forces are equal.

Regardless of how it happens, once mass begins to transfer from one star to the other, the evolution of one or both is modified. The one that gains mass speeds up its evolution, while the one that loses mass may slow down. This process was discovered because certain binary systems were found in which the *less* massive star was the one that had evolved the farthest, quite contrary to what is expected. The most famous such system is the eclipsing binary known as Algol, or β Persei. This binary system consists of a red giant and a B main sequence star. The B star is more massive but is still on the main sequence, whereas the red giant has already left the main sequence. The "Algol paradox" was solved only when it was realized that the red giant must have been the more massive star initially but later lost mass to its companion.

The most spectacular observational effects in mass-transfer binary systems occur when one star has gone all the way through its evolution and become a white dwarf, neutron star, or black hole and then gains new material from the companion star. Special effects, often including X-ray emission, are observed, and it is under these circumstances that stellar remnants are most easily detected. This will be discussed at some length in the next chapter.

Perspective

We have now seen how stars are born, live their lives, and die, and we know how the sequence of events depends on mass. The least massive stars, those with much less than a solar mass, have been little discussed simply because they evolve so slowly that none has yet had time, since the universe began, to complete its main sequence lifetime. The evolution of these stars has had little effect on the rest of the galaxy, while the much rarer heavy-weight stars are profoundly important because their rapid evolution has enriched the galaxy with heavy elements.

Before we finish discussing stars and turn our attention to larger-scale objects in the universe, we must consider the properties of the stellar remnants left behind as stars die.

Summary

1. Stars evolve as they age. We learn how they evolve from observations of stars in different stages; it is especially helpful to observe stars in clusters, because the stars in a cluster all have the same age and are the same distance from us.
2. From cluster observations we know that stars become red giants or supergiants after leaving the main sequence and that the most massive stars do so the most quickly.
3. The internal structure of a star is governed largely by the balance that must exist everywhere between gravity and pressure; this balance is called hydrostatic equilibrium. The mass of a star dictates what its other properties are, because mass governs the inward gravitational force that must be balanced in hydrostatic equilibrium.
4. The energy in stars is produced by nuclear fusion reactions. Stars on the main sequence produce energy by reactions that convert hydrogen into helium.
5. The length of the hydrogen-burning stage of a star is shortest for the most massive stars and longest for the least massive stars.
6. Stars may lose mass through stellar winds or gain mass through exchange in close binary systems, in either case affecting their evolutions.
7. A star with the mass of the Sun spends about 10^{10} years on the main sequence. When the core hydrogen is gone, the reactions there stop, but they continue in a spherical shell outside the core. The outer layers expand and cool, and the star becomes a red giant.
8. The core of the red giant continues to shrink until it becomes degenerate. It continues to heat up, and when it eventually becomes hot enough, a helium flash occurs, after which the star is powered by the triple-alpha reaction, converting helium to carbon.
9. Following the helium-burning phase, the star may undergo instabilities that cause the ejection of one or more planetary nebulae. The star eventually becomes a white dwarf, with a degenerate interior.
10. A more massive star, in the range of five to ten solar masses, evolves more quickly than the lower-mass star, goes through multiple red giant stages, and ends up either as a white dwarf or a neutron star, depending on how much mass it loses during its evolution.
11. The most massive stars evolve very quickly, going through many nuclear reaction stages before exploding as supernovae, leaving neutron star remnants or becoming black holes.
12. The transfer of mass between stars in close binaries can speed up the evolution of the star that gains mass and slow the evolution of the one that loses it, so that the more evolved star can be the less massive.

Review Questions

1. Explain why observations of star clusters are very important in determining how stars evolve.
2. Suppose two stars in the same cluster have identical apparent magnitudes and spectral types. How do their other properties such as mass, diameter, luminosity, and surface temperature compare? Could you reach the same conclusion for a pair of

stars that are not in the same cluster but which also have identical apparent magnitudes?
3. Explain how the main sequence turnoff is used to estimate the age of a cluster of stars.
4. How do we know that a star must be in hydrostatic equilibrium? What would happen if it were not? Can you think of cases where stars are not in hydrostatic equilibrium?
5. Why do nuclear reactions in a star take place only in the core?
6. Why is star formation best observed in infrared wavelengths?
7. Explain why degeneracy causes the core of a star like the Sun to undergo a helium flash after the hydrogen is used up.
8. Summarize the contrasts between the evolution of a star of one solar mass and the evolution of a star ten times more massive.
9. Discuss the role of stellar winds in the evolution of massive stars.
10. From what you have learned about nuclear reaction stages in stars, can you explain why iron is relatively abundant in the universe?

Additional Readings

Balick, Bruce. 1987. The shaping of planetary nebulae. *Sky and Telescope* 73(2):125.

Bethe, Hans A. 1985. How a supernova explodes, *Scientific American* 252(5):60.

Bok, B. J. 1981. The early phases of star formation. *Sky and Telescope* 227(2):48.

Cohen, M. 1975. Star formation and early evolution. *Mercury* 4(5):10.

Flannery, B. P. 1977. Stellar evolution in double stars. *American Scientist* 65:737.

Herbst, W., and G. E. Assousa. 1979. Supernovas and star formation. *Scientific American* 241(2):138.

Kaler, J. 1981. Planetary nebulae and stellar evolution. *Mercury* 10(4):114.

Lada, C. J. 1982. Energetic outflows from young stars. *Scientific American* 247(1):82.

Loren, R. B., and F. J. Vrba. 1979. Starmaking with colliding molecular clouds. *Sky and Telescope* 57(6):521.

Murdin, Paul, and Lesley Murdin. 1985. *Supernovae*. Cambridge, Mass.: Cambridge University Press.

Schaefer, Bradley E. 1985. Gamma-ray bursters. *Scientific American* 252(2):52.

Seeds, M. A. 1979. Stellar evolution. *Astronomy* 7(2):6.

Seward, Fredrick. 1986. Neutron stars in supernova remnants. *Sky and Telescope* 71(1):6.

Shklovskii, I. S. 1978. *Stars: Their birth, life, and death*. San Francisco: W. H. Freeman.

Sweigart, A. V. 1976. The evolution of red giant stars. *Physics Today* 29(1):25.

Werner, M. W., Becklin, E. E., and Neugebauer, G. 1977. Infrared studies of star formation. *Science* 197:723.

Wyckoff, S. 1979. Red giants: The inside scoop. *Mercury* 8(1):7.

Zeilik, M. 1978. The birth of massive stars. *Scientific American* 238(4):110.

Special Insert

The Great Supernova of 1987

One of the most spectacular and scientifically bountiful events in the history of astronomy was first observed on February 23, 1987. A supernova bright enough to be easily seen with the naked eye was discovered in the Large Magellanic Cloud, providing astronomers the first opportunity in nearly four hundred years to study a supernova so bright. Ian Shelton, a Canadian astronomer, working at Canada's southern hemisphere observatory at La Silla, Chile, noticed the new object on a photographic plate as he was developing it. He rushed outside and was able to see with his own eyes what the photograph had told him: The Large Magellanic Cloud contained a new star, at that time just brighter than fifth magnitude. Later that night, other astronomers in New Zealand and Australia independently discovered the supernova (Fig. 1). While some now refer to it as Supernova Shelton, most astronomers use the more mundane name Supernova 1987a, which indicates that this was the first supernova detected in 1987 (typically a dozen or more are found each year in faraway galaxies).

Astronomers around the world were galvanized by the discovery as the news traveled through the International Astronomical Union's telegraph service, and by telephone and computer networks. NASA's *International Ultraviolet Explorer* (*IUE*; see Chapter 4) was making ultraviolet spectroscopic observations within 14 hours, and ground-based observers everywhere in the southern hemisphere trained their telescopes on it. Radio telescopes also observed the supernova within hours of the discovery. Unfortunately, the event occurred at a time when space astronomy was in a lull between major missions (the *Space Telescope*, for example, still sits on the ground because of delays in the space shuttle program), so it was not possible to obtain all the data that would have been desirable. A Japanese X-ray telescope called *Ginga* had just been launched, however, and it was soon able to make X-ray observa-

(continued on next page)

Figure 1 The Region of the Supernova Before and After the Explosion. At left is a photograph made before the supernova occurred; at right is another, with similar exposure time, made afterwards. At this time the supernova was about a fourth-magnitude object, on the way to its eventual peak brightness of magnitude 2.9. The large size of its image compared to normal stars is due to blurring on the photographic emulsion caused by the brilliance of the supernova; the supernova was actually a point source, no bigger than the images of the other stars. (© 1987 Anglo-Australian Telescope Board)

Special Insert

The Great Supernova of 1987

(continued from previous page) tions. The Soviet *MIR* satellite also observed its X rays. NASA moved quickly to mobilize a number of sounding-rocket and high-altitude aircraft and balloon experiments as well, so there will be substantial coverage of the supernova in wavelengths that do not reach the ground. Perhaps the most spectacular observation of all, however, did not concern electromagnetic radiation—it was the detection of neutrinos from the explosion (to be discussed below).

Spectroscopic data quickly showed that the supernova was of Type II, because hydrogen lines were seen in its spectrum. The expansion velocity implied by the Doppler shifts of these lines was nearly 20,000 km/sec. Recall that standard models assume that Type II supernovae result from the core collapse of massive stars that still have normal (that is, hydrogen-rich) gas in their outer layers. As we will see, however, Supernova 1987A has several characteristics that do not fit the standard interpretation. (There is, in fact, a growing consensus that the classification scheme needs revision, because apparently there are more than just two combinations of explosion mechanism and stellar composition.)

Of the many unique opportunities the supernova is providing astronomers, one of the most im-

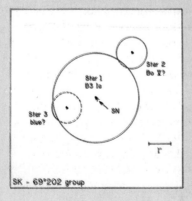

Figure 2 Tracking Down the Progenitor. This photograph and diagram illustrate the relative positions of the three stars seen in pre-supernova photographs to lie near the position of the outburst. The circles represent the uncertainties of the measured positions of each of the three stars. It was eventually determined that Star 1 (Sk −69 202, a B3 I supergiant) is the one that exploded. Note the scale; all three stars lay within just a couple of arcseconds of each other. *(M. Phillips, Cerro Tololo Inter-American Observatory)*

portant is the chance to analyze the star that blew up. Several photographic plates of that region exist in observatory files around the world, so it was possible to examine them to determine what kind of star the progenitor was. This examination led to some initial confusion, because careful study showed that there were no fewer than three stars very close to the position of the explosion (Fig. 2), and for a week or two identification of the star that blew up was uncertain. Finally, a combination of extensive measurements of many plates and ultraviolet data from the *IUE* satellite established beyond reasonable doubt that the progenitor was a twelfth-magnitude blue supergiant (spectral type B3 I), listed in catalogs as Sk −69 202. This discovery was a major surprise for many astronomers, because until then it had been thought that only red supergiants exploded as supernovae of Type II.

The "unusual" nature of the progenitor, once it was established, helped explain some other peculiar features of the explosion. For example, the supernova never reached the luminosity normally expected, and it reached its peak brightness (just brighter than third magnitude) on May 20, some 85 days after its discovery (an unusually long time; supernovae usually reach peak

Special Insert

The Great Supernova of 1987

brightness in 10-20 days). Radio data, on the other hand, showed a rapid decline almost from the beginning, whereas other Type II supernovae typically do not reach their peak radio emission until hundreds of days after the outburst. The *Ginga* and *MIR* satellites initially showed no X rays coming from Supernova 1987A, although this may not be unusual (there is very little information on X-ray emission in the early stages of other supernovae). Now, however, X rays from the supernova have begun to be seen, showing that some of the high-energy radiation from inside the expanding gas cloud is beginning to leak out.

All of these peculiarities fit well with supernova models involving core collapse in a massive star that is not as large as a red supergiant. The small size of the blue (B3 I) supergiant, Sk −69 202, (compared to a red supergiant) accounts for the low initial luminosity, since relatively little surface area is available to radiate light. Another very important measure of the size of the star came from the detection of neutrinos, which escape almost directly from the core as the star collapses. The observed light is emitted later, when the shock front caused by the collapse reaches the star's surface. Thus, the neutrinos reached the Earth before the supernova was visible, but the short (4-hour) delay (equivalent to the time it took for the shock to travel from the star's core to its surface) indicates that the star could not have been as large as a red supergiant. The best fit to all the observations suggests that Sk −69 202 had an initial mass of about 19 solar masses and a diameter roughly 40 times the diameter of the Sun.

As already mentioned, it was a surprise to most astronomers that such a star should explode, although some model calculations had suggested that it could happen. It is still not clear what stage of evolution the star was in: Was it just leaving the main sequence on its way to becoming a red supergiant, or had it already been a red supergiant at least once and then moved back to the left on the H-R diagram as its core contracted and it shed its outer layers in a stellar wind? As yet there is no consensus on the answer to this question, although most supernova theorists seem to be leaning toward the latter point of view. In either case, the explosion itself is thought to have been triggered by the formation of an iron core that then collapsed violently and rebounded, the standard mechanism described in the text. A complicating factor in all these explanations is that the Large Magellanic Cloud has lower overall abundances of heavy elements than "normal" spiral galaxies like our own and most others where Type II supernovae are observed, and lower abundances of heavy elements can alter the evolution of a star. We can be certain that the occurrence of Supernova 1987A will stimulate new efforts to model the evolution of massive stars.

Since Supernova 1987A failed (by a factor of about 15) to reach the magnitude expected for Type II supernova, many astronomers at first assumed that this was a freak event and not any kind of common occurrence. Further analysis showed, however, that many supernovae of comparable brightness could occur in other galaxies without being detected, because they would not stand out so well. Therefore, supernovae like 1987A might actually occur fairly frequently (there have been a small number of other "peculiar" supernovae with some similarities to 1987A).

The detection of neutrinos from Supernova 1987A was a spectacular event for astronomy and for physics. A burst of neutrinos was detected simultaneously by two separate experiments, one in Japan and one in Ohio. This was the first time that extraterrestrial neutrinos had been detected (except for those from the Sun—see Chapter 5), and this event is regarded by some as establishing a new

(continued on next page)

Special Insert

The Great Supernova of 1987

(continued from previous page)
branch of astronomy. As already explained, the timing of the arrival of the neutrinos provided valuable information on the size of the star. However, the neutrinos represent something far more important in terms of understanding the supernovae: The energy the neutrinos carried away from the explosion was greater than the supernova's energy in all other forms by a factor of roughly 100. The visible light emitted over the first several months amounts to some 10^{49} ergs of energy; the kinetic energy of the expanding gas cloud entails perhaps a few times 10^{51} ergs; and the energy released in neutrinos is about 3 or 4 times 10^{53} ergs. These numbers are thought to be typical of Type II supernovae, according to model calculations, but Supernova 1987A provided the first confirmation of the important role of neutrinos. What we see of a supernova, as brilliant as it is, is only a small fraction of the energy produced. Furthermore, even though Supernova 1987A was dimmer than many other supernovae in visible wavelengths of light, its total energy output was comparable to the brightest of supernovae.

Physicists who study elementary particles are as excited as the astronomers by the detection of neutrinos from Supernova 1987A. Detailed analysis of the energy distribution of the neutrinos that were detected, along with the exact timing of their arrival, provides information on their basic properties. As mentioned elsewhere in this text, there has been some debate on the question of neutrino mass. The standard theories of neutrinos say that they are massless, yet controversial experimental data, along with a host of astronomical problems that would be solved, suggest that neutrinos may have a very small mass. The data from the supernova neutrinos are not as complete as would have been desirable (fewer than 20 neutrinos were detected) but the information that was obtained still allows something to be learned about neutrino masses. Preliminary analysis indicates that the mass must be zero or very small, though not necessarily too small to solve the astronomical problems alluded to here (such as the total mass content of the universe and the low observed neutrino emission from the Sun).

The light emitted by the supernova as it continues to shine brightly arises from at least three distinct sources of energy. The most important energy source during the early days of the supernova was heating by shock waves created in the rebound following the initial collapse of the star's core. This heating produced most of the supernova's luminosity at first. A second energy source is radiation from a neutron star, the remnant of the star that exploded. So far there has been no direct confirmation that a neutron star formed, but it seems very likely because neutron star formation is required in the successful models of the explosion and especially because it is the only known way to produce the observed quantity of neutrinos. Another probable source of energy is the decay of radioactive elements produced in the explosion. The most abundant radioactive isotope expected is a form of nickel (^{56}Ni), which quickly decays to a form of radioactive cobalt (^{56}Co) and then more slowly to normal iron (^{56}Fe). Calculations show that an initial mass of about 0.075 solar masses of ^{56}Ni would provide the observed energy from Supernova 1987A. As this isotope decays to iron, the emitted energy decreases in a particular manner (called exponential decay) that should be clearly recognizable in the shape of the supernova's light curve (a graph showing how the brightness changes with time, see Fig. 3). For the first few months the light curve did not follow this shape (probably because other sources of energy dominated), but recent observations show that it is now beginning to do so,

Special Insert

The Great Supernova of 1987

with precisely the rate of decline expected from the decay of ^{56}Co to ^{56}Fe.

The expanding shell of gas from the supernova was so thick for the first several months that there was no chance of seeing into the interior, of actually detecting a neutron star or measuring other properties of the material inside. It is expected that this situation will gradually change as the shell expands and becomes more tenuous. There must be a massive flux of gamma rays and X rays, arising from the radioactive decay of ^{56}Co, trying to make their way out from it, and eventually we may expect this radiation to begin to leak through. Theoretical calculations show that one of the first signs of this thinning will be the detection of X rays, and the very recent detection of X rays by the *Ginga* satellite is therefore an important confirmation of this aspect of the theory. (These are X rays from the radioactive decay; those from the neutron star are not expected to begin escaping for several more months.) Another sign may be the formation of strong ultraviolet emission lines in the outer regions of the expanding cloud, as X rays begin to reach these regions and excite the gas. For this reason, the *IUE* satellite is also monitoring the supernova, and it may have begun to detect some of the expected emission lines.

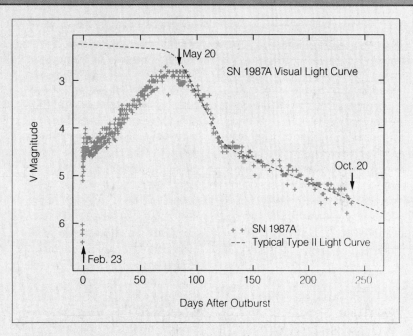

Figure 3 The Light Curve of Supernova 1987A. This graph shows the variation in magnitude of Supernova 1987A since its discovery up to late Fall 1987. The early part of the light curve is very unusual compared to other known supernovae, but once this one reached its peak it began to behave very much like the light curves of other Type II supernovae. The reasons for the difference in the early period is that the star that exploded was smaller than is thought typical, so that it had less surface area to emit light. The later part of the light curve is consistent with energy production by the decay of radioactive nickel (^{56}Ni). *(J. Doggett and R. Fesen, University of Colorado)*

Several gamma ray, X-ray, and ultraviolet telescopes will be launched (by sounding rocket or high-altitude balloon) in Australia in coming months to detect the high-energy emission. Parallel infrared observations are being made in hopes of detecting an "echo" as the light from the supernova explosion reaches outlying interstellar clouds in the vicinity. Such clouds are expected to be present as a result of the earlier mass-losing phases of the progenitor star, particularly if the star had gone through one or more red-supergiant stages. Infrared observations will also provide information on solid dust particles that could form by condensation in supernova outbursts;

(continued on next page)

Special Insert

The Great Supernova of 1987

Figure 4 The Mysterious Glowing Spot. The small spot next to the much larger image of the supernova is the mysterious companion object first detected in late March. (*P. Nisenson, Harvard-Smithsonian Center for Astrophysics*)

(continued from previous page)

such dust permeates interstellar space (see Chapter 9), and it will be helpful to assess how much of it comes from supernovae.

While much of the behavior of Supernova 1987A is now reasonably well understood, there are still some mysteries. The most outstanding of these is the presence of a new companion to the supernova—there is a glowing spot (Fig. 4) displaced by only 0.06 arcseconds from the supernova itself that was not there before the explosion. When it was detected early in the summer of 1987, this mystery spot was very bright (about 10 percent as bright as the supernova itself), but repeat observations later in the summer failed to confirm its presence. At that time the Large Magellanic Cloud was very low on the horizon, so that the observations had to be made through a lot of Earth atmosphere, and this limits the effectiveness of the interferometric measurements originally used to detect the companion (see Chapter 4). A better opportunity to detect it was to occur in November, 1987, and at the time of this writing astronomers were eagerly waiting to see what would be found. It is not known whether the mystery spot is a physical blob of gas that was ejected from the supernova (which would require a speed of nearly forty percent of the speed of light), or whether it is material that was already present in interstellar space and was excited to glow because of radiation (either light or a beam or jet of rapid electrons) from the supernova. Perhaps this mystery will have been solved by the time this text is published.

The occurrence of Supernova 1987A has already had major impact on virtually every field of astronomy. The technical journals are filled with articles on theory or observations of the supernova, and we can expect this to continue for months or years. Over hundreds of years, as the expanding shell evolves into a normal supernova remnant, it is expected to become one of the brightest radio and X-ray sources in the sky, as the ionized gas produces energy by thermal shock and nonthermal processes. In due course, perhaps in months or years, the gaseous remnant may become thin enough to allow astronomers to detect the neutron star, particularly if it is a pulsar (discussed in the next chapter). For decades or centuries the gaseous remnant will be a strong source of visible and ultraviolet emission lines, for centuries it will remain an intense X-ray source, and for millennia it will be a strong radio emitter. Many generations of astronomers will take part in the study of this supernova, and many future astronomy classes will learn about it.

Chapter 8

Stellar Remnants

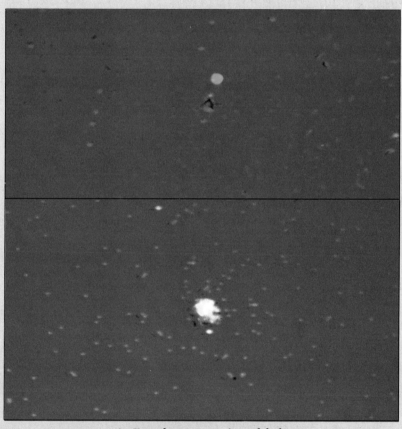

An X-ray burst source in a globular cluster. *(Harvard-Smithsonian Center for Astrophysics)*

Chapter Preview

White Dwarfs, Black Dwarfs
    White Dwarfs, Novae, and Supernovae
Supernova Remnants
Neutron Stars
    Pulsars: Cosmic Clocks
    Neutron Stars in Binary Systems
Black Holes: Gravity's Final Victory
    Do Black Holes Exist?

Our story of stellar evolution is almost complete. We have seen how the properties of stars are measured, and we have learned how they form, live, and die. The only missing link is what is left of a star after its life cycle is completed.

As we saw in the previous chapter, the form of the remnant that remains at the end of a star's life depends on the mass the star had when it finally ran out of nuclear fuel. Three possibilities were mentioned: white dwarf, neutron star, and black hole. Each of these three bizarre objects is discussed in this chapter, with emphasis on the observational properties.

White Dwarfs, Black Dwarfs

Table 8.1 summarizes the properties of a typical white dwarf. An isolated white dwarf is not a very exciting object, from the observational point of view. It is rather dim (Fig. 8.1), and its spectrum is nearly featureless, except for a few very broad absorption lines created by the limited range of chemical elements it contains (Fig. 8.2). The great width of the lines is caused by the immense pressure in the star's outer atmosphere; pressure tends to smear out the atomic energy levels because of collisions between atoms. In effect, the energy levels become broadened, and this broadens the spectral lines when electrons make the transitions between levels.

Additional broadening or shifting of spectral lines may occur because of magnetic effects. If the star had a magnetic field before its contraction to the white dwarf state, that field is not only preserved but also intensified in the process of contraction. Thus a white dwarf may have a very strong magnetic field, and as we have seen previously (Chapters 5 and 6), a magnetic field causes certain spectral lines to be split into two or more parts. In a white dwarf, where the lines are already very broad, this splitting (called the Zeeman effect) usually just makes the lines appear even broader.

The lines in the spectrum of a white dwarf are always shifted toward longer wavelengths from their laboratory positions. This shift is not caused by motion of

Figure 8.1 Sirius B. The dim companion to Sirius was the first white dwarf to be discovered; analysis of its mass, temperature, and luminosity led to the realization that it is very small and dense. *(Palomar Observatory, California Institute of Technology)*

Table 8.1 Properties of a Typical White Dwarf

Mass: 1.0 $M_\odot$
Surface temperature: 10,000 K
Diameter: 0.0015 $D_\odot$ (1.0 $D_\oplus$)
Density: 5×10^5 g/cm^3
Surface gravity: 1.3×10^5 g
Luminosity: 2×10^{31} erg/sec (0.005 $L_\odot$)
Visual absolute magnitude: 11

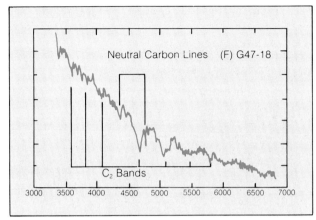

Figure 8.2 A White Dwarf Spectrum. Here we see that white dwarf spectra are without many strong features and that the spectral lines tend to be rather broad. This white dwarf has a large abundance of carbon, as a result of the triple-alpha reaction. *(G. A. Wegner)*

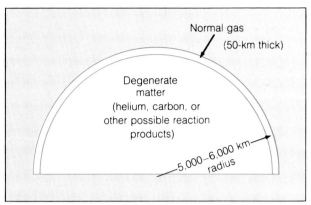

Figure 8.3 The Internal Structure of a White Dwarf. The star's interior is degenerate and would cool very rapidly, except that the outer layer of normal gas acts as a very effective insulator, trapping radiation so that heat escapes only very slowly.

the star; if it were, it would tell us that somehow all the white dwarfs in the sky are receding from us. This redshift is a different type caused by gravitational forces. One of the predictions of Einstein's theory of general relativity is that photons of light should be affected by gravitational fields. White dwarfs, with their immensely strong surface gravities, provide a confirmation of this prediction. In later sections, we will see even more extreme examples of these **gravitational redshifts.**

Left to its own devices, a white dwarf will simply cool off, eventually becoming so cold that it no longer emits visible light. At this point it cannot be seen, and it simply persists as a burnt-out cinder sometimes called a black dwarf.

The cooling process takes a long time, however; several billion years may go by before the star becomes cold and dark. This may seem surprising, since the white dwarf has no source of energy and can only stay hot as long as it retains the heat it contained when it was formed. The outer skin of the white dwarf acts like a very efficient thermal blanket, however, since it consists of gas that is nearly opaque. Radiation cannot penetrate it easily, and because there is no other way of transporting heat away from the interior into space, it takes a long time for the heat to filter out.

The thermal blanket created by the outer atmosphere is very thin; most of the volume of the star is filled with degenerate electron gas (Fig. 8.3), as discussed in the last chapter. The atmosphere, consisting of ordinary gas, may be only about 50 kilometers thick. (Remember that the white dwarf itself is about the size of the Earth, with a radius of some 5000 to 6000 kilometers.) The composition of the interior may be primarily helium if the star never progressed beyond the hydrogen-burning stage; it may be carbon if the triple-alpha reaction was as far as the star got in its nuclear evolution before dying; or it may be some heavier element if the original star was sufficiently massive to have undergone several reaction stages. In any case the mass of the white dwarf must be less than the limit of 1.4 solar masses.

White Dwarfs, Novae, and Supernovae

Ordinarily, a white dwarf cools off and is never heard from again. There are circumstances, however, in which it can be resurrected briefly for another role in the cosmic drama. It has recently become apparent that white dwarfs are central characters in producing spectacular events.

We discussed in the previous chapter the evolution of stars in close binary systems, where mass can transfer from one star to the other. Consider such a system where the star that was originally more massive has gone far enough in its evolution to become a white dwarf. Now the companion begins to transfer matter to

Astronomical Insight 8.1

The Story of Sirius B

Sirius, also known as α Canis Majoris, is the brightest star visible in the heavens. It was discovered in 1844 to be a binary, based on its wobbling motion as it moves through space. (Recall the discussion of astrometric binaries in Chapter 6.) Its period is about fifty years, and the orbital size about 20 AU; analysis of these data using Kepler's third law leads to the conclusion that the companion must have a mass near that of the Sun. At the distance of Sirius (only 2.7 parsecs), a star like the Sun should be easily visible, yet the companion to Sirius defied detection for some time.

Part of the difficulty lay in the overwhelming brightness of Sirius itself, which tended to drown out that of the lesser, very close star. Astronomers were sufficiently intrigued by the mysterious nature of the companion, however, that they made intensive efforts to find it, and finally in 1863 it was seen with an exceptionally fine, newly developed telescope.

It was immediately confirmed that this star was unusually dim for its mass, having an apparent magnitude of 8.7 (the apparent magnitude of the primary star is −1.4) and an absolute magnitude of 11.5, nearly 7 magnitudes fainter than the Sun. Spectra obtained with great difficulty (again due to the brightness of the companion) showed that this object, called Sirius B, has a high temperature similar to that of an O star.

When Henry Norris Russell constructed his first H-R diagram in 1913, he included Sirius B, finding it to lie all by itself in the lower left-hand corner. The only way for the star to have such a low luminosity while its temperature was so high was for it to have a very small radius, and it was deduced at that time that Sirius B was about the size of the Earth. The incredible density was estimated immediately, and it was realized that a totally new form of matter was involved. The term *white dwarf* was coined to describe Sirius B and other stars of this type, which were soon discovered in other binary systems.

It took some time to understand the physics of this new kind of matter, partly because the theory of relativity had to be applied (the electrons in a degenerate gas move at speeds near that of light). The Indian astrophysicist S. Chandrasekhar was the first to solve the problem of the behavior of this kind of gas and by the 1930s was able to construct realistic models of the structure of a white dwarf. From this work came the 1.4-solar-mass limit we have referred to in the text.

Modern observational techniques offer the possibility of detecting white dwarfs more easily. For example, even though Sirius B is about 10 magnitudes (or a factor of 10,000) fainter than Sirius A in visible wavelengths, its greater temperature makes it relatively brighter in the ultraviolet. There is even a wavelength, about 1100 Å, below which the white dwarf is actually brighter than its companion. Hence Sirius B and other white dwarfs have been detected with ultraviolet telescopes. With the development of more sophisticated space instruments, it will become possible to study the properties of many white dwarfs in this way.

the white dwarf (Fig. 8.4), either because the companion has expanded to become a red giant or because it is a hot, luminous star with a rapid stellar wind. The white dwarf may respond to the addition of the new material in a variety of ways, depending on its composition and on the rate at which mass is added. All the possible responses lead sooner or later to an explosive outburst or to many repeated outbursts, because of the degenerate nature of the white dwarf.

If the transferred gas reaches the white dwarf with a high velocity, so that heating occurs due to the energy of impact on the white dwarf's surface, there may be an instant nuclear explosion, enhanced by the fact that the degenerate dwarf material cannot expand and cool when heated. (Recall the discussion of the helium flash in the previous chapter.) In other cases, mass may accrete onto the white dwarf for some time, perhaps decades or centuries, before the new material becomes hot enough to undergo nuclear reactions. The intensity of the outburst depends on how much matter is involved. If enough material is consumed in this explosion, the formerly very dim white dwarf may become more luminous than any ordinary star, reaching an absolute magnitude as bright as -6 to -9 (Fig. 8.5). This is a

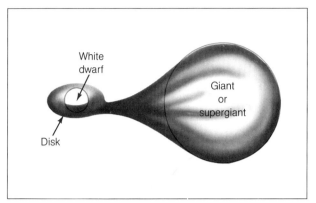

Figure 8.4 Mass Transfer to a White Dwarf. Matter is transferred onto a white dwarf by a red giant companion. Theoretical calculations show that the new material will orbit the white dwarf in a disk but will eventually become unstable and fall down onto the white dwarf. When it does so, nuclear reactions flare up, creating a nova outburst.

Figure 8.5 Nova Cygni 1975. This was one of the brightest novae in recent years. The photograph at left shows the region without the nova; at right is the same region near the time of peak brightness, when the apparent magnitude of the object was near $+1$. *(E.E. Barnard Observatory, photograph by G. Emerson)*

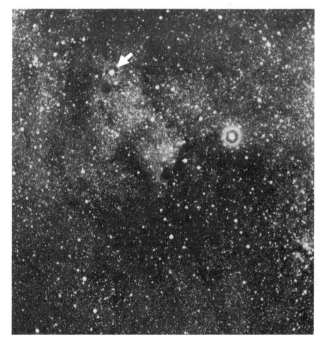

nova. A star that becomes a nova may do so repeatedly, usually with separations of decades between outbursts.

In some binary systems, rather small amounts of matter reach the white dwarf at regular intervals and create relatively minor flare-ups. A general name for such systems is **cataclysmic variables,** and there are a number of specific types. The frequency of the outbursts can be as high as several times a day, but typically it is once every few days.

If material trickles onto the white dwarf at a slow rate, so that not much heating is caused by the impact, the white dwarf can gradually gain mass without any explosion due to the infalling material. In this way a white dwarf can approach the 1.4-solar-mass limit, beyond which degenerate electron gas pressure can no longer support it. As a white dwarf gains mass, it becomes *smaller;* this is another peculiar property of degenerate matter. Thus the white dwarf increases in density. What happens as the white dwarf approaches the mass limit depends on its composition. It may suddenly contract more or less quietly and become either a neutron star or a black hole. If its composition consists of carbon, however, theory shows that nuclear reactions will begin instead, and the degeneracy will cause these reactions to consume much of the white dwarf's mass very rapidly, creating an immense explosion. This is another mechanism for producing a supernova explosion.

In the previous chapter we discussed the supernova caused by the collapse and rebound of a massive star at the end of its nuclear lifetime; now we see that a supernova can also be caused by the addition of mass to a white dwarf. Supernovae created in these two different ways are generally similar in their maximum brightness and the length of time required to become dim again, but they are rather different in many details, such as the types of spectral lines observed and the shape of the **light curve,** a graph showing the brightness variation with time (Fig. 8.6). Supernovae caused by the explosion of massive stars are called **Type II supernovae,** while those produced in the explosion of white dwarfs are **Type I supernovae.**

White dwarfs, then, are most prominent in binary systems where mass transfer from a normal companion takes place. We will see later in this chapter that other forms of stellar remnants are also best observed in these circumstances.

Supernova Remnants

We have learned that stars can explode as supernovae in two different ways, in some cases leaving a stellar remnant (a neutron star or black hole). In all cases there will be another form of matter left over, an expanding gaseous cloud called a **supernova remnant.**

A number of supernova remnants are known to astronomers (Table 8.2). A few, like the prominent Crab Nebula (Fig. 8.7), are detectable in visible light, but many are most easily observed at radio wavelengths (Fig. 8.7), partly because visible light is affected by the interstellar dust, which can conceal a remnant, while radio waves are largely unaffected, and partly because the remnants are intrinsically strong radio emitters. The reason for this is not that they are cold, which would cause their peak thermal emission to occur at long wavelengths, as in the case of the outer planets.

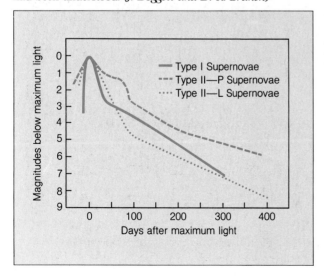

Figure 8.6 Supernova Light Curves. The two types of supernovae are readily distinguished by their spectra and their light curves, which show marked differences. Type I supernovae usually have no hydrogen lines in their spectra and are thought to be the explosions of carbon white dwarfs that have accreted matter. Type II supernovae have hydrogen lines in their spectra and are the explosions that occur in massive stars when all nuclear fuel is gone. As seen here, the peak brightness is slightly different and the rate of decline quite different for the two types, which were distinguished historically before the mechanism for either had been understood. (*J. Doggett and D. R. Branch*)

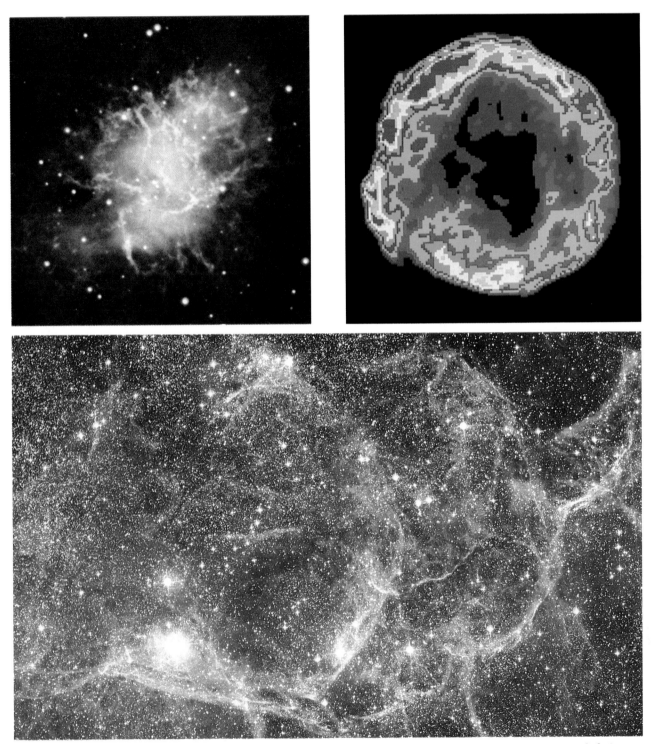

Figure 8.7 Supernova Remnants. At upper left is the Crab Nebula, the remnant of a star that was seen to explode in 1054 A.D. Below is a photograph of part of the Veil Nebula, showing wispy, filamentary structure. At upper right is a radio image of the supernova observed by Tycho Brahe in 1572. (top left: *Lick Observatory photograph*; top right: *National Radio Astronomy Observatory, operated by Associated Universities, Inc., under contract with the National Science Foundation*; bottom: © *1980 Anglo-Australian Telescope Board*)

Table 8.2 Prominent Supernova Remnants

Remnant	Age (yrs)	Distance (pc)
Vela X	10,000	1600
Cygnus loop	20,000	2500
Lupus	975	4000
IC 443	60,000	5000
Crab nebula	932	6500
Puppis A	4000	7200
Cassiopeia A	300	10,000
Tycho's supernova	400	9800
Kepler's supernova	370	20,000

Here a completely different process, called **synchrotron emission,** produces the radio waves.

Synchrotron emission, named for the particle accelerators in which it is produced on Earth, occurs when electrons move rapidly through a magnetic field. The electrons must have speeds near that of light; as they travel through the magnetic field, they are forced to move in a spiraling path, and they emit photons as they do so (Fig. 8.8). The emission occurs over a very broad range of wavelengths, including some visible, ultraviolet, and even X-ray radiation (Fig. 8.9), but these other wavelengths are often not as easily detected as radio, because the radio emission is the strongest.

Supernova remnants that are detected in visible wavelengths usually glow in the light of several strong emission lines, most notably the bright line of atomic hydrogen at a wavelength of 6563 Å, in the red portion of the spectrum. These lines often show large Doppler shifts, indicating that the gas is still moving rapidly as the entire remnant expands outward from the site of the explosion that gave it birth. Often a filamentary structure is seen, suggestive of turbulence but probably modified in shape by magnetic fields.

Remnants are observed today at the locations of several famous historical supernova explosions. The Crab Nebula is the most prominent; it was created in the supernova observed by Chinese astronomers in 1054 A.D. Other supernovae seen by Tycho Brahe (in 1572) and by Kepler and Galileo (1604) also left detectable remnants, although neither is as bright in visible wavelengths as the Crab. Apparently a supernova remnant can persist for 10,000 years or more before becoming too dissipated to be recognizable any longer.

The immense energy of a supernova explosion is comparable to the total amount of radiant energy the Sun will emit over its entire lifetime. Much of this energy takes the form of mass motions as the remnant expands. This kinetic energy heats the expanding gas to very high temperatures. The entire process has a profound effect on the interstellar gas and dust that permeate the galaxy, as we shall see in Chapter 9.

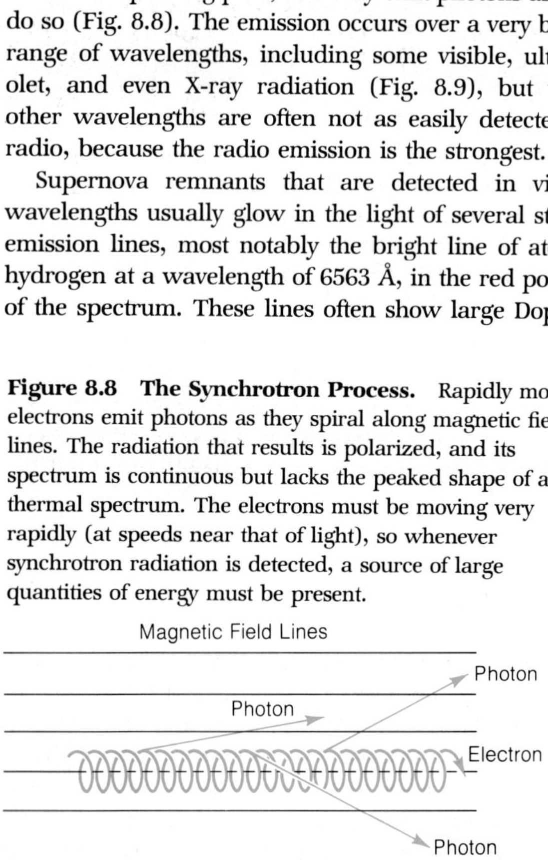

Neutron Stars

We have referred to a neutron star (Table 8.3) as a stellar remnant composed entirely of neutrons that are in a degenerate state similar to that of the electrons in a white dwarf. The pressure created by the degenerate neutron gas is greater than that of a degenerate electron gas, so a star slightly too massive to become a white dwarf can be supported by the neutrons. Again, however, the mass of one of these objects is limited. The limit, which depends on factors such as the rate of rotation, is between 2 and 3 solar masses. Thus the class of stars whose masses at the end of their nuclear

Figure 8.8 The Synchrotron Process. Rapidly moving electrons emit photons as they spiral along magnetic field lines. The radiation that results is polarized, and its spectrum is continuous but lacks the peaked shape of a thermal spectrum. The electrons must be moving very rapidly (at speeds near that of light), so whenever synchrotron radiation is detected, a source of large quantities of energy must be present.

Table 8.3 Properties of a Typical Neutron Star

Mass: 1.5 $M_\odot$
Radius: 10 km
Density: 10^{14} g/cm^3
Surface gravity: 7×10^9 g
Magnetic field: 10^{12} earth's field
Rotation period: 0.001 to 100 sec

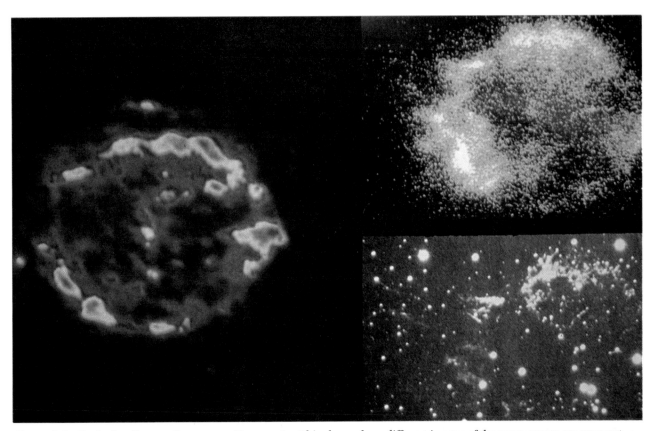

Figure 8.9 Emission from a Supernova Remnant. This shows three different images of the same supernova remnant Cassiopeia A. At left is a radio image, at upper right an X-ray image, and at lower left a visible-light image. The remnant emits at all these wavelengths by the synchrotron process; it is less obvious in visible light because there are many other, brighter sources in the field of view. (left: *National Radio Astronomy Observatory, operated by Associated Universities, Inc., under contract with the National Science Foundation;* top right: *Harvard-Smithsonian Center for Astrophysics;* bottom right: *K. Kamper and S. van den Bergh*)

lifetimes are between 1.4 and about 2 or 3 solar masses become neutron stars.

The structure of a neutron star (Fig. 8.10) is even more extreme than that of a white dwarf. All of the mass is compressed into an even smaller volume (now the radius is about 10 kilometers), and the gravitational field at the surface is immensely strong. The layer of normal gas that constitutes its atmosphere is only centimeters thick, and beneath it may be zones of different chemical composition resulting from previous shell-burning episodes, each zone only a few meters thick at the most. Inside these surface layers is the incredibly dense neutron gas core, which takes the form of a crystalline lattice. The temperature throughout is very high, but because the surface area is so small, a neutron star is very dim indeed. It may have a magnetic field that is also much more intense than that of a white dwarf, created by the compression of the star's original field.

Pulsars: Cosmic Clocks

The properties of neutron stars were predicted theoretically several decades ago, but until 1967 it was not expected that they could be observed because of their low luminosities. In that year, however, an accidental discovery by a radio astronomer established a new class of objects called **pulsars,** which are now believed to be neutron stars. As it turned out, pulsars are only

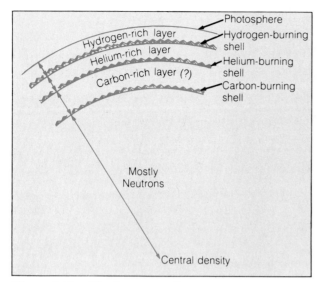

Figure 8.10. The Structure of a Neutron Star. This diagram shows that the outer regions of a neutron star may consist of thin layers of various elements that were produced by nuclear reactions during the star's lifetime. These outer layers are thought to have a rigid crystalline structure because of the intense gravitational field of the neutron star.

one of two distinct ways in which neutron stars can be detected; the other is described in the next section.

Pulsars are radio sources that flash on and off very regularly, and with a high frequency (Fig. 8.11). One of the most rapid of them, the one located in the Crab nebula, repeats itself 30 times a second, and at least a few others are now known to flash even more rapidly than the Crab pulsar (one of them at the incredible frequency of 885 times per second!). The slowest known pulsars have cycle times of more than a minute. In every case, the pulsar is only "on" for a small fraction of each cycle.

Figure 8.11 The Radio Emission from a Pulsar.
Here, plotted on a time scale, is the radio intensity from a pulsar. The radio emission is weak or nonexistent except for a very brief flash once each cycle (in some cases there is a weaker flash between each adjacent pair of strong flashes).

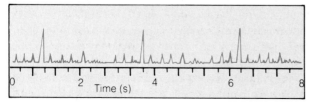

When pulsars were discovered, there was a great deal of excitement and a lot of speculation, including the suggestion that they were beacons operated by an alien civilization. Once the initial shock of discovery wore off, however, a number of more natural explanations were offered by astronomers who sought to establish the identity of the pulsars. It was well known that a variety of stars pulsate regularly, alternately expanding and contracting, but none were known to do it so rapidly. Theoretical studies showed that such quick variations as those exhibited by the pulsars should occur only in very dense objects, denser even than white dwarfs. This led astronomers to think of neutron stars.

Enough was known from theoretical calculations, however, to rule out rapid expansion and contraction of neutron stars as the cause of the observed pulsations. The calculations showed that the vibration period of such an object would be even shorter than the observed periods of the pulsars.

A second possibility, that the rapid periods were produced by rotation of the pulsating objects, was considered. As in the case of physical contraction and expansion, the rotation hypothesis ruled out ordinary stars and white dwarfs. In order to rotate several times per second, a normal star or white dwarf would have to have a surface velocity in excess of the speed of light, a physical impossibility since it would be torn apart by rotational forces before approaching such a rotational velocity. A neutron star, on the other hand, could rotate several times per second and remain intact.

When the Crab pulsar was discovered, a great deal of additional information became available. This pulsar was found to be gradually slowing its pulsations, something that could best be explained if the pulses were linked to the rotation of the object. The rotation could slow as the pulsar gave up some of its energy to its surroundings.

This discovery cleared up another mystery. The source of energy that powers the synchrotron emission from supernova remnants had been unknown; now it was suggested that the slowdown of the rotating neutron star could provide the necessary energy, either by magnetic forces exerted on the surrounding ionized gas or by transfer of energy from the pulsar to the surrounding material through the emission of radio waves.

The remaining question was how the pulses were created by the rotation. Evidently a pulsar acts like a lighthouse, with a beam of radiation sweeping through space as it spins. The question was why a rotating neutron star should emit a beam from just one point on its surface.

The most probable explanation is based on the strong magnetic fields that neutron stars are likely to have (Fig. 8.12). If a neutron star has a strong field, electrons from the surrounding gas are forced to follow the lines of the field, hitting the surface only at the magnetic poles. The result is an intense beam of electrons traveling along field lines, especially concentrated near the magnetic poles of the neutron star, where the field lines are crowded together. The rapidly moving electrons emit synchrotron radiation as they travel along the field lines, creating narrow beams of radiation from both magnetic poles of the star. If the magnetic axis of the star is not aligned with the rotation axis, these beams will sweep across the sky in a conical pattern as the star rotates. If the Earth happens to lie in the direction intersected by one of the beams, we see a flash of radiation every time the beam sweeps by us. Synchrotron radiation is normally emitted over a very broad range of wavelengths, suggesting that pulsars may "pulse" in parts of the spectrum other than the radio. Indeed, this is the case; the Crab pulsar has been identified as a visible-light and X-ray pulsar (Fig. 8.13).

While the details are still somewhat vague, the rotating neutron star model seems to be the best explanation of the pulsars. Since special conditions (nonalignment of the magnetic and rotation axes, and the need for the beam to cross the direction toward the Earth) are required for a neutron star to be seen from Earth as a pulsar, it follows that there should be many neutron stars that do not manifest themselves as pulsars. In the next section we will see how some of these nonpulsating neutron stars are detected.

Neutron Stars in Binary Systems

Earlier we made a general statement that stellar remnants are often most easily observed when they are in binary systems. We have already seen that a white dwarf in a binary can flare up violently if it receives new matter from its companion star. A neutron star reacts similarly in the same circumstances.

If a neutron star is in a binary system where the companion object is either a hot star with a rapid wind

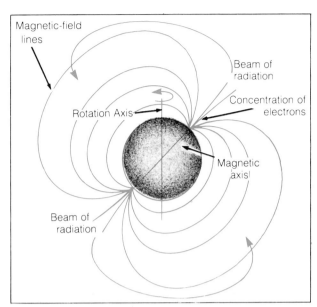

Figure 8.12 The Pulsar Mechanism. Here we see a rapidly rotating neutron star, with its magnetic axis out of alignment with its rotation axis. Synchrotron radiation is emitted in narrow beams from above the magnetic poles, where charged particles, constrained to move along the magnetic field lines, are concentrated. These beams sweep the sky as the star rotates, and if the Earth happens to lie in a direction covered by one of the beams, we observe the star as a pulsar.

or a cool giant losing matter due to its active chromosphere and low surface gravity, some of the ejected mass will reach the surface of the neutron star. Very little of it falls directly down onto the surface; instead, much of it swirls around the neutron star, forming a disk of gas called an **accretion disk** (Fig. 8.14). The individual gas particles orbit the neutron star like microscopic planets and only fall inward as they lose energy in collisions with other particles. The accretion disk acts as a reservoir of material, slowly feeding it inward toward the neutron star.

The disk is very hot because of the immense gravitational field close to the neutron star. As gas in the accretion disk slowly falls closer to the neutron star, gravitational potential energy is converted into heat. The gas in the disk reaches temperatures of several million degrees, hot enough to emit X rays. Hence a neutron star in a mass-exchange binary system is likely to be an X-ray source, and a number of such systems have

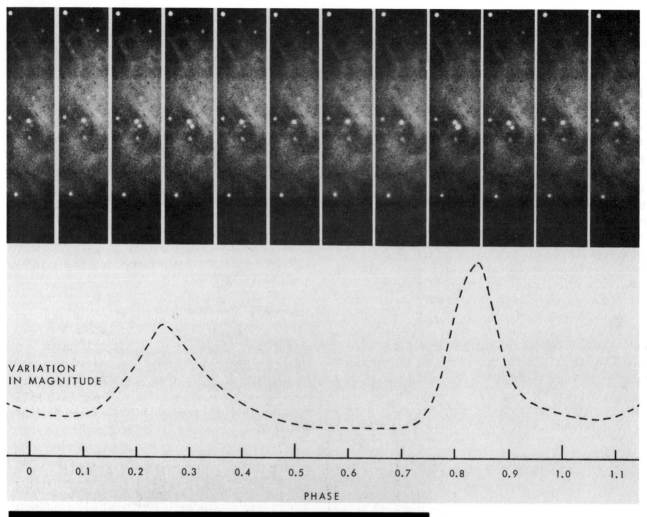
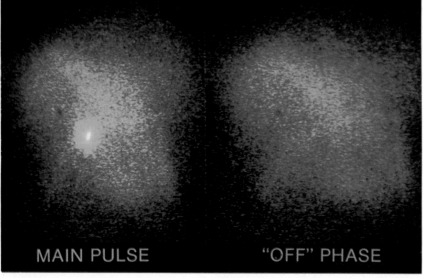

Figure 8.13 Visible and X-ray Flashes from the Crab Pulsar. At top is a sequence of visible-light photographs, accompanied by a light curve, showing how the pulsar in the Crab Nebula flashes on and off during its 0.03-second cycle. Below is a pair of X-ray images, showing the pulsar "on" and "off" in different parts of its cycle. (top: *National Optical Astronomy Observatories;* bottom: *Harvard-Smithsonian Center for Astrophysics*)

been found in the past decade (Fig. 8.15), since the advent of X-ray telescopes launched on rockets or satellites. Often it is known that an X-ray object is part of a binary system because the X rays are periodically eclipsed by the companion star. Eclipses are made likely by the proximity of the two stars (mass exchange would not occur unless it were a close binary) and the fact that the mass-losing companion is likely to be a large star, either an upper main sequence star or a supergiant.

The so-called **binary X-ray sources** (a few are listed in Table 8.4) are among the strongest X-ray emitting objects known. Many of them are most likely neutron stars, although some may be black holes, which would also produce X-ray emission as material fell inward.

Neutron stars can emit X rays in a slightly different way, which also occurs in mass-exchange binaries. If the infalling material trickles down onto the neutron star in a steady fashion, there is continuous radiation of X rays, as we have just seen. If, however, the material falls in sporadically and arrives in substantial quantities every now and then, a major nuclear outburst occurs each time (Fig. 8.16), in close analogy to the nova process involving a white dwarf. This apparently happens in some cases, producing random but frequent X-ray outbursts. The intensity of the outburst,

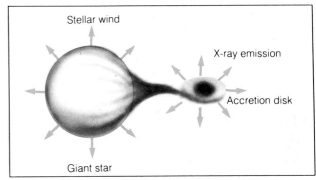

Figure 8.14   An Accretion Disk.   Here a giant star losing mass through a stellar wind has a neutron star companion. Material that is trapped by the neutron star's gravitational field swirls around it in a disk, which is so hot due to compression that it glows at X-ray wavelengths. A nearly identical situation can arise when the compact companion is a black hole.

as in the case of a nova, depends on how much matter has fallen in and been consumed in the reactions. The neutron star binary systems where this occurs are called **bursters.** One important contrast with novae is that the flare-up of a burster occurs much more rapidly, lasting only a few seconds. Another is that most of the emission occurs only in very energetic X rays, so these objects do not show up in visible light.

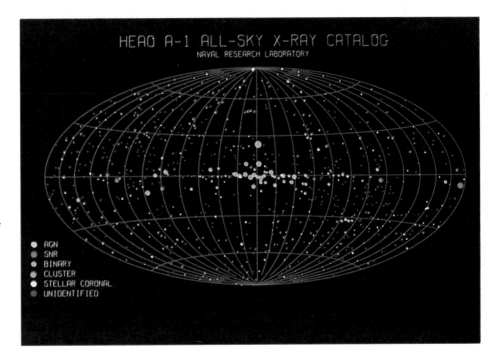

Figure 8.15   An X-ray Map.   This shows the locations of X-ray sources cataloged by the *Einstein Observatory,* an orbiting X-ray satellite that operated in the early 1980s. Most of the sources here are binary systems in which one member is a compact stellar remnant such as a neutron star or a black hole. *(Smithsonian Astrophysical Observatory)*

Table 8.4 Selected Binary X-ray Sources

Source	Period (Days)	Mass of System ($M_\odot$)	Nature of Stars
Cygnus X-1	5.6	40	O9Ib supergiant; probable black hole
Centaurus X-3	2.09	20–25	BOIb-III giant; pulsar (neutron star)
Small Magellanic Cloud X-1	3.89	15–25	BOIb supergiant; pulsar (neutron star)
Vela X-1	8.95	20–30	BO.5Ib supergiant; probable neutron star
Hercules X-1	1.70	2–5	HZ Hercules (A star); pulsar (neutron star)
A0620-00	0.32	11	K dwarf; probable black hole (companion mass $\simeq 10\ M_\odot$)
Large Magellanic Cloud X-3	1.70	15	B3 main sequence star; probable black hole, mass $\simeq 9\ M_\odot$

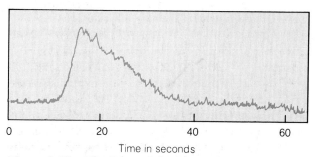

Figure 8.16. The X-ray Light Curve of a Burster. This schematic illustration of the X-ray intensity from one of these sources shows how rapidly the emission flares up and then drops off. Like the nova process, these outbursts are thought to be due to material falling onto the surface of a neutron star and igniting brief episodes of nuclear reactions.

Black Holes: Gravity's Final Victory

In the last chapter we saw what happens to a massive star at the end of its life: it falls in on itself, and no barrier—neither electron degeneracy nor neutron degeneracy—can stop it.

The gravitational field near the surface of a collapsing star grows in strength, as the mass of the star becomes concentrated in an ever-smaller volume. This gravitational field has important effects in the near vicinity of the star, although at a distance it remains unchanged from what it was before the collapse. Close to the star, though, the structure of space itself is distorted, according to Einstein's theory of general relativity. Einstein discovered that accelerations due to changing motion and those due to gravitational fields are equivalent, and from this it follows that space must be curved in the presence of a gravitational field, so that moving particles follow the same path that they would follow if they were being accelerated (see Astronomical Insight 3.3). This applies to photons of light, just as it does to other kinds of particles.

The effects of the curvature of space are not normally noticeable, except under very careful observation or when considering very large distances. Later in this text, when we discuss the universe as a whole and its overall structure, we will consider the latter situations. For now we confine ourselves to local regions where space can be distorted by very strong gravitational fields (Fig. 8.17).

Let us consider a photon emitted from the surface of a star as it falls in on itself (Fig. 8.18). If the photon is emitted at any angle away from the vertical, its path will be bent over further. If the gravitational field is strong enough, the photon's path may be bent over so far that it falls back onto the stellar surface. Photons emitted straight upward follow a straight path, but they lose energy to the gravitational field, which causes their wavelengths to be redshifted. (We discussed this phenomenon in the case of white dwarfs.) When the gravitational field has become strong enough, the photon loses all its energy and cannot escape. Another way of stating this is that the velocity required for escape from the star exceeds that of light. At the point when this happens the star becomes invisible to outside observers because no light from it can reach them.

When the gravitational field of a star becomes strong enough to trap photons, its radius is called the **Schwarzschild radius,** after the German astrophysicist who first calculated its properties some sixty years

Figure 8.17. Geometry of Space Near a Black Hole. Einstein's theory of general relativity may be interpreted in terms of a curvature of space in the presence of a gravitational field. Here we see how this curvature varies near a black hole.

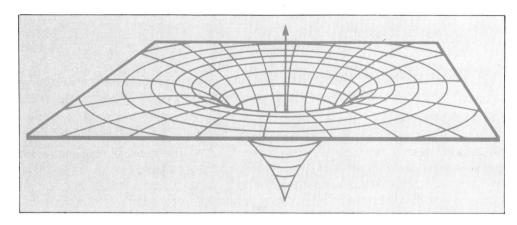

ago. The Schwarzschild radius depends only on the mass of the star that has collapsed. For a star of 10 solar masses, it is 30 kilometers; a star of twice that mass would have twice the Schwarzschild radius, and so on (Table 8.5).

When a collapsing star has shrunk inside its Schwarzschild radius, it is said to have crossed its **event horizon,** because an outside observer cannot see it or anything that happens to it after that point. We have no hope of ever seeing what happens inside the event horizon, but since no force is known that could stop the collapse, we assume that it continues. The mass becomes concentrated in an infinitesimally small region at the center, which is called a **singularity,** because mathematically it is a single point.

What happens to the matter that falls into a singularity is a subject for speculation, but we can be more concrete in discussing what happens just outside the event horizon. From the outsider's point of view, the time sense is distorted by the extreme gravitational field and acceleration in such a way that the collapse would seem to slow gradually, coming to a halt just as the star reached its event horizon. This slowdown, however, is most significant in the last moments before the star's disappearance, and by then the star would be essentially invisible anyway, as most escaping photons are diverted into curved paths or redshifted into the infrared or beyond. The star would seem to disappear rather quickly, despite the stretching of the collapse in time due to relativistic effects. To an observer unfortunate enough to be falling into the black hole, the fall

Figure 8.18 Photon Trajectories from a Collapsing Star. Light escapes in essentially straight lines in all directions from a normal star (left), whose gravitational field is not sufficient to cause large deflections. At an intermediate stage of collapse (center), photons emitted in a cone nearly perpendicular to the surface can escape, while others cannot. Those emitted at just the right angle go into orbit around the star, while those emitted at greater angles fall back onto the stellar surface. After collapse has proceeded to within the Schwarzschild radius, no photons can escape.

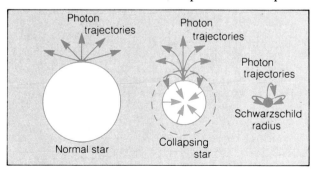

Table 8.5 Schwarzschild Radii for Various Objects

Object	Schwarzschild Radius
Sun	3 km
Earth	0.9 cm
150-lb human	5×10^{-23} cm
Jupiter	2.9 m
Star of 50 $M_\odot$	150 km
Typical globular cluster	5×10^4 km
Nucleus of a galaxy	10^6 to 10^8 km (10^{-7}–10^{-5} pc)
Massive cluster of galaxies	10^{15} km (100 pc)
The universe	10^{26} km (10^{13} pc)

would go very quickly; the observer's sense of time would not be distorted by relativistic effects in the same way as that of the distant observer. (The infalling observer would suffer serious discomfort due to tidal forces.)

Mathematically, a black hole can be described completely by three quantities: its mass, its electrical charge, and its spin. The mass, of course, is determined by the amount of matter that collapsed to form the hole, plus any additional material that may have fallen in later. The electrical charge depends on the charge of the material from which the black hole formed; if it contained more protons than electrons, for example, it would end up with a net positive charge. Because particles with opposite charges attract each other and those with like charges repel, it is thought that electrical forces would maintain a fairly even mixture of particles during and after the formation of the black hole, so that the overall charge would be nearly zero. To illustrate this, imagine that a black hole was formed with a net negative charge. If there were ionized gas around it afterward (as there likely would be, with some of the matter from the original star still drifting inward), the negative charge of the hole would repel additional electrons, preventing them from falling in, while protons would be accelerated inward. In time enough protons would be gobbled up to neutralize the negative charge.

The spin of a black hole is not so easily explained, however. It stands to reason that if the star were spinning before its collapse, as it most likely would have been, the rotation would speed up greatly as the star shrank. A high spin rate actually shrinks the event horizon, allowing an outside observer to see closer in to the singularity residing at the center. It is even possible mathematically, although it presents a physical dilemma, to have sufficient spin that there is no event horizon, so that whatever is in the center is exposed to view. A **naked singularity,** as this has been dubbed, would not produce the usual gravitational effects of a black hole, and it would be possible to blunder into one without any forewarning.

For the most part, it is assumed that the spin rate is never so high that a naked singularity can form, and in fact, black hole properties are usually specified by mass alone, neglecting the effects of charge and spin. It is assumed that our main hope of detecting a black hole is by its gravitational effects, which are determined entirely by the mass.

Before turning to a discussion of how to find a black hole, we should mention that there may be other kinds of black holes formed by processes other than the collapse of individual massive stars. The others include "mini" black holes, very small ones postulated to have formed under extreme density conditions early in the history of the universe, and supermassive black holes, thought possibly to inhabit the cores of large galaxies and to have formed from thousands or millions of stellar masses coalescing in the center. These other possibilities are discussed in later chapters; for now, we turn to the hunt for stellar black holes.

Do Black Holes Exist?

The mass that goes into a black hole during its formation still exists there, hiding inside its event horizon. Even though no light can escape, its gravitational effects persist. The gravitational force of the star is exactly the same as it was before the collapse, except at points so close that they would have been inside the original star.

Our best chance of detecting a black hole, then, is to look for an invisible object whose mass is too great to be anything else. Even if we do find such a thing, it is really only a circumstantial argument, because the conclusion that it is a black hole relies on the theory that neither a white dwarf nor a neutron star can survive if the mass is sufficiently great.

The best opportunity to determine the mass of an object is when it is in orbit around a companion star, where Kepler's third law can be applied to find the masses. The search for black holes, then, leads us to examine binary systems, looking for invisible but massive objects.

This search is facilitated by another property of black holes: If they find new material to pull into themselves, the trapped matter forms an accretion disk, just as we described in the case of a neutron star in a mass-transfer binary. We have already seen that such a disk becomes so hot that it emits X rays. Probably some of the binary X-ray sources (see Table 8.4), therefore, are due to black holes in binary systems, rather than neutron stars. The best way to distinguish between the two possible types of remnants in these systems is to determine the mass of the invisible companion to see whether it is too great to be a neutron star.

The determination of the masses in a binary system

is difficult, if not impossible, if only one of the two stars can be seen, because the information on their velocities and on the inclination angle of the orbit is incomplete. The fact that many X-ray binaries are eclipsing systems (at least the X-ray source is eclipsed; see Fig. 8.19) helps us determine the tilt of the orbit in some cases. The orbital period is usually easily determined from the eclipse frequency as well, but the orbital velocities needed to deduce the sizes of the orbits are often very difficult or impossible to measure. The unseen star, of course, emits little or no light, so there is no hope of measuring Doppler shifts in its spectral lines; the normal companion star is often a hot giant or supergiant, and these stars usually have very broad spectral lines whose velocities cannot be measured accurately. Nevertheless, sometimes enough information can be derived to at least place limits on the mass of the invisible companion, and in one such case, the X-ray binary called Cygnus X-1 (Fig. 8.20), it appears that the collapsed object has to contain at least eight solar masses. This object is one of three leading black hole candidates, the other two being the sources LMC-X3 and A0620-00 (included in Table 8.4).

While the jury is still out on the question of the existence of black holes, the favorable circumstantial evidence is strong. The principle of **Occam's razor** (the simplest explanation of the observed facts is most likely correct) makes acceptance of the existence of black holes natural, there being no simpler way to explain what happens when a massive star collapses. Most astronomers today have adopted the concept of black holes and not only believe that they exist but also consider them integral parts of our universe with numerous important roles.

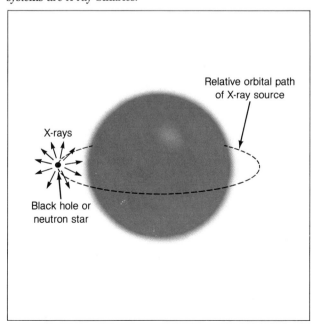

Figure 8.19 Eclipses of an X-ray Source in a Binary System. Because the separation of the two stars in a mass-transfer system is always very small, and because the mass-losing star is usually very large (a red giant or a hot giant or supergiant with a stellar wind), eclipses occur easily. If the mass-receiving star is a neutron star or black hole, and therefore an X-ray emitter, the X rays are likely to be eclipsed periodically by the companion. Hence many such systems are X-ray binaries.

Figure 8.20 Cygnus X-1. This X-ray image obtained by the orbiting *Einstein Observatory* shows the intense source of X rays at the location of a dim, hot star that is thought to be a normal star that is losing mass to its invisible black hole companion. *(Harvard-Smithsonian Center for Astrophysics)*

Perspective

We have examined the corpses of stars to see how they decay and have come away with our minds filled with wonder at the novel forms matter can take. We have seen that the nature of stellar remnants depends entirely on how much mass is left when stars run out of nuclear fuel and collapse, the three possibilities being white dwarfs, neutron stars, and black holes. Stars that die alone, no matter what final form they take, are not likely to be detected again (except for the closest white dwarfs and the neutron stars that happen to appear to us as pulsars), while those that exist in binary systems may be reincarnated in spectacular fashion if mass exchange takes place.

Having learned all we can about stars as individuals, we are ready to move on to larger scales in the universe, to examine galaxies and ultimately the universe itself.

Summary

1. A white dwarf gradually cools off, taking billions of years to become a cold cinder.
2. If new matter falls onto the surface of a white dwarf, for example, in a binary system where the companion star loses mass, there are several possible results. The white dwarf may become a cataclysmic variable if the matter arrives in small amounts with high energy; or it may become a nova, if matter builds up slowly to the point where a larger quantity is involved in the explosion; or it may exceed the white dwarf mass limit and become a neutron star or black hole or explode as a Type I supernova.
3. Massive stars are likely to explode as Type II supernovae when all possible nuclear reaction stages have ceased. The supernova explosion creates an expanding cloud of hot, chemically rich gas known as a supernova remnant.
4. In some cases a remnant of two to three solar masses is left behind in the form of a neutron star consisting of degenerate neutron gas.
5. A neutron star is too dim to be seen directly in most cases, but it may be observed as a pulsar (depending on the alignment of its magnetic and rotation axes and our line of sight), and in a close binary system it may become a source of X-ray emission.
6. Some neutron stars that receive new material in clumps flare up occasionally as X-ray sources called bursters.
7. If the final mass of a star exceeds two or three solar masses, it will become a black hole at the end of its nuclear reaction lifetime.
8. The immensely strong gravitational field near a black hole traps photons of light, rendering the black hole invisible.
9. A black hole may be detected by its gravitational influence on a binary companion or by the X rays it emits if new matter falls in, as in a close binary system where mass transfer takes place. Evidence that the unseen companion is a black hole (rather than a neutron star) can be established only by analysis of the orbits to determine the mass of the unseen object.

Review Questions

1. Why is the final mass a star has, rather than the mass it begins with, the important criterion for determining what form of remnant the star will leave? Why should the final mass be different from the initial mass for a given star?
2. Using information from Chapter 3, compare the surface gravity of the Sun with that of a white dwarf with the same mass but only 0.01 of the radius. Do the same calculation for a neutron star, with twice the mass of the Sun but only 10^{-6} of the Sun's radius.
3. Summarize the techniques by which white dwarfs can be detected from the Earth.
4. Use Wien's Law to calculate the temperature at which a cooling white dwarf becomes essentially invisible to the human eye, assuming that to occur when the wavelength of maximum emission has shifted to the infrared wavelength of 10,000Å.
5. Despite the similarity of their names, novae and supernovae are completely different phenomena. Explain the differences.
6. Discuss the effect of supernova explosions on the chemical composition of the galaxy.
7. Compare the internal structure of a neutron star with that of a white dwarf.

8. If a certain pulsar has a much longer period than another, which is more likely to be surrounded by a detectable supernova remnant? Why?
9. When a star collapses to become a black hole, the gravitational field close to it becomes very strong. Why does this not affect the orbit of the companion star, in the case where the black hole is a member of a binary system?
10. Both neutron stars and black holes can be detected as X-ray sources under certain circumstances (in mass-transfer binary systems). How can we expect to distinguish between these two kinds of stellar remnants in such cases?

Additional Readings

Anderson, L. 1976. X-rays from degenerate stars. *Mercury* 5(5):2.
Bethe, H. A. 1985. How a supernova explodes. *Scientific American* 252(5):60.
Black, D. L. 1976. Black holes and their astrophysical implications. *Sky and Telescope* 50(1):20 (Part 1); 50(2):87 (Part 2).
Clark, David H. 1985. *The Quest for SS433*. Penguin Books.
Greenstein, G. 1985. Neutron stars and the discovery of pulsars. *Mercury* (2):34 (Part 1); (3):66 (Part 2).
Gursky, H., and E. P. J. van den Heuvel. 1975. X-ray emitting double stars. *Scientific American* 232(3):24.
Helfand, D. J. 1978. Recent observations of pulsars. *American Scientist* 66:332.
Kafatos, M., and A. G. Michalitsianos. 1984. Symbiotic stars. *Scientific American* 251(1):84.
Kaufmann, W. 1974. Black holes, worm holes and white holes. *Mercury* 3(3):26.

Kawaler, S. D., and D. E. Winget. 1987. White dwarfs: Fossil stars. *Sky and Telescope* 74(2):132.
Lewin, W. H. G. 1981. The sources of celestial X-ray bursts. *Scientific American* 244(5):72.
Schaefer, B. E. 1985. Gamma-ray bursters. *Scientific American* 252(2):52.
Shaham, Jacob. 1987. The oldest pulsars in the universe. *Scientific American* 256(2):50.
Schramm, D., and W. Arnett. 1975. Supernovae. *Mercury* 4(3):16.
Seward, F. 1986. Neutron stars in supernova remnants. *Sky and Telescope* 71(1):6.
Smarr, L., and W. H. Press. 1978. Spacetime: black holes and gravitational waves. *American Scientist* 66:72.
Thorne, K. S. 1974. The search for black holes. *Scientific American* 233(6):32.
Van Horn, H. M. 1973. The physics of white dwarfs. *Physics Today* 32(1):23.
Wheeler, J. C. 1973. After the supernova, what? *American Scientist* 61:42.
The following articles are all related to Supernova 1987A.
Helfland, David. 1987. Bang: The supernova of 1987. *Physics Today* 40(8):24.
Schorn, Ronald A., 1987. Neutrinos from hell. *Sky and Telescope* 73(5):477.
⸻. 1987. SN1987A: Watching and waiting. *Sky and Telescope* 74(1):14.
⸻. 1987. A supernova in our backyard. *Sky and Telescope* 73(4):382.
⸻. 1987. Supernova 1987A after 200 days. *Sky and Telescope* 74(5):477.
⸻. 1987. Supernova 1987A's fading glory. *Sky and Telescope* 74(3):258.
⸻. 1987. Supernova shines on. *Sky and Telescope* 73(5):470.
⸻. 1987. A surprising supernova. *Sky and Telescope* 73(6):582.
Time. March 23, 1987. 129(12):60.

The lighter Side of Gravity by J. V. Narlikar

"A closed door usually leads you to an open one."
"If an open door is what you truly desire."

Section III

The Milky Way

In this section we will explore the great conglomeration of stars to which our Sun belongs. The Milky Way, a spiral galaxy, is itself a dynamic entity with its own structure and evolution governed by the complex interactions of stars with each other and with the interstellar gas and dust. While this system of some 10^{11} individual stars is infinitely more complicated than a single star and its life story is correspondingly more difficult to unravel, great progress has nevertheless been made in piecing together the puzzle.

In the first of the two chapters of this section, we will examine the overall properties of our galaxy. We will learn how its size and shape and the Sun's location within it were learned from observations carried out from our position within the great disk, where our view is obscured by interstellar haze. We will discuss in this chapter how the various parameters describing the galaxy were derived from observation, and we will learn of some immense and fascinating mysteries that remain. Included in this chapter is a description of the interstellar medium of gas and dust, which contains only a fraction of the galactic mass but plays an essential role in its evolution.

The second chapter on our galaxy ties together the disparate data gathered and examined in this chapter with what we have learned previously about stellar processes and tells the story of the formation and evolution of our galaxy. We will see how the major features of the Milky Way arose from a sequence of developments dictated by simple laws of physics. While there remain some mysteries, such as the quantity of mass in the far reaches of the galactic halo and the traces of violent activity at the core of the galaxy, most of the story can be told.

Having probed the Milky Way, we will then be ready to move outward to examine and understand the countless galaxies in the universe.

Chapter 9

Structure and Organization of the Milky Way

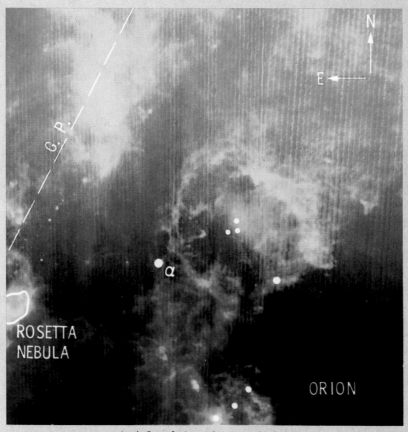

An infrared view of Orion, observed by the *IRAS* satellite. *(NASA/JPL)*

Chapter 9 Structure and Organization of the Milky Way

Chapter Preview

Variable Stars as Distance Indicators
The Structure of the Galaxy and the Location of the Sun
Galactic Rotation and Stellar Motions
The Mass of the Galaxy
The Interstellar Medium
Spiral Structure and the 21-Centimeter Line
The Galactic Center: Where the Action Is
Globular Clusters Revisited
A Massive Halo?

Table 9.1 Properties of the Milky Way

Mass: 1.4×10^{11} $M_\odot$ (interior to Sun's position)
Diameter of disk: 30 Kpc
Diameter of central bulge: 10 Kpc
Sun's distance from center: 10 Kpc (possibly less)
Diameter of halo: 100 Kpc (very uncertain)
Thickness of disk: 1 Kpc (at Sun's position)
Number of stars: 4×10^{11}
Typical density of stars: 20 stars/pc^3 (solar neighborhood)
Average density of interstellar matter: 10^{-24} g/cm^3 (roughly equal to one hydrogen atom/cm^3)
Luminosity: 2×10^{10} $L_\odot$
Absolute magnitude: -20.5
Orbital period at Sun's distance: 2.5×10^8 years

We have discussed stars as individuals, as though they live in a vacuum, isolated from the rest of the universe. For the purposes of analyzing the structure and evolution of stars, this is a suitable approach, but if we want to understand the full stellar ecology, the properties of individuals must be discussed in the context of the larger environment.

Even a casual glance at the nighttime sky shows that the stars tend to be grouped rather than randomly distributed. The most obvious concentration is the **Milky Way** (Fig. 9.1), a diffuse band of light stretching from horizon to horizon, clearly visible only in areas well away from city lights. To the ancients the Milky Way was merely a cloud; Galileo with his primitive telescope recognized it as a region of great concentration of stars. In modern times we know the Milky Way as a galaxy, the great pinwheel of billions of stars to which the Sun belongs (Table 9.1). The hazy streak across our sky is a cross-sectional, or edge-on, view of the galaxy from a point within its disk.

The overall structure of the Milky Way resembles a phonograph record with a large central bulge, the center of which is called the **nucleus** (Figs. 9.2 and 9.3). Nearly all of the visible light is emitted by stars in the plane of the disk, although the galaxy also has a **halo**, a distribution of stars and star clusters centered on the nucleus but extending well above and below the disk. The most prominent objects in the halo are the **globular clusters**, very old, dense clusters of stars characterized by their distinctly spherical shape.

To envision the size of the Milky Way requires a new unit of distance. In Chapter 6 we discussed stellar distances in terms of parsecs, and we found that the nearest star is about 1.3 parsecs from the Sun. To expand to the scale of the galaxy, we speak in terms of **kiloparsecs**, or thousands of parsecs. The visible disk of the Milky Way is roughly 30 kiloparsecs (abbreviated Kpc) in diameter, and the disk is a few hundred parsecs thick. Light from one edge of the galaxy takes about 100,000 years to travel across to the far edge.

Our galaxy is so large that we must discuss new methods of measuring distance, for even the main sequence fitting technique, the most powerful we have discussed so far, fails when star clusters are too distant and faint to allow determination of the individual stellar magnitudes and spectral types.

Variable Stars as Distance Indicators

It was stressed in Chapter 6 that it is always possible to determine the distance to an object if both its absolute and apparent magnitudes are known. In that chapter we discussed spectroscopic parallaxes, where the absolute magnitude of a star is derived from its spectral class, and main sequence fitting, where a cluster of stars is observed and its main sequence determined so that it can be fitted to the standard H-R diagram main sequence. Now we will learn about a special type of star whose absolute magnitude can be determined from its other properties, thus allowing it to be used to measure distances.

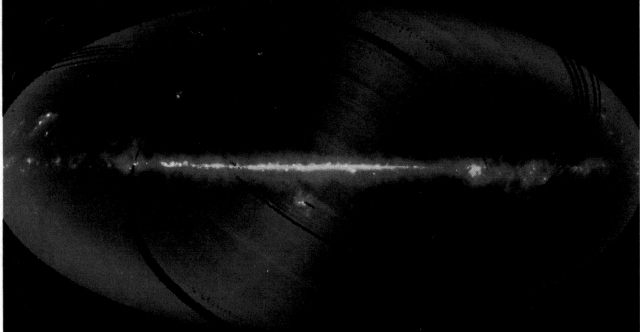

Figure 9.1 The Milky Way. The top mosaic shows a major portion of the cross-sectional view of our galaxy that we see from Earth. Beneath is a composite infrared view from the *IRAS* satellite. *(top: Mt. Wilson and Las Campanas Observatories, Carnegie Institution of Washington; bottom: NASA)*

In Chapter 8 we briefly mentioned variable stars, including those that pulsate regularly (Table 9.2). These stars physically expand and contract, changing in brightness as they do so. One of the first such stars to be discovered was δ Cephei, which is sufficiently bright to be seen with the unaided eye. Following the discovery of δ Cephei in the mid-1700s, other similar stars were found, and these as a class became known as **Cepheid variables,** named after the prototype. Another type of pulsating variable, the **RR Lyrae stars,** was found to exist primarily in globular clusters and to have pulsation periods of less than a day, whereas the Cepheids have periods ranging from 1 to 100 days.

It was discovered in 1912, through observations of variable stars in the Magellanic Clouds (two small, relatively nearby galaxies), that there is a definite relationship between the pulsation period of these stars and their luminosities. For the Cepheids, the period increases with increasing luminosity (or decreasing absolute magnitude), while the RR Lyrae stars all have

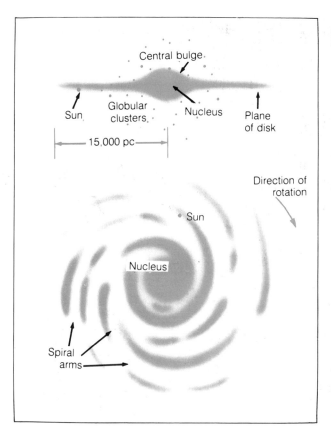

Figure 9.2 The Structure of Our Galaxy. These sketches illustrate the modern view of the Milky Way.

Figure 9.3 A Spiral Galaxy Similar to the Milky Way. This galaxy (NGC 7331) probably resembles our own, as seen from afar. *(Palomar Observatory, California Institute of Technology)*

Table 9.2 Types of Pulsating Variables

Type of Variable	Spectral Type	Period (days)	Absolute Magnitude	Change of Magnitude
δ Cephei (Type I)	F, G supergiants	3 to 50	−2 to −6	0.1 to 2.0
W Virginis (Type II Cepheids)	F, G supergiants	5 to 30	0 to −4	0.1 to 2.0
RR Lyrae	A, F giants	0.4 to 1	0.5 to 1.2	0.6 to 1.3
RV Tauri	G, K giants	30 to 150	−2 to −3	up to 3
δ Scuti	F giants	0.1 to 0.2	2	0.1
β Cephei (β Canis Majoris)	B giants	0.1 to 0.2	−3 to −5	0.1
Long-period variables (Mira)	M supergiants	80 to 600	+2 to −3	3 to 7

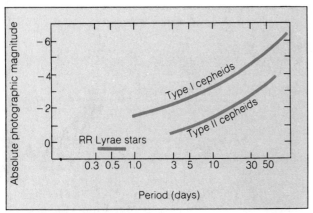

Figure 9.4 The Period-Luminosity Relationship for Variable Stars. This diagram shows how the pulsation periods for Cepheid and RR Lyrae variables are related to their absolute magnitudes. The fact that there are two types of Cepheids, with somewhat different relationships, was not recognized at first, and this led to some early confusion about distance scales.

about the same luminosity. This discovery had a profound implication: When a star is recognized as a variable belonging to one of these classes, its absolute magnitude can be determined simply by measuring its period and using the established period-luminosity relation (Fig 9.4). Once the absolute magnitude is known, the distance can be found by comparing the absolute and apparent magnitudes (average values must be used, since the magnitudes actually vary with the pulsations).

Because Cepheids are giant stars, they are very luminous and can be observed at great distances. Thus they are very powerful tools for measuring distances beyond those reached by other techniques we have discussed. The fact that there are two different types of Cepheids with slightly different period-luminosity relations (Fig. 9.4) was not discovered for some time, and for a while this caused confusion in establishing the size scale of our galaxy.

The Structure of the Galaxy and the Location of the Sun

Because the solar system is located within the disk of the Milky Way, we have no easy way of getting a clear view of where we are in relation to the rest of the gal-

Figure 9.5 Harlow Shapley. Shapley's work on the distances to globular clusters was a key step in the determination of the size of the Milky Way. Shapley also played a prominent role in the discovery of galaxies outside our own (see Chapter 19). Much of his observational work was done before 1920, when Shapley was a staff member at the Mount Wilson Observatory. He later became director of the Harvard College Observatory. (*Harvard College Observatory*)

axy. All we see is a band of stars across the sky, which tells us that we are in the plane of the disk. It is not so easy to determine where we are with respect to the edge and center of the disk.

Early in this century astronomers did not know that our galaxy is permeated with a very rarefied medium of gas and tiny solid particles in interstellar space, and this contributed to the difficulty in determining the structure of the galaxy and our location in it. The haze of tiny solid particles, called **interstellar dust,** obscures our view of distant stars, making it appear as though the number of stars in space decreases as we look farther and farther away. This led one early researcher, the Dutch astronomer J. C. Kapteyn, to conclude that we are at the center of the galaxy because our Sun seemed to be in the densest region. But other techniques soon showed that we are far from the center of the galaxy.

One of these techniques, which was employed by Harlow Shapley (Fig. 9.5), made use of the globular clusters, the spherical star clusters found outside the confines of the galactic disk. These clusters tend to contain RR Lyrae variables, so it was possible for Shapley to determine their distances and hence their locations with respect to the Sun and the disk of the Milky Way. He found that the globular clusters are arranged in a spherical volume centered on a point several thousand parsecs from the Sun (Fig. 9.6), and he argued that this point must represent the center of the galaxy. It would not make physical sense for the globular clusters to be concentrated around any location other than the center of the entire galactic system.

Shapley's conclusion, first published in 1917, was not widely accepted initially, but other supporting evidence was found in the 1920s. Two scientists, Jan Oort of Holland (who will be mentioned in connection with the origin of comets; see Chapter 19) and Bertil Lindblad of Sweden, carried out careful studies of the motions of stars in the vicinity of the Sun. What they found was that these motions could best be understood if the Sun and the stars around it were assumed to be orbiting a distant point; that is, they found systematic, small velocity differences between stars, similar to those between runners on a track who are in the inside and outside lanes (Fig. 9.7). It appeared from these studies that the Sun is following a more-or-less circular path about a point several thousand parsecs away, which indicates that the center of the galaxy is located at that distant center of rotation. This supported Shapley's view of the galaxy, although there were still uncertainties about how far the Sun was from the center.

The discovery of the general interstellar dust medium in 1930 laid to rest the apparent conflict between the findings of Kapteyn and those of Shapley, Oort, and Lindblad, and the only major question remaining was to find the true size of the galaxy. This required refinement of the period-luminosity relation for variable stars, which, as noted earlier, was confused for a while by the fact that there are two types of Cepheids, which had not been recognized as distinct. Eventually a consensus was reached that the Sun is about 10 kiloparsecs from the center of a disklike galaxy whose total diameter is about 30 kiloparsecs.

More recently, evidence has emerged that in fact the Sun is a little closer to the center, perhaps only 8 kiloparsecs out. We see that the problem of determining where we are in our galaxy is not necessarily completely solved even today. There are also significant uncertainties in determining the extent of the galaxy beyond the Sun's position, as we will see in later sections of this chapter.

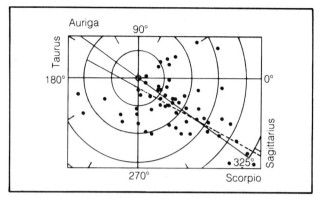

Figure 9.6 Shapley's Measurements of Globular Clusters. This is one of Shapley's original figures illustrating the distribution of globular clusters in the galaxy. The Sun is at the point where the axes intersect, and each circle centered on that point represents an increase in distance of 10,000 parsecs. Note that the distribution of globulars is centered some 10 to 20 kiloparsecs from the Sun. *(Estate of H. Shapley. Reprinted with permission.)*

Figure 9.7 Stellar Motions Near the Sun. Stars just inside the Sun's orbit move faster than the Sun, whereas those farther out move more slowly. Analysis of the relative speeds (as inferred from measurements of Doppler shifts) and distances of stars like these led to the realization that the Sun and stars near it are orbiting a distant galactic center.

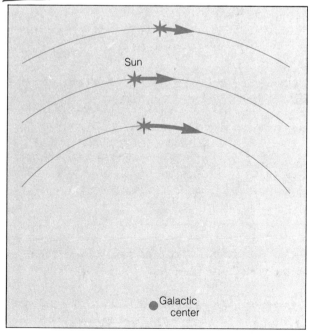

Galactic Rotation and Stellar Motions

An important result of the work of Lindblad and Oort was the development of an understanding of the overall motions in the galaxy. Oort's analysis was especially useful in this regard; he showed not only that the Sun and the stars near it are orbiting the distant galactic center, but also that the rotation of the galaxy is differential, meaning that each star follows its own orbit at its own speed. Thus the portion of the galaxy in the vicinity of the Sun does not act like a rigid disk but like a fluid, with each star moving as an independent particle.

This is not true of the inner portion of the galaxy. There the entire system does rotate like a rigid object, like a record on a turntable. The stars in the inner part of the galaxy are subject to the combined gravitational forces of all the stars around them, and they are not free to follow Keplerian orbits about the galactic center. In the region where rigid-body rotation is the rule, the speeds of individual stars increase with distance from the center, whereas they decrease with distance in the outer portion (Fig. 9.8). In the outer reaches of the galaxy, each star orbits the central portion of the galaxy with little influence from its neighbors, and the stellar orbits are approximately described by Kepler's laws.

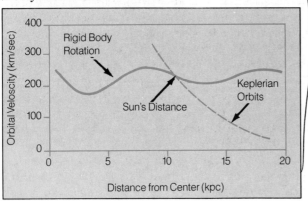

Figure 9.8 The Rotation Curve for the Milky Way. This diagram shows how the stellar orbital velocities vary with distance from the center of the galaxy. The fact that the curve does not simply drop off to lower and lower velocities beyond the sun's orbit, as it would if all the mass of the galaxy were concentrated at its center, indicates that there is a lot of mass in the outer portions of the galaxy. (Data from M. Fich)

This is why the orbital speeds decrease with distance from the center in this part of the galaxy. There is an intermediate distance just inside the Sun's orbit where a transition between rigid-body and Keplerian orbits occurs, and it is here that stars have the greatest orbital velocities. The Sun, which is near this peak position, travels at about 250 kilometers per second, taking roughly 250 million years to make one complete circuit about the galaxy. In its 4- to 5-billion-year lifetime, the Sun has completed fifteen to twenty orbits.

Individual stellar motions do not necessarily follow precise circular orbits about the galactic center. What we have described so far is the overall picture that develops from looking at the composite motions of large numbers of stars. If we look at the individual trees instead of the forest, we find that each star in the great disk has its own particular motion, which may deviate slightly from the ideal circular orbit. These individual motions are comparable to the paths of cars on a freeway, where the overall direction of motion is uniform, but there is a bit of lane-changing here and there.

In the galaxy, the deviation in the motion of a star from a perfect circular orbit is called the **peculiar velocity** of the star; in the case of the Sun, it is called the **solar motion.** The Sun has a velocity of about 20 kilometers per second with respect to a circular orbit, in a direction about 45° from the galactic center and slightly out of the plane of the disk. Most peculiar velocities of stars near the Sun are comparable, amounting to only minor departures from the overall orbital velocity of about 250 kilometers per second. In Chapter 10 we will discuss the **high-velocity stars,** which deviate strongly from the circular orbits followed by most stars in the Sun's vicinity.

The Mass of the Galaxy

Once the true size of the Milky Way was determined, it became possible to estimate its total mass. This could be achieved by measuring the star density in the vicinity of the Sun and then assuming that the entire galaxy has about the same average density. But a much simpler and more accurate technique became possible with the discovery that the stars in our region of the galaxy approximately obey Kepler's laws.

Kepler's third law, in the more complete form developed by Newton, expresses the relationship among the period, the size of the orbit, and the sum of the masses of the two objects in orbit about each other:

$$(m_1 + m_2)P^2 = a^3,$$

where m_1 and m_2 are the two masses (in units of the sun's mass), P is the orbital period (in years), and a is the semi-major axis (in AU). If we consider the Sun to be one of the two objects and the galaxy itself to be the other, we can use this equation to determine the mass of the galaxy. As we have already seen, the orbital period of the Sun is roughly 250 million years: $P = 2.5 \times 10^8$ years. The orbit is nearly circular, with the galactic nucleus at the center, so the semimajor axis is approximately equal to the orbital radius, that is, $a = 10 \text{ Kpc} = 2 \times 10^9$ AU. Now we can solve Kepler's third law for the sum of the masses:

$$m_1 + m_2 = a^3/P^2 = 1.3 \times 10^{11} \text{ solar masses.}$$

Since the mass of the Sun (1 solar mass) is inconsequential compared to this total, we can say that the mass of the galaxy itself is about 1.3×10^{11} solar masses. The Sun is slightly above average in terms of mass, so we conclude that the total number of stars in the galaxy must be 3 to 4×10^{11}, that is, a few hundred billion.

This method refers only to the mass inside the orbit of the Sun; the matter that is farther out has little effect on the Sun's orbit. Thus, when we estimate the mass of the galaxy by applying Kepler's third law in this way, we neglect all the mass that lies farther out. To overcome this problem, radio measurements of interstellar gas in the outer portions of the galaxy have recently been used to determine the orbital velocity (and therefore the orbital period) of material in the outer reaches of the galaxy. It was found that the velocity does not decrease as rapidly with distance as had previously been thought, and this in turn led to the conclusion that quite a bit of mass lies beyond the Sun's orbit. In a later section we will discuss the possibility that most of the mass of the galaxy lies in the halo.

The Interstellar Medium

It has been known for many decades that there is material in space between the stars. We have mentioned the haze of interstellar dust that permeates interstellar space, and photographs show very obvious concentrations of dark or glowing clouds here and there in the galaxy (Figs. 9.9 and 9.10). These clouds are regions of relatively high density, whereas the lower-density material filling most of the volume of the galaxy is virtually transparent. In all, the interstellar gas and dust comprise some 10 to 15 percent of the mass of the galactic disk and play very important roles in its evolution. Stars form from interstellar clouds, as we have seen, and later in life, many stars return some of their substance to the interstellar medium. Because of this, the interstellar medium is studied in detail by astronomers interested in learning more about the lives and deaths of stars and of the galaxy itself.

The interstellar medium is a mixture of gas and dust, with a wide range of physical conditions (Table 9.3). The interstellar dust makes distant stars appear dimmer than they otherwise would and therefore complicates distance determinations based on apparent magnitude measurements. The tendency of the dust to make stars appear dimmer is called **interstellar extinction**. Because the particles tend to block out short-wavelength light more effectively than the longer wavelengths, red light penetrates the interstellar medium more easily than blue light (Fig. 9.11). Thus distant stars appear not only dimmer but also redder than they otherwise would. This trend of increasing extinction with decreasing wavelength continues throughout the spectrum, so that the extinction in ultraviolet wavelengths is very severe, but it is minimal in the infrared portion of the spectrum (Fig. 9.12). Thus it is virtually impossible to observe the inside of even a moderately dense interstellar cloud with an ultraviolet telescope, but we can see deep inside the densest clouds at infrared and radio wavelengths. Infrared observations are a particularly effective technique for observing stars in the process of formation, as discussed in Chapter 7.

Interstellar grains are formed in the material ejected by various kinds of aged or dying stars. Red giant and supergiant stars, for example, are known to form grains in their outer atmospheres, and grain formation seems also to take place in planetary nebulae, novae, and supernova outbursts. Grains form in a condensation process that occurs whenever the proper combination of pressure and temperature prevails. The formation of dust grains in the atmosphere of a red giant is analogous to the formation of frost on the lawn on a cool, humid morning.

Most of the mass in interstellar space is in the form of gas particles rather than grains. The gas is observed by means of absorption lines formed in the spectra of

Figure 9.9   An Emission Nebula.   This dense cloud of interstellar gas and dust is being heated by hot stars embedded within it, causing the gas to glow. Much of the emission is due to hydrogen atoms, which produce strong emission at 6563 Å, accounting for the red color seen here. *(© 1984 Anglo-Australian Telescope Board)*

Table 9.3   Interstellar Medium Conditions

Component	Temperature	Density (particles/cm^3)	State of Gas	How Observed
Coronal gas (intercloud)	10^5–10^6 K	10^{-4}	Highly ionized	X-ray emission, UV absorption lines
Warm intercloud gas	1000 K	0.01	Partially ionized	21-cm emission, UV absorption lines
Diffuse clouds	50–150 K	1–1000	Hydrogen in atomic form, others ionized	Visible, UV absorption lines, 21-cm emission
Dark clouds	20–50 K	10^3–10^5	Molecular	Radio emission lines, IR emission and absorption lines

Figure 9.10 A Dark Cloud. The dark, patchy region here is a dense, cold interstellar cloud, the kind of region where stars can form. The interior temperature may be as low as 20 K, and the density roughly 10^5 particles per cubic centimeter, a high vacuum by earthly standards, but quite dense for the interstellar medium. Some emission nebulosity is present in this region also. (© 1984 Anglo-Australian Telescope Board)

Figure 9.11 Scattering of Light by Interstellar Grains. The grains tend to absorb or deflect blue light more efficiently than red, so a star seen through interstellar material appears red.

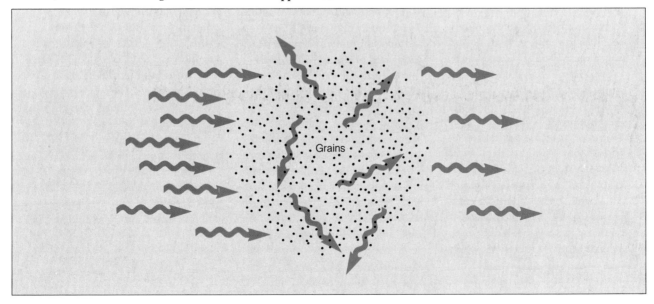

Astronomical Insights 9.1

The Violent Interstellar Medium

Once it was realized (in the 1920s and 1930s) that there is a pervasive interstellar medium of gas and dust, the notion developed that this was a cold, inactive place where nothing much ever changed. Now we know that picture to be false, because it was based on the very limited view of the interstellar medium that was available before the advent of space astronomy.

Because of the physical conditions that prevail in the interstellar medium, only a minute, insignificant portion of it can be observed in visible wavelengths of light. Before ultraviolet, radio, infrared, and X-ray technologies became available to astronomers, little was known about the material between the stars. The most active and important types of interstellar material were invisible.

Thus a great revolution in our knowledge of interstellar conditions has taken place in the past twenty years or so as telescopes have been developed for space observations. The first hints that the interstellar medium is actually a hot, violent place came in the early 1970s, when new ultraviolet and X-ray instruments were launched above the Earth's obscuring atmosphere. The ultraviolet telescope/spectrometer called *Copernicus* discovered in the mid-1970s that very highly ionized gases exist between the stars. These gases require temperatures of several hundred thousand degrees to achieve their high degree of ionization. (The most prominent example is oxygen with five of its eight electrons removed.) At about the same time, rocket-borne X-ray telescopes found that the interstellar medium emits a diffuse glow of X rays, which indicate the presence of gas at a temperature of about one million degrees. Meanwhile, spectroscopic observations in radio and ultraviolet wavelengths revealed clouds here and there in space that are moving at enormous velocities, well in excess of 100 kilometers per second. It became clear that the old view of a placid, unchanging, cold, and dark interstellar medium was incorrect.

It also became clear that a great deal of energy must be injected into interstellar space to create the observed high temperatures and velocities. It is now believed that the sources of this energy are supernova explosions and hot stars with rapid stellar winds. Supernova explosions occur on average only once every 30 to 50 years in our galaxy but release as much energy in an instant as our Sun will produce in its entire lifetime. Hot stars with rapid stellar winds, produce similar quantities of energy in a few million years. These outbursts from massive stars create huge "bubbles" of superheated gas, which expand and join together, creating a galactic network of superhot, low-densitiy gas that occupies much of the volume of the galactic disk. The high-velocity clouds that are observed are thought to be material swept up by the gas expanding outward from supernovae and hot stars with rapid winds.

We point out in the text that the interstellar medium is one half of a cosmic recycling process. In this process, material becomes concentrated in stars, where the chemical composition is altered by nuclear reactions, and is then dispersed through supernova explosions and stellar winds into interstellar space, where it forms the raw material for the formation of new stars. Now we find that the outpouring of matter from stars provides not only the material but also the energy for the formation of stars. Compression of dense clouds by the expanding material from supernovae and stellar winds apparently is responsible for triggering the collapse of interstellar clouds to form new stars. Thus the violence of the interstellar medium may have been a surprise when it was discovered, but it is actually necessary for the vitality of our galaxy.

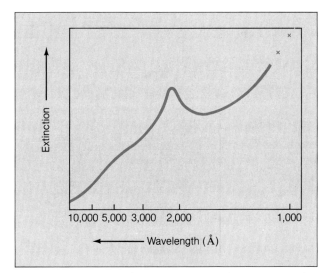

Figure 9.12 Interstellar Extinction. The extinction of starlight by interstellar dust is greatest at short wavelengths, especially in the ultraviolet, and decreases toward longer wavelengths, being very small in the infrared and virtually nonexistent in radio wavelengths.

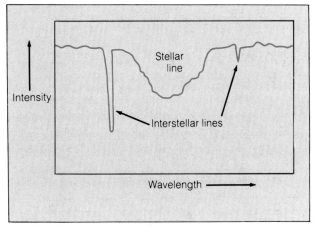

Figure 9.13 Interstellar Absorption Lines. Gas atoms and ions in space absorb photons from distant stars, creating absorption lines in the spectra of the stars. Here we see some interstellar absorption lines in ultraviolet wavelengths, where most of these lines are found. The interstellar lines can be distinguished from the star's own spectral lines because they are narrower, usually have a different Doppler shift (because the star and the cloud producing the line move at different velocities), and represent different states of ionization.

distant stars (Fig. 9.13), mostly in the ultraviolet portion of the spectrum, or by means of various kinds of emission lines. Dense clouds that are heated by nearby hot stars glow by producing emission lines, primarily of hydrogen (Fig. 9.14). The strongest emission line of hydrogen lies at 6563 Å, in the red portion of the spectrum, so color photographs of these **emission nebulae** or **H II regions** show a vivid red appearance (Fig. 9.15).

Radio emission lines are also produced by interstellar gas. Again, hydrogen atoms play an important role, emitting at a wavelength of 21 centimeters. Since hydrogen is the most abundant element, observations of the 21-centimeter emission line are very useful for determining the structure of the galaxy (discussed in the next section). Molecules in the densest interstellar clouds also form radio emission lines at wavelengths ranging from a few millimeters to several centimeters. Molecules, like atoms, have definite energy states, and photons are emitted or absorbed when the molecules change energy states. The energy levels that produce lines in the radio portion of the spectrum are related to the rotational energy of the molecules; being an extended object, a molecule can rotate, and the energy of rotation can have only certain values, just as the energy

level of an electron orbiting an atom can have only certain values. Thus a spinning molecule can slow its spin only by making a sudden drop to a lower energy state, and in doing so it emits a photon whose energy corresponds to the loss of rotational energy of the molecule. Each kind of molecule has its own unique set of rotational energy states and, therefore, its own unique spectrum of radio emission or absorption lines.

Over one hundred different molecular species have been identified in dense "dark" interstellar clouds (see Appendix 12), and more are found frequently as radio telescope technology improves. By far the most abundant molecule is hydrogen (H_2), followed by carbon monoxide (CO). Ironically, hydrogen molecules do not have strong radio emission lines, and their presence is inferred indirectly, except in more rarefied clouds, where they can be observed through ultraviolet absorption lines. The less dense interstellar clouds have few molecules but are composed of a mixture of atoms and ions. Ionization in interstellar space occurs primarily through the absorption of ultraviolet photons rather than collisions between atoms, as is the case in a star's interior or atmosphere.

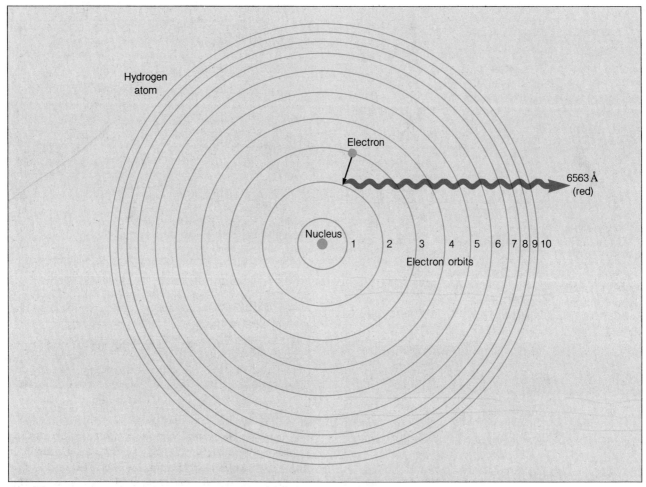

Figure 9.14 Hydrogen Lines from an Emission Nebula. In heated interstellar gas, electrons and protons combine to form hydrogen atoms, usually with the electron initially in an excited state. It then drops down to the lowest state, emitting photons at each step. The strongest emission in visible wavelengths corresponds to the jump from level 3 to level 2 and has a wavelength of 6563 Å. Thus emission nebulae appear red.

The range of physical conditions in interstellar space is enormous. The densest interstellar clouds have densities of perhaps 10^4 to 10^5 particles per cubic centimeter. (Compare this with the Earth's atmospheric density of about 2×10^{19} particles per cubic centimeter!) These clouds are very cold, with temperatures typically between 20 and 50 K. The less dense "diffuse clouds" have densities of 1 to 1000 particles per cubic centimeter and temperatures of 50 to 150 K. Most of the volume of space is filled by even more tenuous material, with a density as low as 10^{-4} particles per cubic centimeter and a temperature as high as 100,000 to 1 million K!

The enormously energized hot intercloud medium is heated by the blast waves from supernova explosions and the rapid outflows from hot stars with winds. The space between the obvious clouds is a violent place filled with superheated gas. The clouds themselves are often in motion, pushed around by the same forces that provide the energy for the intercloud heating. In the next section, we will learn more about how the interstellar material is distributed in the galaxy.

Figure 9.15 An Emission Nebula. The red color of this cloud of hot gas is due to emission by hydrogen atoms. This is a portion of the Rosette nebula. (© *1984 Anglo-Australian Telescope Board*)

Spiral Structure and the 21-Centimeter Line

So far we have spoken of the galactic disk as though it were a uniform, featureless object, whereas we know that this is not a completely accurate picture. The Milky Way is a spiral galaxy, and if we could see it face-on, we would see the characteristic pinwheel shape normally found in galaxies of this type (Fig. 9.16).

Figure 9.16 Spiral Structure. This galaxy has prominent spiral arms, as ours does, because hot, luminous young stars tend to be concentrated in the arms. (The overall appearance of this galaxy is not exactly like the Milky Way, which does not have the bar-like central region seen here.) (© *1977 Anglo-Australian Telscope Board*)

There is a common misconception about the spiral structure in the Milky Way (and other spiral galaxies); namely, that there are few stars between the visible spiral arms. In reality, the density of stars between the arms is nearly the same as it is in the arms. The most luminous stars, however, the young, hot O and B stars, tend to be found almost exclusively in the arms. The interstellar gas and dust tend to be concentrated in the arms, so young stars tend to be concentrated there also. Because these are the brightest stars, their presence in the arms makes the spiral structure stand out.

The fact that we live in a spiral galaxy was not easily discovered, again because we are located within it and see only a cross-sectional view. It was not until 1951 that investigations of the distribution of luminous stars revealed traces of spiral structure in the Milky Way, and even that technique was limited to a small portion of the galaxy. The obscuration caused by interstellar material limits our view of even the brightest stars to a local region, about a thousand parsecs from the Sun at most.

> **Astronomical Insight 9.2**
>
> ## 21-Centimeter Emission from Hydrogen
>
> To understand how a hydrogen atom can emit radiation at a wavelength of 21 centimeters, we need to take a closer look at the structure of the atom.
>
> A hydrogen atom consists of only a proton, which forms the nucleus, and a single electron in orbit about it. We have already learned that the electron can occupy a variety of possible orbits, each corresponding to a different energy state, and that a photon of light is absorbed by the electron if it moves from a low level to a higher one or is emitted if the electron drops from a high level to a lower one. The wavelength of the photon is related to the energy difference between the two electron energy levels; the greater the difference, the shorter the wavelength, and vice versa.
>
> But the energy level structure of the electron is really more complicated than previously described. The electron and the proton are both spinning, and the energy of the electron depends on whether it is spinning in the same or the opposite direction as the proton. If both spin in the same (parallel) direction, the energy state is slightly greater than if they spin in opposite (antiparallel) directions.
>
> As before, the electron can change from one state to the other by either emitting or absorbing a photon. The energy difference is so small, however, that the wavelength of the photon is very much longer than that of visible light. It is 21.1 centimeters.
>
> Hydrogen atoms in space would normally tend to have the electron in the lowest possible energy state, with its spin anti-parallel to that of the proton. Occasionally an atom will collide with another, however, causing the electron to jump to the higher state with the spin parallel to that of the proton. The electron will then spontaneously reverse itself, seeking to return to the lowest energy state, and when it does so, it emits a photon of 21.1-centimeter wavelength. The probability that the electron will make the downward transition is very low, and the electron may remain in the upper state for as long as 10 million years before spontaneously dropping back down. There are so many hydrogen atoms in space, however, that at any given instant, many photons are being emitted. Hence radio telescopes capable of receiving this wavelength can trace the locations of hydrogen clouds throughout the galaxy and beyond.

A major advance in measuring the structure of the galaxy occurred in the same year, when the 21-centimeter radio emission of interstellar hydrogen atoms was first detected. It had been predicted that hydrogen should emit radiation at this wavelength, and the Americans E. M. Purcell and H. I. Ewen were the first to detect this emission, using a specially built radio telescope.

One great advantage of the hydrogen 21-centimeter emission for measuring galactic structure is its ability to penetrate the interstellar medium to great distances. Whereas the brightest stars can be detected at distances of at most a few hundred parsecs, hydrogen clouds in space can be "seen" by radio telescopes from all the way across the galaxy, at distances of several thousand parsecs. Because hydrogen gas is the principal component of the interstellar medium and tends to be concentrated along the spiral arms, observations of the 21-centimeter radiation can be used to trace the spiral structure throughout the entire Milky Way (Fig. 9.17).

When a radio telescope is pointed in a given direction in the plane of the galaxy, it receives 21-centimeter emission from each segment of spiral arm in that direction. Because of differential rotation, each arm has a distinct velocity from the others that are closer to the center or farther out. Therefore, instead of a single

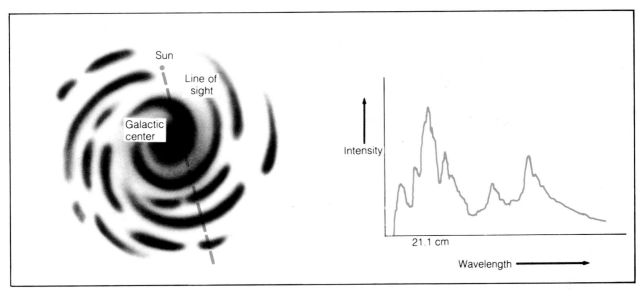

Figure 9.17 21-Centimeter Observations of Spiral Arms. At left is a schematic diagram of the galaxy, showing the direction in which a radio telescope might be pointed to produce a 21-centimeter emission-line profile like the one sketched at right. There are many components of the 21-centimeter line, each corresponding to a distinct spiral arm, and each at a wavelength reflecting the Doppler shift between the velocity of that arm and the earth's velocity.

emission peak at a wavelength of exactly 21.1 centimeters, what we see is a cluster of emission lines near this wavelength but separated from each other by the Doppler effect. By combining measurements such as these with Oort's mathematical analysis of differential rotation, it was possible to reconstruct the spiral pattern of the entire galaxy.

The pattern is much more complex than in some galaxies, where two arms are seen elegantly spiraling out from the nucleus. Instead, the Milky Way consists of bits and pieces of a large number of arms, giving it a definite overall spiral form, but not a smoothly coherent one. As we will learn in Chapter 11, the differences between our type of spiral structure and the regular appearance of some other spiral galaxies may reflect different origins for the arms.

The Galactic Center: Where the Action Is

The center of our galaxy is a mysterious region, forever blocked from our view by the intervening interstellar medium (Fig. 9.18). From Shapley's work on the distribution of the globular clusters, as well as from Oort's analysis of stellar motions, it was known by the 1920s that the center of our great pinwheel lies in the direction of the constellation Sagittarius. There we find immense clouds of interstellar matter and a great concentration of stars. It is one of the richest regions of the sky to photograph, although the best observations can be made only from the Southern Hemisphere.

The central portion of the galaxy consists of a more-or-less spherical bulge, populated primarily by relatively cool stars. The absence of hot, young stars implies that the central bulge has had relatively little recent star formation, something we will have to deal with in our discussion of the history of the galaxy (Chapter 10).

At the center of the bulge lies the nucleus of the galaxy, a small region best located by radio observations. While we cannot see into the central portion of the nucleus in visible light, we can probe the region at longer wavelengths where the interstellar extinction is not such a problem. Radio and, more recently, infrared observations have revealed some interesting features of the galactic core (Fig. 9.19).

The first clue that something unusual was taking place there came from 21-centimeter observations of hydrogen, which revealed a turbulent mixture of clouds moving about at high speed. They also showed that one of the inner spiral arms, about 3 kiloparsecs

Figure 9.18 The Galactic Center. The dark regions are dust clouds in nearby spiral arms. Because of obscuration, photographs such as this do not reveal the true galactic center, but only nearby stars and interstellar matter in the plane of the disk and the outer portions of the central bulge of the galaxy. (© 1980 Anglo-Australian Telescope Board)

from the center, is expanding outward at a velocity of more than 100 kilometers per second. This is suggestive of some sort of explosive activity at the center that has forced matter to move outward.

More recent infrared observations of emission lines from hot gas clouds have shown that interstellar clouds near the center of the nucleus are moving quite rapidly in their orbits, which in turn implies (through the use of Kepler's third law) that they are orbiting a very massive central object. Radio observations made with the *Very Large Array* reveal a remarkable arc of gas extending about 200 parsecs above the galactic plane, apparently held in place by a magnetic field (Fig. 9.20). The origin of this magnetic field is mysterious; it has been suggested that a dynamo created by the enigmatic central object may be responsible.

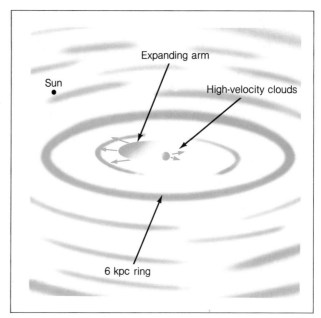

Figure 9.19 Activity at the Galactic Center. A variety of evidence for energetic motions associated with the center of our galaxy is indicated here.

Space observations show that a very small object precisely at the center of the galaxy is emitting an enormous amount of energy in the form of X rays. This object is less than 1 parsec in diameter, yet it appears to be responsible for all the violent and energetic activity described. From the observed stellar velocities, the mass of the central object has been estimated at roughly 10^6 solar masses. This is only a small fraction of the total mass of the galaxy, but it is much greater than the mass of any known object within it and must therefore represent some new kind of astronomical entity. The best explanation that has been offered by astronomers is that a very massive black hole resides at the core of our galaxy. Its gravitational influence would be responsible for the rapid orbital motions of nearby objects, and matter falling in would be compressed and heated, which would account for the X-ray emission. How such an object formed is unknown, but the evidence indicates that at some time during the formation and evolution of the Milky Way great quantities of matter were compressed into a very small volume at its center. Such an intense buildup of matter may have resulted from frequent collisions among stars that caused vast numbers of them gradually to coalesce into a single, massive object at the center. The prospect that such

Figure 9.20 A Radio Image of the Galactic Center. This image, made at a wavelength of 6 centimeters (where thermal continuum emission is measured, rather than a spectral feature such as the 21-centimeter line), shows an arc of gas extending some 200 parsecs above the plane of the galactic disk. *(F. Yusef-Zadeh, M. Morris, and D. Chance)*

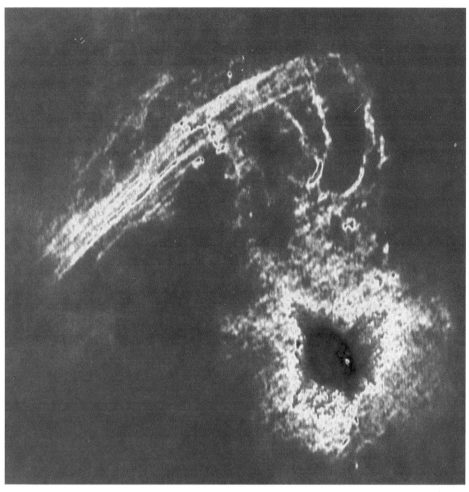

a beast inhabits the core of our galaxy is a bizarre one, yet it may tie in with some of the strange happenings that have been observed in the nuclei of other galaxies (see Chapter 13).

Globular Clusters Revisited

We have not yet said much about the globular clusters, the gigantic spherical conglomerations of stars that orbit the galaxy. A large globular cluster can be an impressive sight when viewed with a small telescope, and a number of them are popular objects for astronomical photography (Fig. 9.21). Well over one hundred of these clusters have been cataloged; no doubt many more are obscured from our view by the disk of the Milky Way.

A single cluster may contain several hundreds of thousands of stars and have a diameter of 10 to 20 parsecs. The mass is usually in the range of several hundred thousand to a few million solar masses. To contain so many stars in a relatively small volume, globular clusters must be very dense compared to the galactic disk. The average distance between stars in a globular cluster is only about 0.1 parsec (recall that the nearest star to the Sun is more than 1 parsec away). If the Earth orbited a star in a globular cluster, the nighttime sky would be a spectacular sight, with hundreds of stars brighter than the first magnitude.

An H-R diagram for a globular cluster (Fig. 9.22) is rather peculiar looking compared to the familiar diagram for stars in the galactic disk. The main sequence is almost nonexistent, having stars only on the extreme lower portion. On the other hand many red giants and

Figure 9.21 A Globular Cluster. This cluster, designated Hodge II, actually resides in one of the Magellanic Clouds, which are small galaxies just outside the Milky Way. (© 1984 Anglo-Australian Telescope Board)

a number of blue stars lie on a horizontal sequence extending from the red giant region across to the left in the diagram. Putting all these facts together points to a very great age for globular clusters. As we saw in Chapter 7, in a cluster H-R diagram the point where the main sequence turns off toward the red giant region indicates the age of the cluster: the lower the turn-off point, the older the cluster. Using this technique to date globular clusters leads typically to age estimates of 14 to 16 billion years, comparable to the accepted age of the galaxy itself. Thus globular clusters are among the oldest objects in the galaxy. As we shall see, their presence in the galactic halo provides important data on the early history of the galaxy.

The sequence of stars extending across the globular cluster H-R diagram from the red giant region to the left is called the **horizontal branch**. It is most likely that these are stars that have completed their red giant stages (as discussed in Chapter 7) and are moving to the left on the diagram, perhaps on their way to becoming white dwarfs.

Some globular clusters have been found to contain X-ray sources in their centers (Fig. 9.23). In a few cases these are X-ray "bursters," described in Chapter 8 as

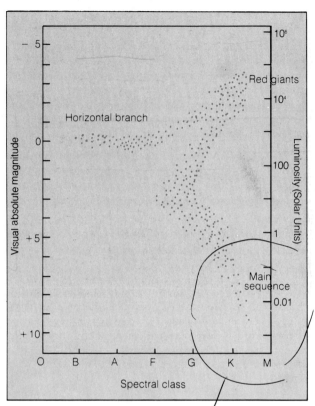

Figure 9.22 The H-R Diagram for a Typical Globular Cluster. The main sequence branches off at a point near the bottom, indicating the great age of the cluster.

being neutron stars that are slowly accreting new matter on their surfaces, probably in binary systems where the companion star is losing mass. In other cases, however, the X-ray source does not have the recognizable properties of a binary system and may instead be caused by a single object at the center of the cluster. This object could be a giant black hole, formed as stars in the cluster collided and fell together in the core.

A Massive Halo?

The halo of the Milky Way galaxy has traditionally been envisioned as a very diffuse region, populated by a scattering of dim stars and dominated by the giant globular clusters. There was little or no evidence of any substantial amount of interstellar material, and the halo was assumed not to contain a significant fraction of the galaxy's mass.

Figure 9.23 A Globular Cluster X-ray Source. Several globular clusters have been found to contain intense X-ray sources at their centers. This X-ray image obtained with the *Einstein Observatory* clearly shows the central source of a globular cluster. *(Harvard-Smithsonian Center for Astrophysics)*

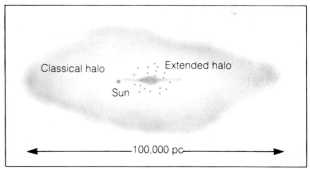

Figure 9.24 The Extensive Halo of the Milky Way. This illustrates the shape and scale of the very extended galactic halo, whose presence has been inferred from the shape of the galactic rotation curve (see Fig. 9.8) and from absorption lines formed by interstellar gas.

Some of these ideas are changing as a result of very recent discoveries, and now it is believed that the halo of the galaxy may be much more extensive and massive than previously thought (Fig. 9.24). We mentioned in an earlier section of this chapter that radio observations have revealed an unexpectedly high amount of mass in the outer portions of the disk, indicating that the galaxy is more extensive than previously believed. At the same time, other evidence (to be described in later chapters) led to a general expectation that many galaxies may have large quantities of matter in their halos. Finally, in the early 1980s, ultraviolet observations revealed a large amount of very hot, rarefied interstellar gas in the halo of our galaxy, probably extending to distances of several thousand parsecs above and below the plane of the disk. This gas is very turbulent, with clouds traveling at speeds of several hundred kilometers per second, and is highly ionized, indicating a temperature of at least 100,000 K or more.

At present we cannot say much about the origin or total quantity of material in the halo of our galaxy. It is quite possible that in addition to the gas now known to be present, there are enough stars, too dim to be detected, to add significantly to the total mass of the galaxy. Some astronomers believe that as much as 90 percent of the galaxy's mass may be in the halo. This means that the total mass of the galaxy is more like 10^{12} solar masses than the 10^{11} or so that we estimated from applying Kepler's third law to stars in the Sun's vicinity.

The *Space Telescope,* to be launched later in the 1980s, should be able to settle this question by making more sensitive measurements of both the gas and the stars in the halo.

Perspective

We have discussed the anatomy of the galaxy, particularly its overall structure and motions. The solar system is located some 10 kiloparsecs from the center of a flattened rotating disk containing a few hundred billion stars. Some 10 to 15 percent of the mass of the disk is in the form of interstellar gas and dust, the interstellar medium serving as the source of new stars and the repository of material ejected by aging stars. The structure of the disk has been unraveled with the help of radio and infrared observations. It consists of spiral arms—delineated by bright, hot stars and interstellar gas—and a central nucleus that contains relatively dim red stars and a mysterious, energetic gremlin that stirs up the core region. Many questions remain about the structure and extent of our galaxy; some will be answered with the launch of the *Space Telescope.*

We have yet to discuss the workings of the galaxy; we have seen what it is like, but have said little about why. This will be the subject of the next chapter.

Summary

1. The Milky Way is a spiral galaxy consisting of a disk with a central nucleus and a spherical halo where the globular clusters reside. The disk is about 30 kiloparsecs in diameter.
2. Distances within the Milky Way can be determined by measuring variable star periods and applying the period-luminosity relation.
3. The true size of the Milky Way and the Sun's location within it were difficult to determine because the view from our location within the disk is obscured by interstellar extinction.
4. Star counts seem to indicate that the Sun is in the densest part of the Milky Way, but measurements of the distribution of globular clusters and analysis of stellar motions show that the Sun is some 10 kiloparsecs from the center of the galaxy. The discrepancy was resolved when it was discovered that interstellar extinction affects the star counts.
5. The inner part of the galactic disk rotates rigidly, while the outer parts are fluid, with each star following its own individual orbit approximately described by Kepler's laws.
6. Individual stellar orbits in the Sun's vicinity are generally in the plane of the disk and are nearly circular. Deviations from perfect circular motion by stars are called peculiar motions (or the solar motion in the case of the Sun).
7. The mass of the galaxy, determined by the application of Kepler's third law to the Sun's orbit, is roughly 10^{11} solar masses. This technique ignores any mass that resides in the halo or in the galactic plane outside the Sun's position.
8. Between 10 and 15 percent of the mass of the disk is in the form of interstellar matter, including both dust particles and gas. This medium ranges from dark, cold clouds, where the gas is molecular and where stars form, to more diffuse clouds and to a very hot intercloud medium with temperatures as high as one million K. The interstellar material is stirred up by supernovae, which account for the very hot regions and for motions of interstellar clouds.
9. The spiral structure of our galaxy is most easily and directly measured through radio observations of the 21-centimeter line of hydrogen atoms in space. The spiral pattern is complex, with many segments of spiral arms.
10. A variety of evidence indicates that there is chaotic, energetic activity associated with the central core of our galaxy. The data show that there is a compact, massive object there, and the best explanation is that it is a massive black hole.
11. The globular clusters that inhabit the halo of the galaxy are very old and therefore provide information on the early history of the galaxy.
12. There may be large quantities of mass in the galactic halo in the form of interstellar gas, dim, cool stars, or both. It appears that as much as 90 percent of the mass of the galaxy may be in the halo.

Review Questions

1. Recalling what you learned about stellar parallaxes as distance determinators in Chapter 6, calculate the distance to a star whose parallax angle is $\pi = 0''.001$ (about the smallest angle that can be measured with current techniques). Compare the distance to such a star with the diameter of the galactic disk, and discuss the practicality of using parallax measurements to probe the structure of our galaxy.
2. Suppose a Cepheid variable is found to have a period of 10 days. From the period-luminosity relation, its average absolute magnitude is found to be $M = -4$. Its average apparent magnitude is $m = +16$. What is its distance? How does this compare with the diameter of the galactic disk?
3. Compare Shapley's reasoning, that the center of the galaxy should be the place about which the globular clusters orbit, with that of Aristarchus, who determined for similar reasons that the Sun, not the Earth, is at the center of the solar system.
4. Why do the most rapid circular orbits around the galaxy occur at the distance from the center where rigid-body rotation gives way to fluid rotation?
5. The fact that interstellar dust makes a distant star appear redder than it otherwise would means that the star's measured $B - V$ color index will be larger than it should be. How do you think the error could be detected; that is, how could an as-

tronomer tell whether the color index has been affected by interstellar dust?
6. Explain why the interstellar medium must have the same chemical composition as the Sun.
7. Summarize the reasons why the 21-centimeter line of hydrogen is the best tool we have for tracing the spiral structure of the galaxy.
8. Suppose the orbit of a star lying 20 kiloparsecs from the galactic center were analyzed to determine the mass of the galaxy. How would you expect the result to differ from what was found from analysis of the Sun's orbit?
9. Summarize the evidence for the existence of a massive black hole at the center of the Milky Way.
10. We have seen that the globular clusters are comparable in age to the galaxy itself. What implications does this have for the chemical composition of the stars in these clusters?

Additional Readings

Bok, B. J. 1972. Updating galactic spiral structure. *American Scientist* 60:708.

———. 1981. The Milky Way galaxy. *Scientific American* 244(3):92.

Bok, B. J., and P. Bok. 1981. *The Milky Way*. Cambridge, Mass.: Harvard University Press.

de Boer, K. S., and B. D. Savage. 1982. The coronas of galaxies. *Scientific American* 247(2):54.

Geballe, T. R. 1979. The central parsec of the galaxy. *Scientific American* 241(1):52.

Gingerich, O., and B. Welther. 1985. Harlow Shapley and the Cepheids. *Sky and Telescope* 70(6):540.

Herbst, W. 1982. The local system of stars. *Sky and Telescope* 63(6):574.

King, I. R. 1985. Globular clusters. *Scientific American* 252(6):78.

Mullholland, D. 1985. The beast at the center of the galaxy. *Science 85* 6(7):50.

Sanders, R. H., and G. T. Wrixon. 1974. The center of the galaxy. *Scientific American* 230(4):66.

Seeley, D., and R. Berendzen. 1978. Astronomy's great debate. *Mercury* 7(3):67.

Weaver, H. 1975/6. Steps towards understanding the large-scale structure of the Milky Way. *Mercury* 4(5):18 (Part 1); 4(6):18 (Part II); 5(1):19 (Part III).

Chapter 10

The Formation and Evolution of the Galaxy

A spiral galaxy (NGC253), showing its stellar populations. *(© 1980 Anglo-Australian Telescope Board)*

Chapter Preview

Stellar Populations and Elemental Gradients
Stellar Cycles and Chemical Enrichment
The Care and Feeding of Spiral Arms
Galactic History

The galaxy that we observe today is the end product of some 10 to 20 billion years of evolution. A variety of processes have shaped it, and many have left unmistakable evidence of their action. In this chapter we will discuss the major influences on the galaxy and the developments that they have made.

We have laid the groundwork that will enable us to understand one of the most elegant concepts in the study of astronomy: the interplay between stars and interstellar matter and the long-term effects of this cosmic recycling on the evolution of the galaxy.

Stellar Populations and Elemental Gradients

To begin this story, we focus first on the overall structure of the galaxy (Fig. 10.1) and the types of stars found in various regions. We have already seen that most of the young stars in the galaxy tend to lie along the spiral arms; indeed, it is the brilliance of the hot, massive O and B stars and their associated H II regions that makes the arms stand out from the rest of the galactic disk. We have also noted that in the nuclear bulge of the galaxy, as well as in the globular clusters and the halo in general, there are few young stars and relatively little interstellar material.

Careful scrutiny has revealed a number of distinctions between the stars that lie in the disk, particularly those in the spiral arms, and the stars in the nucleus and halo. Analysis of the chemical compositions of the stars has shown that those in the halo and nucleus tend to have relatively low abundances of heavy elements such as metals, while stars in the vicinity of the Sun (and in the spiral arms in general) contain these elements in greater quantities. Nearly all stars are dominated by hydrogen, of course, but in the halo stars the heavy elements represent an even smaller trace than in the spiral arm stars. The relative abundance of heavy

Figure 10.1   A Spiral Galaxy Similar to the Milky Way. This vast conglomeration of stars and interstellar matter is following its own evolutionary course, just as individual stars do. Note that the spiral arms tend to be blue in color, while the central bulge is yellow. These colors indicate the types of stars that dominate in the different parts of the galaxy. (© 1980 Anglo-Australian Telescope Board)

elements in a halo star may typically be a factor of one hundred below that found in the Sun. If iron, for example, is 10^{-5} as abundant as hydrogen in the Sun, it may be only 10^{-7} as abundant in a halo star.

A picture has emerged in which the galaxy has two distinct groups of stars, those in the halo and nuclear bulge on one hand, and those in the spiral arms on the other (Fig. 10.2). These two groups have been designated **Population II** (for the stars in the halo) and **Population I** (those in spiral arms) (Table 10.1). The Sun is thought to be typical of Population I stars, and its chemical composition is usually adopted as representative of the entire group.

Almost every time an attempt is made to classify astronomical objects into distinct groups, however, the boundaries between them turn out to be a little vague. There are usually intermediate objects whose properties fall between categories, and the stellar populations are no exception to this. For example, the stars in the disk of the galaxy that do not fall strictly within the

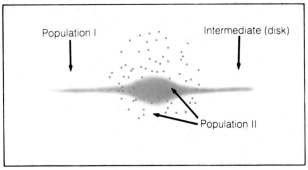

Figure 10.2 The Distribution of Populations I And II. This cross-section of the galaxy shows that Population II stars lie in the halo and central bulge, whereas Population I stars inhabit the spiral arms in the plane of the disk. Intermediate population stars are distributed throughout the disk.

Table 10.1 Stellar Populations

	Population I	Population II
Age	Young to intermediate	Old
Distribution	Disk, spiral arms	Halo, central bulge
Composition	Normal metals	Low metals
Constituents	Disk, arm stars	Low-metal stars
	O, B stars	High-velocity stars
	Interstellar matter	Globular clusters
	Type I Cepheids	Type II Cepheids
		RR Lyrae stars
	Type II supernovae	Type I supernovae
		Planetary nebulae
	The Sun	

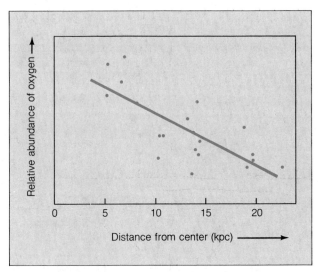

Figure 10.3 An Abundance Gradient. This diagram shows how the abundance of oxygen (relative to hydrogen) varies with distance from the center of the galaxy. More stellar generations have lived and died in the dense regions near the center, so nuclear processing is further advanced in the central region than it is farther out. (*Data from Blair, W. P., R. P. Kirschner, and R. A. Chevalier 1982*, Astrophysical Journal 254:50)

spiral arms have intermediate properties between those of Population I and Population II and are often called simply the **disk population.** Furthermore, there are gradations within the two principal population groups, and we speak therefore of "extreme" or "intermediate" Population I or Population II objects. It is much more accurate to view the stellar populations as a smooth sequence of stellar properties represented by very low heavy element abundances at one end and sunlike abundances at the other.

The differences in chemical compositions from one part of the galaxy to another provide astronomers with very important clues in their efforts to piece together the history of the galaxy. Differences that occur gradually over substantial distances are referred to as **gradients,** and the variations of stellar composition from one part of the galaxy to another are therefore referred to as **abundance gradients.** There is a gradient of increasing heavy element abundances from the halo to the disk (Fig. 10.3). Within the disk itself there is a similar gradient from the outer to the inner portions. (This does not include the stars in the nuclear bulge, which, as we have pointed out, tend to have low abundances of heavy elements.)

Most of the stars near the Sun belong to Population I; they are disk stars orbiting the galactic center in approximately circular paths like the Sun. There are a few nearby Population II stars here and there, and they are distinguished by a number of properties in addition to their compositions. The most easily recognized of these properties is that their motions depart drastically from those of Population I stars. As a rule, Population II stars do not follow circular paths in the plane of the disk, instead having highly elliptical orbits that are randomly oriented (Fig. 10.4), much like cometary orbits in the solar system. Thus most of the Population II stars that are seen near the Sun are just passing through the disk from above or below it. The Sun and the other Population I stars in its vicinity move in their orbits at speeds around 250 kilometers per second, so the Sun moves very rapidly with respect to the Population II stars that pass through its neighborhood in a perpendicular direction. From our perspective here on the Earth, these Population II objects are therefore classified as **high-velocity stars.**

High-velocity stars were discovered and recognized as a distinct class even before the differences between Population I and Population II were enumerated. It was only later that these objects were equated with Population II stars following randomly oriented orbits in the galactic halo.

The systematic differences between the stellar populations arose from the differing conditions when they formed. In this regard it is important to recall that globular clusters, which are representative of Population II, are among the oldest objects in the galaxy, while Population I includes young, newly formed stars. We will follow this discussion further in a later section.

Stellar Cycles and Chemical Enrichment

The fact that stars in different parts of the Milky Way have distinctly different chemical makeups is readily explained in terms of stellar evolution. Recall (from Chapter 7) that as a star lives its life, nuclear reactions gradually convert light elements into heavier ones. The first step occurs while the star is on the main sequence, when hydrogen nuclei in its deep interior are fused into helium. Later stages, depending on the mass of the star, may include the fusion of helium into carbon and possibly the formation of even heavier elements by the addition of further helium nuclei. The most massive stars, which undergo the greatest number of reaction stages, also form heavy elements in the fiery instants of their deaths in supernova explosions.

All of these processes work in the same direction: They act together gradually to enrich the heavy element abundances in the galaxy. Material is cycled back and forth between stars and the interstellar medium, and with each passing generation a greater supply of heavy

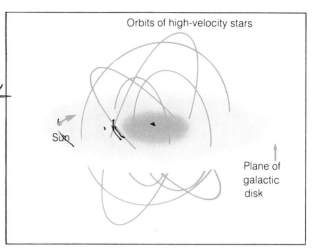

Figure 10.4 The Orbits of High-Velocity Stars. These stars follow orbits that intersect the plane of the galaxy. When such a star passes near the Sun, it has a high velocity relative to us because of the Sun's rapid motion along its own orbit.

elements is available. As a result, stars formed where there has been a lot of stellar cycling in the past are born with higher quantities of heavy elements than those formed where little previous cycling has occurred. We know that Population II stars are very old because they formed out of material that had not yet been chemically enriched.

Population I stars like the Sun condensed from interstellar material that had previously been processed in stellar interiors and in supernova explosions. As a rule, therefore, they formed at moderate-to-recent times in the history of the galaxy and are not as ancient as Population II objects.

Besides elemental abundance variations with age, there can also be variations with location. As noted in the preceding section, there is a distinct abundance gradient within the disk of the galaxy, the stars nearer the center having higher abundances of heavy elements than those farther out. This is a reflection of enhanced stellar cycling near the center, rather than age difference. The central portion of the disk is where the density of stars and interstellar material is highest, and in this region there has been relatively active stellar processing because star formation has proceeded at a greater rate than in the less dense outer portions of the disk. Over the lifetime of the galaxy, more generations of stars have lived and died in the central region than in the rarefied outer reaches.

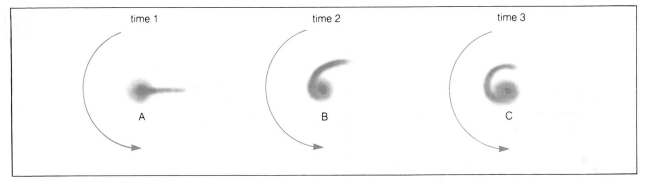

Figure 10.5 The Windup of Spiral Arms. If spiral arms were simple streamers of material attached to a rotating galaxy, they would, within a few hundred million years, wind tightly around the nucleus. The galaxy is much older than that, so some other explanation of the arms is needed.

The Care and Feeding of Spiral Arms

In Chapter 9 we described the structure of the galaxy and described how its spiral nature was discovered and mapped, but we said nothing about how the spiral arms formed nor why they persist. Both questions are important, for it is clear that if the arms were simply streamers trailing along behind the galaxy as it rotates, they should have wrapped themselves tightly around the nucleus (Fig. 10.5) like string being wound into a ball. Because the galaxy is old enough to have rotated at least forty to fifty times since it formed, something must be preventing the arms from winding up, or they would have done so long ago.

Maintaining the spiral arms—a distinct question from that of forming them in the first place—is a very complex business, and we do not yet fully understand how it is done. The first successful theory on this appeared in 1960 and is still being refined. The essence of this theory is that the large-scale organization of the galaxy is imposed on it by wave motions. We have understood waves as oscillatory motions created by disturbances, and we know that waves can be transmitted through a medium over long distances while individual particles in the medium move very little. In the case of water waves, for example, a floating object simply bobs up and down as a wave passes by, whereas the wave itself may travel a great distance. Here we are distinguishing between the wave and the medium through which it moves.

The waves that apparently govern the spiral structure in our galaxy are not transverse waves, like those in water, but are compressional waves similar to sound waves and to certain seismic waves (the P waves; see Chapter 16 for an elaboration on the different types of waves). In this case the wave pattern consists of alternating regions of high and low density. When a compressional wave passes through a medium, the individual particles vibrate back and forth along the direction of the wave motion. There is no motion perpendicular to that direction, in contrast with water waves.

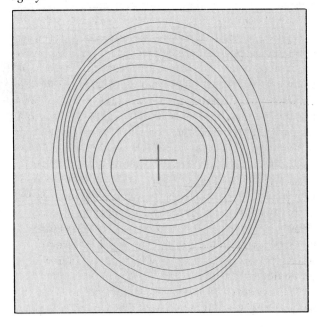

Figure 10.6 The Density Wave Theory. This drawing depicts the manner in which circular orbits are deformed into slightly elliptical ones by an outside gravitational force. The nested elliptical orbits are aligned in such a way that there are density enhancements in a spiral pattern. This pattern rotates at a steady rate and does not wind up more tightly.

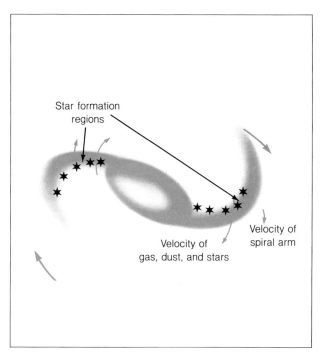

Figure 10.7 The Effect of a Spiral Density Wave on the Interstellar Medium. As the wave moves through the medium, material is compressed, leading to enhanced star formation. The young stars, H II regions, and nebulosity along the density wave are what we see as a spiral arm.

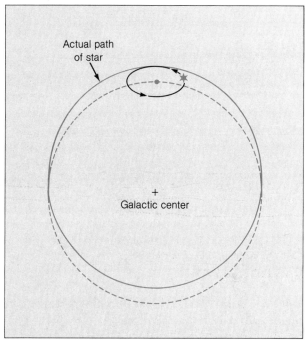

Figure 10.8 Epicyclic Orbits. The oscillatory motions created by spiral density waves cause individual stars to have orbits about the galaxy that are not perfect circles but ovals. The shape of the orbit may be viewed as a combination of a large circle and a small circle (epicycle). *(Adapted from Shu, F. 1982, The Physical Universe (San Francisco: University Science Press), Fig. 12.12, p. 267)*

The theory of galactic spiral structure that invokes waves as a means of maintaining the spiral arms is called the **density wave theory**. This theory supposes that a spiral wave pattern centered on the galactic nucleus creates a pinwheel of alternating dense and relatively empty regions (Fig. 10.6). The density waves have more effect on the interstellar medium than on stars, so the spiral arms are characterized primarily by concentrations of gas and dust. These lead, in turn, to concentrations of young stars, because star formation is enhanced in regions where the interstellar material is compressed (Fig. 10.7).

In its simplest form, the wave pattern is double, that is, there are just two spiral arms emanating from opposite sides of the nucleus. There are galaxies that have such a simple spiral structure, but many, including the Milky Way, are more complicated. The density wave theory allows the possibility of more arms if the waves have a shorter "wavelength," or distance between them.

The waves rotate about the galaxy at a fixed rate that is constant from the inner portions to the outer edge. Thus, while the outer portions of the arms appear to trail the rotation, in fact they move all the way around the galaxy in the same time period that the inner portions do. The waves are essentially rigid, in strong contrast with the motions of the stars. Just as in the case of water waves, the motion of the spiral density waves is quite distinct from the motions of the individual particles (stars) in the medium through which the waves travel. In the Milky Way, individual stars orbit the galaxy at their own speeds. In the inner part of the galactic disk, out to the Sun's position, the stars move faster than the density waves. This means that as a star circles the galaxy, every so often it will overtake and pass through a region of high density as it penetrates a spiral arm. Because their motion is slowed slightly when they are in a density wave, stars tend to become concentrated in the arms, just as cars on a highway become jammed together at a point where traffic flow is constricted.

The motion of an individual star about the galaxy is not precisely circular. The spiral density waves impose

Astronomical Insight 10.1

A Different Kind of Spiral Arm

Recently an alternative to the spiral density wave theory has been proposed to explain the existence and persistence of the spiral arms. In this picture, the differential rotation of the galaxy, along with the great extent of some regions of star formation, plays a key role.

There are regions in the galaxy where very large complexes of interstellar gas and dust are found. Here star formation seems to take place by the sequential process described in Chapter 7, in which the stellar winds and supernova explosions of massive, short-lived stars give rise to new generations in the near vicinity, as shock waves compress the adjacent clouds. In several cases these hotbeds of star formation are many tens of parsecs in extent. On this size scale, the differential rotation of the galaxy will distort the star-forming region. The part of the vast cloudy region nearer the galactic center is pulled ahead of the outer part, and the entire complex is stretched into a curved segment that resembles a portion of a spiral arm.

In addition to its shape, such a region has other characteristics of spiral arms, most notably a concentration of interstellar matter and hot, young stars.

In due course a giant dark cloud will be consumed by star formation, the luminous stars in it will die out, and the entire region will fade into obscurity. A spiral arm segment created in this way is therefore only a temporary feature of the galaxy.

If such segments were continually being formed and dissipated, at any given moment a galaxy would have plenty of spiral structure. The arms would not be permanent but would be constantly forming and dissipating. This is especially likely in the cases of galaxies with a rather chaotic spiral structure, consisting of many bits and pieces of arms, rather than a small number of smooth, complete ones in a simple overall pattern. From what we know of the structure of the Milky Way, our own galaxy is a strong candidate for this process.

---

slight oscillatory motions on the stars as they pass through the waves, with the result that the orbit of a star can be represented as an epicyclic motion (Fig. 10.8)

We have noted that the spiral arms stand out mainly because hot, luminous young stars are found exclusively in the arms due to the enhancement of star formation there where the interstellar gas is compressed. According to the density wave theory, the most active stellar nurseries should be located on the inside edges of the arms, where stars and gas catch up with and enter the compressed region. This is difficult to check observationally in our own galaxy, but it seems to be true of others.

The most luminous stars have such short lifetimes that they evolve and die before having sufficient time to pull ahead of the arm where they were born; therefore, we find no bright, blue stars between the arms. Less luminous stars, which live billions instead of only millions of years, can survive long enough to orbit the galaxy many times, and these stars are spread almost uniformly throughout the galactic disk, between the arms as well as in them. The Sun, for example, is old enough to have circled the galaxy some fifteen to twenty times, and therefore has passed through spiral arms and the intervening gaps on several occasions. It is only coincidence that the Sun is presently in a spiral arm; it is not necessarily the one in which it formed.

The initial formation of the density wave that is the cause of the spiral arms is a separate question. A rotating disk of particles, such as the disk of a galaxy like ours, will naturally form spiral density waves if disturbed by an outside force (Fig. 10.9). This suggests that the spiral density waves in our galaxy were triggered by the gravitational influence of neighboring galaxies, most likely the Magellanic Clouds, a pair of small galaxies that orbit the Milky Way.

Another possible mechanism for creating spiral density waves, with no help from neighboring galaxies, occurs in certain galaxies that have noncircular central bulges. In these **barred spiral galaxies** (see Chapter 11) the asymmetric shape of the central region can create the gravitational disturbance needed to initiate spiral density waves. Some calculations suggest that the barred spiral structure is the more natural form for a disklike galaxy to take and that the presence of a massive, extended halo is what prevented our own galaxy from becoming a barred spiral instead of the "normal" spiral it is. If so, other normal spiral galaxies may also have massive halos, whereas the barred spirals may not. This will be discussed further in the next chapter.

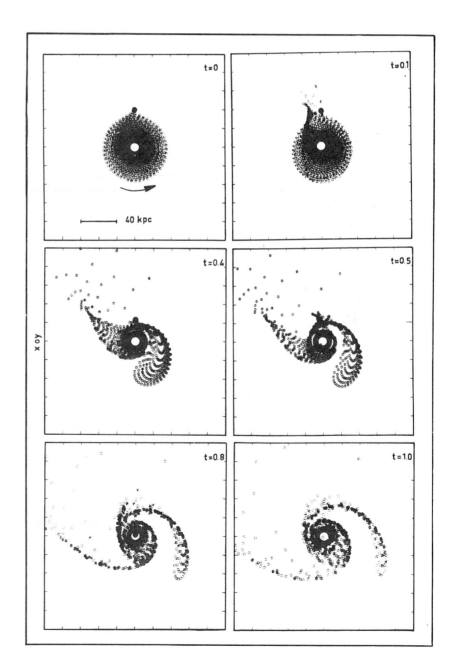

Figure 10.9 The Creation of Spiral Structure. This sequence of computer-generated models shows how a rotating disk forms spiral density waves when subjected to a gravitational force.

Galactic History

We are now in a position to tie together all the diverse information on the nature of the Milky Way and develop a picture of its formation and evolution.

The pertinent facts that must be explained include the size and shape of the galaxy, its rotation, the distribution of interstellar material, elemental abundance gradients, and the dichotomy between Population I and Population II stars in terms of composition, distribution, and motions. The task of fitting all the pieces of the puzzle together is made easier by the fact that we can reconstruct the time sequence, knowing that as a rule stars with high abundances of heavy elements were formed more recently than those with lower abundances.

The oldest objects in the galaxy are in the halo, which is dominated by the globular clusters but contains a large (but unknown) number of isolated, dim, red stars as well. Estimates based on the main sequence turnoff in the H-R diagrams for globular clusters indicate ages of between 14 and 16 billion years, and we conclude that the age of the galaxy itself is comparable.

The spherical distribution of the halo objects about the galactic center demonstrates that when they formed, the galaxy itself was round. Evidently the progenitor of the Milky Way was a gigantic spherical gas cloud consisting almost exclusively of hydrogen and helium. Very early, perhaps even before the cloud began to contract (Fig. 10.10), the first stars and globular clusters formed in regions where localized condensa-

Figure 10.10 Formation of The Galaxy. The pregalactic cloud, rotating slowly, begins to collapse (a). The stars formed before and during the collapse have a spherical distribution and noncircular, randomly oriented orbits (b). The collapse leads to a disk with a large central bulge, surrounded by a spherical halo of old stars and clusters (c). The disk flattens further, and eventually forms spiral arms (d). Stars in the disk have relatively high heavy element abundances because they formed from material that had been through stellar nuclear processing.

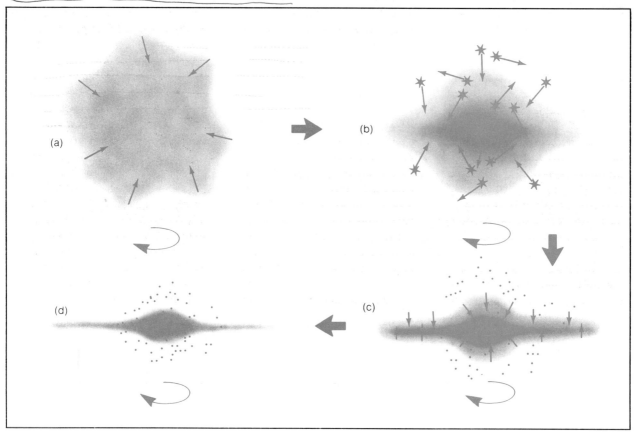

Astronomical Insight 10.2

The Missing Population III

It was explained in the text that the two populations of stars should be viewed as a range of stellar characteristics such as age and metal content rather than a strict separation of two completely distinct classes of stars. The most extreme Population II stars, those with the lowest metal abundances and the greatest ages, represent one end of the range, while the youngest stars in spiral arms, containing the highest metal abundances, represent the other end of the range.

The fact that extreme Population II stars contain *some* heavy elements, even though it may be only 1 percent of the metal content of Population I stars, has proven difficult to explain. In the picture of galactic formation and evolution presented in this chapter, the Population II stars are portrayed as the first stars to form from the collapsing cloud of gas that was to become the galaxy. This cloud of gas is assumed to be primordial, that is, never involved in stellar formation and processing. Therefore the cloud should have had a composition consisting almost entirely of hydrogen and helium, with *no* heavy elements. The question, then, is why the oldest stars we observe, those belonging to extreme Population II, have the metal contents that they do. Even these stars must have formed from material that had been processed through stars, becoming somewhat enriched with heavy elements.

The missing population of stars formed from completely unprocessed material and having virtually zero heavy element abundances has become known as **Population III**. Several astronomers have spent considerable time and effort in the unsuccessful attempt to find such stars.

No widely accepted explanation for the lack of Population III stars has been devised, but there have been suggestions. One is that before the pregalactic cloud collapsed, there was a brief episode of star formation and nuclear processing in which a number of very massive, short-lived stars were formed and quickly evolved to explode in supernovae, creating heavy elements. These heavy elements then became part of the gas that collapsed to form the disk, so that stars formed during and after the collapse (the oldest stars we now observe) contain some heavy elements and represent what we know today as Population II. None of the original Population III stars remain because they were all too massive and short-lived to survive for long.

The chief difficulty in this and other pictures in which an early star-formation episode occurred is to explain how star formation became so active at a time when the pregalactic material was still widely dispersed in a giant, tenuous cloud. Despite our lack of understanding of *how* this occurred, it seems probable that it did, since we find no Population III stars. Furthermore, there is independent evidence that galaxies do undergo periods of intense star formation and processing; such episodes are observed in some other galaxies, called **starburst galaxies**. Therefore, it is quite possible that our galaxy went through such a time early in its history. Several studies are under way today in an effort to understand how and why this may have occurred.

tions occurred. These stars contained few heavy elements and were distributed throughout a spherical volume with randomly oriented orbits about the galactic center. Eventually the entire cloud began to fall in on itself, and as it did, star formation continued to occur, so many stars were born with motions directed toward the galactic center. These stars assumed highly elongated orbits, accounting for the motions observed today in Population II stars.

Apparently the pregalactic cloud was originally rotating, for we know that as the collapse proceeded, a disklike shape resulted. Rotation forced this to happen, just as it did in the case of the contracting cloud that was to form the solar system (see Chapter 15). Rotational forces slowed the contraction in the equatorial plane, but not in the polar regions, so that material continued to fall in there. The result was a highly flattened disk. Stars that had already formed before the disk took shape retained the orbits in which they were born; stars are unaffected by the fluid forces (such as viscosity) that caused the gas to continue collapsing to form a disk.

Stars that were formed after the disk had developed had characteristics unlike those of their predecessors in at least two respects: They contained greater abundances of heavy elements, because by this time some stellar cycling had occurred, enriching the interstellar gas; and they were born with circular orbits lying in the plane of the disk. These are the primary traits of Population I stars.

Since the time of the formation of the disk, there have been additional but relatively gradual changes. Further generations of stars have lived and died, continuing the chemical enrichment process (particularly in the inner regions of the disk) and creating the chemical abundance gradient mentioned earlier. Apparently the enrichment process was once more rapid than it is today, because the present rate of star formation is too slow to have built up the quantities of heavy elements that are observed in Population I stars. At some time in the past, probably just when the disk was forming, there must have been a period of intense star formation, during which the abundances of heavy elements in the galaxy jumped from almost none to nearly the present level. A large fraction of the galactic mass must have been cooked in stellar interiors and returned to space in a brief intense episode of stellar cycling that has not been matched since.

We have just about recounted all the events that led to the present-day Milky Way, except for the formation of the spiral arms. It is not known when this took place, but it was probably soon after the formation of the disk itself. We see relatively few disk-shaped galaxies without spiral structure; this observation leads to the conclusion that a disk galaxy does not exist long in an armless state.

It is difficult to guess exactly what may become of the galaxy in the future. It appears that some sort of balance has been reached in the grand recycling process between stars and the interstellar material, so that the interstellar medium is being replenished by evolving and dying stars about as rapidly as it is being consumed by stellar births. Therefore we do not expect the interstellar material to disappear gradually, terminating star formation, as it apparently has in certain other types of galaxies.

Perspective

We have now seen our galaxy as an active, dynamic entity. Individual stars orbit its center in individual paths, yet the overall machinery is systematically organized. The majestic spiral arms rotate at a stately pace, while stars in interstellar matter pass through them. We have come to understand the active processes of stellar cycling and chemical enrichment, which still occur. In a constant turnover of stellar generations, the deaths of the old give rise to the births of the new, while the violence of their death throes energizes a chaotic interstellar medium.

The lessons we have learned by examining our own Milky Way will be remembered as we move out into the void, probing the distant galaxies.

Summary

1. Stars in the galactic halo have low abundances of heavy elements and are referred to as Population II stars, while those in the spiral arms of the disk have "normal" compositions and are called Population I stars.
2. There are gradients of increasing heavy element abundance from halo to disk and from the outer to the inner portions within the disk.
3. Population II stars have randomly oriented, highly elliptical orbits, while Population I stars have

nearly circular orbits that lie in the plane of the disk. When passing through the disk near the Sun's location, Population II stars are seen as high-velocity stars.
4. The variations in heavy element abundance from place to place within the galaxy reflect variations in age: Very old stars, formed before stellar nuclear reactions had produced a significant quantity of heavy elements, have low abundances of these elements, while younger stars were formed after the galactic composition had been enriched by stellar evolution.
5. Spiral arms are probably density enhancements produced by spiral density waves, which rotate about the galaxy while stars and interstellar material pass through them. Because the interstellar gas is compressed in these density waves, star formation tends to occur there, explaining why young stars are found predominantly in the spiral arms.
6. The existence and characteristics of Population I and II stars can be explained in a picture of galactic evolution that begins with a spherical cloud of gas that has little or no heavy elements at first. Population II stars are those that formed while the cloud was still spherical or just beginning to collapse and had no significant quantities of heavy elements. Population I stars are those that formed later, after the galaxy had collapsed to a disk and stellar evolution had produced some heavy elements.
7. The spiral arms formed after the disk had been created by the collapse of the original cloud. The spiral density waves that maintain the arms probably started as a result of gravitational disturbances, possibly due to nearby galaxies.

Review Questions

1. Explain why the globular clusters have the lowest heavy element abundances of any objects in the galaxy.
2. The intermediate disk population of stars, which uniformly fill the disk of the galaxy without being confined to the spiral arms, have lower heavy element abundances than extreme Population I stars, which have formed recently in the spiral arms. Explain how this has come about.
3. Suppose a star is observed to have its spectral lines shifted to the red. The strong line of hydrogen whose rest wavelength is 6563 Å is found to lie at a wavelength of 6567.38 Å. What is the line-of-sight velocity of this star (see Chapter 4)? Is it a Population I or a Population II star?
4. Explain why the evolution of massive stars, rather than that of the much more common low-mass stars, has been the principal contributor to the enrichment of heavy elements in the galaxy.
5. Why are O and B stars in the galaxy found only in the spiral arms?
6. Explain the contrast between the orbital motions of stars circling the galaxy in the disk and the motions of the spiral density waves that are thought to be responsible for the spiral arms.
7. If the mass of the galaxy is 2×10^{11} solar masses, use Kepler's third law to determine the orbital period of a globular cluster whose semimajor axis is 20 kiloparsecs. (Remember to convert this into astronomical units.)
8. How was the formation of our galaxy similar to the formation of the solar system (Chapter 15)? Does the solar system have the equivalent of a halo?
9. How do we know that the interstellar material that pervades space has, at some time in the past, been processed through stellar interiors?
10. Summarize the various roles played by supernovae in the evolution of the galaxy.

Additional Readings

Bok, B. J. 1981. Our bigger and better galaxy. *Mercury* 10(5):130.
Bok, B. J., and P. Bok. 1981. *The Milky Way*. Cambridge, Mass.: Harvard University Press.
Burbidge, G., and E. M. Burbidge. 1958. Stellar populations. *Scientific American* 199(2):44.
Iben, I. 1970. Globular cluster stars. *Scientific American* 223(1):26.
Larson, R. B. 1979. The formation of galaxies. *Mercury* 8(3):53.
Shu, F. H. 1973. Spiral structure, dust clouds, and star formation. *American Scientist* 61:524.
———. 1982. *The physical universe*. San Francisco: W. H. Freeman.
Weaver, H. 1975/1976. Steps towards understanding the spiral structure of the Milky Way. *Mercury* 4(5):18 (Part 1); 4(6):18 (Part 2); 5(1):19 (Part 3).

Section IV

Extragalactic Astronomy

We are ready now to move out of the confines of our own galaxy and explore the universe beyond. The Milky Way is just one of billions of galaxies in the cosmos, and we will find that while our own system is typical of a certain class of these objects, galaxies have a wide variety of shapes, sizes, and peculiarities.

The first chapter of this section describes both the observational properties of galaxies beyond the Milky Way and their distribution in space. We will rely here on what was learned about our own galaxy in the previous section, for many aspects of its structure and evolution are also characteristic of other galaxies, particularly other spirals. While we will not again discuss the evolution of individual galaxies in great detail, we will assume that the forces that have shaped the Milky Way are at work in other situations as well. Our examination of the distribution of galaxies will bring us to our first discussions of the universe as a whole, and we will pay particular attention to groupings and clusters and the possibility that these are in turn organized into larger associations. We will see in the process that galaxies in clusters can interact with each other in ways that modify their individual properties and influence the nature of the clusters in which they reside.

One question addressed in this chapter is of critical importance to our later discussions of the universe as a whole: It concerns the uniformity of the distribution of matter. The implications of this are not discussed until Chapter 14, but the observational evidence is cited in Chapters 11 and 12.

Chapter 12 describes two major aspects of the universe, and in examining them we will, for the time being, direct our attention away from individual objects to the universe itself. The two profound observational discoveries are the expansion of the universe and the cosmic background radiation that fills it. Both are legacies of the fiery origin of the universe, and their observed properties teach us something of its early history. Here we will see clearly for the first time that the universe itself is a dynamic, evolving entity. It had a beginning, has been changing since, and will continue to develop.

With the perspective gained from Chapter 12, in Chapter 13 we will turn our attention back to individual objects whose properties can best be appreciated in the context of an expanding universe. A variety of peculiar galaxies, especially the quasi-stellar objects, have fantastic properties that were discovered by astronomers only because the nature of the universal expansion was already known. These objects, on the frontiers of the observable universe, have the potential to tell us a great deal about its history. An important theme here is the realization that very distant objects are only seen as they were long ago because of light travel time. This will prove to be useful because it allows us to probe early times in the history of the universe, but it is also a hindrance, because of the difficulty of comparing distant objects with nearby ones. The quasi-stellar objects represent a fundamental component of the early universe, and therefore the mysteries they present for astronomers are among the most important currently under study.

The last chapter of this section describes the current state of cosmology, the science of the universe as a whole. Here we tie together all the diverse information gained from the preceding chapters in assessing the overall structure of the universe, and we especially emphasize the question of its future. If there is a single premier question in modern astronomy, this is it, and we are on the verge of knowing the answer.

Having completed this section, we will be ready to turn our attention to small-scale objects in the universe, the planets, and the solar system.

Chapter 11

Galaxies Upon Galaxies

The Fornax cluster of galaxies.
(© 1984 Royal Observatory,
Edinburgh)

Chapter Preview

The Hubble Classification System
Measuring the Properties of Galaxies
The Origins of Spirals and Ellipticals
Clusters of Galaxies
    The Local Group
    Rich Clusters: Dominant Ellipticals and Galactic
        Mergers
    Cluster Masses
Superclusters
The Origins of Clusters

We have spoken of our galaxy as one of many, a single member of a vast population that fills the universe. Given all that we have learned about humankind's ordinary position in the cosmos, this is no surprise. Having traced our painful progression from the geocentric view to the realization that we occupy an insignificant planet orbiting an ordinary star in an obscure corner of the galaxy, we should be surprised if we found that our galaxy held any kind of unique status in the larger environment of the universe as a whole. It doesn't.

Despite the seeming inevitability of this idea, the actual proof that our galaxy is not alone was some time in coming and only arrived after considerable debate and controversy. The so-called **nebulae**—dim, fuzzy objects scattered throughout the sky—have been known since the early days of astronomical photography, but it was not until the mid-1920s that they were demonstrated to be galaxies, rather than closer objects such as gas clouds or star clusters. The proof of their galactic nature was announced in 1924, when the American astronomer Edwin Hubble (Fig. 11.1) reported that he had found Cepheid variables in the prominent Andromeda galaxy (until then known as the Andromeda nebula, since its true nature was not known) and used the period-luminosity relation to show that this nebula was entirely too distant to be within our own galaxy.

The Hubble Classification System

Having established that the nebulae were truly extragalactic objects, Hubble began a systemic study of their

Figure 11.1 Edwin Hubble. Hubble's discovery of Cepheid variables in the Andromeda nebula led to the unambiguous conclusion that this object lies well beyond the limits of the Milky Way and must therefore be a separate galaxy. Hubble later made important discoveries about the properties of galaxies and what they tell us about the universe as a whole (see Chapter 12). *(Mt. Wilson and Las Campanas Observatories, Carnegie Institution of Washington)*

properties. The most obvious basis for establishing patterns among the various types was to categorize them according to shape. Hubble did this, designating the spheroidal nebulae **elliptical galaxies** (Fig. 11.2), as distinct from the **spiral galaxies** (Figs. 11.3 and 11.4). Within each of the two general types, Hubble established subcategories on the basis of less dramatic gradations in appearance (Table 11.1). The ellipticals displayed varying degrees of flattening and were sorted out according to the ratio of the long axis to the short axis (Fig. 11.5), with designations from E0 (spherical) to E7 (the most flattened). The number following the letter E is determined from the formula $10(1 - b/a)$, where a is the long axis and b is the short axis, as measured on a photograph.

Among the spirals, Hubble based his classification on the tightness of the arms and the compactness of the

Figure 11.2 An Elliptical Galaxy. These smooth, featureless galaxies are probably more common than spirals. *(National Optical Astronomy Observatories)*

Figure 11.3 A Spiral Galaxy. This is NGC 4321, a beautiful example of a face-on spiral galaxy. During the early years of the twentieth century, there was considerable controversy among astronomers over the true nature of objects like this. *(J. D. Wray, University of Texas)*

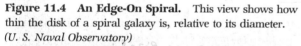

Figure 11.4 An Edge-On Spiral. This view shows how thin the disk of a spiral galaxy is, relative to its diameter. *(U. S. Naval Observatory)*

nucleus (Fig. 11.6). The types ranged from Sa (tight spiral, large nucleus) to Sc (open arms, small nucleus). The Milky Way is probably an Sb in this system, intermediate in both characteristics, although it is difficult to be sure, because we cannot get an outsider's view of what our galaxy looks like.

Hubble recognized a variation of the spiral galaxies in which the nucleus has extensions on opposing sides, with the spiral arms emanating from the ends of the extensions. He called these galaxies **barred spirals** (Fig. 11.7) and assigned them the designations SBa through SBc, using the same criteria as before in establishing the a, b, and c subclasses (Fig. 11.8).

Following Hubble's original work on galaxy classification, an intermediate class called the S0 galaxies has been recognized. These appear to have a disk shape but no trace of spiral arms.

Hubble arranged the types of galaxies in an organization chart that has become known as the "tuning fork" diagram (Fig. 11.9). Because there are two types of spirals, Hubble chose not to force all the types into a single sequence but to split it into two branches. It was thought for a time that this diagram represented

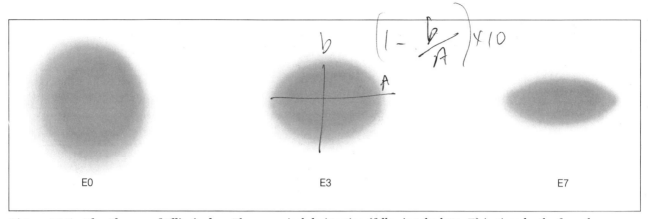

Figure 11.5 The Shapes of Ellipticals. The numerical designation (following the letter E) is given by the formula $(1 - b/a) \times 10$, where b is the short axis and a the long axis of the galaxy image. Here are three examples: the E0 galaxy has $a = b$ (that is, it is circular); the E3 galaxy has $a = 1.43b$; and the E7 has $a = 3.3b$. The E7 galaxy is the most highly elongated of the ellipticals.

an evolutionary sequence, but when later studies showed that all types of galaxies contain old stars, it became clear that one type of galaxy does not evolve into another. The tuning fork diagram is not an age sequence after all, and the differences in galactic type must be explained in some other way.

Most of the galaxies listed in catalogs are spirals, which are about evenly divided between normal and barred spirals. Only about 20 to 30 percent of the listed galaxies are ellipticals, and a comparable number are S0 galaxies. The remaining few percent are called irregular galaxies because they do not fit into the normal classification scheme. Because there are probably many small, dim elliptical galaxies that are usually not sufficiently prominent to appear in catalogs, it seems likely that ellipticals actually outnumber spirals in the universe. This certainly is the case in dense clusters of galaxies, as we will see. Table 11.2 lists a few prominent galaxies whose properties indicate the typical values for galaxies of different types.

Although the irregular galaxies are, by definition, misfits, it has proven possible to find some systematic characteristics even in their case. Most have a hint of spiral structure, although they lack a clear overall pattern, and are designated as **Type I irregulars.** The rest, a small minority, simply do not conform in any way to the normal standards and are assigned the classification of **Type II irregulars** or peculiar galaxies (for example, see Fig. 11.10). The Magellanic Clouds are both Type I irregulars.

Figure 11.6 An Assortment of Spirals. This sequence shows spiral galaxies of several subclasses. *(Palomar Observatory, California Institute of Technology)*

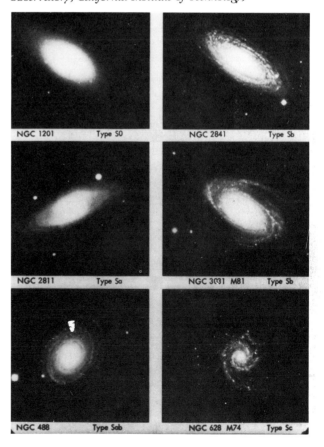

Table 11.1 Types of Galaxies

Type	Designation	Characteristics
Elliptical	E0–E7	Spheroidal shape; subtype determined by the formula $10(1 - b/a)$, where a is the long axis and b is the short axis of the galaxy.
Dwarf elliptical	dw E	Spheroidal shape, very low mass and luminosity.
S0	S0	Disklike galaxies with no spiral structure.
Spiral	Sa–Sc	Disklike galaxies with spiral arms; subtype determined by relative size of nucleus and openness of spiral structure; Sa refers to large nucleus and tightly wound arms, whereas Sc refers to small nucleus and open spiral arms.
Barred spiral	SBa–SBc	Spiral galaxies with elongated, barlike nuclei; subtypes determined the same way as for spiral galaxies.
Irregular I	Ir I	Disklike galaxies with evidence of spiral structure, but not well organized.
Irregular II	Ir II	Galaxies that do not fit into any of the other types; some Ir II galaxies have been found to be normal spirals heavily obscured by interstellar gas and dust.

 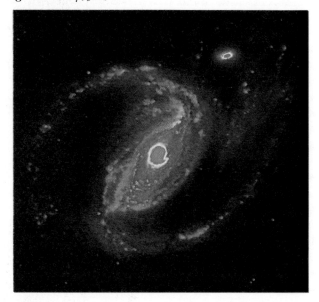

Figure 11.7 Barred Spirals. At left is M83, a well-known example of a spiral galaxy with a barlike structure through the nucleus; at right is a computer-enhanced view of another barred spiral. (left: © *1980 Anglo-Australian Telescope Board*; right: *photo by H. C. Arp, image processing by J. J. Lorre, Image Processing Laboratory, JPL*)

Measuring the Properties of Galaxies

In order to probe the physical nature of the galaxies, we must first determine their distances, because without this information, such fundamental parameters as masses and luminosities cannot be deduced.

We have already mentioned one technique for distance determination that can be applied to some galaxies: the use of Cepheid variables. These stars are sufficiently luminous to be identified as far away as a few million parsecs (**megaparsecs,** abbreviated **Mpc**), which is sufficient to reach the Andromeda nebula and several other neighbors of the Milky Way. It is not adequate, however, for probing the distances of most galaxies, so other techniques had to be developed.

Recall from our discussions of stellar distance determinations (Chapter 6) that we can always find the distance to an object if we know both its apparent and absolute magnitudes. This is the principle behind the spectroscopic parallax method, the main sequence fitting technique, and even the Cepheid variable period-luminosity relation. Any object whose absolute magnitude is known from its observed characteristics can be referred to as a **standard candle,** and an assortment of standard candles are used in extending the distance scale to faraway galaxies.

The most luminous stars are the red and blue supergiants, which occupy the extreme upper regions of the H-R diagram. These can be seen at much greater distances than Cepheid variables and therefore are important links to distant galaxies. The absolute magnitudes of these stars are inferred from their spectral classes, just as in the spectroscopic parallax technique, although these stars are so rare that there is substantial uncertainty in assuming that they conform to a standard relationship between spectral class and luminosity. In a variation of this technique, it is simply assumed that there is a fundamental limit on how luminous a star can be and that in a collection of stars as large as a galaxy, there will always be at least one star at this limit. Then to determine the distance to a galaxy, it is necessary only to measure the apparent magnitude of the brightest star in it and to assume that its absolute magnitude is at the limit, which is about $M = -8$. This technique extends the distance scale by about a factor of 10 beyond what is possible using Cepheid variables, that is, to 10 or more megaparsecs.

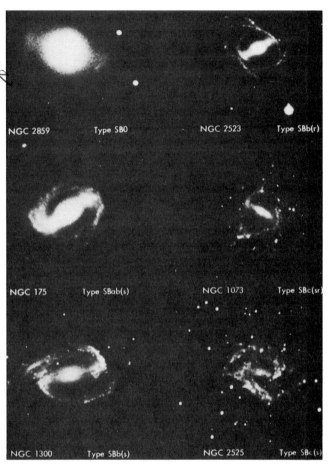

Figure 11.8 Barred Spirals. Roughly half of all spiral galaxies have central elongations, or bars, from which the spiral arms emanate. *(Palomar Observatory, California Institute of Technology)*

Other standard candles come into play at greater distances. These include supernovae (Fig. 11.11), which at peak brightness always reach the same absolute magnitude. (Care must be exercised in making measurements, because there are two distinct types of supernovae with different absolute magnitudes.) Supernovae can be observed at distances of hundreds of megaparsecs and are therefore very useful distance indicators; the major drawback of this method is that we get an opportunity to measure distances only to galaxies that happen to have supernovae occur in them. The ultimate standard candle, useful to distances of thousands of megaparsecs, is the brightest galaxy in a

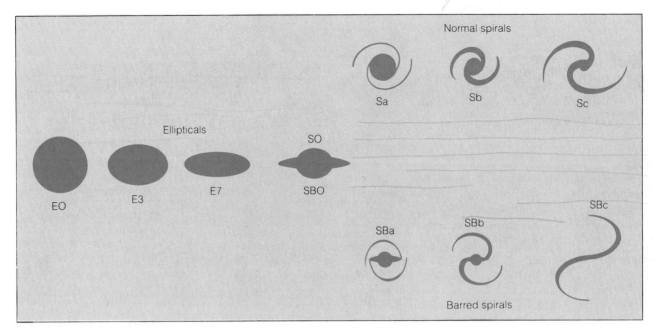

Figure 11.9 The Tuning-Fork Diagram. This traditional manner of displaying the galaxy types was originally devised by Hubble. For quite some time this was thought to be an evolutionary sequence, although the imagined direction of evolution was reversed at least once. Now it is known that in the normal course of events galaxies do not evolve from one type to another.

Table 11.2 Selected Bright Galaxies

Galaxy	Type	Angular Diameter (arcmin)	Diameter (kpc)	Distance (kpc)	Apparent Magnitude*	Absolute Magnitude*	Mass (M.)
Large Magellanic Cloud	Ir I	460	7	55	0.1	−18.7	1×10^{10}
Small Magellanic Cloud	Ir I	150	3	63	2.4	−16.7	2×10^{9}
Andromeda (M31)	Sb		16	700	3.5	−21.1	3×10^{11}
M33	Sc	35	6	730	5.7	−18.8	1×10^{10}
M81	Sb	20	16	3200		−20.9	2×10^{11}
Centaurus A	E0p	14	15	4400	7	−20	2×10^{11}
Sculptor system	dw E	30	1	85	7	−12	3×10^{6}
Fornax system	dw E	40	2	170	7	−13	2×10^{7}
M83	SBc	10	12	3200	7.2	−20.6	—
Pinwheel (M101)	Sc	20	23	3800	7.5	−20.3	2×10^{11}
Sombrero (M104)	Sa	6	8	1200	8.1	−22	5×10^{11}
M106	Sb	15	17	4000	8.2	−20.1	1×10^{11}
M94	Sb	7	10	4500	8.2	−20.4	1×10^{11}
M82	Ir II	8	7	3000	8.2	−19.6	3×10^{10}
M32	E2	5	1	700	8.2	−16.3	3×10^{9}
Whirlpool (M51)	Sc	9	9	3800	8.4	−19.7	8×10^{10}
Virgo A (M87)	E1	4	13	13,000	8.7	−21.7	4×10^{12}

*The apparent and absolute magnitudes represent the light from the entire galaxy. Because the light from a galaxy comes from a large area on the sky, a galaxy of a given apparent magnitude is not as easily seen by the eye as a star of the same magnitude. Even though several galaxies in the list are brighter than sixth magnitude, only the two Magellanic Clouds and the Andromeda galaxy are easily visible to the unaided eye.

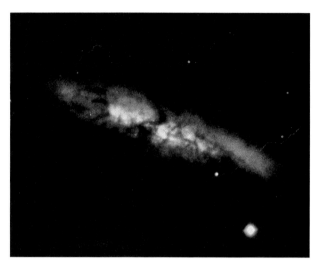 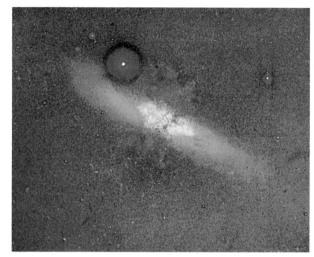

Figure 11.10 A Peculiar Galaxy. Not all galaxies fit the standard classifications. This is M82, a well-known example of a galaxy with an unusual appearance. For a time, it was thought that the nucleus of this galaxy was exploding, but more recent analysis has indicated that the odd appearance is due simply to a very extensive region of interstellar gas and dust surrounding the center of a spiral galaxy that is undergoing an episode of intense star formation. The righthand image uses computer-enhanced colors to highlight the gas and dust. (left: *Lick Observatory photograph*; right: *photo by H. C. Arp, image processing by J. J. Lorre, Image Processing Laboratory, JPL*)

cluster of galaxies. As in the case of the brightest star in a galaxy, astronomers assume that the brightest galaxy in a cluster always has about the same absolute magnitude. (Actually, experience has shown that the brightest galaxy in a cluster may not be so standard from one cluster to another and that the second- or third-ranked galaxy is a better standard candle.)

A new technique called the **Tully-Fisher method**, in which the luminosity of a galaxy is inferred from its mass, shows great promise for spiral galaxies. The mass is not measured directly; what is measured instead is the rotational velocity of the galaxy, which depends on its mass. The rotational velocity is estimated from the width of the 21-centimeter radio emission line from the galaxy. The Doppler effect broadens the line according to the maximum spread of velocities from one side of the galaxy's disk to the other (Fig. 11.12). Thus the luminosity (hence the absolute magnitude) of a spiral galaxy can be inferred from the width of its 21-centimeter emission line, so that its distance can be found by measuring the apparent magnitude and comparing it with the absolute magnitude. This technique works best when infrared magnitudes are used, because infrared light is relatively unaffected by interstellar dust within the galaxy being observed. The Tully-

Figure 11.11 A Supernova in a Distant Galaxy. Before the distinction between novae and supernovae became clear, these flare-ups (arrow) contributed to the controversy over the nature of the nebulae. Now that these occurrences in galaxies are known to be supernova explosions, comparisons of their apparent and assumed absolute magnitudes provide distance estimates. (*Palomar Observatory, California Institute of Technology*)

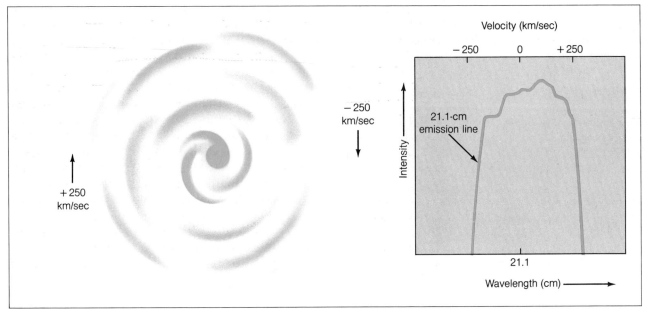

Figure 11.12 Rotation Velocity and the Width of the 21-Centimeter Line. This shows how the rotation velocity of a spiral galaxy is related to the width of the 21-centimeter emission line from hydrogen gas in the galaxy. Because of the Doppler effect, 21-centimeter photons from one side of the galaxy are shifted toward shorter wavelengths, those from the other side are shifted toward longer wavelengths, and those from the central regions are not shifted. The result is a 21-centimeter line that is broadened according to how rapidly the galaxy is rotating. Because the rotation speed is governed by the central mass of a galaxy, the width of the line is related to the mass of the galaxy and can be used to estimate its absolute magnitude for distance determination.

Fisher method is now being used in efforts to improve the intergalactic distance scale, which has very important implications for our understanding of the universe as a whole.

It is important to keep in mind how uncertain *all* ga-lactic distance determination methods are. To assume that all objects in a given class are identical in basic properties such as luminosity is always a risky business, especially when such assumptions are applied to objects as distant as external galaxies or as rare as the brightest star in a galaxy. There is little else that can be done, however, so we must simply recognize the inherent limitations in accuracy and take them into account. The uncertainties in distance determinations carry over to our measurements of other properties of galaxies.

The masses of nearby galaxies can be measured in the same manner as the mass of our own galaxy: by applying Kepler's third law to the orbital motions of stars or gas clouds in the outer portions. All that is required is to measure the orbital velocity at some point well out from the center and to determine how far from the center that point is (which in turn requires knowledge of the distance to the galaxy). Then Kepler's third law leads to

$$M = a^3/P^2,$$

where M is the mass of the galaxy (in solar masses), and a and P are the semimajor axis (in AU) and the period (in years) of the orbiting material at the observed point. (See the more complete discussion in Chapter 3 to remind yourself how this equation was developed.)

There are two difficulties in using this technique: (1) It is hard to isolate individual stars in distant galaxies and then measure their velocities using the Doppler shift, and (2) both the distance and the orientation of the galaxy must be known before the true orbital velocity and semi-major axis can be determined. The orientation can usually be deduced for a spiral galaxy, since it has a disk shape whose tilt can be seen, and the distance can be estimated using one of the methods just outlined. In most cases the orbital velocities are measured at several points within a galaxy from the center out as far as possible, and the data are plotted on a **rotation curve** (Fig. 11.13), which is simply a diagram showing the variation of orbital velocity with distance from the center. Doing this is useful for several reasons: It provides a means of checking whether the observations go sufficiently far out in the galaxy to reach the region where the orbiting material follows Kepler's third law; it helps in determining the orientation of the galaxy; and it ensures that the velocities have been measured as far out as possible from the center. The most effective means of obtaining velocity data on the outer portions of a spiral galaxy is to measure the 21-centimeter emission from hydrogen, which can be detected at greater distances from the center than visible light from stars can be measured.

The technique of measuring rotation curves can best be applied to spiral galaxies, where there is a disk with stars and interstellar gas orbiting in a coherent fashion. In elliptical galaxies there is no such clear-cut overall motion, and a slightly different technique must be used. The individual stellar orbits are randomly oriented in the outer portions of an elliptical galaxy, so that there is a significant range of velocities within any portion of the galaxy's volume. This range of velocities, called the **velocity dispersion,** is greatest near the center of the galaxy, where the stars move fastest in their orbits, and smaller in the outer regions. Furthermore, the greater the mass of the galaxy, the greater the velocity dispersion (at any distance from the center), so a measurement of this parameter can lead to an estimate of the mass of a galaxy. The velocity dispersion is deduced from the widths of spectral lines formed by groups of stars in different portions of a galaxy (Fig. 11.14); the greater the internal motion within the region observed, the greater the widths of the spectral lines, due to the Doppler effect (because some stars are moving toward the Earth and others away from it).

Kepler's third law can sometimes be used in a different way entirely in determining galactic masses. There are double galaxies here and there in the cosmos, orbiting each other exactly like stars in binary systems. In these cases, Kepler's third law can be applied

Figure 11.14 The Velocity Dispersion for an Elliptical Galaxy There is no easily measured overall rotation of an elliptical galaxy, so the mass cannot be estimated from a rotation curve. Instead the average velocites of stars at a known distance from the center are used. Light from a small area of the galaxy's image is measured spectroscopically. The random motions of the stars in the observed part of the galaxy broaden the spectral lines by the Doppler effect, and the amount of broadening is a measure of the average velocity of the stars in the observed region. At a given distance from the center of the galaxy, the higher the average velocity, the greater the mass contained inside that distance.

Figure 11.13 A Schematic Rotation Curve for a Galaxy. This rotation curve illustrates the fact that in most spiral galaxies, the rotation velocity does not drop off in the outer regions, as it would if most of the mass were concentrated at the center of the galaxy. The data on rotation speed in the outermost regions comes from radio observations.

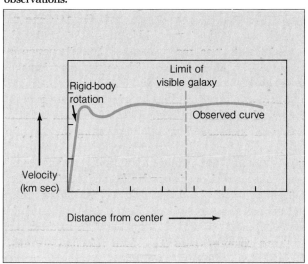

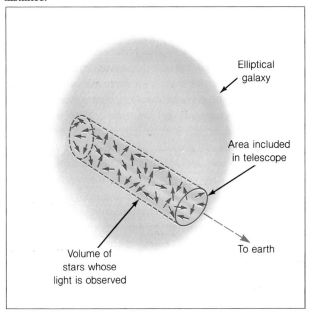

to derive an estimate of the combined mass of the two galaxies. The uncertainties are even more severe than in the case of a double star, however, because the orbital period of a pair of galaxies is measured in hundreds of millions of years. Thus, the usual problems of not knowing the orbital inclination or the distance to the system are compounded by inaccuracies in estimating the orbital period, something that is usually well known for a double star. Still, this technique is useful and has one major advantage: It takes into account *all* the mass of a galaxy, including whatever part of it is in the outer portions, beyond the reach of the standard rotation curve or velocity disper-sion measurements. Interestingly, galactic masses estimated from double systems are generally much larger than those based on measurements of internal motions within galaxies, possibly indicating that most galaxies have extensive halos containing large quantities of matter. As noted in Chapter 9, there is independent evidence that our own galaxy has a massive halo, perhaps containing as much as 90 percent of the total mass.

Once the distance to a galaxy has been established, its luminosity and size can be deduced directly from the apparent magnitude and apparent diameter. Both quantities are found to vary over wide ranges, with luminosities as low as 10^6 and as high as 10^{12} times that of the Sun, and diameters ranging from about 1 to 100 kiloparsecs.

Elliptical galaxies generally display a wider range of luminosities and sizes than the spirals. The latter tend to be more uniform, with luminosities usually between 10^{10} and 10^{12} solar luminosities and diameters between 10 and 100 kiloparsecs. The smallest elliptical galaxies are called **dwarf ellipticals.** These may be very common, but they are too dim to be seen at great distances, so we can only say for sure that there are many of them near our own galaxy and the Andromeda galaxy.

It appears that galaxies of a given type tend to be fairly uniform in other properties, just as stars of a given spectral type are the same in other ways. One quantity often used by astronomers to characterize galaxies is the **mass-to-light ratio** (M/L), which is simply the mass of a galaxy divided by its luminosity in solar units. Values of the mass-to-light ratio typically range from 5 to 200. Any value larger than 1 means that the galaxy emits less light per solar mass than the Sun; that is, the galaxy is dominated in mass by stars that are dimmer than the Sun. Even the smaller mass-to-light ratios that are observed for galaxies are much larger than 1; a value of M/L = 50, for example, means that 50 solar masses are required to produce the luminosity of one Sun. Interestingly, counts of stars in the Sun's part of our galaxy indicate a mass-to-light ratio near 1, whereas we expect (based on observations of similar galaxies) that it is much larger than 1 for our galaxy as a whole. This indicates that significant mass in galaxies like ours must be in forms that are very difficult to see, perhaps not in the form of normal stars at all. It is possible that the invisible mass resides in the halos of galaxies, in some unseen form. Recall (from Chapter 10) that this appears to be true of our own galaxy. The whole question of the nature of the unseen "dark matter" in galaxies is currently an area of intense interest in astronomy.

Elliptical galaxies tend to have the largest mass-to-light ratios, consistent with our earlier statement that these galaxies are relatively deficient in hot, luminous stars. Spirals, on the other hand, which contain some of these stars, have lower mass-to-light values. It is worth stressing again, however, that even the low mass-to-light ratios for these galaxies are much greater than 1, indicating that they too are dominated in mass by very dim stars or dark matter (including, of course, interstellar gas and dust, but this usually accounts for less than 25 percent of the mass of spiral galaxies). Hence a spiral galaxy, with all its glorious bright blue disk stars, actually has far more dim red ones.

The colors of galaxies can also be measured by using filters to determine the brightness at different wavelengths. In general the spirals are not as red as the ellipticals, again indicating that the latter galaxies contain a higher fraction of cool, red stars. There are also color variations within galaxies: In a spiral, for example, the central bulge is usually redder than the outer portions of the disk where most of the young, hot stars reside.

Both the mass-to-light ratios and the colors of galaxies are indicators of the relative content of Population I and Population II stars. Recall that Population I stars tend to be younger, and included in this group are all the bright, blue O and B stars. By contrast, Population II objects are old and include only cool, relatively dim main sequence stars and red giants. Therefore a red overall color, along with a high mass-to-light ratio, implies that Population II stars are dominant, while a low mass-to-light ratio and a bluer color means that some Population I stars are included. Thus elliptical galaxies seem to be almost entirely Population II objects, while spirals contain a mixture of the two populations.

This dichotomy between the two types of galaxies is

found also in comparing the interstellar matter content of spirals and ellipticals. Photographs of spirals, especially those seen edge-on, often clearly show the presence of dark dust clouds, and face-on views typically reveal a number of bright H II regions. Neither shows up on photographs of elliptical galaxies. Radio observations of the 21-centimeter line of hydrogen bear this out; emission is usually present from spirals but is weak or absent in ellipticals.

The Origins of Spirals and Ellipticals

In attempting to explain the differences between spiral and elliptical galaxies, astronomers have suggested several types of theories, none of which is fully developed at present.

The weight of all the evidence cited in the previous sections is that spiral galaxies are still dynamic, evolving entities, with active star formation and recycling of material between stellar and interstellar forms, whereas elliptical galaxies have reached some sort of equilibrium in which these processes are not taking place at a significant rate. Because both types of galaxies contain old stars, we know that there are no significant age differences between the two types; we cannot simply conclude that the ellipticals are older and have run out of gas.

One of the earliest suggestions was that rotation is responsible for the division between the two types. We learned in the previous chapter that our own galaxy is thought to have formed from a rotating cloud of gas that flattened into a disk as it contracted. The rotation of the cloud was the cause of the disk formation, so perhaps elliptical galaxies are the result of contracting gas clouds that did not rotate rapidly enough to form disks. It is not clear how this would account for the lack of interstellar material and star formation, however.

One problem with this idea is that elliptical galaxies can and do rotate, yet they have not formed disks. Therefore rotation cannot be the full explanation. Apparently the key is whether a disk forms before most of the gas in the contracting cloud has been consumed by star formation (Fig. 11.15). If all the gas is converted into stars before the collapse has proceeded very far, the result is an elliptical galaxy. If the initial rate of star formation is not so great, the cloud has time to form a disk while it still contains a large quantity of interstellar gas and dust. The reason that the timing is so important is that stars, once formed, act as individual particles and continue to orbit the galaxy without forming a disk. In contrast, gas acts as a fluid and settles into a disk. (In essence the question is whether there is viscosity that will dissipate energy and allow the material to sink into the plane of a disk; stars don't encounter each other frequently enough to create a fluid viscosity, but gas particles do.)

While these ideas represent progress toward understanding why some galaxies are ellipticals and others are spirals, there remain substantial questions. It is not clear why the initial star formation rate should differ from one galaxy to another, for example, although it has been suggested that it may be related to the density of the cloud from which the galaxy contracts.

A secondary question is why some spirals are barred and others are not. Theoretical studies suggest that the presence or absence of a massive halo may be the determining factor. Without a halo, a disk of stars will be subject to an instability that naturally causes its central region to become elongated and barlike. Once a bar has formed, the asymmetry of the gravitational force it exerts on the surrounding disk is sufficient to create a spiral density wave, so that a barred spiral galaxy results. If, on the other hand, a massive halo is present, its gravitational influence inhibits the formation of a bar. Spiral density waves often form anyway, due to the gravitational influence of another galaxy nearby, and cause a normal spiral to form.

Clusters of Galaxies

While galaxies may be considered the largest single objects in the universe (if indeed an assemblage of stars orbiting a common center can be viewed as a single object), there are yet larger scales on which matter is organized. Galaxies tend to be located in clusters (Fig. 11.16), rather than being distributed uniformly throughout the cosmos, and the clusters in turn have an uneven distribution, with concentrations of clusters referred to as **superclusters.** It may be that superclusters are the largest entities in the universe.

Clusters of galaxies range in membership from a few (perhaps half a dozen) to many hundreds or even thousands (Appendix 13). Just as stars in a cluster orbit a common center, so galaxies in a group or cluster are gravitationally bound together and follow orbital paths about a central point. In a large, rich cluster,

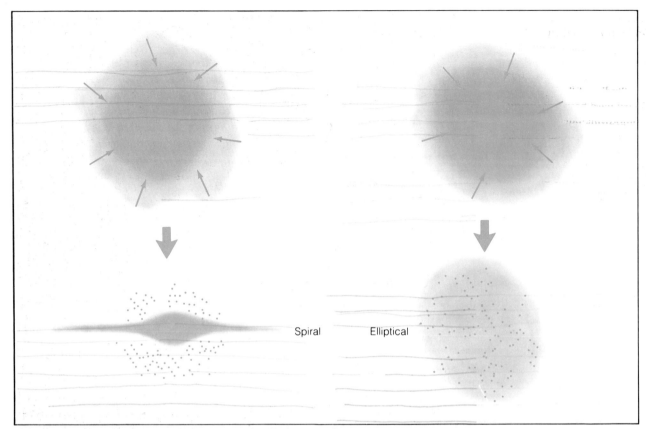

Figure 11.15 The Origins of Spirals and Ellipticals. Both types of galaxies begin as giant gas clouds that collapse gravitationally. If collapse to a disk occurs before all the gas is converted into stars, the result is a spiral galaxy. If the gas is entirely used up in star formation before a disk forms, the result is an elliptical gallaxy. The rate of star formation relative to the rate of collapse to a disk is probably determined by the initial density of the cloud, and may be influenced by the rotation rate.

Figure 11.16 A Portion of a Cluster of Galaxies in the Constellation Hercules. Many different galaxy types are seen here. *(Palomar Observatory, California Institute of Technology)*

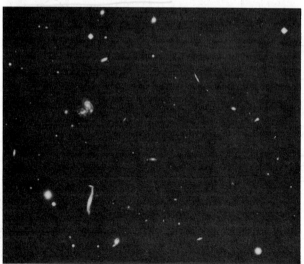

frequent encounters between galaxies during the cluster formation era cause the members to take on a smooth, spherical distribution; in a small group like the one to which the Milky Way belongs, the arrangement of individual galaxies is more haphazard, creating an amorphous overall appearance.

The Local Group

The Milky Way belongs to a small group of galaxies known as the Local Group. This group consists of about thirty members arranged in a random distribution. Despite the relative proximity of these galaxies to us, it has been difficult to ascertain their properties in some cases because of obscuration by our own galactic disk. It is not even possible to say with certainty how many members the Local Group has.

Among the member galaxies are three spirals, two of which—the Andromeda galaxy (Figs. 11.17 and 11.18) and the Milky Way—are rather large and lu-

Figure 11.17 The Andromeda Galaxy. This large spiral, comparable to the Milky Way, is the most distant object visible to the unaided eye. Studies of the Andromeda galaxy have been very useful in improving our understanding of the structure and evolution of our own galaxy. *(Palomar Observatory, California Institute of Technology)*

Figure 11.18 New Views of the Andromeda Galaxy. Here are an X-ray image (left) and an infrared picture (right) of our neighbor spiral galaxy. The X-ray view reveals associations of stars with hot coronae and binary X-ray sources; the infrared image shows the locations of cold dust clouds. *(left: Harvard-Smithsonian Center for Astrophysics; right: NASA)*

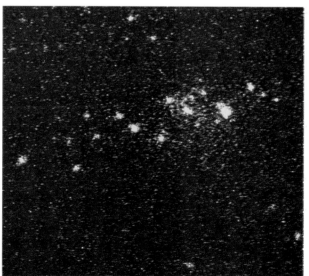
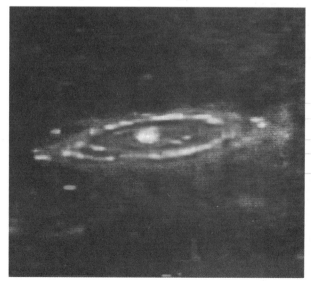

minous. They are probably the brightest and most massive galaxies in the entire cluster. Two large galaxies discovered in the 1970s through infrared observations (Fig. 11.19) were thought for a while to be members of the Local Group. They would have significantly altered the constitution of the cluster, but more recent evidence indicates that both are too distant to be members.

Most of the other members are ellipticals, many of them dwarf ellipticals (Fig. 11.20). There may be additional members of this type undetected so far because of their faintness. Four irregular galaxies (Fig. 11.21)—two of them are the Large and Small Magellanic Clouds (Fig. 11.22)—are also found within the Local Group. In addition there are globular clusters that are probably distant members of our galaxy but lie so far away from the main body of the Milky Way that they appear isolated. A list of known members of the Local Group is given in Appendix 13.

The Local Group is about 800 kiloparsecs in diameter and has a roughly disklike overall shape, with the Milky Way located a little off center. Beyond the outermost portions of the Local Group there are no conspicuous external galaxies for a distance of some 1100 kiloparsecs.

The Magellanic Clouds and the Andromeda galaxy (also known commonly as M31, its designation in the widely cited Messier Catalog; see Appendix 15) have been particularly well studied because of their proximity and prominence and because of what they can tell us about galactic evolution and stellar processing.

The Large and Small Magellanic Clouds appear to the unaided eye as fuzzy patches, easily visible only on

Figure 11.19 Maffei 1 and 2. This infrared photograph reveals two large galaxies (the fuzzy images, upper right and lower right) that were for a while considered possible members of the Local Group. More extensive analysis has shown, however, that they probably are not. (The circular marks are overexposed images of closer stars.) *(H. Spinrad)*

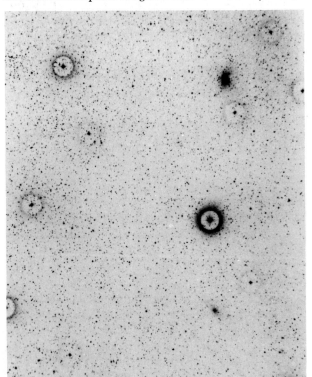

Figure 11.20 The Sculptor Dwarf Galaxy. This loose conglomeration of stars lies about 83 kiloparsecs from the Sun. Such a small, dim galaxy would not be noticed if it were much farther away, but the number of systems of this type in the Local Group leads to the assumption that they may be very common in the universe. *(P. W. Hodge, photo taken at the Boyden Observatory)*

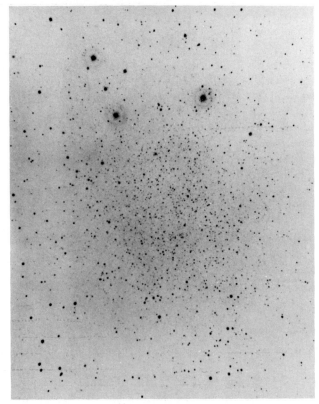

Figure 11.21 IC1613. This is an irregular galaxy, a member of the Local Group. *(P. W. Hodge, photo taken at the Lick Observatory)*

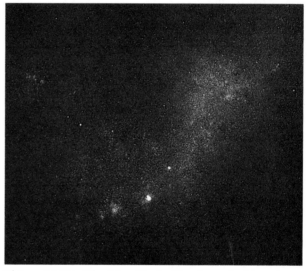

Figure 11.22 The Magellanic Clouds. The large cloud is shown in the upper photo, and the small cloud in the lower picture. Our nearest neighbors, these galaxies orbit as satellites of the Milky Way. *(top: © 1984 Royal Observatory, Edinburgh; bottom: R. J. Dufour, Rice University)*

dark, moonless nights. They lie near the south celestial pole, and can therefore be seen only from the Southern Hemisphere. Their name originated from the fact that the first Europeans to see them were Ferdinand Magellan and his crew, who made the first voyage around the world in the early sixteenth century.

The Magellanic Clouds are considered to be satellites of the Milky Way. They have lesser masses and follow orbits about it, taking several hundred million years to make each circuit. Lying between 50 and 65 kiloparsecs from the Sun, both are Type I irregulars. They contain substantial quantities of interstellar matter and are quite obviously the sites of active star formation, with many bright nebulae and clusters of hot, young stars. Measurements of the colors of these galaxies and of the spectra of some of their brighter stars indicate that they have five to ten times lower heavy element abundances than Population I stars in our galaxy. This seems to indicate that the Magellanic Clouds have not undergone as much stellar cycling and recycling as the Milky Way. These and other Type I irregular galaxies may generally be viewed as galaxies in extended adolescence, not yet having settled down into mature disks. In the case of the Magellanic Clouds, the reason for the unrest is probably the Milky Way's tidal forces.

The great spiral of M31, the Andromeda galaxy, is the most distant object visible to the unaided eye, lying some 700 kiloparsecs from our position in the Milky Way. All that the eye can see is a fuzzy patch of light, even if the telescope is used, but when a time-exposure photograph is taken, the awesome disk stands out (although the spiral arms are still difficult to see because the disk is viewed from an oblique angle). The Andromeda galaxy is so large, extending over several degrees across the sky, that full portraits can be obtained only by using relatively wide-angle telescope optics. (Most large telescopes have extremely narrow fields of view.)

The Andromeda galaxy is probably very similar to our own galaxy and thus has taught us quite a bit about the nature of the Milky Way. The two stellar populations were first discovered through studies of stars in Andromeda, and the effects of stellar processing on the chemical makeup of different portions of the galaxy are better determined for Andromeda than for our galaxy. The study of this galaxy and other members of the Local Group has been very important in the development of distance-determination techniques for more distant galaxies and clusters.

Rich Clusters: Dominant Ellipticals and Galactic Mergers

In contrast with small groups such as the Local Group, many clusters of galaxies are much larger and contain hundreds or even thousands of members (Fig. 11.23). The density of galaxies in such a cluster is relatively high, and therefore there are numerous close encounters between galaxies as they follow their individual orbits about the center. When two galaxies pass close by each other, they exert mutual gravitational forces that can have profound effects. These close encounters apparently were very frequent in the formation era of rich clusters, accounting for the smooth overall distribution of galaxies, with the greatest density toward the center.

By contrast, small groups or clusters rarely reach this state and retain a less regular appearance.

The central regions of large, rich clusters are more highly dominated by elliptical and S0 galaxies (Fig. 11.24) than are small groups or isolated galaxies. Near the center of a rich cluster, some 90 percent of the galaxies may be ellipticals or S0s, whereas about 60 percent of noncluster galaxies are spirals. This contrast is probably a direct result of the frequent near-collisions between galaxies in dense clusters. When two galaxies have a close encounter, the tidal forces they exert on each other stretch and distort them. Under some circumstances any interstellar matter in them can be pulled out and dispersed (Fig. 11.25). The outer regions, such as the halos, can also be stripped away. The net effect is like what happens to rocks in a tumbler; the galaxies are gradually ground down into smooth remnants. A spiral galaxy subjected to these cosmic upheavals may assume the form of an elliptical. In time most of the spirals in a cluster, particularly those in the dense central region, may be converted into ellipticals and S0s.

The frequent gravitational encounters between galaxies during cluster formation have another interesting effect: They cause a buildup of galaxies right at the

Figure 11.24 Galaxy Types in a Rich Cluster. In dense, highly populated clusters of galaxies, nearly all of the members in the central portion are ellipticals or S0s, and there is often a giant, dominant elliptical at the center. Only in the outer portions are many spirals seen.

Figure 11.23 A Portion of a Rich Cluster of Galaxies. Most of the objects in this photograph are galaxies. This cluster lies in the constellation Hercules. *(National Optical Astronomy Observatories)*

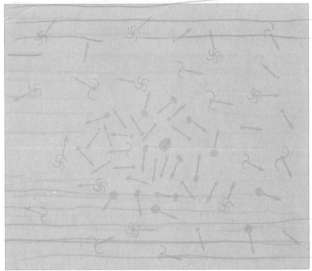

center of the cluster. When two galaxies orbiting within a cluster come together, one always gains energy and moves to a larger orbit, while the other (the more massive of the two) loses energy and drops closer to the center. Thus a gradual sifting process—like the differentiation that occurs inside some planets as the heavy elements sink toward the core—gradually builds up a dense central conglomeration of galaxies at the heart of the cluster. There these galaxies may actually merge, the end result being a single gigantic elliptical galaxy (Figs. 11.26 and 11.27) that continues to grow larger as new galaxies fall in.

Another distinctive characteristic of some rich clusters of galaxies is the existence of a very hot gaseous medium filling the spaces between the galaxies. This **intracluster gas** was discovered through X-ray observations (Fig. 11.28), which showed that the temperature of the gas is as high as a hundred million degrees, much hotter even than the highly ionized gas in the interstellar medium in our galaxy. This observation raised the possibility that the general intergalactic void is filled with such gas, although if it is, the density outside of clusters must still be rather low, or X-ray data would have revealed its presence. (The significance of a possible general intergalactic medium is discussed in Chapter 14.)

A different type of X-ray measurement indicates that the hot intracluster gas originates in the galaxies themselves, rather than entering the cluster from the intergalactic void. Spectroscopic measurements made at X-ray wavelengths have revealed that the gas contains iron, a heavy element, in nearly the same quantity (relative to hydrogen) as the Sun and other Population I stars. Such a high abundance of iron could only have been produced in nuclear reactions inside stars. Therefore this intracluster gas must once have been involved in part of the cosmic recycling that goes on in galaxies, as stars gradually enrich matter with heavy elements before returning it to space. How the gas was expelled into the regions between galaxies is not clear, but it may have been swept out during near-collisions be-

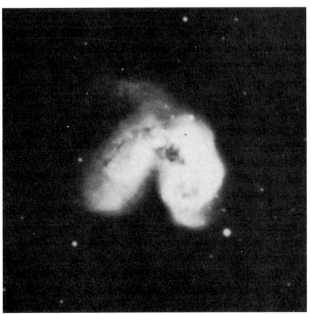

Figure 11.25 A Collision Between Galaxies. This shows a pair of galaxies undergoing a near-collision. At right is a computer simulation of their interaction, showing how the present appearance was created. This type of interaction between galaxies is believed responsible in some cases for the removal of interstellar matter, converting spiral galaxies to elliptical or S0 galaxies. *(Photograph from the Palomar Observatory, California Institute of Technology; computer simulation courtesy A. Toomre and J. Toomre)*

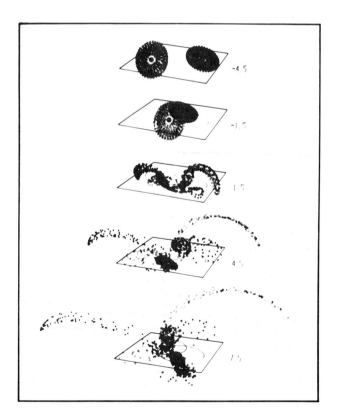

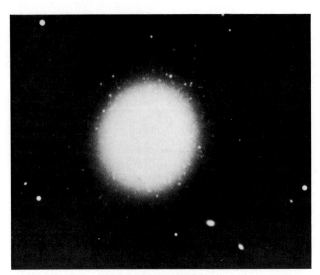

Figure 11.26 A Giant Elliptical Galaxy. This is M87, a well-known example of a dominant central galaxy in a rich cluster. This particular galaxy is discussed further in Chapter 13. (*National Optical Astronomy Observatories*)

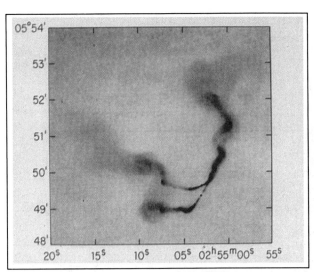

Figure 11.27 A Radio Image of a Galactic Merger. This is a radio map of a giant elliptical galaxy with two nuclei (the bright points at lower center, each having a pair of wispy gaseous jets emanating from it; the jets are discussed in Chapter 13). Evidently this galaxy is in the process of forming from the merger of two galaxies whose centers have not yet completely merged. (*National Radio Astronomy Observatory, operated by Associated Universities, Inc., under contract with the National Science Foundation*)

tween galaxies, or it may have been ejected in galactic winds created by the cumulative effect of supernova explosions and stellar winds.

Cluster Masses

There are two distinct methods for measuring the mass of a cluster of galaxies, and both are quite uncertain. This is a crucial problem, as we shall see in Chapter 14, because of the importance of knowing how much mass the universe contains.

The simpler and more straightforward of the two methods is to estimate the masses of the individual galaxies in a cluster, using techniques described earlier in this chapter, and add them up. In many cases, particularly for distant clusters where it is impossible to measure the rotation curves or velocity dispersions of individual galaxies, the only way to estimate a cluster's mass is to measure its brightness and then use a standard mass-to-light ratio to derive the mass. This is inaccurate because it depends on knowing the cluster's distance from us and because it assumes that the galaxies adhere to the usual mass-to-light ratios for their types. It also neglects any matter in the cluster that may lie between the galaxies.

The second method (Fig. 11.29) is similar to the velocity dispersion technique used to estimate masses of elliptical galaxies. The mass of a cluster is estimated from the orbital speeds of galaxies in its outer portions; the faster they move, the greater the mass of the cluster. This method has the advantage that it measures all the mass of the cluster, whether it is in galaxies or between them, but it has the disadvantage that the necessary velocity measurements, particularly for a very distant cluster, are difficult. Furthermore, the technique is valid only if the galaxies are in stable orbits about the cluster; if the cluster is expanding or some of the galaxies are not really gravitationally bound to it, the results will be incorrect. In rich clusters at least, the smooth overall shape and distribution gives the appearance of a bound system, and this technique is probably valid. It may not be valid for small clusters such as the Local Group. This method always leads to an estimated cluster mass that is much greater than that derived by adding up the masses of the visible galaxies, probably because that technique neglects intracluster gas and extended galactic halos.

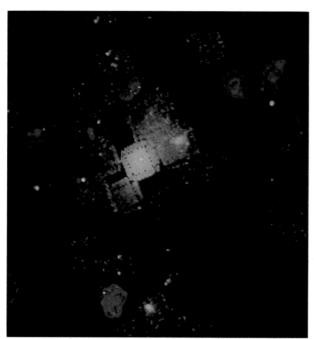

Figure 11.28 Intracluster Gas in the Virgo Cluster. This is an X-ray image from the *Einstein Observatory*, showing emission from the entire central portion of this cluster of galaxies. The X rays are being emitted by very hot gas that fills the space between galaxies. The X-ray emission is shown in blue; radio images of galaxies in the cluster are shown in red. *(X-ray data from the Harvard-Smithsonian Center for Astrophysics; radio data and image preparation by the National Radio Astronomy Observatory, operated by Associated Universities, Inc., under contract with the National Science Foundation)*

Superclusters

We turn now to consider the overall organization of the matter in the universe. We noted earlier that clusters of galaxies may represent the largest scale on which matter in the universe is clumped, but that there is some evidence for a higher-order organization, even in the case of the Local Group (Fig. 11.30). It appears that clusters of galaxies are themselves concentrated in certain regions and are commonly referred to as **superclusters**.

The reality of superclusters has been difficult to establish, and for a while many astronomers were not convinced. Today, however, there are few doubts that

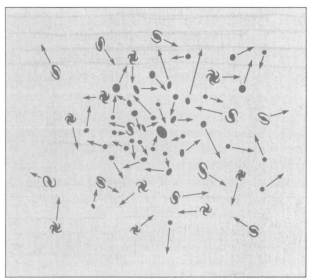

Figure 11.29 Galaxy Motions within a Cluster. Galaxies move randomly within their parent cluster. In a large cluster, where the overall distribution of galaxies is uniform, analysis of the average galaxy velocities is used to estimate total mass of the cluster, just as the velocity dispersion method is used to measure the masses of elliptical galaxies.

clusters of galaxies tend to be grouped, although there is still disagreement on the significance of this. The best evidence lies in the distribution of rich clusters (some 1500 of these have been cataloged), which clearly tend to congregate in certain regions, with relatively empty space in between.

The uncertainty that remains has to do with whether the groupings are random occurrences or whether they reflect a fundamental unevenness in the distribution of matter in the universe. These possibilities are usually tested by comparing the apparent clustering of clusters with what would be expected from coincidence if they were actually distributed in a random fashion (after all, in any random scattering of objects, there will always be occasional accidental groupings). Mathematical experiments of this type appear to show that the observed superclusters could be random concentrations of clusters and therefore may not have profound significance for the universe as a whole. On the other hand, very recent observations appear to show that a large number of clusters of galaxies may be linked into an enormous filamentary supercluster spanning a major portion of the sky. Previously unrecognized because it

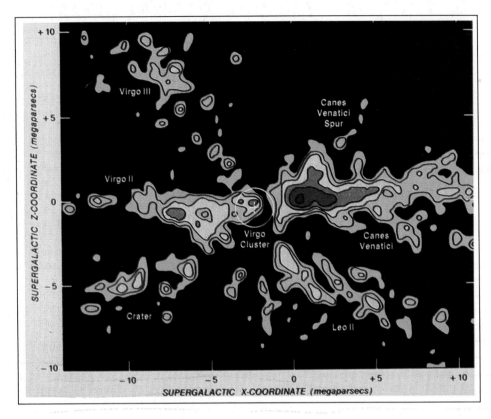

Figure 11.30 **The Local Supercluster.** This is an artist's concept of the cluster of galaxy clusters to which the Local Group belongs. *(Illustration from* Sky and Telescope *magazine by Rob Hess, © 1982 by the Sky Publishing Corporation; also with the permission of R. B. Tully)*

crosses the plane of the Milky Way, where our view of distant galaxies is obscured, this supercluster, if real, is far too large to have been the result of random groupings of clusters. This supercluster may show that there is a basic unevenness in the distribution of matter in the universe, a possibility whose significance is discussed in Chapter 14.

The Origins of Clusters

The basic problem of galaxy formation bears a great resemblance to that of star formation: We know that collapse of the initial cloud will occur only if a sufficiently high density of matter exists in a volume sufficiently small that the mutual gravitational attraction of the gas particles overcomes their natural tendency to fly about as free individuals. The Jeans criterion for collapse tells us that in the early days of the universe, condensations that formed would have had masses of only about 10^5 solar masses, too small to form normal galaxies. Other physical processes that have been ana-

lyzed lead to masses much larger than individual galaxies, so there is a problem in understanding how individual galaxies formed. One possibility is that the initial condensations had masses comparable to clusters of galaxies and that subsequent fragmentation led to the formation of individual galaxies.

Another suggestion is that massive stars formed early in the history of the universe (calculations show that relatively small condensations could have led to star formation even before galaxies formed) and that the supernova explosions resulting from their deaths could have created sufficiently strong shock waves in the universal gas to cause it to fragment and collapse into galaxies. This idea is similar to the sequential star formation theory discussed in Chapter 7.

Whatever the initial cause, the clouds that began to collapse were probably much more massive than individual galaxies (around 10^{15} solar masses according to some calculations). Just as in the case of star formation, when these overly massive condensations developed, they soon fragmented (Fig. 11.31) because, as their densities increased, smaller volumes of the gas were

Astronomical Insights 11.1

Holes in the Universe

One of the fundamental properties of the universe is thought to be its homogeneity, or uniformity from one location to another. We discuss in this chapter the hierarchy of galaxies, clusters, and superclusters with the tacit assumption that the universe as a whole is more-or-less uniformly filled with these objects. As we will see in Chapter 14, the assumption of homogeneity is commonly made in the construction of models of the universe, but it is an assumption that is sometimes questioned. If the cosmos is not uniformly filled with matter, if the distribution is somehow unbalanced or uneven, it may have important implications about the formation and history of the universe.

Astronomers make a great effort to measure the distribution of matter. The technique most often used is to choose one or more specific, representative regions of the sky and carefully search in those regions for all galaxies brighter than some predetermined limit. Distances to all the detected galaxies are then estimated (usually using the Hubble expansion law; see Chapter 12) in order to develop a three-dimensional picture of each sampled region of the sky. From this it is possible to determine the overall distribution of galaxies and discover whether any large-scale clumpiness or unevenness exists.

From such a study the astronomical world recently received news of a large portion of space that is apparently devoid of visible matter. In the course of an extensive survey of faint galaxies, a group of astronomers found a region on the sky some 35° across where there appeared to be no galaxies at distances between 240 million parsecs (Mpc) and 360 Mpc. At that distance, 35° of angular diameter corresponds to a linear dimension of 180 Mpc, so it appeared that there was a void over 100 Mpc across in all dimensions, centered some 300 Mpc away.

Statistical studies show that random distributions of galaxies should produce voids, or empty zones, no larger than about 20 Mpc across. Gravitational effects, in which clumps in the distribution of matter are enhanced by the attraction of galaxies for each other, are thought capable of producing voids up to 35 Mpc in diameter. An empty space as large as 100 Mpc across would have profound implications, for at the present time there is no known way to produce such a hole in space, unless the universe is fundamentally inhomogeneous. Therefore the announcement of the discovery of a hole in space immediately evoked a great deal of interest.

Since the original announcement of the existence of a void in the distribution of galaxies, several researchers have sought to verify the discovery. In a very recent announcement, a group of astronomers reported that the overall distribution of galaxies is a network of bubble-shaped voids, with the galaxies between distributed in arclike sheets or filaments (see Fig. 11.32). This conclusion was based on the most complete and far-reaching sampling of galaxies yet carried out, involving some 1100 galaxies in a large strip of sky near the north polar region of our own galaxy (observing in this direction avoids the obscuration of interstellar gas and dust within the Milky Way).

If the nonuniform distribution of galaxies in this recent survey is representative of the universe as a whole, it will have profound implications for theories of the early stages of the universe and of galaxy formation. The discovery of a cellular distribution of galaxies implies that the universe was somehow nonuniform from the beginning and that galaxies simply formed where the concentrations of mass were highest.

One theory, published before the recent discovery, suggests that galaxy formation was triggered

(continued on next page)

> ## Astronomical Insights 11.1
>
> ### Holes in the Universe
>
> *(continued from previous page)* by the explosion of very massive primordial stars in supernovae. These massive stars were thought to have formed from concentrations of matter in the very young universe before galaxies formed. (The development of this theory was motivated in part by the need to explain how some heavy elements could have been present when galaxies formed.) A consequence of this theory was the prediction that galaxies should have formed in curved sheets as the result of expanding shock waves from the supernova explosions, creating arclike structures much like those that have now been found. This match between theory and observation appears to be a nice confirmation of the theory, except that the predicted structures are much smaller than those observed. Still, the idea that galaxies formed originally from the debris left over from primordial, gigantic explosions is a tempting explanation for the bubblelike distribution of galaxies in the universe implied by the recent observations.

able to collapse on their own. An initial giant cloud thus broke apart into many small ones, each on its way to becoming an individual galaxy. The entire group was still held together gravitationally, so the galaxies that formed remained in orbit about a common center, and the result was a cluster of galaxies.

The view just outlined does not answer the question of the origin of superclusters. It may be that the supercluster-sized concentrations of matter formed first and then fragmented to form clusters and finally individual galaxies. This view is sometimes referred to as the "top-down" model since it postulates that the objects at the top of the hierarchy of sizes formed first, and the smaller ones next. Statistical studies leave open the possibility that the "bottom-up" view, in which the smallest objects (individual galaxies) formed first, could be more appropriate. In this model early turbulence or some other factor allowed individual galaxies to form first, and then random gravitational influences of galaxies on each other caused groupings into clusters and, ultimately, superclusters.

It is possible that neither view is correct, that the first objects to form had the masses of ordinary clusters of galaxies, which then fragmented to form individual galaxies, but which also formed into superclusters by random gravitational attraction. The recent discoveries about large-scale linkages of superclusters into a universal filamentary structure tend to support the top-down viewpoint, however, because these largest systems, as noted earlier, simply cannot be the result of random groupings of clusters. A very recent reaffirmation of this possibility arose from a new survey of galaxies in which distances and positions of *all* galaxies in selected large portions of the sky were measured, so that in effect a three-dimensional map could be constructed. This map shows not only filamentary struc-

Figure 11.31 Fragmentation of a Primordial Cloud.
At left is a gigantic cloud containing the mass of many galaxies. As it collapses it breaks up into fragments, which then collapse individually and form galaxies. This is one scenario for the origin of galaxies; in another, individual clouds form first then merge to form galaxies of varying size and mass.

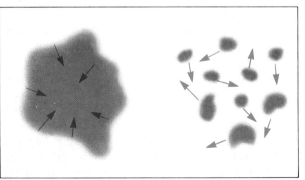

Figure 11.32 Arclike Superclusters. This shows the distribution of galaxies in a region of the sky that has recently been mapped completely, showing that superclusters are arranged in curved filamentary structures. This is evidence that the distribution of matter in the universe is not uniform, but that there is a large-scale structure. *(M. Geller, J. Huchra, M. Kurtz, and V. de Lapparent, Harvard-Smithsonian Center for Astrophysics)*

tures, but arclike chains of superclusters, as though the early universe had some sort of a bubblelike structure (Fig. 11.32). This has enormous implications for studies of the evolution of the universe and will be discussed in Chapter 14.

Perspective

We have at last completed our tour of the universe, having discussed nearly all the forms and types of organization that matter can take. We must still deal with a few specific kinds of objects that have not yet been described, but the overall picture of the universe in its present state is now more-or-less complete.

We are ready to tackle questions having to do with the nature of the universe itself, its overall properties, and its dynamic nature, and we will do so before examining some of the peculiar objects that are clues to the past.

Summary

1. Galaxies are categorized by shape in two general classes: spirals and ellipticals. There are also a substantial number of SO galaxies (disk-shaped, but with no spiral structure) and irregular galaxies.
2. Distances to galaxies are measured using a variety of standard candles such as Cepheid variables, extremely luminous stars, supernovae, and galaxies of standard types. A recent promising technique uses a correlation between 21-centimeter line width and luminosity to determination absolute magnitude and distance.
3. Masses of spiral galaxies are determined through the application of Kepler's third law to the outer portions, where the orbital velocity and period are determined from a rotation curve. The internal velocity dispersion is used to determine the masses of elliptical galaxies. Galactic masses can also be determined from the application of Kepler's third law to binary galaxies.
4. Galactic luminosities and diameters vary quite widely among elliptical galaxies but not so widely among spirals.
5. While all galaxies are dominated by Population II stars, spirals tend to contain a greater proportion of Population I stars, have substantial quantities of interstellar matter, and generally seem to be in a state of continuous evolution and stellar cycling. Ellipticals, on the other hand, have few or no Population I stars, contain little or no cold interstellar

matter, and generally do not seem to have active stellar cycling at the present time.

6. Spiral galaxies appear to originate from rotating gas clouds that flatten into disks before all the gas is used up in star formation. Ellipticals seem to result when star formation consumes all of the gas before collapse to a disk occurs.
7. Many galaxies are members of clusters rather than being randomly distributed throughout the universe.
8. The Milky Way is a member of a cluster called the Local Group, which has about 35 members.
9. The nearest neighbors to the Milky Way are the Magellanic Clouds, both Type I irregulars.
10. The Andromeda galaxy, a huge spiral of type Sb, is similar in size and general properties to the Milky Way, and the two are among the most prominent members of the Local Group.
11. The dominance of elliptical galaxies in rich clusters is probably the result of the conversion of spirals into ellipticals by the tidal forces of other galaxies and by drag created by intracluster gas. In many rich clusters there is a giant elliptical galaxy at the center that probably formed from the merger of several galaxies that settled there as a result of collisions.
12. Masses of clusters can be determined from the sum of the masses of individual galaxies or by the internal velocity dispersion of the galaxies in the cluster.
13. Clusters of galaxies tend to be grouped into aggregates called superclusters, but it is not certain whether these are random concentrations formed by mutual gravitational attraction or fundamental inhomogeneities of the universe.
14. Clusters of galaxies formed due to an uneven distribution of matter at some point early in the history of the universe, but it is not known how this clumpiness originally developed. It is not known whether the largest structures formed first and then fragmented or the smallest objects (individual galaxies) formed first and then congregated to form larger groupings.

Review Questions

1. A supernova of one type has an absolute magnitude of $M = -19$ at peak brilliance. Assuming such objects can be observed to apparent magnitudes as faint as $m = +24$, how far away can supernovae of this type be used as distance indicators?
2. Explain how our knowledge of the distance to a far-away galaxy depends on how well we know the Sun-Earth distance.
3. In a spiral galaxy with a mass-to-light ratio of 50, it takes 50 solar masses to produce each solar luminosity of energy that the galaxy emits. What does this tell us about the type of star that is most common in such a galaxy?
4. Would you expect elliptical galaxies to have the same proportion of heavy elements as spiral galaxies? Explain.
5. Contrast the evolution of an elliptical galaxy with that of the Milky Way.
6. If a dwarf elliptical galaxy has an absolute magnitude of $M = -15$ and can be detected with an apparent magnitude as faint as $m = +20$, how far away can these galaxies be found? Compare this distance with the diameter of the Local Group and with the distance of the Virgo cluster, a moderately large cluster some 15 megaparsecs distant.
7. Discuss the similarities and contrasts between a rich cluster of galaxies and a globular cluster of stars.
8. Why is it inaccurate to estimate the mass of a cluster of galaxies by using standard mass-to-light ratios for the individual members?
9. Suppose the temperature of the intracluster gas in a rich cluster of galaxies is 100 million K (10^8 K). At what wavelength does this gas emit most strongly? (Review the discussion of Wien's Law in Chapter 4.)
10. Discuss the parallels between the formation of individual stars and that of individual galaxies.

Additional Readings

de Boer, K. S., and Savage, B. D. 1982. The coronas of galaxies. *Scientific American* 247(2):54.

Geller, M. J. 1978. Large-scale structure of the universe. *American Scientist* 66:176.

Gorenstein, P., and Tucker, W. 1978. Rich clusters of galaxies. *Scientific American* 239(5):98.

Groth, E. J., Peebles, P. J. E., Seldner, M., and Soneira, R. M. 1977. The clustering of galaxies. *Scientific American* 237(5):76.

Hirshfeld, A. 1980. Inside dwarf galaxies. *Sky and Telescope* 59(4):287.

Hodge, P. W. 1981. The Andromeda galaxy. *Scientific American* 44(1):88.

Larson, R. B. 1977. The origin of galaxies. *American Scientist* 65:188.

Mathewson, D. 1985. The clouds of Magellan. *Scientific American* 252(4):106.

Mitton, S. 1976. *Exploring the galaxies.* New York: Scribner.

Rubin, V. C. 1983. Dark matter in spiral galaxies. *Scientific American* 248(6):96.

Sandage, A., Sandage, M., and Kristian, J., eds. 1976. *Galaxies and the universe.* Chicago: University of Chicago Press.

Strom, S. E., and Strom, K. M. 1977. The evolution of disk galaxies. *Scientific American* 240(4):56.

Talbot, R. J., Jensen, E. B., and Dufour, R. J. 1980. Anatomy of a spiral galaxy. *Sky and Telescope* 60(1):23.

Toomre, A., and Toomre, J. 1973. Violent tides between galaxies. *Scientific American* 229(6):38.

van den Bergh, S. 1976. Golden anniversary of Hubble's classification system. *Sky and Telescope* 52:410.

Chapter 12

Universal Expansion and The Cosmic Background

A microwave receiver used to measure the anistropy of the cosmic background radiation.
(George Smoot, Lawrence Berkeley Laboratory)

Chapter Preview

Hubble's Great Discovery
Hubble's Constant and the Age of the Universe
Redshifts as Yardsticks
A Cosmic Artifact: The Microwave Background
The Crucial Question of the Spectrum
Isotropy and Daily Variations

A recurrent theme throughout our study of astronomy has been the dynamic nature of celestial objects. Everything we have studied, from planets to stars to galaxies, has turned out to be in a constant state of change. Now that we are prepared to examine the entire universe as a single entity, we should not be surprised to see this theme maintained. That the universe itself is evolving, with a life story of its own, has been accepted by most astronomers, but there was doubt in the minds of some until very recently. In this chapter we will study the evidence.

Hubble's Great Discovery

Even before it was established unambiguously that the nebulae were really galaxies, a great deal of effort went into observing them. Their shapes were studied carefully, and their spectra were analyzed in detail. Typically the spectrum of a galaxy resembles that of a moderately cool star, with many absorption lines (Fig. 12.1). Because the spectrum represents the light from a huge number of stars, each moving at its own velocity, the Doppler effect causes the spectral lines to be rather broad and indistinct. Nevertheless, it is possible to analyze these lines in some detail, and one of the early workers in this field, V. M. Slipher, discovered during the decade before the Shapley-Curtis debate in 1920 that the nebulae tend to have large velocities, almost always directed away from the solar system. The spectral lines nearly always were found to be shifted toward the red, and this tendency was most pronounced for the faintest of the nebulae. Slipher measured velocities as great as 1800 kilometers per second.

Following his work, others continued to study spectra of nebulae, with heightened interest after Hubble's demonstration in 1924 that these objects were undoubtedly distant galaxies, comparable to the Milky Way in size and complexity. Hubble and others estimated distances to as many nebulae as possible by using the Cepheid variable period-luminosity relation for the nearby ones and other standard candles, such as novae and bright stars, for the more distant ones.

In 1929 Hubble made a dramatic announcement: The speed with which a galaxy moves away from the earth is directly proportional to its distance. If one galaxy is twice as far away as another, for example, its velocity is twice as great. If it is ten times farther away, it is moving ten times faster.

The implications of this relationship between distance and velocity are enormous. It means that the universe itself is expanding, its contents rushing outward at a fantastic pace, and that the entire universe was once concentrated at a single point. All the galaxies in the cosmos are moving away from all the others (Fig. 12.2).

To envision why the velocity increases with increasing distance, it is perhaps best to resort to a common analogy. Imagine a loaf of bread that is rising. If the dough has raisins sprinkled uniformly through it, they will move farther apart as the dough expands. Suppose that the raisins are 1 centimeter apart before the dough begins to rise, and consider what has happened one hour later, after it rises to the point where adjacent raisins are 2 centimeters apart. A given raisin is now 2 centimeters from its nearest neighbor, but 4 centimeters away from the next one over, 6 centimeters from the next one, and so on. The distance between any pair of raisins has doubled. From the point of view of any one of the raisins, its nearest neighbor had to move away from it at a speed of 1 centimeter per hour, the next raisin farther away had to move at 2 centimeters per hour, the next one at 3 centimeters per hour, and so on. If we let the dough continue to rise to the point where the adjacent raisins are 3 centimeters apart, we find that all the distances between pairs have tripled over what they were to start with, and the speeds needed to accomplish this are again directly proportional to the distance between a given pair. The farther away a raisin is to begin with, the farther it must move in order to maintain the regular spacing during the expansion, and therefore the greater its velocity. In the same way, the galaxies in the universe must increase their separations from each other at a rate proportional to the distance between them (Fig. 12.3), or else they would become bunched up. By observing the velocities

260 Section IV Extragalactic Astronomy

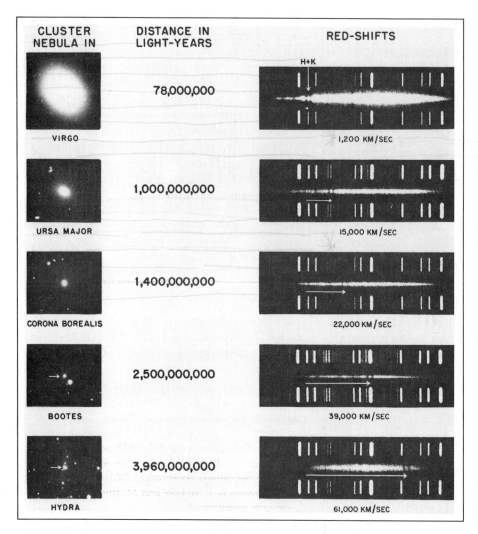

Figure 12.1 Spectra of Galaxies. These examples illustrate the broad absorption lines characteristic of spectra of large groups of stars such as galaxies. It was noticed before 1920 that the spectral lines tend to be shifted toward the red in galaxy spectra, indicating that galaxies as a rule are receding from us. *(Palomar Observatory, California Institute of Technology)*

of galaxies, astronomers are keeping watch on the raisins in order to see how the loaf of bread is coming along.

It is important to realize that it doesn't matter which raisin we choose to watch; from any point in the loaf, all other raisins appear to move away with speeds proportional to their distances from that point. Thus we do not conclude that our galaxy is at the center of the universe; from *any* galaxy, it would appear that all others are receding. It is also important to realize that the raisin-bread analogy has serious shortcomings: It suggests that the universe can be viewed as a simple three-dimensional object, with an edge and a center, which is inappropriate (this is discussed further in Chapter 17).

Following Hubble, a number of other astronomers have extended the study of galaxy motions to greater and greater distances, with the same result: As far as the telescopes can probe, the galaxies are moving away from each other at speeds proportional to their distances. Expansion is a major feature of the universe we live in and one that must be taken into account as we seek to understand its origins and its fate.

It is often convenient to display the data showing the universal expansion in a plot of velocity versus distance (as Hubble originally did; see Fig. 12.4), or on a plot

Chapter 12 Universal Expansion and the Cosmic Background   261

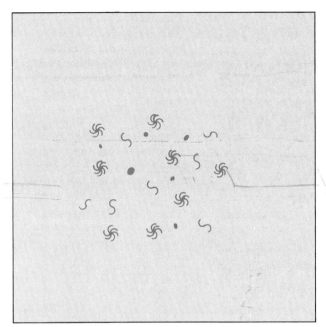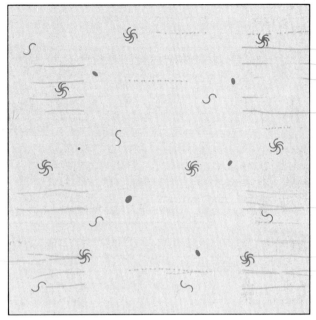

Figure 12.2 The Expanding Universe. These drawings show a number of galaxies at one time (left), and again at a later time (right). All of the spacings between galaxies have increased due to expansion of the universe.

Figure 12.3 Velocity of Expansion as a Function of Distance. The upper row represents seven galaxies, and the lower row represents the same galaxies one billion years later, when the distances between adjacent galaxies have doubled from 1 to 2 megaparsecs. From the viewpoint of an observer in *any* galaxy in the row, the recession velocity of its nearest neighbor is 1 megaparsecs per billion years; of its next nearest neighbor, 2 megaparsecs per billion years; of the next, 3 megaparsecs per billion years; and so on. The velocity is proportional to the distance in all cases, for any pair of galaxies.

Figure 12.4 The Hubble Law. This early diagram prepared by Hubble and M. Humason shows the relationship between galaxy velocity and distance. The slope of the relation has been altered since, with addition of data for more galaxies and improvements in the measurement of distances, but the general appearance of this figure is the same today. *(Estate of E. Hubble and M. Humason. Reprinted with permission)*

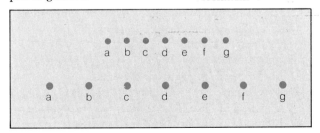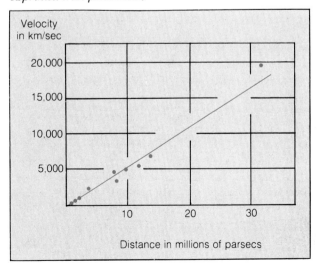

of redshift versus apparent magnitude (Fig 12.5), since increasing apparent magnitude (that is, decreasing brightness) indicates increasing distance.

The galaxies within a cluster have their own orbital motions, which are separate from the overall expansion (Fig. 12.6). As the universe expands, the clusters move apart from each other, while the individual galaxies within a cluster do not. The motions of galaxies within a cluster are relatively unimportant for very distant galaxies that are moving away from us at speeds of thousands of kilometers per second, but for nearby ones the orbital motion can rival or exceed the motion due to the expansion of the universe. The Andromeda nebula, for example, is actually moving *toward* the Milky Way at a speed of about 100 kilometers per second, whereas at its distance of 700 kiloparsecs from us, the universal expansion should give it a velocity away from us of about 40 to 70 kilometers per second.

Hubble's Constant and the Age of the Universe

The relation discovered by Hubble can be written in simple mathematical form:

$$v = Hd,$$

where v is a galaxy's velocity of recession and d is its distance in megaparsecs. The Hubble constant, H, is given in units of kilometers per second per megaparsec (km/sec/Mpc).

The value of H is difficult to establish (Table 12.1). It is done by collecting as large a body of data as possible on galactic velocities and distances and then deducing the value of H that best represents the relation-

Figure 12.6 Local Motions. Although there is a systematic overall expansion of the universe, individual galaxies within clusters, and even clusters within superclusters, have random individual motions. Hence within the Local Group the galaxies are not uniformly receding from each other.

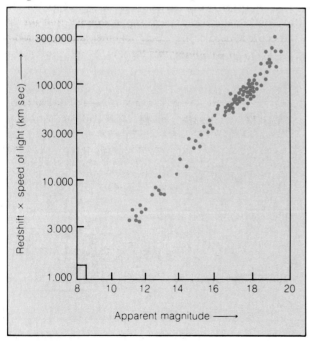

Figure 12.5 A Modern Version of the Hubble Law. Often, apparent magnitude is used on the horizontal axis instead of distance, because the two quantities are related, particularly if the diagram is limited to galaxies of the same type, as this one is. The small rectangle at the lower left indicates the extent of the relationship as Hubble first discovered it; today many more galaxies, much dimmer, have been included. (*Adapted from J. Silk, 1980,* The Big Bang [*San Francisco: W. H. Freeman*])

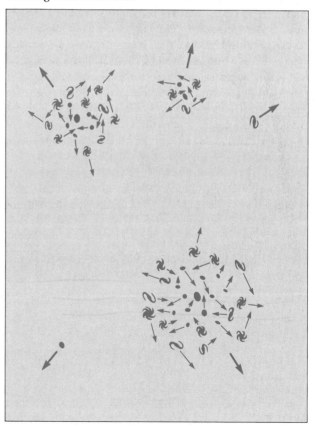

Table 12.1 Example Measurements of the Hubble Constant

Value of H (km/sec/Mpc)	Observer	Date	Implied Age of Universe (years)
540	Hubble	1930	1.8×10^9
260	Baade	1949	3.8×10^9
180	Humason et al.	1955	5.4×10^9
75	Sandage	1956	1.3×10^{10}
56	Sandage and Tammann	1974	1.8×10^{10}
100	van den Bergh	1975	1.0×10^{10}
80	Tully and Fisher	1977	1.2×10^{10}
100	de Vaucouleurs et al.	1979	1.0×10^{10}
95	Aaronson	1980	1.0×10^{10}
65	Mould et al.	1980	1.5×10^{10}
50	Sandage and Tammann	1982	2.0×10^{10}
82	Aaronson and Mould	1983	1.2×10^{10}

Source: Information from Rowan-Robinson, M. 1985, *The Cosmological Distance Ladder* (New York:W.H. Freeman).

ship between distance and velocity. Hubble did this first, finding $H = 540$ kilometers per second per megaparsec, meaning that for every megaparsec of distance, the velocity increased by 540 kilometers per second. The standard candles were not very well established in Hubble's day, when it was still news that the nebulae were distant galaxies, and this turned out to be an overestimate. The best modern values for H are between 50 and 100 kilometers per second per megaparsec. Until very recently, a consensus had been developing that the best value is 55 kilometers per second per megaparsec, but a new distance determination for a number of galaxies using the Tully-Fisher method (Chapter 11) has now suggested a value close to 90 kilometers per second per megaparsec. Today a value near 90 kilometers per second per megaparsec seems to be gaining acceptance, but many astronomers are still at work refining this. The precise value of H has extremely important implications for our understanding of the universe and its expansion, and intense research effort is being devoted to refining the estimates.

If the universe is expanding, it follows that all the matter in it used to be closer together than it is today (Fig. 12.7). If we follow this logic to its obvious conclu-

Figure 12.7 Backtracking the Expansion. If the universe began as a single point, the present rate of expansion provides information on how long ago the expansion began. With considerable uncertainty due to the fact that the expansion rate has probably not been constant, this leads to an estimated age for the universe of between 10 and 20 billion years.

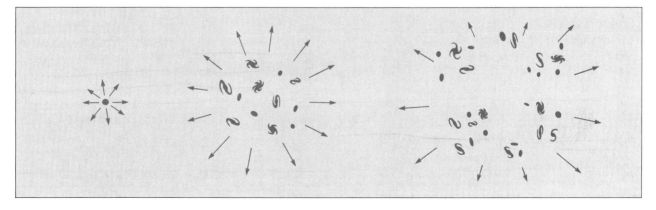

sion, we find that the universe was once concentrated in a single point, from which it has been expanding ever since.

From the rate of expansion, we can calculate how long ago the galaxies were all together in a single point. From the simple expression

time = distance/velocity,

we find

age = d/Hd = $1/H$.

The age of the universe is equal to $1/H$, if the expansion has been proceeding at a constant rate since it began. As we will see in Chapter 14, this assumption of a constant expansion rate is not strictly valid (expansion was more rapid in the beginning), but it gives a useful estimate of the age for now.

If we adopt, for the sake of discussion, a value for H of 75 kilometers per second per megaparsec, the age of the universe is 4.6×10^{17} seconds = 1.5×10^{10} years (to carry out this calculation, first you must convert the value of a megaparsec into kilometers).

Figure 12.8 Finding Distances from The Hubble Law. This diagram shows how the distance to a galaxy can be deduced directly from the Hubble law, if the redshift, hence the velocity, is measured.

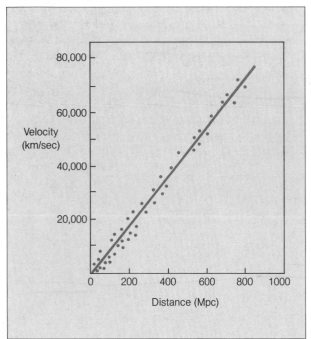

A simple calculation based on the observed expansion rate has led to a profound conclusion: Our universe is about 15 billion years old. This fits in with what we have learned about galactic ages. The Sun, for example, is thought to be about 4.5 billion years old, and the most aged globular clusters are apparently 12 to 18 billion years old.

It is interesting to consider what happens to our estimate of the age of the universe if we choose other values of H (Table 12.1). If Hubble's value of 540 kilometers per second per megaparsec had been correct, the age would have been calculated as less than 2 billion years, and if the recently suggested value of about 90 kilometers per second per megaparsec is found to be correct, the age is 11 billion years. In either case, there is a conflict with the estimated ages of globular clusters; it will be interesting to see how this conflict is resolved if indeed a value for H of 90 kilometers per second per megaparsec becomes widely accepted. Age estimates based on $1/H$ are probably too large, because they do not take into account the fact that the expansion rate at the earliest times was somewhat more rapid than it is today. This further complicates things if a value of 90 kilometers per second per megaparsec is found to be best. Whatever the correct ages are, this situation illustrates how uncertain some of the most important measurements in astronomy can be.

There are many other important questions about the universe that depend on the value of H, and this is the reason so much effort is being expended to define its value as accurately as possible. One of the principal reasons for development of the *Space Telescope* is to probe the most distant possible galaxies to determine their distances and velocities so that this information can be used to help find the correct value of H.

Redshifts as Yardsticks

The expansion of the universe has a very practical benefit for astronomers concerned with the properties of distant galaxies: It provides a means of determining their distances (Fig. 12.8). If we know the value of H, we can find the distance to a galaxy simply by measuring its velocity as indicated by the Doppler shift of its spectral lines. The distance is given by

$d = v/H$

Thus, if we adopt H = 75 kilometers per second per megaparsec, a galaxy found to be receding at 7500 ki-

Astronomical Insight 12.1

The Distance Pyramid

Having reached the ultimate distance scales, we can pause to look back at the progression of steps that got us here. As noted in the text, the use of Hubble's relation to estimate distances to the farthest galaxies depends critically on the value of H, which can only be determined from knowledge of the distances to a representative sample of galaxies. Those distances in turn depend on various standard candles, themselves calibrated through the (use of other methods such as the Cepheid variable period-luminosity relation) that are applicable only to the nearest galaxies. Our knowledge of these close extragalactic distances rests on the known distances to objects within the Milky Way, such as star clusters containing variable stars, which are used to calibrate the period-luminosity relation. The distances to these clusters in turn are known from more local techniques such as spectroscopic parallaxes and main sequence fitting, and these ultimately depend on the distances to the nearest stars, derived from trigonometric parallaxes.

Thus the entire progression of distance scales, all the way out to the known limits of the universe, depends on our knowledge of the distance from the Earth to the Sun, the basis of trigonometric parallax measurements. At every step of the way outward, our ability to measure distances rests on the previous step. If we revise our measurement of local distances, we must accordingly alter all our estimates of the larger scales, which will affect our perception of the universe as a whole. This elaborate and complex interdependence of distance determination methods is known as the **distance pyramid** and is sometimes depicted in graphic form as shown above.

A very important step in the sequence shown above is represented by the Hyades cluster, a galactic star cluster some 44 parsecs away. The distance to this cluster is determined from the **moving cluster method**, in which a combination of Doppler

(continued on next page)

Distance	Key Object	Method
10^{-6} pc	Sun	↓ Radar, planetary motions
10^{-5}		
10^{-4}		
10^{-3}		
10^{-2}		Trigonometric parallax
10^{-1}		
1	α Centauri	
10^{1}		
10^{2}	Hyades cluster	↓ Moving cluster method
10^{3}		Main sequence fitting
10^{4}	Limits of Milky Way	↓
10^{5}	Magellanic Clouds	
10^{6}	Andromeda galaxy	↓ Cepheid variables
10^{7}		↓ Brightest stars
10^{8}	Virgo cluster	
10^{9}		
10^{10}		↓ Brightest galaxies

Astronomical Insight 12.1

The Distance Pyramid

(continued from previous page) shift measurements and proper motion measurements (see Chapter 6) reveal the true direction of cluster motion, so that a comparison of angular motion and true velocity then gives the distance to the cluster. Because much of the rest of the distance pyramid depends on the application of this technique to the Hyades cluster, great care has been taken in the analysis, and every so often it is redone. Whenever the distance to the Hyades is altered slightly, the impact spreads as astronomers judge how this affects other distance scales in the universe. One of the things the *Space Telescope* may accomplish is to make direct parallax measurements of stars in the Hyades, providing a new and important measurement of the distance to the cluster and thus revising once again our understanding of the distance scale of the universe.

---

lometers per second, for example, has the distance

$$d = \frac{7500 \text{ km/sec}}{75 \text{ km/sec/Mpc}} = 100 \text{ Mpc}.$$

This technique can be applied to the most distant and faint galaxies, where it is nearly impossible to observe any standard candles. It is important to remember, however, that the use of Hubble's relation to find distances is only as accurate as our value of H, and the value that we use is derived in turn from distance determinations that depend on standard candles. Furthermore, because the expansion rate of the universe has not been constant, for very distant galaxies it is necessary to know what the rate was in the past in order to determine the distance accurately.

A Cosmic Artifact: The Microwave Background

Since Hubble's discovery of the universal expansion, many astronomers and physicists have explored its significance, both for what it may tell us about the future and for the information it provides about the past. A number of scientists, beginning in the 1940s with George Gamow and his associates, have carried out extensive studies of the nature of the universe in its earliest days. It was quickly realized that if all the matter was originally compressed into a small volume, it must have been very hot, since any gas heats up when compressed. The fiery conditions inferred for the early stages of the expansion have led to the term **big bang**, which is the commonly accepted title for all theories of the universal origin that start with an expansion from a single point.

Gamow's primary interest was in analyzing nuclear reactions and element production during the big bang (the modern understanding of this is described in Chapter 14), but in the course of his work he came to another important realization: The universe must have been filled with radiation when it was highly compressed and very hot, and this radiation should still be with us. At first, when the temperature was in the billions of degrees, there was γ-ray radiation, but later, as the universe cooled, X rays and then ultraviolet light filled it. During the first million years or so after the expansion started, the radiation was constantly being absorbed and re-emitted (scattered) by the matter in the universe, but eventually (when the temperature had dropped to around 4000 K) protons and electrons combined to form hydrogen atoms, and from that time on there was little interaction between the matter and the radiation. In due course the matter organized itself into galaxies and stars, while the radiation has simply continued to cool as the universe has continued to expand (Fig. 12.9). The intensity and spectrum of the radiation depend only on the temperature of the universe and are described by the laws of thermal radiation (see Chapter 4).

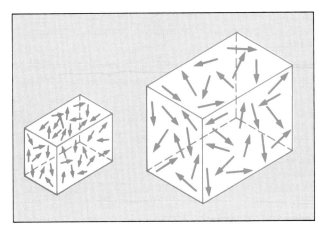

Figure 12.9 The Cooling of the Universal Radiation. At left is a box representing the early universe, filled with intense gamma-ray radiation. At right is the same box, greatly expanded. The radiation is still there, but its wavelength has increased, and the temperature represented by its thermal spectrum has decreased.

Figure 12.10 The Radio Receiver That Discovered The 3-Degree Background Radiation. Bell Laboratories scientists Robert Wilson, left, and Arno Penzias in front of the horn-reflector antenna with which they found a persistent noise that turned out to be the remnant radiation from the big bang. (*Courtesy of AT&T*)

Gamow and others made rough calculations showing that the present temperature of the radiation filling the universe should be very low indeed, about 25 K. More recently R. H. Dicke and his colleagues independently carried out new calculations, which predicted an even lower temperature, about 5 K. Using Wien's law, it is easy to calculate the most intense wavelength of such ra-diation: If the temperature is 5 K, the value of λ_{max} is 0.06 centimeter, in the microwave part of the radio spectrum.

Gamow's principal interest was in element production during the big bang, rather than in the remnant radiation, and in the late 1940s the technology needed to detect the radiation was not yet developed. Hence no attempt was made in Gamow's time to search for the radiation, and the idea was forgotten until Dicke's work. The remnant radiation was eventually detected in 1965 quite by accident, when radio astronomers at Bell Laboratories found a persistent source of background noise while testing a new antenna (Fig. 12.10).

The implication soon became clear: The "noise" was the universal radiation that the theorists had predicted should exist. This was one of the most significant astronomical observations ever made, for it provided a direct link to the origins of the universe. More than a piece of circumstantial evidence, this was a real artifact, the kind of hard evidence that carries weight in a court of law.

The Crucial Question of The Spectrum

There remained questions about the interpretation of the radiation, however, and doubt in the minds of some who did not accept the big bang theory of the origin of the universe. If the radiation was really the remnant of the primeval fireball, it was expected to fulfill certain conditions, and intensive efforts were made to see whether it did.

One important prediction was that the spectrum of a simple glowing object whose emission of radiation is due only to its temperature should have the characteristic shape of a thermal spectrum, as explained in Chapter 4. If, on the other hand, the source of the radiation were something other than the remnant of the initial universe, such as distant galaxies or intergalactic gas, it could have a different spectrum. Thus the shape of the spectrum—its intensity as a function of wavelength—was a very important test of the interpretation of its origin.

Some of the early results were confusing, due to unforeseen complications, but now the situation seems to have been resolved. The spectrum does indeed follow the shape expected (Fig. 12.11), and the radiation is considered to be a relic of the big bang. The most difficult part of the task of measuring the spectrum was

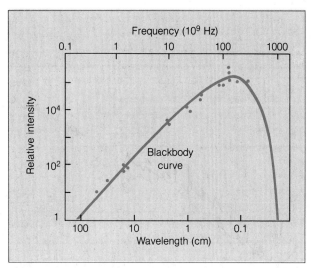

Figure 12.11 The Measured Background Spectrum. This curve shows the spectrum of the cosmic background radiation, plotted as a function of wavelength (a frequency scale is also given at the top). The points are measurements; the solid line represents a thermal spectrum for a temperature of just under 3 K. (© *West Publishing Co.*)

observing the intensity at short wavelengths, around 1 millimeter or less, where the earth's atmosphere is especially impenetrable, but this was finally accomplished with high-altitude balloon instruments.

The net result of all the effort expended in determining the spectrum is that the temperature of the radiation is about 2.7 K, and its peak emission occurs at a wavelength of 1.1 millimeters. The radiation is commonly referred to as the **microwave background** or **3-degree background radiation**.

The care that went into establishing the true nature of the background radiation reflects the importance of what it has to tell us about the universe. As we will see in Chapter 14, there are alternatives to the big bang theory, and those who support other theories have sought other explanations of the microwave radiation. Its close adherence to the spectrum expected of the radiation from the cosmic fireball envisioned in the big bang theory has presented grave difficulties for these alternative points of view.

Isotropy and Daily Variations

Another expectation of the background radiation, in addition to the nature of its spectrum, is that it should fill the universe uniformly, with no preference for any particular location or direction. A medium or radiation field that has no preferred orientation but looks the same in all directions is said to be **isotropic**. Some of the early observations of the 3-degree background radiation were designed to test its isotropy, for again, failure to meet this expectation could imply some origin other than the big bang.

In simple terms, a test of isotropy is just a measurement of the radiation's intensity in different directions to see whether or not it varies. From the time it was discovered, the microwave background showed the high degree of isotropy expected. The issue has been pressed, however, for a variety of reasons. One possible alternative explanation of the radiation, for example, was that it arises from a vast number of individual objects, such as very distant galaxies, so closely spaced in the sky that their combined emission appears to come uniformly from all over. To test this idea, a group of astronomers attempted to see whether the radiation is patchy on very fine scales, as it would be if it came from a number of point sources. So far no evidence of any clumpiness has been found, and the theory of big bang origin for the radiation has not been threatened.

There is a kind of subtle nonuniformity that is expected of the microwave radiation even if it is a remnant of the big bang. Because of the Doppler shift, the observed intensity of the radiation as seen by a moving observer will vary with direction, depending on whether the observer is looking toward or away from the direction of motion. If the observer looks ahead, toward the direction of motion, a blueshift occurs, with the peak of the spectrum being shifted slightly toward shorter wavelengths (Fig. 12.12). In terms of temperature, this means that the radiation looks a little hotter when viewed in this direction. If the observer looks the other way, there is a slight redshift, and the measured temperature is cooler.

The Earth is moving in its orbit about the Sun. The Sun, in addition, is orbiting the galaxy, the galaxy is moving along its own path through the Local Group, and the Local Group has its own motion within the local supercluster. All of these motions combined represent a velocity of the Earth with respect to the frame of reference established by the background radiation, a velocity that should produce slight differences in the radiation temperature if the radiation is viewed in different directions. Because of the Earth's rotation, we should be able to point our radio telescope straight up and alternately see high and low temperatures, as our

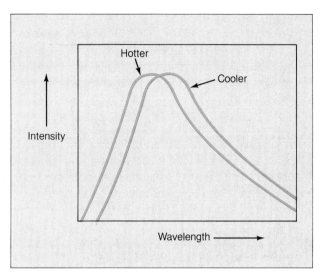

Figure 12.12 The Effect of a Doppler Shift on The Cosmic Background Radiation. If the observer on the Earth has a velocity with respect to the radiation, this shifts the peak of the spectrum a little and affects the temperature that observer deduces from the measurements. Motion of the Earth results in a daily cycle of tiny fluctuations in the observed temperature, known as the twenty-four-hour anisotropy.

Figure 12.13 The *Cosmic Background Explorer*. This satellite, to be launched in 1989, will make the most precise and extensive measurements yet of the microwave background radiation. (*NASA, courtesy of J. C. Mather*)

telescope points toward and then, 12 hours later, away from the direction of motion. Thus there should be a daily variation cycle in the radiation, referred to as the **24-hour anisotropy**. (*Anisotropic* is the term for a nonuniform medium.)

The actual change in temperature from one direction to the other is very small and was not successfully measured until very recently. The observed temperature difference from the forward to the rearward direction is only a few thousandths of a degree, and very sophisticated technology was required in order for this to be measured. The data show that the Earth is moving with respect to the background radiation at a speed of about 400 kilometers per second. When the Sun's known orbital velocity about the galaxy is taken into account, this implies that the galaxy itself is moving at 600 to 800 kilometers per second with respect to the background. This motion must be a combination of the motion of the galaxy in its orbit within the Local Group, the motion of the Local Group within the supercluster to which it belongs, and possibly the motion of the supercluster itself.

The use of the 3-degree background radiation as a tool for deducing local motions with respect to the radiation is predicated on the assumption that the radiation itself is fundamentally isotropic. At present there is no reason to doubt this, but the possibility remains that it is not, that the universe itself is somehow lopsided. If it is, it will be very difficult indeed to measure its asymmetry, because of the complexity of the motions that our observing platform, the Earth, undergoes.

Future observations are being planned to improve our measurements of the 3-degree background. A satellite called the *Cosmic Background Explorer* (Fig 12.13) is due to be launched in 1989 and will provide, from its vantage point above the Earth's atmosphere, the best data yet on the spectrum and isotropy of the background radiation.

Perspective

We have learned now that the universe, like all the subordinate objects within it, is a dynamic, evolving entity. One of the grandest stories in astronomy has been the unfolding of the concept of universal expansion. An idea forced on astronomers by the evidence of the redshifts, it has led directly to the big bang picture, which in turn tells us the age of the universe, explains the origin of some of the elements, and is verified through the presence of the microwave background radiation.

Before we finish the story by studying modern theories of the universe and their implications for its future, we must take a closer look at some of the objects that inhabit it. There is a breed of strange and bizarre entities whose very nature seems to be tied up with the evolution of the universe, and which therefore will provide us with a few additional bits of evidence for our final discussion.

Summary

1. It was discovered early in this century that galaxies tend to have redshifted spectra, implying that they are moving away from us.
2. Hubble discovered in 1929 that the velocity of recession is proportional to distance, demonstrating that the universe is expanding.
3. The rate of expansion, expressed in terms of Hubble's constant H, is probably between 50 and 100 kilometers per second per megaparsec.
4. The fact that the universe is expanding implies that it originated from a single point, and the rate of expansion, if constant, tells us that it began some 10 to 20 billion years ago. This may be considered to be the age of the universe.
5. The relationship between velocity and distance allows the distance to a galaxy to be estimated from its velocity, through the Doppler effect. This method of determining distance extends to the farthest limits of the observable universe.
6. The fact that the universe began in a single point implies that it was very hot and dense initially, and this in turn implies that the early universe was filled with radiation. As the expansion has proceeded, this radiation has been transformed into microwave radiation with a thermal spectrum corresponding to a temperature of about 3 K. The cosmic background radiation was discovered in 1965.
7. To distinguish a primordial origin for the background radiation from other possible origins (such as numerous galaxies spread throughout the universe), it is necessary to measure the spectrum to see whether it truly is a thermal spectrum. Observations to date are consistent with the assumption that it is.
8. Another important test is to determine whether the radiation is isotropic or whether it might be unevenly intense in different directions. Careful observations have revealed no evidence of patchiness or unevenness that might imply that the radiation arises in a large number of individual sources or that the early universe was nonuniform in any way.
9. There is an anisotropy in the radiation, however, created by the Doppler effect due to the Earth's motion. By comparing the radiation intensity in the forward and backward directions, it has been possible to determine the Earth's velocity with respect to the radiation. The Earth's motion is a combination of its orbital motion about the Sun, the Sun's orbital motion about the galaxy, the galaxy's motion within the Local Group, the Local Group's motion within the local supercluster, and possibly the motion of the local supercluster with respect to the reference frame of the radiation.

Review Questions

1. Using the information on the Doppler shift in Chapter 4, determine the speed of recession of a galaxy whose ionized calcium line (assume the rest wavelength is 3933.00 Å) is observed at a wavelength of 3936.93 Å.
2. Suppose three galaxies are situated initially so that the distance from galaxy A to galaxy C is twice the distance from A to B. After a period of time, the distance between galaxies A and B has doubled. Now how does the distance between A and C compare with that between A and B?
3. To help you understand how the age of the universe is determined from the rate of expansion, determine the age of a supernova remnant whose outermost portions are expanding away from the center at a rate of 1000 kilometers per second if its radius is 10 parsecs. (You will have to convert parsecs to kilometers to do this.) Express your answer in years.
4. If the past expansion of the universe was more rapid than the present expansion, the correct age is less than the value calculated by assuming that

H has been constant at the present value. Suppose the correct present value of H is 75 kilometers per second per megaparsec, but the average value over the entire history of the universe has been 85 kilometers per second per megaparsec. Calculate the age of the universe using this value.

5. Using information from Chapter 11, summarize the kinds of observations needed in order to determine the value of H, the Hubble constant.

6. Suppose the position of the line of ionized calcium is observed in the spectra of three galaxies. The rest wavelength of this line is 3933 Å, and in the three galaxies it is observed at wavelengths of 3936 Å, 4028 Å, and 3942 Å. Rank the three galaxies in order of increasing distance. (As an additional exercise, you may want to calculate the distances to the galaxies, but you need not do that simply to rank them.)

7. If the Hubble constant is $H = 75$ kilometers per second per megaparsec, calculate the distance to a galaxy whose ionized calcium line is observed at a wavelength of 4028 Å, instead of the rest wavelength of 3933 Å.

8. What was the wavelength of maximum emission for the universal radiation field when hydrogen atoms formed from protons and electrons, when the temperature was 4000 K (recall Wien's law, Chapter 4)?

9. Explain why it is necessary, when attempting to observe the cosmic background radiation, to cool the telescope and instruments to a low temperature.

10. Discuss the possible ways in which the cosmic background radiation could be anisotropic (apart from the 24-hour anisotropy), and the implications if it is.

Additional Readings

Abell, G. 1978. Cosmology—the origin and evolution of the universe. *Mercury* 7(3):45.

Barrow, J. D., and J. I. Silk. 1980. The structure of the early universe. *Scientific American* 242(2):118.

Layzer, D. 1975. The arrow of time. *Scientific American* 233(6):56.

Muller, R. A. 1978. The cosmic background radiation and the new aether drift. *Scientific American* 238(5):64.

Shu, F. H. 1982. *The physical universe.* San Francisco: W. H. Freeman.

Silk, J. 1980. *The big bang: The creation and evolution of the universe.* San Francisco: W. H. Freeman.

Weinberg, S. 1977. *The first three minutes.* New York: Basic Books.

Chapter 13

Peculiar Galaxies, Explosive Nuclei, and Quasars

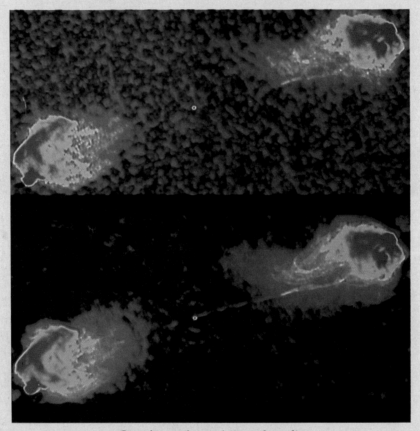

Steps in creating an image of a radio galaxy from *VLA* observations. *(National Radio Astronomy Observatory, operated by Associated Universities, Inc., under contract with the National Science Foundation)*

Chapter 13 Peculiar Galaxies, Explosive Nuclei, and Quasars 273

Chapter Preview

The Radio Galaxies
Seyfert Galaxies and Explosive Nuclei
The Discovery of Quasars
The Origin of the Redshifts
The Properties of Quasars
Galaxies in Infancy?

In discussing the characteristics of galaxies in the preceding chapters, we overlooked a variety of objects, some of them galaxies and some possibly not, that have unusual traits. As in many other situations in astronomy, the so-called peculiar objects, once understood, have quite a bit to tell us about the more usual ones.

To appreciate fully the bizarre nature of some of these astronomical oddities, it was best to delay their introduction until this point when we have essentially completed our survey of the universe. We know about the expansion and the big bang, and we are just in the process of tying our knowledge together. In this chapter we will uncover a number of vital clues in the cosmic puzzle.

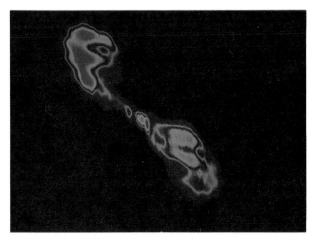

The Radio Galaxies

Most of the first astronomical sources of radio emission to be discovered, other than the Sun, were galaxies. Most of these, when examined optically, have turned out to be large ellipticals, often with some unusual structure (Fig. 13.1). The first of these objects to be detected, and one of the brightest, is called Cygnus A (named under a preliminary cataloging system in

Figure 13.1 Centaurus A. This is a giant elliptical radio galaxy, showing a dense lane of interstellar gas and dust across the central region. At top is a visible-light photograph; in the center is a radio image; and at the bottom is an X-ray image. Note that the radio and X-ray emissions come from the lobes above and below the visible galaxy. (top: © 1980 Anglo-Australian Telescope Board; middle: *National Radio Astronomy Observatory, operated by Associated Universities, Inc., under contract with the National Science Foundation*; bottom: *Harvard-Smithsonian Center for Astrophysics*)

which the ranking radio sources in constellations are listed alphabetically). This galaxy has a strange double appearance, and after sufficient refinement of radio observing techniques (see the discussion of radio interferometry in Chapter 4) it was eventually found that the radio emission comes from two locations on opposite sides of the visible galaxy and well separated from it (Figs. 13.2 and 13.3). High resolution observations with the *Very Large Array* have revealed remarkable detail in some radio galaxies (Fig. 13.4).

This double-lobed structure is a common feature of the so-called **radio galaxies,** those with unusually strong radio emission. Most, if not all, ordinary galaxies emit radio radiation, but the term *radio galaxy* is applied only to the special cases where the radio intensity is many times greater than the norm. The core of the Milky Way is a strong radio source as viewed from the Earth, but it is not in a league with the true radio galaxies. Typical properties of radio galaxies are listed in Table 13.1.

While the double-lobed structure is standard among elliptical radio galaxies, the visual appearance varies quite a bit. We have already noted the double appearance of Cygnus A; another bright source, the giant elliptical Centaurus A (the object shown in Fig. 13.1), appears to have a dense band of interstellar matter bisecting it, and it has not one but two pairs of radio lobes, one much farther out from the visible galaxy than the other. Other giant elliptical radio galaxies have other kinds of strange appearances. One of the most famous is M87, also known as Virgo A, which has a jet that protrudes from one side and is aligned

Figure 13.2 A Radio Image of a Radio Galaxy. Here the lighter areas represent maximum radio brightness. The galaxy as seen in visible wavelengths does not even appear here; this image is completely dominated by the double side lobes. *(National Radio Astronomy Observatory, operated by Associated Universities, Inc., under contract with the National Science Foundation)*

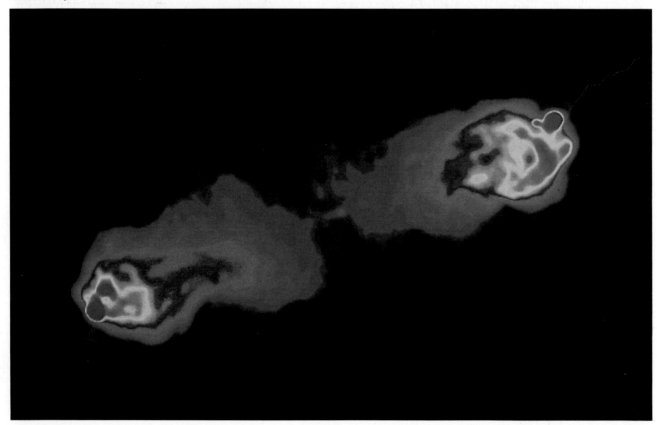

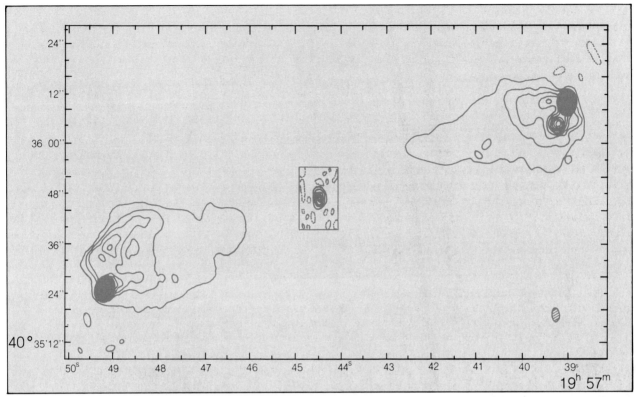

Figure 13.3 A Map of a Radio Galaxy. The intensity contours here represent radio emission from the two side lobes of the radio galaxy Cygnus A. The blurred image at the center illustrates the size of the visible galaxy compared to the gigantic side lobes. *(Mullard Radio Astronomy Observatory, University of Cambridge)*

with one of the radio lobes (Fig. 13.5). Careful examination shows that this jet actually consists of a series of blobs that appear to have been ejected from the core of the galaxy in sequence. Many galaxies have been discovered to have radio-emitting jets. These include galaxies already known to be peculiar, such as Centaurus A (Fig. 13.1) and at least one spiral galaxy (Fig. 13.6). In some cases double jets are seen on opposite sides of a galaxy (Fig. 13.7).

The size scale of the radio galaxies can be enormous. In some cases the radio lobes or jets extend as far as 5 megaparsecs from the central galaxy. Recall that the diameter of a large galaxy is only one-tenth of this distance and that the Andromeda galaxy is less than one megaparsec from our position in the Milky Way.

The first radio galaxies were detected in the 1940s, and by the 1950s studies of the radio spectra of these objects were in progress. It was discovered that they are emitting by the synchrotron process (mentioned in Chapter 8), which requires a strong magnetic field and a supply of rapidly moving electrons. The electrons are forced to follow spiral paths around the magnetic field lines, and as they do so they emit radiation over a broad range of wavelengths. The characteristic signa-

Table 13.1 Properties of Radio Galaxies

Galaxy type: Most often elliptical or giant elliptical.
Radio luminosity compared to optical luminosity: 0.01 to 10.
Radio source shape: Either double-lobed or compact central source; often jets are seen that emit throughout the spectrum.
Variability: Intensity variations in times as short as days may occur in the compact radio sources.
Nature of spectrum: Usually synchrotron or inverse Compton spectrum, usually polarized.

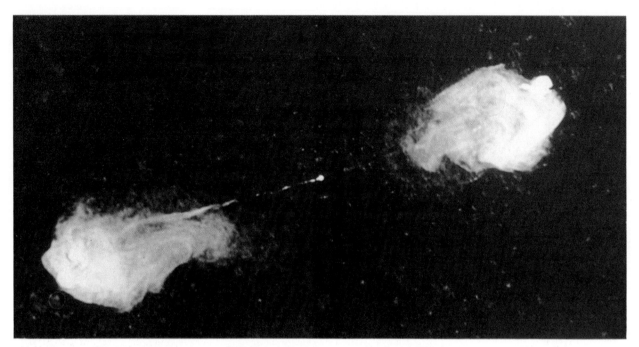

Figure 13.4 Details of Radio Lobes. This image of Cygnus A was obtained with the *Very Large Array* and shows fine detail suggesting turbulence and flows in the lobes. *(National Radio Astronomy Observatory, operated by Associated Universities, Inc., under contract with the National Science Foundation)*

Figure 13.5 A Giant Elliptical Radio Galaxy with A Jet. This is M87, also known as Virgo A, showing its remarkable linear jet of hot gas. At left are short-exposure photographs, specially processed to maximize the visibility of the jet and its lumpy structure. On a longer-exposure photograph (right), this galaxy looks like a normal elliptical, the light from the jet being drowned out by the intensity of light from the galaxy. *(H. C. Arp)*

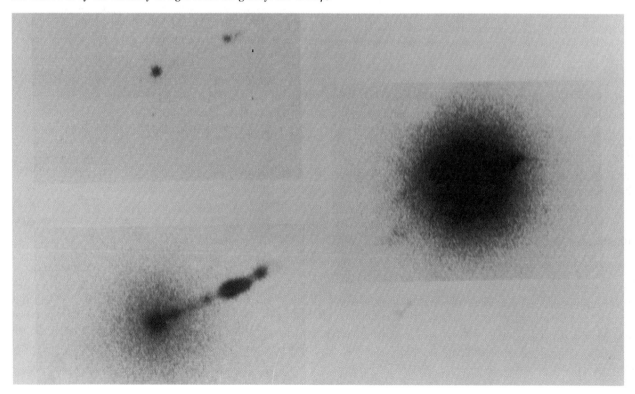

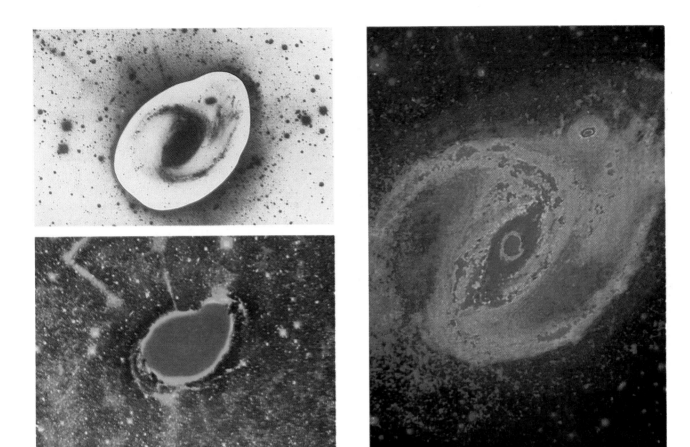

Figure 13.6 Jets in an Unusual Galaxy. Here are three views of the barred spiral galaxy NGC1097, showing its two remarkable linear jets. At top left is a black-and-white photograph that has been processed to enhance the visibility of the jets. At lower left is a false-color version. An enlarged view of the inner part of the galaxy is shown at right. *(H. C. Arp)*

ture of synchrotron radiation, in contrast with the thermal radiation from hot objects such as stars, is a sloped spectrum with no strong peak at any particular wavelength. The radiation from a synchrotron source is also polarized. Whenever an object is found to be emitting by the synchrotron process, observers immediately conclude that some highly energetic activity is taking place to produce the rapid electrons that are required to create the emission.

In some cases it appears that a different process that also requires a supply of rapidly moving electrons is responsible for the radio emission. If there is ordinary thermal radio emission from a galaxy, the presence of electrons moving near the speed of light can alter the spectrum of the radio emission as energy is transferred from the electrons to the radiation. This process is called **inverse Compton scattering,** and it has the same implication as synchrotron radiation: There must be a source of vast amounts of energy to produce the rapidly moving electrons required to create the radiation.

The source of this energy in radio galaxies is a mystery, although some interesting speculation has been fueled by certain characteristics of the galaxies. The double-lobed structure indicates that there are two clouds of fast electrons on either side of the visible galaxy, apparently created by energetic particles that have been ejected from the galactic core, perhaps by an explosive event that expelled material symmetrically in opposite directions. This impression is heightened by the many cases where jets are seen protruding from the core (Figs. 13.6 and 13.7) and by the galaxies that have more than one pair of lobes, aligned in the same direction but with one farther out than the other. In these galaxies it seems that successive outbursts have taken place, each one ejecting material in opposite directions.

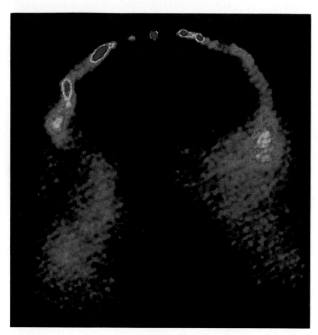

Figure 13.7 Dual Radio Jets. While visible jets are rare, radio images often reveal jetlike structures. Here is a radio image of a galaxy with a pair of jets emanating from opposite sides and curving away from the galaxy, apparently because the galaxy is moving through an intergalactic gas medium. *(National Radio Astronomy Observatory, operated by Associated Universities, Inc., under contract with the National Science Foundation)*

One method envisioned by astronomers for producing prodigious amounts of energy from a small volume such as the core of a galaxy is to have a giant black hole there surrounded by an accretion disk. The general idea is that superheated gas that has been compressed by the immense gravitational field of the black hole escapes along the rotation axes and is channeled into two opposing jets. It is probably better to view this process not as explosive, but as a more-or-less steady expulsion of gas. It has been suggested that *all* radio galaxies have jets, that this is the only mechanism for forming the double radio lobes. The fact that the jets are not *observed* in all radio galaxies is thought to be an effect of how the jets are aligned with respect to our line of sight. The emission from rapidly moving particles is confined to a narrow angle in the forward direction; therefore an observer to the side of a jet will see little of its radiation.

The theory that massive black holes are the energy sources in radio galaxies has received some observational support. Recently it was reported that evidence for a central massive object has been found in the heart of M87, the radio galaxy with the well-known visible jet. Based on measurements of the velocity dispersion near the center of the galaxy, it was deduced that a large amount of mass must be confined in a very small volume at the core, and a black hole seemed the best explanation. Subsequent observations, however, have failed to confirm the reportedly high velocities of stars in the inner portions of the galaxy, and the case for a black hole in M87 has been weakened. Further efforts to find direct evidence of supermassive black holes are being made.

Seyfert Galaxies and Explosive Nuclei

While it is true that the strongest radio emitters among galaxies are giant ellipticals, they are by no means the only ones with evidence for explosive events in their cores. Some spiral galaxies also display such behavior. Even the Milky Way is not immune, as we learned in Chapter 9, which summarized the evidence for the existence of a compact, massive object at the center of our galaxy. There are other spiral galaxies with much more pronounced violence in their nuclei. These galaxies as a class are called Seyfert galaxies (Fig. 13.8; Table 13.2), after the astronomer who discovered and cataloged many of them in the 1940s.

Seyfert galaxies have the appearance of ordinary spirals, except that the nucleus is unusually bright and blue in color, in contrast with the red color of most normal spiral galaxy nuclei. About 10 percent of them are radio emitters, with spectra indicating that the synchrotron process is at work. The radio emission usually comes from the nucleus rather than from the double side lobes. The spectrum of the visible light from a Seyfert nucleus typically shows emission lines, something completely out of character for normal galaxies. The emission lines, formed in an ionized gas, are sometimes very broad, indicating velocities of several thousand kilometers per second in the gas that produces the emission. Evidently the cores of these galaxies are in extreme turmoil, with hot gas swirling about and tremendous amounts of energy being generated.

Figure 13.8 A Seyfert Galaxy. This photograph of NGC 4151, a well-studied example, shows the enormous intensity of light from the nucleus compared to the rest of the galaxy. *(Palomar Observatory, California Institute of Technology)*

Few spirals show such effects, and it is not clear whether the Seyfert behavior is a phase they all pass through at some time or whether only a few act in this manner. Later in this chapter we will discuss evidence supporting the first of these possibilities.

The Discovery of Quasars

In 1960 spectra were obtained of two starlike, bluish objects (Fig. 13.9) that had been found to be sources of radio emission. No radio stars were known, and astronomers were very interested in these two objects, called 3C48 and 3C273 (their designations in a catalog of radio sources that had recently been compiled by the radio observatory of Cambridge University in England). The spectra of the two objects, which had several strong emission lines, defied understanding for some three years. The objects became known as **quasars** (short for "quasi-stellar radio sources"), or **quasi-stellar objects** (sometimes simply **QSO**'s) because of their starlike appearance but nonstellar spectra.

It was eventually realized that the strong emission lines in the two quasars were lines of the well-known Balmer series due to hydrogen, but shifted toward far longer wavelengths than those measured in the labo-

Table 13.2 Properties of Seyfert Galaxies

They are spiral galaxies.
Their luminosities are comparable to the brightest normal spirals.
They have bright, compact blue nuclei with X-ray emission.
Their nuclei show emission lines of highly ionized gas.
About 10 percent are radio sources.
Most are variable in times of days or weeks.
Emission from the nucleus is synchrotron radiation.
Some have radio jets.

Figure 13.9 Quasi-Stellar Objects. These starlike objects are quasars, whose spectra are quite unlike those of normal stars. As explained in the text, these objects are probably more distant, and therefore more luminous, than any normal galaxies. *(Palomar Observatory, California Institute of Technology)*

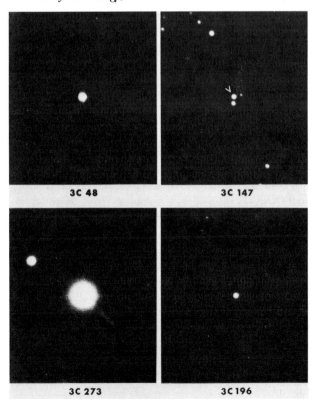

ratory. The shifts of the lines in 3C273 indicated a redshift of 16 percent, that is, the object emitting these lines must be moving away from the earth at 16 percent of the speed of light, or 48,000 kilometers per second! In the case of 3C48, the shift was even greater, about 37 percent, corresponding to a velocity of 111,000 kilometers per second. (These velocities actually have to be corrected for relativistic effects, as described in Appendix 14, and the correct values are 44,000 kilometers per second for 3C273 and 91,000 kilometers per second for 3C48.)

Since the early 1960s, several thousand additional quasars have been discovered. Only a small fraction are radio sources, and they may differ from the first two in other ways; but they invariably have highly redshifted emission lines, and in very many cases they also have weak absorption lines. In a number of quasars the redshift is so huge that spectral lines whose rest wavelengths are in the ultraviolet portion of the spectrum are shifted all the way into the visible region. The grand champion today is a quasar with redshift of 443 percent, so that the strongest of all the hydrogen lines, whose rest wavelength is 1216 Å, is shifted all the way to 6566 Å. This quasar is *not* traveling away from us at 4.43 times the speed of light, however; for very large velocities, the Doppler shift formula has to be modified in accordance with relativistic effects. The velocity of this quasar is 93.4 percent of the speed of light, still an enormous speed.

The Origin of the Redshifts

In order to understand the physical properties of the quasars, it is essential to discover the reason for the high redshifts. We have already tacitly adopted the most obvious explanation, that they are due to the Doppler effect in objects that are moving very rapidly, but we must still ascertain the nature of the motions. Furthermore, one alternative explanation was suggested that has nothing to do with motions at all.

One consequence of Einstein's theory of general relativity (verified by experiment) is that light can be redshifted by a gravitational field. Photons struggling to escape an intense field lose some of their energy in the process, and as this happens, their wavelengths are shifted toward the red. We were exposed to this concept in Chapter 8, when we discussed the behavior of light near black holes, which have such strong gravitational fields that no light can escape at all. For a while it was considered possible that quasars are stationary objects sufficiently massive and compact to have large gravitational redshifts.

This suggestion has now largely been ruled out for two reasons:

1. There is no known way for an object to be compressed enough to produce such strong gravitational redshifts without falling in on itself and becoming a black hole. If enough matter is squeezed into such a small volume that its gravity produces redshifts as large as those in some quasars, no known force could prevent this object from collapsing further. A neutron star does not have as large a gravitational redshift as those found in quasars, and we already know that the only possible object with a stronger gravitational field than that of a neutron star is a black hole.

2. Even if such a massive, yet compact object could exist, its spectral lines would be smeared out (as in a white dwarf, but more extreme), and light emitted from slightly different levels in the object would have different gravitational redshifts, again causing spectral lines to be broadened.

For both of these reasons we are forced to accept the Doppler shift explanation for the redshifts. The problem then is to explain how such large velocities can arise. One possibility is that the quasars are relatively nearby (by intergalactic standards) and are simply moving away from us at very high speeds, perhaps as the result of some explosive event. This "local" explanation requires some care: To accept it, we must explain why no quasars have ever been found to have blueshifts. In other words, if quasars are nearby objects moving very rapidly, it is not easy to see why none of them are approaching us. It might be possible to argue that they originated in a nearby explosion (at the galactic center, perhaps) a long time ago, so any that happened to be aimed toward us have had time to pass by and are now receding (Fig. 13.10). There are serious difficulties with this picture, though, primarily in the amount of energy that would have been required to get all these objects moving at the observed velocities, an amount that would dwarf the total light output of the galaxy over its entire lifetime.

We are left with one alternative, which still poses problems but is perhaps more acceptable than the others. Let us consider the possibility that the quasars are

Astronomical Insight 13.1

The Redshift Controversy

In the text we presented a standard set of arguments demonstrating that quasar redshifts are cosmological, due to the expansion of the universe, and implying that the qua-sars must be very distant objects. While it is true that most astronomers accept this viewpoint, it is not universally adopted, and there are those who favor other explanations of the redshifts.

The evidence cited by the opponents of the cosmological interpretation consists primarily of cases where quasars are found apparently associated with galaxies or clusters of galaxies but do not have the same redshift as the galaxies. It is argued in these cases that the quasar is physically associated with the galaxy or cluster of galaxies and is therefore at the same distance, so its redshift cannot be cosmological.

The evidence favoring these arguments can be quite striking. When a quasar is found within a cluster of galaxies, there is a natural tendency to think of it as physically a member of the group. It can be argued from statistical grounds that the chances of accidental alignments between galaxies and quasars are so low that the observed associations between these objects cannot be coincidental. There are even cases where long-exposure photographs appear to show gaseous filaments connecting a galaxy with a quasar that has a different redshift. This would seem to prove that the redshift cannot be cosmological.

This radical view of quasars leaves many important questions open. The most difficult of these is how to explain the redshifts, if they really are not cosmological, for the arguments cited in the text against gravitational and local Doppler redshifts are still valid. No satisfactory explanation of the quasar redshifts has been offered by the opponents of the cosmological viewpoint.

The counterarguments center on statistical calculations. One point is that the calculations showing that the quasar-galaxy associations have a low probability of occurring by chance are made after the fact. That is, after a few such associations were found, arguments were made that this was a very unlikely thing to occur by chance alignments of objects randomly distributed over the sky. This is somewhat akin to arguing that the chances of being dealt a certain combination of cards in a game are very low. This is true before the deal, but meaningless after the fact. Based on this and other analogies, many scientists think it is incorrect to argue that chance associations of quasars and galaxies are improbable. More to the point, they argue, quasars associated with galaxies are more likely to be discovered than those that are not; therefore, it is natural to find a disproportionately high number of quasars that happen to be aligned on the sky with galaxies. This is an example of a **selection effect,** because the process by which quasars are found is not perfectly random and thorough; the attention of astronomers is naturally concentrated on regions where there are many galaxies, so those regions are more thoroughly searched for quasars.

Probably the most telling argument against the noncosmological redshifts for quasars arises from the fact (noted in the text) that a number of quasars have now been found to be embedded in galaxies or within clusters of galaxies that share the same redshift. In fact recent observations show that *every* quasar that is sufficiently nearby for an associated galaxy to be detected is indeed embedded within a galaxy that has the same redshift as the quasar. There has not yet been an exception to this, and the data amount to virtual proof that quasars are the cores of galaxies and that their redshifts are cosmological.

The controversy is by no means over, however. Adherents of the noncosmological interpretation of the redshifts continue to find remarkable cases of apparent galaxy-quasar connections where the redshifts do not match. Perhaps observations made in the near future, with the *Space Telescope* or other planned instruments, will provide unequivocal evidence supporting one side or the other.

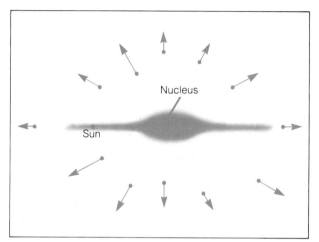

Figure 13.10 A "Local" Hypothesis for Quasar Redshifts. If quasars were very rapidly moving objects ejected from our galactic nucleus some time ago, they could all be receding from our position in the disk by now. This would explain the fact that only redshifts are found, but there is no known mechanism for providing the energy required to produce such great speeds.

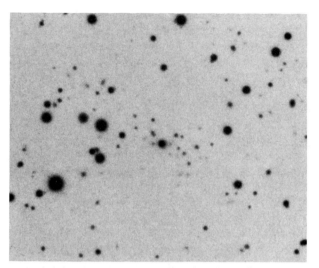

Figure 13.11 A Quasar in a Cluster of Galaxies. The image at the center of this very long exposure photograph is a quasar. The much fainter objects around it are galaxies, forming a cluster of which the quasar is apparently a member. The galaxies and the quasar have the same redshift, indicating that they lie together at the same distance. (H. Spinrad)

very distant objects moving away from us with the expansion of the universe. In this view they are said to be at "cosmological" distances, meaning that they obey Hubble's relation between distance and velocity, just as galaxies do. We find, if we adopt this assumption, that 3C273 is some 590 megaparsecs away, and 3C48 is over 1200 megapar-secs distant. (These distances are based on a value of the Hubble constant $H = 75$ km/sec/megaparsec and the assumption that the rate of expansion of the universe has been constant, something that is probably not true; this will be discussed further in Chapter 14.)

The best evidence that the quasars are at cosmological distances is the fact that some have been found in clusters of galaxies (Fig. 13.11) with the same redshift as the galaxies. This was an extremely difficult observation to make, because at the distances where the quasars exist, the much fainter galaxies are very hard to see, even with the largest telescopes.

Supporting evidence for the cosmological redshift interpretation is provided by the **BL Lac objects.** Named for the prototype, an object called BL Lacertae that was once thought to be a variable star, these are nuclei of elliptical galaxies with very bright central cores. The nucleus of a BL Lac object displays many of the properties of a quasar, with radio synchrotron emission, variable brightness, and enormous luminosity (but it lacks emission lines). The surrounding galaxy shows a normal absorption line spectrum with a redshift consistent with its apparent distance, which is undoubtedly cosmological. Generally the redshifts of BL Lac objects are relatively small (by quasar standards), and they are apparently similar to nearby quasars. The implications of the fact that they are embedded in galaxies will be discussed later in this chapter.

Distances of hundreds or thousands of megaparsecs, inferred for the quasars with very large redshifts, have enormous implications. One is that the light that we receive from them has traveled toward us for many billions of years. When we look at a quasar, we are looking far into the past history of the universe, and we must keep this in mind as we attempt to interpret them. The fact that quasars are seen only at high redshifts, meaning great distances, indicates that they were more common in the early days of the universe than they are now. This will prove to be a very important clue to their true nature. It has been speculated that there is a limit to quasar redshifts; that is, no quasars can be found with a redshift greater than this limit. The implication is fantastic, for it means that we are looking back to a time before the universe had organized itself into such objects as galaxies and quasars.

Another important implication of the great distances to quasars is that they are the most luminous objects known. Their apparent magnitudes (the brightest, 3C273, is a thirteenth-magnitude object) combined with their tremendous distances, imply that their luminosities are far greater than even the brightest galaxies, by factors of hundreds or even thousands. As we will see, explaining this astounding energy output is one of the central problems of modern astronomy.

The Properties of Quasars

Over fifteen hundred quasars have been cataloged, and more are being discovered all the time. Many are being studied with care, primarily by means of spectroscopic observations. Within the past few years, it has even become possible to observe quasars at ultraviolet and X-ray wavelengths with the use of satellite observatories such as the *International Ultraviolet Explorer* and the *Einstein Observatory*. The faintness of the quasars makes all these observations difficult, but their importance makes the effort worthwhile. As a result of all the intensive work being done on quasars, a great deal is now known about their external properties, even though their origin and source of energy remain mysterious.

The blue color that characterizes the first quasars discovered is a general property of all quasars (Table 13.3), but the radio emission is not. Only about 10 percent are radio sources, contrary to the early impression that all are. (This type of misunderstanding is known as a **selection effect,** because the process of selecting quasars was based at first on a certain assumption about their characteristics. For a while, the only method used to look for quasars was to search for objects with radio emission, so naturally all those that were found were radio sources.) The radio emission, and usually the continuous radiation of visible light as well, shows the characteristic synchrotron spectrum.

Apparently *all* quasars emit X rays, again by the synchrotron process, according to the data collected by the *Einstein Observatory*, the most sensitive X-ray satellite yet launched. Hence X-ray observations may be a reliable technique for finding new quasars, particularly those of relatively low redshift, which are relatively nearby and have stronger X-ray emissions than the more distant ones.

Some photographs of quasars reveal evidence of structure instead of a single point of light. The most notable of these is 3C273, which played such a key role in the initial discovery of quasars. This object shows a linear jet extending from one side (Fig. 13.12), closely resembling the one emanating from the giant radio galaxy M87. Perhaps this means that similar processes are occurring in these two rather different objects.

Many quasars vary in brightness, usually over times of several days to months or years (although in one

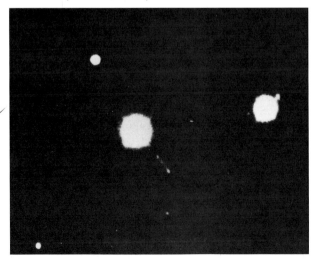

Figure 13.12 A Quasar with a Jet. This is 3C273, one of the first two quasars discovered. This visible-light photograph reveals a linear jet very much like those seen in many radio galaxies. The radio structure of quasars is usually double-lobed, also similar to radio galaxies. *(National Optical Astronomy Observatories)*

Table 13.3 Properties of Quasars

They are characterized by large redshifts.
Their spectra are dominated by emission lines of highly ionized gas.
Their optical luminosities are 100 to 1000 times those of normal galaxies (if they are at cosmological distances).
About 10 percent are radio sources.
They appear as compact, blue objects, with X-ray emission.
Many are variable in times of days or weeks.
Their emission is due to synchrotron radiation.
Some have radio or optical jets.

case, variations were seen over just a few hours). This variability is very important, for it provides information on the size of the region in the quasar that is emitting the light. This region cannot be any larger than the distance light travels in the time over which the intensity varies. There is no way an entire object that might be hundreds or thousands of light-years across can vary in brightness in the time of a month or so. The light travel time across the object would guarantee that we would only see changes over times of hundreds or thousands of years, as the light from different parts of the object reached us (Fig. 13.13). Therefore the observed short-term variations in quasars tell us that the fantastic energy emitted by these objects is often produced within a volume no more than a light-month, or about 0.03 parsec, in diameter. This obviously places stringent limitations on the nature of the emitting object.

The spectra of quasars have already been described in broad outline, but there is a great deal of detail as well (Fig. 13.14). All quasars have emission lines, generally of common elements such as hydrogen and helium, and often of carbon, nitrogen, and oxygen. (Some of the latter elements have strong lines only in the ultraviolet and therefore are best observed in cases where the redshift is sufficiently large to move these lines into the visible portion of the spectrum.) In many cases, the emission lines are very broad, which shows that the gas that forms them has internal motions of thousands of kilometers per second. The degree of ionization tells us that the gas is subjected to an intense radiation field, which continually ionizes the gas by the absorption of energetic photons.

Many quasars have absorption lines in addition to the emission features. Strangely enough, the absorption lines are usually at a different redshift (almost always smaller) than the emission lines, which indicates that the gas creating the absorption is not moving away from us as rapidly as the gas producing the emission. Further confounding the issue is the fact that many quasars have multiple absorption redshifts; that is, they have several distinct sets of absorption lines, each with its own redshift, and each therefore representing a distinct velocity. This shows that there are several absorbing clouds in the line of sight. There are two possible origins for these clouds (Fig. 13.15):

1. The quasar itself, which is moving at the high speed indicated by the redshift of the emission lines, may be ejecting clouds of gas, some of which happen to be aimed toward the Earth. These clouds therefore have lower velocities of recession than the quasar itself, so the redshift of the absorption lines they form is less than the redshift of the emission lines formed in the quasar.

Figure 13.13 The Implication of Time Variability for The Size of the Emitting Object. This illustrates why an object cannot appear to vary in less time than it takes for light to travel across it. As a simple case, this spherical object is assumed to change its luminosity instantaneously. On Earth we observe the first hint of this change when light from the nearest part of the object reaches us, but we continue to see the brightness changing gradually as light from more distant portions reaches us.

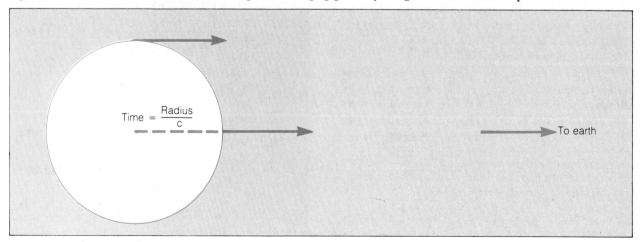

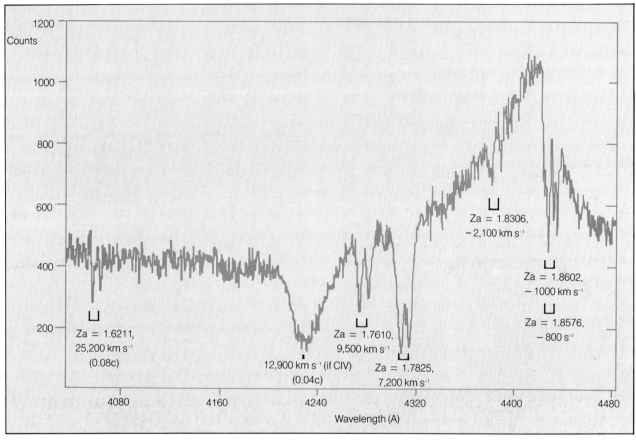

Figure 13.14 A Quasar Spectrum. This spectrum shows several features typical of quasar spectra: a broad emission line centered near 4,400 Å; broad absorption line near 4,240 Å; and narrow absorption lines at several wavelengths. All features seen here are due to carbon that has lost three electrons (designated CIV). Redshifts are indicated. *(Data from Weymann, R. J., R. F. Carswell, and M. G. Smith, 1981, Annual Reviews of Astronomy and Astrophysics 19:41)*

2. The absorption lines may arise in the gaseous disks and halos of galaxies that happen to lie between us and the more distant quasar. Each galaxy has a recession velocity determined by the expansion of the universe, and because the intervening galaxies are not as far away as the quasars, they are not moving as fast and therefore have lower redshifts. If this interpretation of the absorption lines is correct, as is thought to be true in at least some cases, their analysis should provide important information on the nature of the extensive halos thought to surround many galaxies.

In some cases the absorption spectra of quasars show only lines due to hydrogen and none due to heavier elements. These lines apparently arise in gas that has not been enriched at all by stellar processing, and it is thought that these might be intergalactic clouds that have never formed into galaxies, so that no stars have ever formed in them. These clouds may be material remnants of the big bang; if so, they can tell us much about the exact amount of nuclear processing that occurred in the early stages of the expansion.

Galaxies in Infancy?

A variety of theories, some of them rather fanciful, have been proposed to explain the quasars. One idea has gradually become widely accepted, however, and we will restrict ourselves to discussing that one hypothesis, keeping in mind that there are other suggestions.

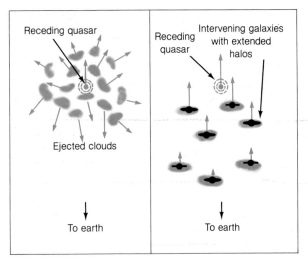

Figure 13.15 Two Explanations of Quasar Absorption Lines. In one scenario (left), the absorption lines are formed by clouds of gas ejected from the quasar. Clouds ejected directly toward the Earth have lower velocities of recession than the quasar itself. In the other more likely view (right), the absorption lines are formed in the extended halos of galaxies that happen to lie between us and the quasar. The galaxies are closer to us than the quasar and therefore have lower velocities of recession and smaller redshifts.

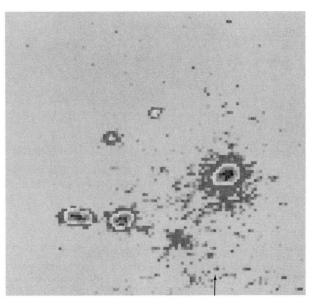

Figure 13.16 A Quasar Embedded in a Galaxy. This photograph, obtained with an electronic detector, contains several star images, a galaxy (slightly elongated object lower left), and a quasar (right). The quasar image is extended, just as the galaxy image is, and the fuzzy region surrounding it is indeed a galaxy. In every case where a quasar is close enough to us that a galaxy would be bright enough to be detected, one has been, indicating that all quasars lie at the centers of galaxies. *(M. Malkan)*

The prevailing interpretation of the nature of quasars was inspired by the fact that they existed only in the long-ago past, by their association with the BL Lac objects (which clearly are galaxies), and by their resemblance to the nuclei of Seyfert galaxies. The statement that they existed only in the past is based on the observation that all are very far away, so that the light travel time ensures that we are seeing things only as they were billions of years ago, not as they are today. The resemblance to Seyfert nuclei is striking: Both are blue; both are radio sources in about 10 percent of the cases; both vary on similar time scales; and both have very similar emission line spectra. Seyferts lack the complex absorption lines often seen in the spectra of quasars, but this would be expected if the quasar absorption lines are formed in the halos of intervening galaxies; since Seyfert galaxies are not so far away, there are likely to be many other galaxies along their lines of sight. The nuclei of Seyferts also differ from quasars in the amount of energy they emit, being considerably less luminous.

The picture that is developing is that quasars are young galaxies, with some sort of youthful activity taking place in their centers. In this view, the Seyfert galaxies are descendants of quasars, still showing activity in their nuclei, but with diminished intensity. If we carry this idea another step, we are led to the suggestion that normal galaxies like the Milky Way are later stages of the same phenomenon; recall the mildly energetic (by quasar standards) activity in the core of our galaxy.

The notion that quasars may be infant galaxies is supported by some rather direct evidence. As noted earlier, some quasars have been found associated with clusters of galaxies, showing that they can be physically located in the same region of space at the same time. In addition, there are a few examples where careful observation has revealed a fuzzy, dimly glowing region surrounding a quasar (Fig. 13.16). This is probably a galaxy with a quasar embedded in its center. It is likely that all quasars are located in the nuclei of galaxies, which are so much fainter that they usually cannot be seen. A short-exposure photograph of a Seyfert galaxy (Fig. 13.17) shows only the nucleus, which is very

Astronomical Insight 13.2

The Double Quasar and Gravitational Lenses

Recently two quasars were discovered, very near each other in the sky, with properties so nearly identical that it has been concluded that they are in fact two images of the same quasar. If so, astronomers have found the first example of a **gravitational lens,** something predicted to be a possibility on the basis of Einstein's theory of general relativity. Since a gravitational field can bend light rays, it is possible for such a field to act like a lens. In particular, calculations have shown that the gravitational field of a galaxy can bend the light from a distant object around it. If the galaxy were perfectly aligned with the more distant object, we would see either a single point image, as if the galaxy were not there (only if we were precisely at the focal point of the lens), or, more likely, a circular ring-shaped image. In the more probable event that the galaxy is slightly off-center between us and the distant light source, we will see two or more distinct images, on either side of the intervening galaxy. The double quasar is thought to be such a case, the two images reaching us by coming around opposite sides of the gravitational lens formed by a galaxy between us and the quasar. Careful photographic studies revealing what appears to be a galaxy between the two images of the quasar help support this interpretation.

Since the discovery of the first gravitational lens, other apparent cases have been found, all involving quasars seen behind galaxies. Sometimes more than two images of the quasar are seen. One of the best tests of whether the multiple images really are of the same quasar would be to see whether they vary with time in the same fashion, but this is complicated by the fact that the light from the different images

(continued on next page)

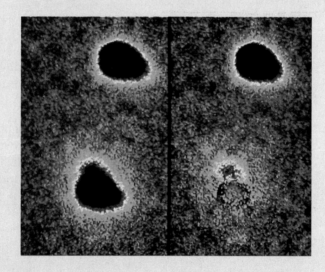

Institute for Astronomy and Planetary-Geosciences Data Processing Facility, University of Hawaii, courtesy of A. Stockton.

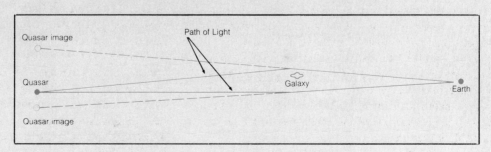

This diagram illustrates how an intervening galaxy can act as a gravitational lens, forming two images of a distant quasar.

Astronomical Insight 13.2

The Double Quasar and Gravitational Lenses

(continued from previous page) follows different paths to us. The length of one path may be longer than the other, so when the quasar changes in brightness, we see this at different times for the separate images. Therefore, we have to rely on similarities in redshift, apparent magnitude, and spectral details in determining whether the multiple quasars are images of the same object.

As we look to greater and greater distances, so that we see more and more distant objects, the sky becomes filled with images of galaxies and quasars. It stands to reason that we should find more and more cases of galaxies lying nearly in front of distant quasars, and we can expect gravitational lenses to show up more and more. This seems to be happening, as new telescopes

and detectors reveal ever fainter objects. Eventually we may reach the point where there are so many multiple quasar images due to gravitational lenses that we will not know how many quasars we really are seeing. It will be as though the universe is screening our view beyond a certain distance, by distorting the picture with many images of each object beyond that distance.

Figure 13.17 Three Exposures of a Seyfert Galaxy. At left is a short-exposure photograph of the Seyfert galaxy NGC 4151, in which only the nucleus is seen, resembling the image of a star. The center and right-hand images are longer exposures of the same object, revealing more of its outer, fainter structure. This sequence illustrates how a quasar, which is even more luminous than a Seyfert galaxy nucleus, can appear as a starlike image even though it is embedded in a galaxy. *(Photographs from the Mt. Wilson Observatory, Carnegie Institution of Washington, composition by W. W. Morgan)*

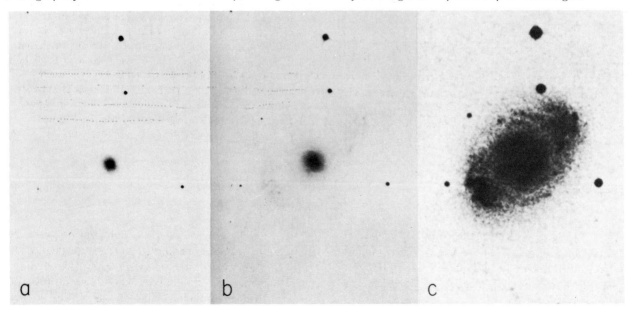

much brighter than the surrounding galaxy, and it is clear that if we observed one of these galaxies from so far away that it was near the limit of detectability, all we would see would be a blue starlike object resembling a quasar.

While we can make a strong case that quasars are young galaxies, we are still far from understanding all of their properties. The main mystery remaining is the source of the tremendous energy of quasars and active galactic nuclei. A number of possibilities have been raised, but only one seems viable today.

This possibility, which keeps coming up in situations where large amounts of energy seem to be coming from small volumes of space, is that quasars are powered by massive black holes (Fig. 13.18). In the chaotic early days of collapse, when a galaxy was just forming out of a condensing cloud of primordial gas, it may be that a great deal of material collected at the center. If many stars formed there, frequent collisions could have caused them to settle in more and more tightly, until they coalesced to form a black hole. The gravitational influence of such a massive object would have then stirred up the surroundings, creating the high electron velocities required to fuel the synchrotron emission. Infalling matter would have formed an accretion disk around the black hole, from which X rays would be emitted, in analogy with stellar black holes and neutron stars in binary systems. In this picture the source of energy in quasars is the same as in the radio galaxies described earlier in this chapter. Even the existence of jets of hot gas is the same; such jets have been observed emanating from several quasars.

Whatever the origin and power source of the quasars, we will certainly learn a lot about the nature of matter and the early history of the universe when we are able to answer all the questions about them. The *Space Telescope*, with its broad wavelength coverage and great sensitivity, will provide invaluable information on these fundamentally important objects.

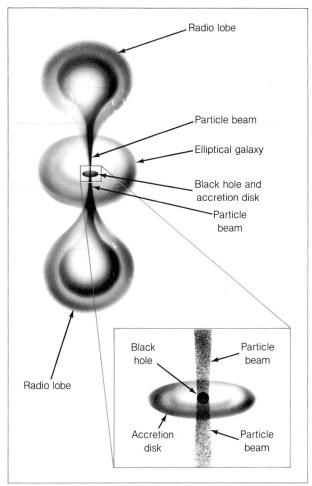

Figure 13.18 A Geometrical Model for a Quasar, Seyfert Galaxy, or a Radio Galaxy. All of these objects have certain features in common that fit the picture shown here. A central object (most likely a supermassive black hole with an accretion disk) ejects opposing jets of energetic charged particles. These jets produce synchrotron radiation and build up double radio-emitting lobes on either side of the central object. These lobes extend well beyond the confines of the galaxy in which the central object is embedded.

Perspective

In this chapter we have learned of new and wondrous things: the mighty radio galaxies and the enigmatic quasars. Along the way we have gained several important bits of information on the universe itself. It is time to tackle the fundamental question of the origin and fate of the cosmos.

Summary

1. Radio galaxies emit vast amounts of energy in the radio portion of the spectrum. These galaxies, often giant ellipticals, commonly show structural peculiarities. The radio emission is nonthermal (usually synchrotron radiation) and in most cases is pro-

duced from two lobes on opposite sides of the visible galaxy.
2. The source of energy in a radio galaxy is unknown but appears to be concentrated at the core.
3. Seyfert galaxies are spiral galaxies with compact, bright blue nuclei that produce emission lines characteristic of high temperatures and rapid motions.
4. Several point sources of radio emission found in the early 1960s looked like blue stars and were therefore called quasi-stellar objects, or quasars.
5. Quasars have emission line spectra with very large redshifts, corresponding to velocities that are a significant fraction of the speed of light. If these redshifts are cosmological, quasars are on the frontier of the universe, are extremely luminous, and are seen as they were billions of years ago.
6. Quasars are sometimes radio sources, usually emit X rays, are compact and blue in color, and in some cases vary over times of only a few days. This implies that their tremendous energy output arises in a volume only a few light-days across.
7. Many quasars have absorption lines at a variety of redshifts (always smaller than the redshift of the emission lines); these absorption lines may be created by matter being ejected from the quasars or, more likely, by intervening galactic disks and halos.
8. The most satisfactory explanation of the quasars is that they are very young galaxies. This interpretation is supported by the facts that they are seen only at great distances (that is, they existed long ago) and that they resemble the nuclei of Seyfert galaxies. Photographs that have revealed galaxies surrounding quasars confirm this suggestion.
9. The source of the energy that powers quasars is a premier mystery of modern astronomy. The most probable explanation is that a quasar has a massive black hole at its core; this can produce enough energy (from the infall of matter) to account for the great luminosity of quasars as well as their time variability. By inference, similar objects may exist in the nuclei of galaxies such as Seyfert galaxies, radio galaxies, and the Milky Way.

Review Questions

1. Summarize the contrasts between radio galaxies and the radio emission from the Milky Way.
2. In radio galaxies with more than one pair of radio-emitting lobes on either side, the alignment between the inner and outer lobes is generally quite precise. What does this tell us about the source of the lobes?
3. Suppose that emission lines are found in the spectrum of a quasar at wavelengths of 2189 Å, 2790 Å, and 5040 Å. If these are identified as the Lyman-alpha line of hydrogen (rest wavelength 1216 Å), the three-times ionized carbon line (rest wavelength 1550 Å), and the strong line of ionized magnesium (2800 Å), what is the redshift of this quasar? How far away is it, if Hubble's constant has the value $H = 75$ kilometers per second per megaparsec? (Note: You must use the relativistic redshift equation.) If the apparent magnitude of this quasar is $m = +16$, what is its absolute magnitude?
4. Explain, in your own words and using your own sketch, why the time scale of variation in light from a source such as a quasar places a limit on the diameter of the emitting region.
5. Compile a list of similarities and contrasts between quasars and Seyfert galaxies.
6. Why are quasar absorption lines *always* observed at smaller redshifts than the emission lines? Would this necessarily be true if the redshifts of the emission lines were not cosmological?
7. From what you have learned about quasars in this chapter and what you learned about the *Space Telescope* in Chapter 4, describe some important observations of quasars that might be made with the *Space Telescope*.
8. If quasars are young galaxies, how does this help demonstrate the difficulty of using faraway galaxies as standard candles in estimating large distances in the universe?
9. Explain why nearly all of the first quasars detected are radio sources, but only about 10 percent of *all* quasars are.
10. Explain why the material in intergalactic clouds may provide clues for astronomers about conditions very early in the life of the universe.

Additional Readings

Blandford, R. D., M. C. Begelman, and M. J. Rees. 1982. Cosmic jets. *Scientific American* 246(5):124.
Chaffee, F. H. 1980. The discovery of a gravitational lens. *Scientific American* 243(5):60.

Disney, M. J., and P. Vèron. 1977. BL Lac objects. *Scientific American* 237(2):32.

Downs, A. 1986. Radio galaxies. *Mercury* XV(2):34.

Feigelson, E. D., and E. J. Schreier. 1983. The X-ray jets of Centaurus A and M87. *Sky and Telescope* 65(1):6.

Ferris, T. 1984. The radio sky and the echo of creation. *Mercury* XIII(1):2.

Hamilton, D., W. Keel, and J. F. Nixon. 1978. Variable galactic nuclei. *Sky and Telescope* 55(5):372.

Hazard, C. and S. Mitton, eds. 1979. *Active galactic nuclei.* Cambridge, England: Cambridge University Press.

Kaufmann, W. 1978. Exploding galaxies and supermassive black holes. *Mercury* 6(5):78.

Lawrence, J. 1980. Gravitational lenses and the double quasar. *Mercury* 9(3):66.

Margon, B. 1983. The origin of the cosmic X-ray background. *Scientific American* 248(1):104.

Osmer, P. S. 1982. Quasars as probes of the distant and early universe. *Scientific American* 246(2):126.

Shipman, H. L. 1976. *Black holes, quasars, and the universe.* Boston: Houghton-Mifflin.

Silk, J. 1980. *The big bang: The creation and evolution of the universe.* San Francisco: W. H. Freeman.

Strom, R. G., G. K. Miley, and J. Oort. 1975. Giant radio galaxies. *Scientific American* 233(2):26.

Tyson, T., and M. Gorenstein. 1985. Resolving the nearest gravitational lens. *Sky and Telescope* 70(4):319.

Wyckoff, S., and P. Wehinger. 1981. Are quasars luminous nuclei of galaxies? *Sky and Telescope* 61(3):200.

Chapter 14

Cosmology: Past and Future of the Universe

An artist's view of the *Hubble Space Telescope* in operation, probing the structure of the universe. *(NASA)*

Chapter 14 Cosmology: Past, Present, and Future of the Universe

Chapter Preview

Underlying Assumptions
Einstein's Relativity: Mathematical Description of the Universe
Open or Closed: The Observational Evidence
Total Mass Content
The Deceleration of the Expansion
The History of Everything
What Next?

Given the time and opportunity, humans through the ages have devoted themselves to speculation on the grandest scale of all. Poets, philosophers, and theologians have approached the question of the origin and future of the universe in countless ways. So have astronomers, with the exception that a somewhat restrictive set of rules is followed: The answers that are accepted must not violate known laws of physics. By retaining this requirement, scientists attempt to approach the problem in an objective, verifiable manner.

There are difficulties in maintaining this idealized posture, however, and we shall try to point them out clearly. Because the universe in which we live is, as far as we know, unique, we have no opportunity to check our hypotheses by comparison with other examples. Furthermore, no matter how thoroughly and rigorously we trace the evolution of the universe by application of known physical laws, there will always be fundamental questions that are beyond the scope of physics. As a result, even the most careful and objective scientists reach a point where they have to make certain unverifiable assumptions, and at that point they become philosophers or theologians. In this chapter we shall restrict ourselves to the questions that can in principle have objective, verifiable answers.

Technically, the study of the universe as it now appears is **cosmology**; that is really what the preceding thirteen chapters have all been about. The study of its origins is **cosmogony**, and this word applies to the big bang as well as to the earlier theories on the origin of the solar system, once thought to be the entire universe. In practice, the general subject of the nature of the universe and its evolution are lumped under the heading of cosmology, and so we now embark on a pursuit of this subject.

Underlying Assumptions

Even to begin to study the universe as a whole, we must make certain assumptions. These can be tested in principle, although it is not clear that there is a practical way to do so, and therefore in making these assumptions the astronomer is on the verge of acting as a philosopher.

A central rule traditionally set forth by cosmologists is that the universe must look the same at all points within it. This does not mean that the appearance of the heavens should be identical everywhere, but it means that the general structure—the density and distribution of galaxies and clusters of galaxies (Fig. 14.1)—should be constant. This assumption states that the universe is **homogeneous**.

A related but slightly different assumption has to do with the appearance of the universe when viewed in different directions. The assumed property in this case

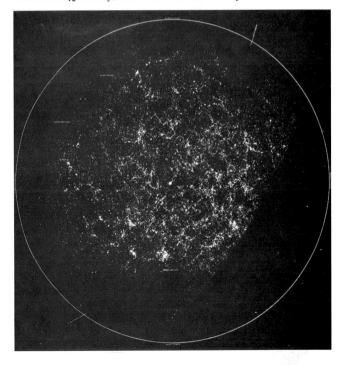

Figure 14.1 The Cosmological Principle. The universe is assumed to be homogeneous and isotropic, meaning that it looks the same to all observers in all directions. Here is a segment of the universe, filled with galaxies. Their distribution, while not identical from place to place, is similar throughout. (© 1978, 1979 by The Co-Evolution Quarterly, Box 428, Sausalito, California)

is **isotropy**; that is, the universe must look the same to an observer, no matter in which direction the telescope is pointed. We have already encountered this concept in connection with the cosmic microwave background radiation. Again, the assumption of isotropy does not imply identical constellations of galaxies and stars in all directions, just comparable ones.

Both of these assumptions are thought to apply only on the largest scales, larger than any obvious clumpiness in the universe, such as clusters or superclusters of galaxies. It should also be pointed out here, although it is not discussed until later in the chapter, that both assumptions have been questioned by some modern theories and observations of the structure of the universe. It is even thought surprising in some current theories that the universe should be as homogeneous as it is. Hence the assumptions of homogeneity and isotropy should now be viewed more as an observation that must be explained than as a postulate of cosmological theory.

The statement that the universe is homogeneous and isotropic is often referred to as the **cosmological principle**. It is often stated in terms of how the universe looks to observers; that is, the universe looks the same to all observers everywhere. It is difficult to verify the cosmological principle. We can test the assumption of isotropy by looking in all directions from Earth, and indeed we do not find any deviations. On the other hand, we cannot test the homogeneity of the universe by traveling to various other locations to see whether things look any different. What we can do is count very dim, distant galaxies to see whether their density appears to be any greater or less in some regions than in others, but even this is complicated by the fact that we are looking back in time to an era when the universe was more compressed and dense than it is now. The 3-degree background radiation gives us another tool for testing both homogeneity and isotropy, and it also appears to satisfy the cosmological principle. The best we can say for now is that the available evidence supports this principle but does not completely rule out the possibility of subtle anisotropies or inhomogeneities.

Einstein's Relativity: Mathematical Description of the Universe

The simple act of making assumptions about the general nature of the universe by itself tells us very little about its past or its future. It elucidates certain properties of the universe as it is today, but to tie this into a quantitative description, to develop a theory capable of making predictions that can be tested, requires a mathematical framework. In developing such a mechanism for describing the universe, the astronomer seeks to reduce the universe to a set of equations and then to solve them, much as a stellar structure theorist studies the structure and evolution of stars by constructing numerical models.

The most powerful mathematical tool for describing the universe was developed by Albert Einstein (Fig. 14.2). His theory of general relativity represents the properties of matter and its relationship to gravitational fields. Within the context of his theory, Einstein developed a set of relations called the **field equations** (Fig. 14.3) that express in mathematical terms the interaction of matter, radiation, and gravitational forces in the universe. Although other scientists have developed and explored the consequences of alternatives to general relativity, most research in cosmology today involves finding solutions to Einstein's field equations and testing these solutions with observational data.

The basic premise of general relativity, called the **equivalence principle**, is that acceleration due to a gravitational field is indistinguishable from accelera-

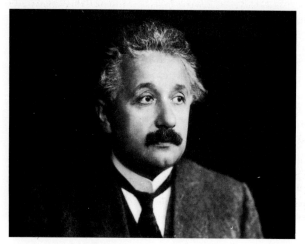

Figure 14.2 Albert Einstein. Among Einstein's great contributions was the development of a mathematical formalism to describe the interaction of gravity, matter, and energy in the universe with the nature of space-time. This framework, general relativity, has withstood all observational and experimental tests applied to it so far. *(The Granger Collection)*

Astronomical Insight 14.1

The Mystery of the Nighttime Sky

A very simple question, first asked over two centuries ago and discussed at length in the early 1800s by the German astronomer W. Olbers, leads to profound consequences. The question is: Why is the sky dark at night?

In an infinite universe filled with stars, every line of sight, regardless of direction, should intersect a stellar surface. The nighttime sky should therefore be uniformly bright, not dark, as the star images in the sky would literally overlap each other. The fact that this is not so presents a paradox worth pondering.

One suggested solution is that interstellar extinction can diminish the light from distant stars sufficiently to make the sky appear black. Careful consideration shows, however, that this is not so. Light that is absorbed by dust in space causes the grains to heat up. If the universe were really filled with starlight as suggested by Olbers' paradox, the grains would become so hot that they would glow, and we would still have a uniformly bright nighttime sky.

Another suggested explanation has to do with the expansion of the universe. The redshift due to the expansion, if we look sufficiently far away, becomes so great that the light never gets here in visible form. It becomes redshifted to extremely large wavelengths, losing nearly all of its energy. This solution also works in an infinitely old universe, so that even if there is no beginning to the universe, there is still a horizon.

Yet another solution to the paradox turns out to be more important. To understand this solution, we must realize that the universe is not really infinite, at least from a practical point of view. At sufficiently great distances we expect to see no galaxies or stars, because we will be looking back to a time before the universe began. Thus, even though the universe has no edge, there is a horizon beyond which stars and galaxies do not exist. Therefore, this argument goes, all lines of sight do not have to intersect stellar surfaces; many reach the darkness beyond the horizon of the universe, and we have a dark nighttime sky.

Another factor is that the source of energy available to produce light—that is, all the stars in the universe—is limited. If we calculate the total energy that possibly can be produced by all stars everywhere, we find that it is too little to make our sky appear bright. The universe simply does not contain enough energy.

The simple question of why the sky is dark at night, if analyzed fully, could have led to the conclusion that the universe has a finite age or a finite size a full century before these things were learned from other observational and theoretical developments.

---

tion due to a changing rate or direction of motion. One way to visualize this is to imagine you are inside a compartment with no windows (Fig. 14.4). If this compartment is on the Earth's surface, your weight feels normal because of the Earth's gravitational attraction. If, however, you are in space and the compartment is being accelerated at a rate equivalent to one Earth gravity, your weight will also feel normal. There is no experimental way to tell the difference between the two situations, short of opening the door and looking out.

One consequence of the equivalence of gravity and acceleration is that an object passing near a source of gravitational pull (that is, any other object having mass) undergoes acceleration, and therefore follows a curved path. In a universe containing matter this means that all trajectories of moving objects are

EINSTEIN: Einheitliche Feldtheorie von Gravitation und Elektrizität 415

Jnabhängig von diesem affinen Zusammenhang führen wir eine kontravariante Tensordichte $\mathfrak{g}^{\mu\nu}$ ein, deren Symmetrieeigenschaften wir ebenfalls offen lassen. Aus beiden bilden wir die skalare Dichte

$$\mathfrak{H} = \mathfrak{g}^{\mu\nu} R_{\mu\nu} \qquad (3)$$

und postulieren, daß sämtliche Variationen des Integrals

$$\mathfrak{J} = \int \mathfrak{H} dx_1 dx_2 dx_3 dx_4$$

nach den $\mathfrak{g}^{\mu\nu}$ und $\Gamma^{\alpha}_{\mu\nu}$ als unabhängigen (an den Grenzen nicht variierten) Variabeln verschwinden.

Die Variation nach den $\mathfrak{g}^{\mu\nu}$ liefert die 16 Gleichen

$$R_{\mu\nu} = 0, \qquad (4)$$

die Variation nach den $\Gamma^{\alpha}_{\mu\nu}$ zunächst die 64 Gleichungen

$$\frac{\partial \mathfrak{g}^{\mu\nu}}{\partial x_\alpha} + \mathfrak{g}^{\sigma\nu}\Gamma^{\mu}_{\sigma\alpha} + \mathfrak{g}^{\mu\sigma}\Gamma^{\nu}_{\alpha\sigma} - \delta^{\nu}_{\alpha}\left(\frac{\partial \mathfrak{g}^{\mu\beta}}{\partial x_\beta} + \mathfrak{g}^{\sigma\beta}\Gamma^{\mu}_{\sigma\beta}\right) - \mathfrak{g}^{\mu\nu}\Gamma^{\beta}_{\alpha\beta} = 0. \qquad (5)$$

Wir wollen nun einige Betrachtungen anstellen, die uns die Gleichungen (5) durch einfachere zu ersetzen gestatten. Verjüngen wir die linke Seite von (5) nach den Indizes ν, α bzw. μ, α, so erhalten wir die Gleichungen

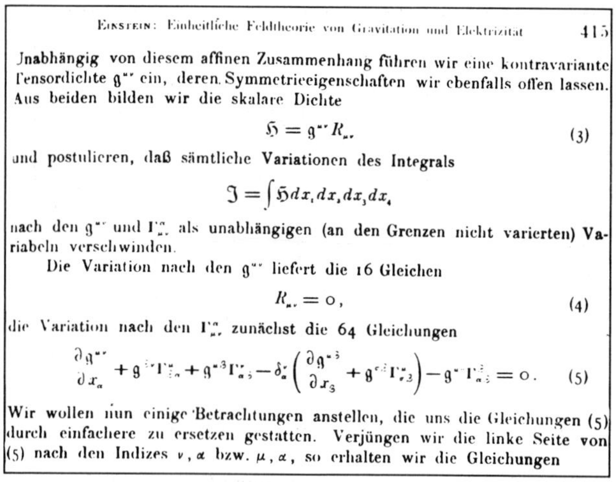

Figure 14.3 The Field Equations. Here are portions of the equations that describe the physical state of the universe. To a large extent, the science of cosmology involves finding solutions to these equations. *(The Granger Collection)*

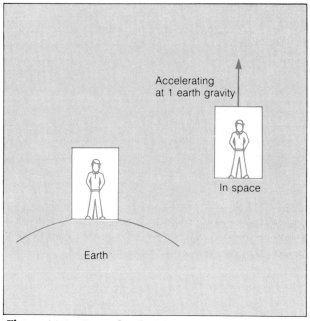

Figure 14.4 General Relativity. The person in the enclosed room has no experimental or intuitive means of distinguishing whether he is motionless on the surface of the Earth, or in space accelerating at a rate equivalent to one Earth gravity. The implication of this is that the acceleration due to motion and that due to gravity are equivalent, which in turn implies that space is curved in the presence of a gravitational field.

curved, and it is often said that space itself is curved. The degree of curvature is especially high close to massive objects (Fig. 14.5), but there is also an overall curvature of the universe, due to its total mass content. The solutions to the field equations specify, among other things, the degree and type of curvature. We will return to this point shortly.

Einstein's solution to the equations, developed in 1917, had a serious flaw in his view: It did not allow for a static, nonexpanding universe. In what he later admitted was the biggest mistake he ever made, Einstein added an arbitrary force called the **cosmological constant** to the field equations, solely for the purpose of allowing the universe to be stationary, neither expanding nor contracting.

Others developed different solutions, always by making certain assumptions about the universe. Some of these assumptions were necessary in order to simplify the field equations so that they could be solved. For example, in 1917 W. de Sitter developed a solution that corresponded to an empty universe, one with no matter in it.

By the 1920s solutions for an expanding universe were found, primarily by the Soviet physicist A. Friedmann and later by the Belgian G. LeMaître, who went so far as to propose an origin for the universe in a hot, dense state from which it has been expanding ever since. This was the true beginning of the big bang idea, and it was developed some three years before Hubble's observational discovery of the expansion. Of course, once it was found that the universe is not static but dynamic, the original need for Einstein's cosmological constant disappeared. It is usually included by modern cosmologists in the field equations, but its value is generally assumed to be zero.

The general relativistic field equations allow for three possibilities regarding the curvature of the universe and three possible futures (Fig. 14.6). A central question of modern studies of cosmology has to do with deciding which possibility is correct. One is referred to as "negative curvature," and an analogy to this is a saddle-shaped surface (Fig. 14.7) that is curved everywhere, has no boundaries, and is infinite in extent. This type of curvature corresponds to what is called an **open universe,** a solution to the field equations in which the expansion never stops. (It does slow down, however, because the gravitational pull of all the matter in it tends to hold back the expansion.)

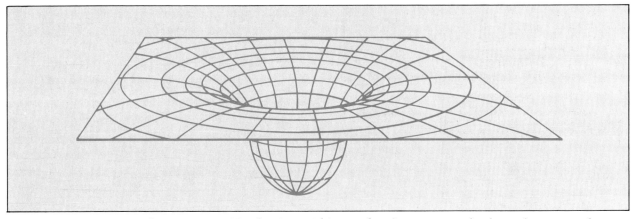

Figure 14.5 Curvature of Space Near a Massive Star. This curved surface represents the shape of space very close to a massive star. The best way to envision what it means to say that space is curved in this way is to imagine photons of light as marbles rolling on a surface of this shape.

Figure 14.6 The Three Possible Fates of the Universe. This diagram shows the manner in which the average distance between galaxies will change with time for the open, flat, and closed universe. In the first case, galaxies will continue to separate forever, although the rate of separation will slow. In the second case, the rate of separation will, in an infinite time, slow to a halt but will not reverse. In the third case, the galaxies eventually begin to approach each other, and the universe returns to a single point.

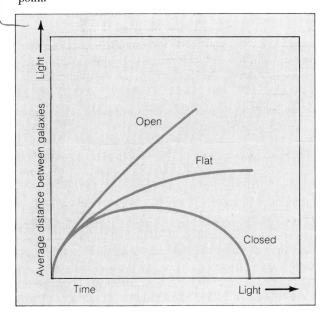

A second possibility is that the curvature is "positive," corresponding to the surface of a sphere (Fig. 14.8), which is curved everywhere, has no boundaries, but is finite in extent. This possible solution to the field equations is called the **closed universe,** and it implies that the expansion will eventually be halted by gravitational forces and will reverse itself, leading to a contraction back to a single point.

The third and last possibility is that the universe is a **flat universe** with no curvature. This corresponds to the case where the outward expansion is precisely balanced by the inward gravitational pull of the matter in

Figure 14.7 A Saddle Surface. This is a representation of the geometry of an open universe, which has negative curvature, is infinite in extent, has no center in space, and has no boundaries.

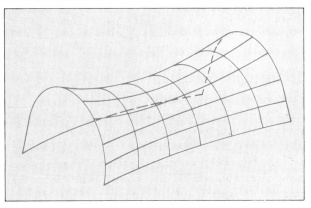

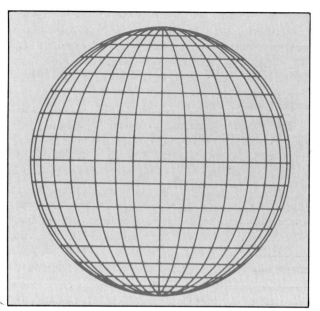

Figure 14.8 The Surface of a Sphere. This surface represents a closed universe, which has positive curvature and a finite extent but no boundaries and no center.

the universe, so that the expansion will eventually come to a stop but will not reverse itself. This balanced, static state requires a perfect coincidence between the energy of expansion and the inward gravitational energy of the combined total mass of the universe. It may therefore seem very unlikely, but we don't know what set of rules governed the amount of matter the universe was born with, so there is no logical reason to rule out this possibility. Most astronomers, however, generally pose the question in terms of whether the universe is open or closed, without specifically mentioning the possibility that it is flat. As we will see in the next section, if a perfect balance has been achieved, it will be very difficult to determine this from the observational evidence, which has large uncertainties.

Open or Closed: The Observational Evidence

Substantial intellectual and technological resources are devoted today to the question of determining the fate of the universe. Theorists are at work developing and refining the solutions to the field equations or seeking alternatives to general relativity that might provide equally or more valid representations. Observers are busily attempting to test the theoretical possibilities by finding situations where competing theories should lead to different observational consequences. This is a difficult job because most of the differences will show up only on the largest scales; to detect them requires observations of the farthest reaches of the universe.

Within the context of general relativity the premier question is whether the universe is open or closed. There are two general observational approaches to this question: determining whether there is enough matter in the universe to produce sufficient gravitational attraction to close it and measuring the rate of deceleration of the expansion, to see whether it is slowing rapidly enough to eventually stop and reverse itself.

Total Mass Content

The total mass in the universe is the quantity that determines whether gravity will halt the expansion. The field equations express the mass content of the universe in terms of density, the amount of mass per cubic centimeter. This is convenient for observers because it is obviously simpler to measure the density in our vicinity than to try to observe the total mass everywhere in the universe. The field equations can be solved for the value of the density that would produce an exact balance between expansion and gravitational attraction, that is, density corresponding to a flat universe. If the actual density is greater than this **critical density,** there is sufficient mass to close the universe, and the expansion will stop. The critical density depends on the value of Hubble's constant H; for a value near 75 kilometers per second per megaparsec, it is calculated to be roughly 8×10^{-30} gram per cubic centimeter, or about 4 protons per cubic meter, a very low value by any earthly standards.

The most straightforward way to measure the density of the universe is simply to count the galaxies in some randomly selected volume of space, add up their masses, and divide the total number of galaxies by the total volume. Care must be taken to choose a very large sample volume, so that clumpiness due to clusters of galaxies is not important. This technique yields very low values for the density, only a few percent of the critical density. It may seem that this would answer the question and that the universe is open, but this method may overlook substantial quantities of mass.

One clue to this overlooked mass arises from determinations of cluster masses based on the velocity dispersion of the galaxies in them (see discussion in Chapter 11). These mass measurements, for reasons still not entirely clear, always yield much higher values than the estimated total mass of the visible galaxies in the cluster. The disparity can be as great as a factor of ten or more, so if the larger values are correct, the average density of the universe comes closer to the critical value. Even the larger values based on velocity dispersion measurements, however, fall short of the amount needed to close the universe. If the mass density is to exceed the critical value, there must be large quantities of matter in some form that has not yet been detected (Fig. 14.9). The hidden matter cannot be inside of clusters of galaxies, because its presence would have been detected by the velocity dispersion measurements.

If there is a lot of dark matter in the universe, it could take several different forms. One possibility that is obvious to us by now is that there may be many black holes in space between clusters. The idea that there may be black holes in intergalactic space has been seriously suggested by the astrophysicist Stephen Hawking, who hypothesized the existence of countless numbers of "mini" black holes, formed during the early stages of the big bang. These would have very small masses, much smaller even than the mass of the Earth, but would be so numerous that they could easily exceed the critical density. Current observational evidence argues against the existence of such objects in great numbers, however.

Another possibility is that neutrinos, the elusive subatomic particles produced in nuclear reactions, have mass. Recall (from Chapter 5) that these particles permeate space, freely traveling through matter and vacuum alike, but that standard theory says they are massless. A recent controversial experiment indicates that this last assertion may not be correct, that they may contain miniscule quantities of mass after all. If so, they are sufficiently plentiful to provide more than the critical density, thereby closing the universe. Further experiments have been performed to determine whether neutrinos have mass, all of them so far being more consistent with the standard notion that these particles are massless. At present we cannot say whether neutrinos provide the matter density required to close the universe. Other kinds of elementary particles have been proposed, however, and physicists today are at work trying to solve this long-standing astronomical puzzle by better understanding the fundamental structure of subatomic particles.

Figure 14.9 Dark Matter in The Universe. Are the isolated galaxies that we see all that there is in the universe? Or are they merely points that happen to glow, embedded in a universal sea of unknown substance, whose mass density overwhelms that represented by the galaxies?

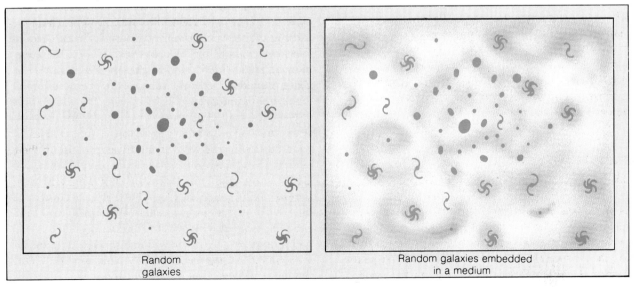

The Deceleration of the Expansion

Another approach to answering the question of whether the universe is open or closed is to compare the present expansion rate with what it was early in the history of the big bang, to see how much slowing, or deceleration, has occurred (Fig. 14.10). The expansion has certainly slowed—the question is how much. If it has only decelerated a little, then we infer that it is not going to slow down enough to stop and reverse itself, but if there has already been a lot of deceleration, then we conclude that the expansion is coming to a halt, and that the universe is closed.

To establish the deceleration rate requires knowing both the present expansion rate and the expansion rate at a time long ago, shortly after the big bang began. Neither quantity is easy to determine: We know that the value of the Hubble constant H, which tells us the present expansion rate, is quite uncertain to begin with. To measure the expansion rate at early times in the history of the universe is even more difficult, for it involves determining the distances and velocities of galaxies or clusters of galaxies that are so distant that we see them as they were when the universe was young. Such objects are now at the very limits of detectability, but it is hoped that the *Space Telescope* will push the frontier back far enough to permit observations of velocities in the early days of the expansion.

We still must rely on standard candles to establish the distance scale, and for such distant objects this procedure becomes even more uncertain than usual. When we look so far back in time, we are seeing galaxies that are much younger than those near us and that may therefore have quite different properties from mature galaxies. For example, their stellar populations, brightest stars and nebulae, and even total galactic luminosities may be rather different from those of nearby galaxies. It may not be valid to assume that the brightest galaxy in a faraway cluster of galaxies has the same absolute magnitude as the brightest galaxies in nearer clusters.

The present evidence on deceleration is that the universe has not slowed very much and therefore is not on its way to stopping and beginning to contract. As we have seen, however, the observational uncertainties are great, and this conclusion is not very firm.

A recently employed indirect technique for measuring the deceleration avoids many of the difficulties of observing extremely distant galaxies. Some light elements were created in the early stages of the expansion, and the amounts that were produced depend on the fraction of the universal energy that was in the form of mass rather than radiation at the time of element formation. The primordial mass density is in turn related to the present-day deceleration of the expansion, so if the primordial density is deduced from the abundances of elements that were created, we can infer the deceleration rate. The strategy, therefore, is to measure the abundance of some element that was produced in the big bang, and from that to derive the deceleration.

The most abundant element (other than hydrogen) produced in the big bang is helium. A measurement of how much helium was created would be a good indicator of the deceleration. The problem with this element, however, is that the amount produced in the big bang does not depend strongly on deceleration. Furthermore, helium is also produced in stellar interiors,

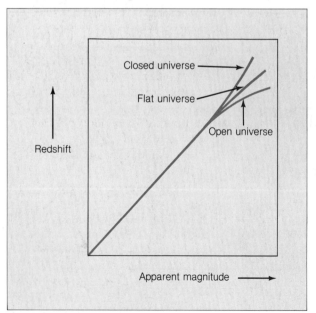

Figure 14.10 The Effect of Deceleration in the Hubble Diagram. This shows the relationship between velocity of recession and distance for the three possible cases: closed (upper curve), flat (middle), or open (lower). Present data seem to favor the open universe, although there is substantial uncertainty, largely because of problems in applying standard candle techniques to galaxies so far away that they are being seen as they were at a young age. The distinction is also made difficult by the subtlety of the contrast between the shapes of the curves.

Astronomical Insight 14.2

The Inflationary Universe

While many of us are sometimes preoccupied with the effects of inflation on the economy, cosmologists have begun to speak in terms of a different kind of inflation, one having to do with the early universe. The big bang cosmology is widely accepted, but some nagging problems remain. One is that it is thought to be very unlikely for a universe starting from a singularity as envisioned in the big bang to be as symmetrical as the observed universe. It seems unlikely that the universe would appear as homogeneous and isotropic as observations tell us it is. As we learned in the text, a great deal of effort has been put into testing this property of the universe, because any departures from homogeneity would provide information on imbalances or asymmetries in the early epochs of the expansion. Another mystery is why the universe should be so close to balance between being open or being closed.

A new kind of expansion model has been developed in which the homogeneity of the universe is easily understood. Calculations have shown that at a certain point very early in the history of the universe, when the temperature was around 10^{27} K, conditions were right for small regions to separate themselves from the rest of the universe and expand very rapidly to much larger sizes. A reasonable analogy would be the creation of bubbles in a liquid, with the bubbles then growing much larger almost instantaneously. In this cosmological model, each "bubble" becomes a universe in its own right, with no possibility of communication with other bubbles.

The major advantage of this so-called **inflationary universe** model is that a universe born of a tiny cell in the early expansion would be expected to remain symmetric and homogeneous as it grew. The reasons for this are somewhat abstract but are related to the fact that the various portions of the tiny region that was to expand into the present universe were so close together that they were in a form of equilibrium with little or no variation in physical conditions. In a larger region, such as in the standard big bang cosmology, no such equilibrium would have existed, and there would be no reason to expect such uniformity throughout the resulting universe.

Another advantage of the inflationary model is that in such a universe, it would be expected that the expansion would go on forever but would continue to slow, approaching a stationary state; that is, a flat universe is the natural result of rapid expansion from a tiny bubble as envisioned in this model. As we have seen in the text, observations tell us that our universe is nearly flat (although perhaps not quite). If it were not so closely balanced, it would not be so difficult to determine whether it is open or closed.

A problem for the inflationary model is its disagreement with the observational data on another kind of particle, one that has yet to be unambiguously detected. The inflationary theory predicts the formation in the early universe of a large number of **magnetic monopoles**, subatomic particles with single magnetic poles instead of the usual pair of opposing poles. Such particles would be very difficult to detect, the best technique being to look for the tiny electrical currents they would briefly create in wires they passed near. So far only one possible detection of a magnetic mono-pole has been reported, while the inflationary theory leads to the expectation that these particles should not be so rare.

The homogeneity of a bubble in the inflationary universe is an advantage in explaining some observations, as we have seen, but it is also a disadvantage. In such a uniform universe, it is very difficult to understand how the matter ever became lumpy

(continued on next page)

Astronomical Insight 14.2

The Inflationary Universe

(continued from previous page) enough to form galaxies and clusters of galaxies. In the text we spoke of possible turbulence in the early universe and other mechanisms having to do with concentrations of matter, but in the inflationary model it is very difficult to see how such concentrations would ever have formed.

First suggested in 1981, the inflationary model is currently under active discussion by cosmologists. There will continue to be refinements to it, as more of the observational constraints are confronted. It will be very interesting to see whether it withstands the test of continued scrutiny.

---

so it is difficult to determine how much of what we see in the universe today is really left over from the big bang.

Better candidates are ^{7}Li, an isotope of the element lithium, and deuterium, the hydrogen isotope that has one proton and one neutron in the nucleus. Trace quantities of these species were also produced in the big bang, and, as far as is presently known, are not made in any other way. To date most efforts have been concentrated on measuring the present-day abundance of deuterium, although in principle ^{7}Li can be measured as well. While deuterium (and lithium) can be destroyed by nuclear processing in stars, it is not produced in that way (even if it were, it would not survive the high temperatures of stellar interiors without undergoing further reactions). The present deuterium abundance in the universe should therefore represent an upper limit on the quantity created in the big bang. Direct measurement (Fig. 14.11) of the amount of deuterium in space became possible in the 1970s, with the launch of the *Copernicus* satellite, which made ultraviolet spectroscopic measurements and was able to observe absorption lines of interstellar deuterium atoms. The abundance of deuterium that was found is sufficiently high to imply a low density, which points to a small amount of deceleration. Thus this test, like the others, indicates that the universe is open. There is some uncertainty, though, because the *Copernicus* data seem to indicate an uneven distribution of interstellar deuterium throughout the galaxy, and this would not be expected if the deuterium really formed exclusively in the big bang. A clumpy distribution of deuterium in the galaxy could indicate that some of it is somehow produced by stars, in which case its abundance is not a strict test of deceleration. A recent measurement of the ^{7}Li abundance in the galaxy is consistent with the deuterium results; that is, the observed quantity of ^{7}Li is also large enough to imply that the universe is open.

It is worth noting that all the observational techniques tried so far to discover whether the universe is

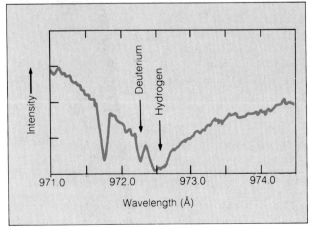

Figure 14.11 An Ultraviolet Absorption Line of Interstellar Deuterium. The abundance of deuterium in space is determined from the analysis of absorption lines it forms in the spectra of background stars. Here we see a weak deuterium line close to a strong absorption feature due to normal hydrogen. The spectrum shown here was obtained with the *Copernicus* satellite. (J. B. Rogerson and D. G. York)

open or closed have pointed toward a fairly close balance. That is, the data do not point to a universe that is closed or open by a wide margin; whatever the correct answer is, we can say that the universe is close to being flat. This is very significant, for such a close balance is not necessarily expected in the big-bang theory and must be regarded in this theory as a coincidence. There are other models, however, such as the very recent **inflationary universe** theory, in which a flat or nearly flat universe is expected (see Astronomical Insight 14.2).

The History of Everything

Having outlined the present state of knowledge and pointed out the basis for current and planned observational tests, we are now at the forefront of modern cosmology. At this point it is useful to review the development of the universe, highlighting some of the significant events. The very fact that we can do this, that we can say with any degree of certainty what conditions were like at the beginning and at points along the way, is a triumph of modern science.

It is impossible to describe the physical universe at the precise moment that the expansion began; it is physical nonsense to deal with infinitely high temperature and density. It is possible, however, to calculate the conditions immediately after the expansion started and at any later time. Many of the most interesting events in the early history of the universe, and the ones currently under the most active investigation, took place at very early times, before even a ten-thousandth of a second had passed. Under the conditions of density and temperature that existed then, matter and the forces that act on it were quite different from anything we can experience, even in the most advanced laboratory experiments. Even the familiar subatomic particles such as protons, neutrons, and electrons could not exist but were replaced by *their* constituent particles.

Current particle physics theory holds that the most fundamental particles are **quarks** and **leptons**. Modern theory also provides a basis for believing that all three and perhaps all four of the fundamental forces in nature (see Astronomical Insight 3.2) may really be manifestations of the same phenomenon. So far it has been possible to show that three of the fundamental forces (the electromagnetic force and the weak and strong nuclear forces) may be manifestations of the same basic interaction. Under the physical conditions that we are used to, these forces behave very differently, but when density and temperature are very high, as they were early in the expansion of the universe, the forces are indistinguish-able. Some aspects of the theory that shows this have been confirmed by laboratory experiments. The theoretical framework connecting the three forces is called **grand unified theory** (**GUT** for short). This name may be somewhat exaggerated, because the fourth force, gravity, has not yet been shown to be unified with the other three, and some theories of gravity indicate that it may not be.

The unification of the other three forces implies that in the very early moments following the beginning of the expansion, until 10^{-35} second had passed, the universe contained only quarks, leptons, and related particles and radiation, and the only forces operating in it were gravity and the unified force that was later to become recognizable as the electromagnetic, strong, and weak forces.

Let us now jump forward in time to an epoch in the early universe when matter was beginning to take on more familiar forms (Figs. 14.12 and 14.13) and the four fundamental forces were already acting as four distinct forces. At 0.01 second after the beginning of the expansion, the temperature was perhaps 100 billion degrees (10^{11} K), and electrons and positrons began to appear. The temperature dropped to 10 billion degrees (10^{10} K) by 1.09 seconds after the start, and by then protons and neutrons were appearing. At this point, most of the energy of the universe was still in the form of radiation. Within a few minutes, conditions became better suited for nuclear reactions to take place efficiently, and the most active stage of element creation began. The principal products, in addition to the helium and deuterium already mentioned, were tritium (another form of hydrogen, with one proton and two neutrons in the nucleus) and the elements lithium (three protons and four neutrons) and, to a minor extent, beryllium (four protons and five neutrons). Nearly all the available neutrons combined with protons to form helium nuclei, a process that was complete within four minutes of the beginning of the expansion. At this point, with some 22 to 28 percent of the mass in the form of helium, the reactions were essentially over, except for some production of lithium and beryllium over the next half hour.

The expansion and cooling continued, but nothing significant happened for a long time after the nuclear

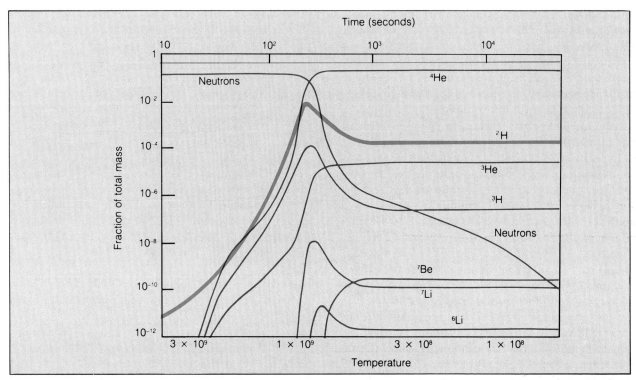

Figure 14.12 Element Formation in the Big Bang. This diagram illustrates the relative rates of formation of various light elements by nuclear reactions during the early stages of the big bang expansion. The production rates are shown as functions of time and temperature. *(Data from Wagoner, R. V. 1973, The Astrophysical Journal 179:343)*

reactions stopped. Eventually the density of the radiation had reduced sufficiently that its energy was less than that contained in the mass; that is, the energy derived from $E = mc^2$ became greater than that contained in the radiation. At that point, it is said, the universe became matter-dominated, rather than radiation-dominated.

The matter and radiation continued to interact, however, because free electrons scatter photons of light very efficiently. The strong interplay between matter and radiation finally ended nearly a million years after the start of the big bang, when the temperature became low enough to allow electrons and protons to combine into hydrogen atoms. The atoms still absorbed and reemitted light, but they did it much less effectively, because they could do so only at a few specific wavelengths. From this time on, matter and radiation went separate ways. The radiation simply continued to cool as the universe expanded, reaching its present temperature of 2.7 K some 10 to 20 billion years after the expansion began.

Sometime in the first billion years or so, the matter in the expanding universe became clumpy and fragmented into clouds and groups of clouds that eventually collapsed to form galaxies and clusters of galaxies. As we noted in Chapter 11, the cause of the clumpiness and fragmentation is still unknown. Once galaxies began to form, the subsequent evolution of matter followed steps already described in the preceding chapters.

Because there is so much uncertainty about how the initial stages of fragmentation took place, there is a great deal of interest in probing back far enough into the past to get some information on this process. To do so will be another of the goals of the *Space Telescope*.

What Next?

To discuss the future of the universe is obviously a speculative venture. We cannot even answer with absolute certainty the basic question whether it is open or closed. There have been attempts to calculate future conditions in the universe, in analogy to the theoretical work on the early stages discussed in the last section. In the case of the future of the universe, however, the uncertainties are much greater, and the following descriptions should be regarded as *very* speculative.

If the universe is open or flat, it will have no definite end; it will just gradually run down. The radiation background will continue to decline in temperature, approaching absolute zero. As stellar processing continues in galaxies, the fraction of matter that is in the form of heavy elements will continue to grow, and the supply of hydrogen, the basic nuclear fuel, will diminish. It is predicted that all the hydrogen should be gone by about 10^{14} years after the birth of the universe, so the universe in which stars dominate has now lived approximately one ten-thousandth of its lifetime. The recycling process between stars and the interstellar medium will continue until this time, but gradually the matter will become locked up in black holes, neutron stars, and white dwarfs. Dead and dying stars will continue to interact gravitationally, eventually colliding often enough in their wanderings about that all planets will be lost (at about 10^{17} years), and galaxies will dissipate as their constituent stars are lost to intergalactic space (by about 10^{18} years). Further speculation shows that a new physical process will take over at later times. The new grand unified theory tells us that the proton, a basically stable particle, may disintegrate in a very low-probability process that occurs on the average once in 10^{32} years for a given proton. When the universe reaches an age of about 10^{20} years, enough protons will begin to evaporate here and there that the energy produced will keep the remnant stars heated, although only to the modest temperature of perhaps 100 K. At an age of over 10^{32} years, most protons will have decayed, and the universe will consist largely of free positrons, electrons, black holes, and radiation (the extremely cold remnant of the big bang). The final stage that has been foreseen occurs at an age of 10^{100} years, when sufficient time has passed for all black holes to evaporate, and nothing is left but a sea of positrons, electrons, and radiation. (Theory says that black holes can disintegrate with a very low probability, meaning that if we wait long enough, eventually they will do so.)

If the universe is closed, then someday, perhaps some 50 billion years from now, the expansion will

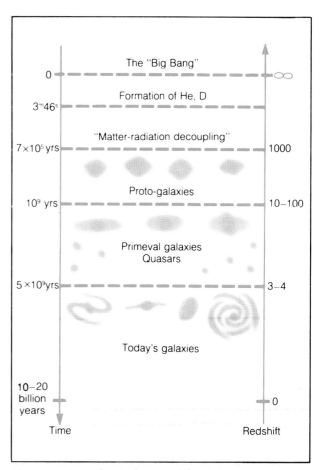

Figure 14.13 The Evolution of the Universe. This diagram shows, as a function of time since the start of the big bang, significant stages in the evolution of the universe. It also shows the types of objects that existed at each stage, and the redshift at which they are (or would be) observed at the present time. *(J. M. Shull)*

stop and be replaced by contraction. The deterioration described above will still take place, until the final moments when the universe once again becomes hot and compressed, entering a new singularity. In some views, purely conjectural and without possibility of verification, such a contraction would be followed by a new big bang, and the universe would be reborn. This concept of an oscillating universe, while pleasing to the minds of many, will not be fulfilled unless the present weight of the evidence favoring an open universe is found to be in error.

Astronomical Insight 14.3

Particle Physics and Cosmology

There is some irony in the fact that the science of the smallest scales, particle physics, is finding its greatest opportunities in cosmology, the science of the largest scales. One of the most rapidly advancing areas in modern particle physics coincides with one of the brightest fields in modern astrophysics, bringing full circle one of the themes of astronomy, the relationship between the science of the atom and that of the cosmos.

In particle physics, scientists are concerned with discovering the nature of matter and energy at its most fundamental level. It is now known that the atom and the major subatomic particles, such as electrons, protons, and neutrons, are actually composed of more basic particles, known in general as elementary particles. These include a wide variety of particles whose names are not familiar to most people. The most modern theories of elementary particles postulate that all matter is made up of basic particles called quarks, which can have several different combinations of properties. Quarks and other particles known only from theory are sought experimentally in **particle accelerators,** enormous machines that can accelerate particles to such high kinetic energies that they are broken down into their constituent elementary particles when they collide. Because there is an equivalence between matter and energy (as expressed in Einstein's theory of general relativity), elementary particles can be created from pure energy in sufficient quantities. Thus the objective in building particle accelerators to probe elementary particles is to inject enough energy into the accelerators to create the particles being sought. Progress is made every time new levels of energy are reached.

Some of the particles that are predicted to exist will require energies so high that it may never be practical to create them in accelerators. Here is where astrophysics, specifically cosmology, may come to the rescue. Energies high enough to create all possible particles certainly existed during the earliest stages of the universal expansion. Thus elementary particle physicists may find some of the answers they seek by studying what is known about the big bang.

We cannot actually observe the composition of the universe at the time of the big bang, but we know that today's universe is the product of that era. By examining the detailed atomic composition of the universe today, particle physicists can hope to deduce the mixture of particles that inhabited the universe during its initial moments. A remarkable degree of success has already been achieved in just this way, as particle theorists have been able to derive with considerable certainty the physical conditions that must have prevailed in early times. The reason we think we know as much as we do about the formation of the elements in the early universe is that the theory of elementary particles is capable of reproducing the observed abundances of the elements, starting from the conditions believed to be present a fraction of a second after the start of the expansion. Now the hope is that the process can be reversed, that particle physicists can learn about elementary particle processes by observing the state of today's universe.

Perspective

We have now completed our survey of the large-scale universe. We have probed the workings of the stars, learned of the majesty of the galaxies, and delved into the secrets of the universe itself.

Now it is time to return to our local neighborhood to gain a better understanding of the tiny realm of the planets. In the next section we discuss the solar system, including the Earth.

Summary

1. In cosmological studies, astronomers usually adopt the cosmological principle, which states that the universe is both homogeneous and isotropic. Existing observational data tend to support this assumption.
2. Einstein's theory of general relativity, which describes gravitation and its equivalence to acceleration, is used to describe the universe as a whole mathematically. In the context of this theory, field equations are written that represent the interaction of matter, radiation, and energy in the universe with the curvature of space-time. Solutions of the field equations amount to definitions of the properties of the universe.
3. Three general solutions to the field equations are considered by modern cosmologists. They correspond to a closed universe (positive curvature), an open universe (negative curvature), and a flat universe (no curvature).
4. A major question in modern astrophysics is whether the universe is closed (expansion eventually to be reversed) or open (expansion never to stop).
5. There are two general types of tests of whether the universe is open or closed: (a) ascertaining whether there is enough mass in the universe to gravitationally halt the expansion; and (b) determining the rate of deceleration of the expansion, to see whether it is slowing enough to stop eventually.
6. The total mass content is measured in terms of the average density of the universe, and the matter that is visible in the form of galaxies is not sufficient to close the universe. Various suggestions have been made concerning other forms in which the necessary mass could exist.
7. The deceleration is measured in two ways: (a) by comparing past and present expansion rates, through observations of very distant galaxies and (b) by inferring the early expansion rate from the present-day abundances of elements that formed only during the big bang.
8. Both the observed total mass content and the inferred deceleration of expansion indicate that the universe is open. This conclusion is not universally accepted, and observational tests are continuing.
9. The early stages of the universal expansion, up to the time when matter and radiation decoupled, can be described quite precisely and certainly by modern physics. Following the initial moment of infinite density and temperature, the first atomic nuclei formed just over three minutes later, and all of the early element production was finished within a few minutes. Matter and radiation decoupled almost a million years later; it is not so well understood how the universe subsequently organized itself into stars and galaxies.
10. The future of the universe appears to have two possibilities. If it is open or flat, it will gradually become cold and disorganized. If it is closed, it will eventually contract, perhaps to a new beginning in another big bang.

Review Questions

1. In Chapter 11 we discussed the fact that galaxies often are grouped in clusters. In this chapter we have asserted that the universe appears to be homogeneous; that is, it has a uniform distribution of matter. Explain how galaxies can be grouped in clusters and at the same time the universe can be homogeneous.
2. Recall from Chapter 12 that there is a 24-hour anisotropy in the 3-degree background radiation. Why does this not violate the assumption adopted here that the universe is isotropic?
3. Are Einstein's field equations more like Kepler's laws of planetary motion or Newton's derivation of Kepler's laws? Explain.
4. Suppose a star ejects a spherical shell of gas as it forms a planetary nebula. Depending on the mass

of the star and the initial velocity of ejection, the shell of gas will either escape completely or fall back onto the star. What determines this, and how is it analogous to the question of whether the universe is open or closed?

5. Why do astronomers not simply measure the density of matter within the well-observed solar neighborhood, within a few hundred parsecs of the Sun, to determine whether there is enough mass to close the universe? What do you think the result would be if this local density were used? (You may want to review Chapter 9 in answering this.)

6. How do we know that the existence of massive black holes in the nuclei of all galaxies would not be sufficient to close the universe?

7. Summarize the difficulties in measuring very distant galaxies in order to infer the early expansion rate of the universe.

8. Why is deuterium a better species than helium to use in inferring the early expansion rate of the universe? Why would either be better than iron, for this purpose?

9. What would it tell us about the early expansion of the universe if it was discovered that substantial quantities of elements such as carbon, nitrogen, and oxygen had been formed in the big bang? Would this imply that the universe is open or closed?

10. Summarize the ways in which the *Space Telescope* should help answer cosmological questions raised in this chapter.

Additional Readings

Barrow, J. D., and J. Silk. 1980. The structure of the early universe. *Scientific American* 242(4):118.

Bartusiak, M. 1985. Sensing the ripples in space-time. *Science 85* 6(3):58.

Dicus, D. A., J. R. Letaw, D. C. Teplitz, and V. L. Teplitz. 1983. The future of the universe. *Scientific American* 248(3):90.

Ferris, T. 1977. *The red limit: The search for the edge of the universe.* New York: William Morrow.

Field, G. B. 1982. The hidden mass in galaxies. *Mercury* 11(3):74.

Gaillard, M. K. 1982. Toward a unified picture of elementary particle interactions. *American Scientist* 70(5):506.

Overbye, D. 1985. The shadow universe (dark matter). *Discover* 6(5):12.

Page, D. N., and M. R. McKee. 1983. The future of the universe. *Mercury* 12(1):17.

Penzias, A. A. 1978. The riddle of cosmic deuterium. *American Scientist* 66:291.

Schramm, D. N. 1974. The age of the elements. *Scientific American* 230(1):69.

Silbar, M. L. 1982. Neutrinos: Rulers of the universe? *Griffith Observer* 46(1):9.

Trefil, J. S. 1978. Einstein's theory of general relativity is put to the test. *Smithsonian* 11(1):74.

——— 1983. How the universe began. *Smithsonian* 14(2):32.

——— 1983. How the universe will end. *Smithsonian* 14(3):72.

Tucker, W., and K. Tucker. 1982. A question of glaxies. *Mercury* 11(5):151.

Weinberg, S. 1977. *The first three minutes.* New York: Basic Books.

Section V

The Solar System

We have studied the large-scale features of the universe, and we have learned what is known of the Sun and stars. With this background, we now turn to a discussion of the small scale, of the planets and other bodies that make up our solar system.

The first of the six chapters in this section describes the formation of the Sun and planets, with some mention of historical developments in the quest to understand our beginnings and a detailed exposition of the current best model of this process. This discussion draws upon an earlier one on star formation (Chapter 7).

Chapter 16 is a description of the Earth and Moon. We devote a fairly intensive discussion to these bodies, the only ones yet visited by humans, because we know the most about them and because they serve as examples for comparison with other bodies. The next chapter, Chapter 17, summarizes and compares what is known of the other terrestrial planets, Venus, Mars, and Mercury. The emphasis is on understanding why these planets, which are much like the Earth in many ways, are so different in other ways.

Chapter 18 treats the outer planets, concentrating on the four gaseous giants, Jupiter, Saturn, Uranus, and Neptune. Three of these four have been visited by space probes from Earth, so we have quite a bit of information to discuss and compare. A brief final section of the chapter is devoted to Pluto, the tiny misfit of a planet.

The asteroids, comets, and meteoroids that populate interplanetary space are the topic of Chapter 19. The discussion of comets is especially thorough, because so much has been learned from the recent apparition of Halley's comet.

Chapter 15

Origins of the Solar System

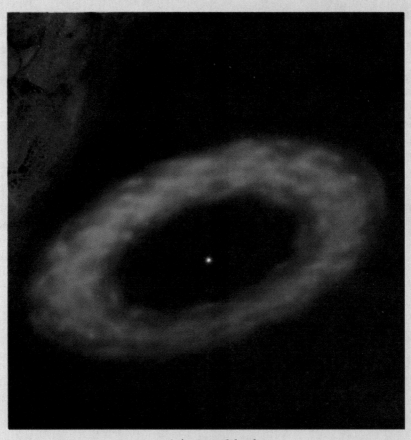

An artist's view of the dust around Vega. *(NASA)*.

Chapter Preview

Overall Properties of the Planetary System
Early Theories of Solar System Formation
A Modern Scenario
Are There Other Solar Systems?

In the cosmic scheme, planets are very minor bodies, small, dim objects difficult or impossible to observe even at the distance of the nearest star. Yet planets may very likely be the only places capable of supporting life, the only locales where temperature and other conditions are moderate.

In this chapter we consider the overall properties of the Sun's family of planets, and we discuss the processes thought to have led to their formation. Along the way we will gain some insight into the question of whether planets orbit other stars and how we might try to detect such planets.

Overall Properties of the Planetary System

The nine planets that are known to us orbit the Sun in nearly circular paths all lying roughly in the same plane. The solar system can be likened to a great disk with the Sun at the center and the planetary orbits arranged concentrically about it (Fig. 15.1). In addition to the planets, there are other solar system bodies: there is a wide belt of **minor planets,** small bodies orbiting largely between Mars (the fourth planet from the Sun) and Jupiter (the fifth); there are comets, whose solar orbits are not confined to the plane of the planetary orbits; and there is a pervasive, but very rarefied, medium of interplanetary gas and dust, concentrated in the plane of the solar system disk.

The distance between successive planets increases as we go outward from the Sun, creating a pattern (Fig. 15.2) similar to ever-widening ripples on a pond where a fish has jumped. This pattern, which Copernicus was able to draw to scale using simple geometric arguments, helped to convince him that the Sun, not the Earth, must be the central object in the solar system. The orbit of each planet is an ellipse (see Chapter 3), but usually an ellipse with a small eccentricity, meaning that it is close to being a circle with the Sun near the center.

The motions of the planets and their satellites are very systematic, with all of the planets and most of the satellites orbiting in the same direction (Fig. 15.2), known as **prograde motion.** The spins of the planets and satellites are also generally in the same direction as the orbital motions. If we look down on the solar system from a point above the north pole of the Earth, these motions are in the counterclockwise direction. The minor planets (also known as **asteroids**) orbit in the prograde direction, while the comets orbit in random directions.

The physical properties of the planets vary markedly, despite the many similarities among them in the character of their motions and the systematic arrangement of their orbits. The innermost four planets (Fig. 15.3) are small, dense, rocky objects, known collectively as the **terrestrial planets.** The next four (Fig. 15.4), starting at roughly five times the Earth's distance from the Sun (that is, five astronomical units, or 5 AU), are much larger and more massive, consist mostly of gas, and are called either the **giant planets** or the **Jovian planets,** after Jupiter, the largest of them all. The outermost planet, Pluto, is unlike any other in the solar system, having many more similarities with the larger moons of the giant planets than with any other planet.

The properties of the nine planets are listed in Table 15.1. We see that the terrestrial planets, besides having higher densities than the gaseous giants, are warmer. Not shown in the table are the significant differences in chemical composition; the terrestrial planets contain relatively small abundances of the lighter-weight elements such as hydrogen and helium, whereas these two elements dominate the composition of the outer planets. The outer planets tend to have ring systems and many more satellites than the terrestrial planets, implying that something about the formation processes for these two groups of planets was very different.

The varying nature of the planets from the inner to the outer solar system, along with the overall shape of the system and the regularities in the motions of the planets and satellites, provides clues we can use in attempting to uncover the story of the birth and evolution of the solar system. The exceptions to the norm also can help us. Here and there are motions that stand out for their defiance of the usual pattern; for example, Venus (the second planet) rotates slowly in the direction

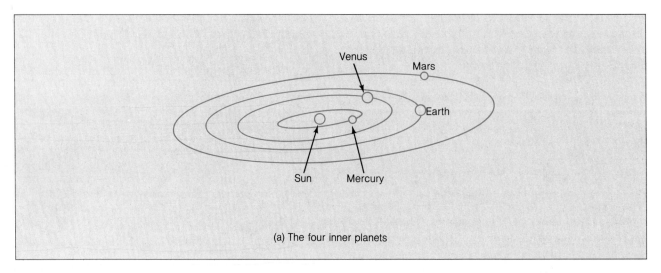

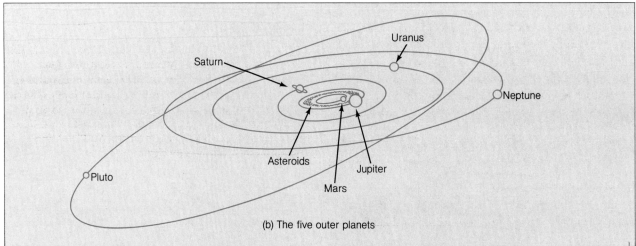

Figure 15.1 The Coalignment of Planetary Orbits. All but one of the nine planets have nearly circular orbits in a common parallel plane. The exception is Pluto, whose orbit is tilted 17 degrees with respect to the ecliptic, and is sufficiently eccentric (elongated) that Pluto is actually closer to the Sun than Neptune at times.

opposite the spins of the other planets (we call this **retrograde motion**); Uranus (the seventh planet) is tipped over so far that its north pole points slightly below the plane of the solar system's disk; some of the satellites in the outer solar system orbit in the retrograde direction; and Pluto has a strangely eccentric and tilted orbit about the Sun.

In the following sections, we will discuss how the solar system came about, beginning with an overview of some early attempts to unravel the story.

Early Theories of Solar System Formation

The first serious theory based on the heliocentric view of the solar system was developed by the French mathematician and philosopher René Descartes, who in 1644 advanced the idea that the solar system formed from a gigantic whirlpool, or vortex, in a universal fluid, with the planets and their satellites forming from smaller eddies. This hypothesis was rather crude, without any clearly specified idea of the nature of the cosmic substance from which the Sun and planets arose, but it did account for the fact that all the orbital motions are in the same direction.

The hypothesis of Descartes was the first of a general type known as **evolutionary theories,** theories in which the formation of the solar system occurred as a natural by-product of the sequence of events that produced the Sun. In an evolutionary theory, no special circumstances are needed to create the planets, other than the fact that the Sun formed. Theories of this type lead to a natural expectation that there may be planets orbiting other stars.

Descartes' idea was further elaborated by the German philosopher Immanuel Kant, who in 1755 used the principles of the recently discovered Newtonian mechanics to argue that a rotating gas cloud would flatten into a disk as it contracted (Fig. 15.5). Kant's theory was called the **nebular hypothesis,** because it invoked formation of the Sun and planets from a interstellar cloud, or nebula. In 1796 Pierre Simon de Laplace, a French mathematician, added to this general idea the notion that as the spinning cloud flattened into a disk, concentric rings of material broke off due to rotational forces, so that at one point the early solar system would have looked very much like the planet Saturn with its rings (Fig. 15.5). Each ring was then supposed to have condensed into a planet.

Soon a problem arose that led to some doubts concerning the validity of the evolutionary models. If the Sun formed from the concentration of a rotating gas

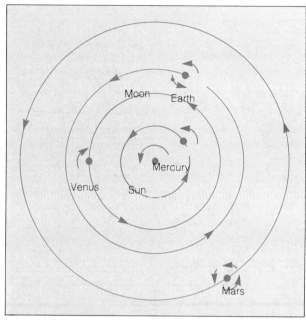

Figure 15.2 Prograde Motions in the Inner Solar System. The rotations and orbital motions of the planets are generally in the same direction. This sketch of the orbits of the inner four planets shows that Venus has retrograde rotation but that the other three planets and their satellites obey the rule. The outer planets (except for Pluto) and nearly all their satellites do also.

Table 15.1 Physical Data for the Planets

Planet	Mass*	Diameter	Density	Surface Gravity	Temperature
Mercury	0.0558	0.382	5.50 g/cc	0.38 g	100–700 K
Venus	0.815	0.951	5.3	0.90	730
Earth	1.000	1.000	5.517	1.00	200–300
Mars	0.107	0.531	3.96	0.38	130–290
Jupiter	318.1	10.79	1.33	2.64	130
Saturn	95.12	8.91	0.68	1.13	95
Uranus	14.54	4.05	1.20	0.89	95
Neptune	17.2	3.91	1.58	1.13	50
Pluto	0.002	0.23	1.1	0.05	40

*The masses and diameters are given in units of the Earth's mass and diameter, which are 5.974×10^{27} grams and 12,734 km, respectively.

Figure 15.3 The Terrestrial Planets. All of these images were obtained from space: Mercury (upper left) and Venus (upper right) were both photographed by *Mariner 10*; Earth (lower left) by one of the *Apollo* missions; and Mars (lower right) by one of the *Viking* orbiters. The terrestrial planets are all relatively small and dense, with rocky surfaces. *(NASA)*

Figure 15.4 The Giant Planets. The images of Jupiter (upper left), Saturn (upper right), and Uranus (lower left) were all obtained at close range by the *Voyager* probes; the image of Neptune (lower right) was obtained from the Earth, using a special filter that isolates spectral features due to methane in the atmosphere. The giant planets are very different from the terrestrial planets, being much larger, far less dense, and having no solid surfaces. *(NASA)*

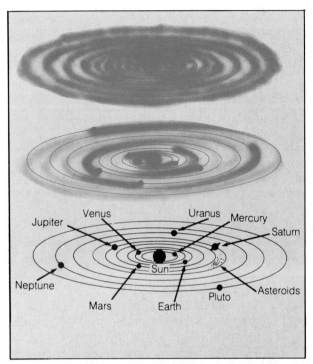

Figure 15.5 The Hypothesis of Kant and Laplace
Descartes's simple vision of a vortex was refined by Kant, who realized that rotation should cause a collapsing cloud to take a disklike shape; and by Laplace, who hypothesized that a rotating disk would form detached rings that could then condense into planets.

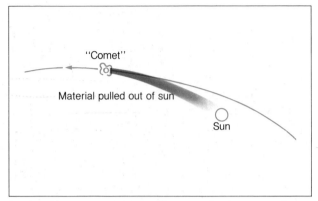

Figure 15.6 The Catastrophe Hypothesis. According to this theory, a passing body would gravitationally draw from the Sun the material that would subsequently form planets.

cloud, then the laws of physics dictate that the Sun should have formed with a rapid spin. For the same reason that ice skaters spin faster when their arms are pulled in close to their bodies, a star should spin faster as it contracts (the **angular momentum,** a measure of the mass, size, and spin rate of an object, must remain constant, so that if the size changes, the spin rate must change to compensate). The Sun's rotation period is about 25 days at the equator, whereas it should have been only a few hours or days if the rotation rate had increased as expected while the Sun condensed. The difficulty is reconciling the Sun's slow spin with the expectations of the evolutionary theories led to the temporary abandonment of those theories by many astronomers.

An alternative class of theories called **catastrophic theories** had been suggested, and it now began to receive increased attention. These theories of the formation of the planetary system envision a solitary Sun to begin with but then invoke some singular, cataclysmic event that disrupted the Sun and formed the planets. The first of these theories was proposed by another Frenchman, Georges Louis de Buffon, who in 1745 suggested that a massive body (which he referred to as a comet, although it was much larger than any cometary nucleus) passed so near the Sun that its gravitational pull forced material out of the Sun (Fig. 15.6), and this gas then condensed to form the planets.

Buffon's idea was largely ignored until the beginning of the twentieth century, when the difficulty with the Sun's slow spin forced scientists to consider alternatives to the evolutionary theories. By 1905, the English astronomers T. C. Chamberlain and F. R. Moulton had elaborated on Buffon's original idea and suggested that another star was the object that passed near the Sun, creating tidal forces that caused matter from inside the Sun to stream out and later condense into planets. Further refinements to the bypassing star hypothesis were soon provided by the British scientists H. Jeffreys and J. H. Jeans.

A disturbing problem with the catastrophic theories was the low probability that the Sun could have collided with another star, given the great distances between stars in our part of the galaxy. Of course, the possibility could not be completely ruled out, for even a low-probability event can occur, but the problem was sufficiently worrisome to lead to the development of an alternative catastrophic theory, one in which the Sun started out as a member of a triple system but was then freed from the gravitational bonds of the other two when the three nearly collided. In the process of expelling the Sun from the triple star system, the gravity of the other stars tore some of the solar material out into a tail that later condensed to form the planets. This theory, developed by H. N. Russell, R. A. Lyttleton, and F. Hoyle, replaced the difficulty of forming the Sun and the planets together with one of forming the Sun and two other stars together. The advantage of this theory was that the slow rotation of the Sun did not have to be produced during its formation but could have developed later, when the Sun nearly collided with the other stars and was ejected.

By the mid-twentieth century the catastrophic theories were becoming quite unwieldy, not only because of the special circumstances that had to be assumed but also because of several fundamental problems. For example, calculations showed that even if some material really were pulled out of the Sun's interior, it would be so hot that it would expand and dissipate into space, rather than condensing to form planets. Another difficulty was how to pull material out of the Sun and give the extracted matter sufficient angular momentum for the planets without also giving it enough speed to escape entirely.

A more modern but quite serious objection to the catastrophic theories concerns the abundance of the hydrogen isotope deuterium. Deuterium is a form of hydrogen whose nucleus contains a proton and a neutron, in contrast with the normal form of hydrogen, whose nucleus consists solely of a proton. It happens that deuterium is destroyed in nuclear reactions that will inevitably occur if the gas containing deuterium is subjected to high enough temperatures, as would be the case inside a star. Even temperatures in the Sun's outer layers are high enough to destroy deuterium. Hence, if the solar system were formed of material pulled out of the Sun, we would expect to find no deuterium in the planets and interplanetary bodies. To the contrary, deuterium is found to be a normal constituent of the solar system, so it must have been present in the primordial material from which the solar system formed (in Chapter 14 we discussed the origin of deuterium in the universe), and thus this material was never part of the Sun. This evidence emphatically rules out all catastrophic theories that suggest that the planets formed from gas pulled out of the Sun or any other star (except possibly a very cool one).

In the 1940s a new type of theory was proposed by the Soviet astronomer O. Y. Schmidt and was developed in some detail by H. Alfven, the Swedish astrophysicist who was later to win a Nobel prize for his work on the behavior of ionized gas in the solar magnetic field. Their idea was that once the Sun had formed, its gravitational field trapped material from the surrounding interstellar medium and this material then formed into planets. This general idea, called the **accretion theory** of solar system formation, never gained much popularity, because the problems with the evolutionary theory began to be solved at about the time the accretion theory was being developed. Thus most research efforts after this time were centered on the evolutionary theory.

Progress was made in the 1940s on the general question of how a contracting cloud could flatten into a disk and then break up into eddies that could form planets. The German physicist C. F. von Weizsäcker showed that the disk would tend to rotate differentially, that is, the inner parts would orbit the center faster than the outer regions, and that this would cause the disk to break up into eddies (Fig. 15.7). The work of von Weizsäcker even showed how the relative sizes of the eddies would vary with the distances of the planets

Figure 15.7 Von Weizsäcker's Theory of a Turbulent Disk. This sketch shows how a rotating disk would form eddies, according to the theory of C. F. von Weizsäcker.

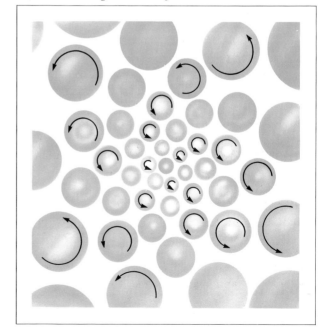

from the Sun and their varying sizes from one to the next. More detailed analyses of a stability of rotating disks were later carried out by a number of others, who showed that rather than forming regular eddies as envisioned by von Weizsäcker, a rotating disk would become clumpy, forming localized regions where the density was high enough to allow material to fall together gravitationally. This process would lead naturally to the breakup of the disk into a number of small, solid bodies. These in turn could later merge together to form the planets.

The breakthrough in understanding the Sun's slow rotation came in the early 1960s with the discovery of the solar wind. Because the Sun continuously expels a stream of charged particles, interplanetary space is filled with them, and the Sun's magnetic field tries to pull them along with it as the Sun rotates (Fig. 15.8). This creates a constant drag on the Sun (imagine trying to spin a pinwheel under water), and this drag, given long enough to act, slows the Sun's spin. If the Sun was born with a rapid rotation, this **magnetic braking** process could easily have slowed it to its present rate in the billions of years that have passed. In fact, it is quite likely that the density of charged particles in space around the Sun was much greater in the early days of the solar system, so the drag created as the Sun's magnetic field tugged on them was probably greater than it is now, and the braking could have taken place in a relatively short time.

With the understanding of the magnetic braking process, most of the major pieces had fallen into place, and all that remained was further refinement of the theory to account for the details. There are still many unanswered questions, but the general idea of how the solar system formed (to be described in the next section) is probably correct.

A Modern Scenario

Since the evolutionary theory postulates that the formation of the planets was a natural by-product of the Sun's formation, we begin our story with a reminder of how stars like the Sun are thought to form (described in greater detail in Chapter 7). Here we will briefly review the star formation process and point out how planets probably arose. In this discussion we will be guided by the need to explain the observed properties of the planets, as described at the beginning of this chapter.

The Sun is thought to have formed from the gravitational collapse of an interstellar cloud (Fig. 15.9), initially a very diffuse object composed chiefly of hydrogen (about 73 percent by mass), helium (about 25 percent), and traces of other common elements (see Appendix 3). The cloud containing all of the mass that was to become the solar system was so diffuse that its density was around 10 to 100 atoms per cm^3 (compare this with the Earth's atmospheric density of more than 10^{19} molecules/cm^3!) and so extended that it was perhaps ten light-years in diameter. The collapse may have occurred because a chance concentration of interstellar material became dense enough for gravity to overcome the pressure that tended to support the cloud, or it may have happened as a result of compression due to a disturbance traveling through the interstellar medium from a violent event such as a supernova explosion.

A combination of theory and observation of the for-

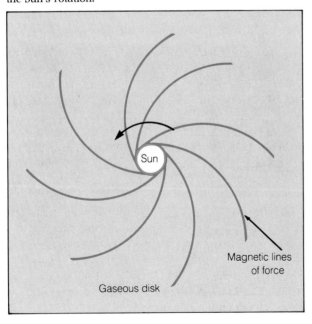

Figure 15.8 Magnetic Braking. The young, rapidly spinning Sun has the magnetic field structure shown, the field lines rotating with the Sun. The early solar system was permeated with large quantities of debris and gas, some of which was ionized and therefore subject to electromagnetic forces. The Sun's magnetic field exerted a force on the surrounding gas, which in turn created drag that slowed the Sun's rotation.

Figure 15.9 Regions of Star Formation. At left is the Orion nebula, in the "sword" of the constellation Orion, and at right is the region near the M supergiant Antares (bright object at right center). In both regions, newly-formed stars are known from infrared observations to lie within the dark, dusty clouds seen here. (left: © *1985 Anglo-Australian Telescope Board*; right: © *1979 Royal Observatory, Edinburgh*)

mation of other stars tells us how the collapse proceeded. The innermost portion of the cloud fell in on itself very quickly, leaving much of the outer material still suspended about the center. The rotation of the cloud sped up as its size diminished, and if the cloud had a magnetic field to begin with (most do), the field was intensified in the central part as a result of the condensation. The core of the cloud began to heat up due to the energy of impact as the material fell in. It eventually began to glow, first at infrared wavelengths (Fig. 15.10) and finally, after a prolonged period of gradual shrinking, at visible wavelengths. Nuclear reactions began in the center when the temperature and pressure were sufficiently high, and powered by these reactions, the Sun began its long lifetime as a star.

The steps in the formation of the Sun described up to this point are fairly well understood and have been observed to be taking place today in many cloudy regions of the galaxy. In a number of stellar nurseries, infrared sources are found embedded inside dark interstellar clouds, indicating that newly formed stars are hidden there, still heating up. The details of planet formation are a little sketchier, however, because we have no way to observe the process as it occurs. All that we know about planet formation has been inferred from observations of the end product: the present-day solar system.

As the central part of the interstellar cloud collapsed to form the Sun, the outer portions were forced into a disk shape by rotation, just as Kant had argued. At this stage there was an embryonic Sun surrounded by a flattened, rotating cloud called the **solar nebula** (Fig. 15.11). The inner portions of the nebula were hot, but the outer regions were quite cold.

 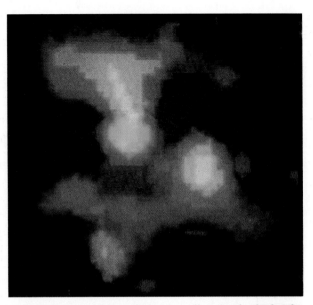

Figure 15.10 Star Formation. At left is seen evidence of new star formation in a dense, dark interstellar cloud. The bright trail at the upper left is a streak of glowing gas being heated by beams of light and shock waves from a newly-forming star that is hidden in the dark cloud. At right is an infrared image revealing a young star inside of another dark cloud; blue indicates the location of the new star, which is hotter than its surroundings (orange). (left: *photograph by B. J. Bok, made at the 4-m telescope at the Cerro Tololo Inter-American Observatory;* right: *R. D. Gehrz, J. Hackwell, and G. Grasdalen, University of Wyoming.*)

Figure 15.11 Steps in the Sun's Formation. An interstellar cloud initially very extended and rotating very slowly, collapses under its own gravitation. This happens most quickly at the center. As the collapse occurs, the internal temperature rises and the rotation rate increases. Eventually the central condensation becomes hot and dense enough to be a star, with nuclear reactions in its core.

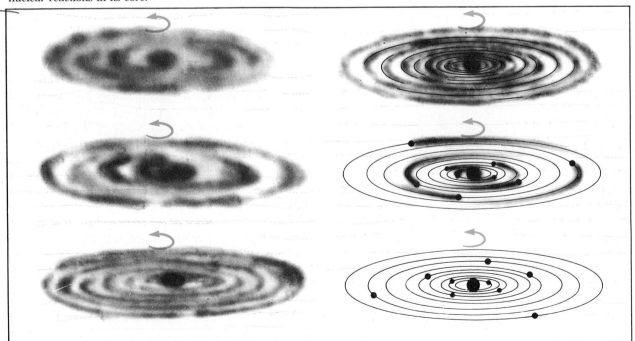

Astronomical Insight 15.1

The Explosive History of Solar System Elements

Even in the evolutionary theory now widely accepted, the development of the solar system required explosive events. The evolution of the universe itself may be viewed as the result of a series of explosions, including the initial one from which the universe has been expanding ever since. A variety of astronomical evidence (discussed thoroughly in Chapter 23) shows that the universe began some 10 to 15 billion years ago in a single point and at some time in the past began to expand outward. The initial conditions were so extreme in temperature and density that the birth of the universe must have been an explosive event, commonly called the **big bang.**

For a brief time during the early stages of the universal expansion, conditions were suitable for nuclear fusion reactions to occur, and the primordial substance was converted from a soup of subatomic particles to one of recognizable elements (primarily hydrogen and helium, with almost no trace of heavier elements). As the universe continued to expand, galaxies and individual stars eventually formed. Just as nuclear reactions in the Sun form a heavier element (helium) from a light one (hydrogen), similar reactions take place inside all stars, gradually enriching the universal content of heavy elements. Stars more massive than the Sun undergo more reaction stages, going from the production of helium to that of carbon, and from carbon to other elements. When these stars die, often explosively, the newly formed heavy elements are returned to interstellar space and are available for inclusion in new stars. Thus during the 5 to 10 billion years between the big bang and the formation of the solar system, the chemical makeup of our galaxy gradually changed, so that roughly 2 percent of the total mass was in the form of elements heavier than helium.

Since this roughly represents the composition of the Sun and the primordial material in the solar system, it might seem that there is no more to the story of the origin of the elements. There are compelling reasons to believe, however, that the formation of the solar system was influenced by at least two local explosive events. Certain atomic isotopes that are present in solar system material are the result of explosive nuclear reactions that must have taken place in relatively recent times (compared to the time that has passed since the big bang). These isotopes, one a form of aluminum (with 26 instead of the normal 27 neutrons and protons) and the other a form of plutonium (with 244 protons and neutrons instead if 242) are radioactive, with half-lives short enough that they would not have been present when the solar system formed unless they had been created relatively recently in explosive events. The plutonium isotope in question probably formed in a stellar explosion called a supernova that occurred a few hundred million years before the solar system formed, whereas the aluminum must have been created in a nearby supernova only a million years or so before the solar system was born.

The second of these inferred explosions may have played a crucial role in triggering the formation of our Sun and planetary system. We have said that the solar system formed from an interstellar cloud that collapsed. What we have not discussed, however, is what caused this collapse. Normally an interstellar cloud is quite stable and will not fall in on itself unless something happens to compress it. Various mechanisms can bring this about, and one of them is a shock wave from an explosive event such as a supernova. Thus our solar system may have been born as the direct consequence of the death of a nearby star in a singular event sometimes known as the "bing bang." We derive from this event not only some minor nuclear isotope abundance anomalies, but also the very existence of the solar system. It is interesting to note that even the evolutionary origin of the solar system discussed in this chapter must have been strongly influenced by what can only be described as catastrophic events.

Throughout the solar nebula, the first solid particles began to form, probably by the growth of the interstellar grains that were mixed in with the gas. For every element or compound, there is a combination of temperatures and pressure at which it condenses or "freezes out" of the gaseous form, in direct analogy with the formation of frost on a cold night on Earth. The **volatile elements** require a very low temperature in order to condense, so these materials tended to stay in gaseous form in the inner portions of the solar nebula but condensed to form ices in the outer portions. The elements that condense easily, even at high temperatures, are called **refractory elements,** and these elements formed the first solid material in the warm inner portions of the solar nebula. This material therefore consists of rocky debris containing only low abundances of the volatile species.

In due course, rather substantial objects resembling asteroids in size and composition were built up, and these objects are referred to as **planetesimals.** Mathematical analysis of the process by which particles orbiting the young Sun collided and built up their sizes shows that the growth of large planetesimals happened rather quickly (in only a few thousand years) and that most of the material in the disk would have gone into large planetesimals. Thus a stage was reached where collisions between large bodies were likely. Because the planetesimals were all orbiting the Sun in a disk, their relative speeds would have been rather small, so that when collisions took place, the colliding bodies could stick together. Even at low speeds (perhaps 10 km/sec typically) enough energy would be released in a collision between large planetesimals to partially vaporize or melt both objects, helping them merge into a single spherical object. In most cases, the planet that was formed in this way rotated in the same direction as the overall rotation of the disk, but it is possible that an off-center collision could alter the direction of spin. This may have happened during the formation of Venus and Uranus, accounting for the slow retrograde rotation of Venus and the large tilt of Uranus. As we will discuss in Chapter 16, many scientists now think that the Moon formed as the result of a grazing collision between the young Earth and a very large planetesimal.

The scenario just described apparently applies to the terrestrial planets, which seem to have formed in two stages: (1) the condensation of refractory elements, leading to the development of planetesimals and (2) the accretion of the planetesimals to form planets. The low quantity of volatile elements that characterizes the terrestrial planets was already established when the planets formed and then was exaggerated by the release of volatile gases during their early histories, when they underwent molten periods.

For the outer planets, there is more uncertainty about the sequence of events that led to their formation. These planets contain a much higher proportion of volatile gases, which would be expected due to the lower temperatures in the outer portions of the solar nebula. Thus, planetesimals that formed there would have contained higher relative abundances of gases like hydrogen and helium, and so would the planets that later formed through the coalescence of these planetesimals.

The most recent theory of the formation of the giant planets suggests that their cores formed from the coalescence of planetesimals, just as the terrestrial planets themselves formed. In the outer solar system, however, where the temperatures were much lower than in the inner portions of the system, the solid objects that formed from the coalescence of planetesimals were able to gravitationally trap gas from large volumes in their vicinities, and these planets therefore accreted extended gaseous atmospheres. Tidal forces due to the Sun played an important role for this scenario; for the inner planets, these forces helped prevent the accretion of gases from the surrounding nebula, but in the outer solar system, farther from the Sun, the tidal forces were weaker and did not prevent the trapping of gases by the giant planets.

This viewpoint is prompted in part by the extensive ring and satellite systems of the outer planets, which resemble miniature solar nebulae. The suggestion is that these planets, in accreting gas from their surroundings, formed disks much like the solar nebula itself. Gravitational and rotational effects would have caused the infalling material to form a disk in the equatorial plane of each planet, and then instabilities in the disk would have caused eddies to form, leading to the coalescence of satellites (Fig. 15.12). Ring systems formed close to the planets, where tidal forces prevented the small particles from merging to form satellites. The rapid rotation rates of the giant planets are explained in this view by the fact that angular momentum conservation would force these planets to spin more rapidly as the disks contracted (and by the fact that there was no magnetic braking process, in contrast with the early Sun).

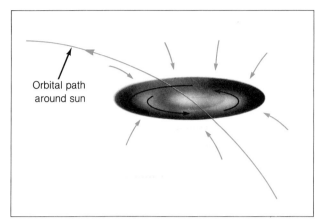

Figure 15.12 Formation of a Giant Outer Planet. Here a rotating disk has formed around a condensation in the outer solar system. Lumps in the disk grow to become satellites, except in the innermost portion where tidal forces prevent this, and a ring system forms instead. The entire system of planet, rings, and moons orbits the Sun.

It is interesting that the satellites and rings of Uranus lie in its equatorial plane, despite the large tilt of the planet. This indicates that Uranus accumulated its disk of icy debris after the event that caused its highly tilted orientation, and this scenario is consistent with the picture of outer planet formation just described. In this view, material trapped by Uranus would be forced into a disk in the equatorial plane of the planet, regardless of the orientation of that plane. Hence the tilt of Uranus must have been established before it trapped the material that was to form its satellites and rings.

The asteroids probably formed as planetesimals similar to those that eventually created the terrestrial planets but were prevented from coalescing into a planet (most likely by the gravitational effects of Jupiter, as discussed in Chapter 19). The comets probably formed farther out, as will also be described in Chapter 19.

Once the planets and satellites were formed, the solar system was nearly in its present state, except that the solar nebula had not totally dissipated. There was still a lot of gas and dust swirling around the Sun, along with numerous planetesimals that had not yet accreted onto planets. The Sun's magnetic field, pulling at ionized gas in the nebula, was very effective during this time in slowing the solar rotation through the magnetic braking process that we have already described. Meanwhile, the remaining planetesimals were floating

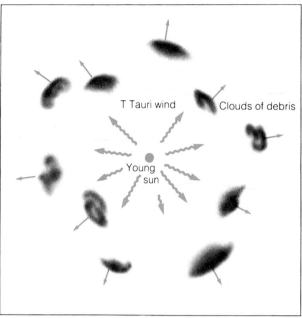

Figure 15.13 The T Tauri Phase. A star like the Sun is thought to go through a phase very early in its lifetime during which it violently ejects material in a high-velocity wind. This wind sweeps away matter left over from the formation process.

about, now and then crashing into the planets and their satellites. Many of the planets and satellites in the solar system still are covered with craters caused by the impacts of interplanetary debris in the early history of the system.

The leftover gas in the solar nebula was dispersed rather early, when the infant Sun underwent a period of violent activity, developing a strong wind that swept the gas and tiny dust particles out into space (Fig. 15.13). Only the larger solid bodies, the newly formed planets and the planetesimals, were left behind. This wind phase has been observed in other newly formed objects called **T Tauri** stars and is apparently in a natural stage in the development of a star like the Sun.

Once the majority of the remaining interplanetary debris was eliminated, either by the T Tauri wind or by accretion onto planetary bodies, the solar system looked much like it does today. The planets went their own ways, either continuing to evolve geologically, as most of the terrestrial planets have, or remaining perpetually frozen in their original condition, as the outer planets apparently have.

Are There Other Solar Systems?

As far as we know, the entire process by which our planetary system formed was a by-product of the Sun's formation, requiring no singular, catastrophic events. We might expect, therefore, that the same processes have occurred countless times as other stars were born, and therefore there may be many planetary systems in the galaxy. Even if we rule out double and multiple solar systems, in which perhaps all the material was either ejected or included in the stars, a vast number of candidates remain.

To detect planets orbiting distant stars is not easy. There are two distinct ways in which this might be done. One would be to directly observe planets in the vicinity of a star; and the other would be to detect the subtle motions of the parent star due to the gravitational influence of its planets. Direct detection is extremely difficult, because even the brightest planet is far dimmer than any star, and thus the star's brilliance is likely to drown out the feeble glow from a nearby planet.

The second method is also difficult but may have a better chance of succeeding. As we learned in Chapter 3, the Sun and a planet orbit their common center of mass, so that both are moving about this point. Because the Sun is so much more massive than any planet, the center of mass is always very close to the center of the Sun, and the Sun's orbit about this point is very small. Nevertheless, the Sun does undergo orbital motion, which could be detected from afar by a sufficiently sensitive technique. Therefore other stars that have planets should also undergo such motion, which we might hope to detect. One way to do this would be to very carefully measure the star's position (with respect to background stars) from time to time, to see whether it changes in a regular manner consistent with orbital motion. Another way is to measure the spectrum of the star's light at different times, to see whether there are regular Doppler shifts that could be due to orbital motion. (These two techniques correspond to the detection of astrometric and spectroscopic binary stars, as described in Chapter 6.) Both techniques are made difficult by the subtlety of the expected motions.

For a long time there was one example cited of a star that was thought to have a planet. A nearby star called Barnard's star was thought to be wobbling due to orbital motion caused by a massive planet, but a recent re-examination of the data has shown that there is no clear-cut indication of such motion. At the moment the case for the existence of a massive planet circling Barnard's star is not well established.

Infrared observations show some promise as a technique for detecting other planetary systems. The *IRAS* satellite mapped most of the sky in far infrared wavelengths and found several stars that are surrounded by disks containing tiny solid particles of the sort throught to be the precursors of planetesimals (Fig. 15.14). Another kind of infrared observation that is possible at ground-based telescopes involves the use of interferometry (see Chapter 4) to detect planets near stars. Even though a star is much brighter than a planet in visible wavelengths, the difference is not so great in the infrared, which is where a cool object such as a planet emits most strongly. The use of interferometry allows very closely spaced objects to be separated. Very recently, this technique revealed the existence of a planetlike body orbiting a nearby cool dwarf star. The data indicate that the dim companion is more massive and warmer than any planet in our solar system but is nevertheless not a star (such objects have been called **brown dwarfs**). It is hoped that continued refinement of infrared interferometric techniques will eventually reveal true planets orbiting nearby stars.

Other kinds of new technology may also provide new information in this area. The *Space Telescope*, for example, will orbit above the Earth's atmosphere and will therefore be free of the distortion and fuzziness that plague ground-based telescopes. It is possible that the *Space Telescope* will allow astronomers to see objects comparable to Jupiter, if they orbit nearby stars. The fine resolution of the *Space Telescope* will also allow more precise measurements of stellar motions than have been achieved from the ground, so this instrument will offer improved chances of detecting planets by that technique as well.

While there is substantial optimism that soon planets will be detected orbiting other stars, there is little hope that any of the techniques discussed here will directly detect small objects similar to the terrestrial planets. We may have to rely on the high probability that if massive, Jupiter-like planets are detected, other, Earth-like planets should also be present. The next few years should bring some very interesting results and speculation on these possibilities.

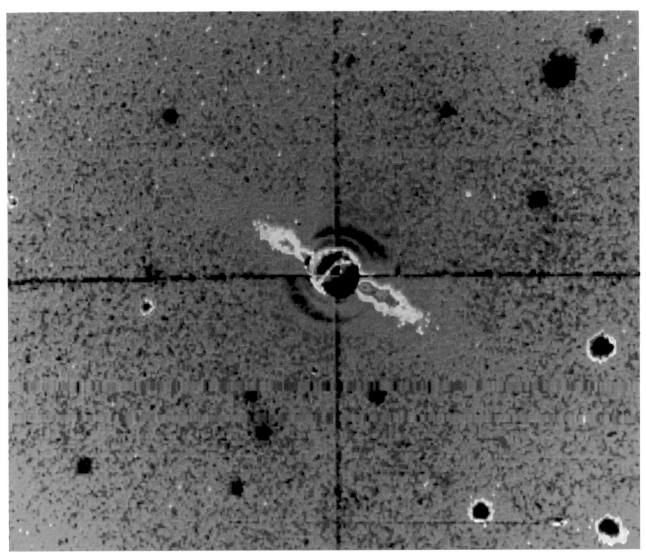

Figure 15.14 A Star with a Preplanetary Disk. This image of the star β Pictoris shows a disk of solid particles edge-on, much like the solar nebula. Infrared data have revealed a number of stars with surrounding material, supporting the suggestion that planetary systems may form around many stars. *(NASA/JPL, courtesy of B. A. Smith)*

Perspective

In this chapter we have outlined a natural, evolutionary process that seems to account for all of the observed properties of the solar system. The orbital characteristics of the planets, their compositions, the Sun's slow rotation, and the nature and motions of the interplanetary material have all been understood. Many details are yet to be worked out, but the overall scenario is probably essentially correct.

With this introduction, we are ready to examine the properties of the planets and other bodies in the solar system in more detail. In the next chapter we will discuss the Earth, our home planet, and its companion, the Moon.

Summary

1. The facts that must be explained by a successful theory of the formation of the solar system include the orbital and spin motions of the planets and satellites, the contrasts between the terrestrial and giant planets, the slow rotation of the Sun, and the existence and properties of the interplanetary bodies.
2. There are two general classes of theories: catastrophic theories and evolutionary theories.
3. Catastrophic theories require some singular event, such as a near-collision between stars, to distort the Sun by tidal forces and pull out matter that can then condense to form the planets.
4. Evolutionary theories postulate that the planets formed as a natural by-product of the formation of the Sun. These theories, although they are simpler, were not accepted for a long time because of their failure to explain the slow spin of the Sun.
5. When magnetic braking was understood in the 1960s, the evolutionary theory became most widely accepted.
6. In the modern theory, the planets formed from condensations in the solar nebula, the flattened disk of gas and dust that formed around the young Sun.
7. The terrestrial planets formed from the coalescence of planetesimals, solid, asteroidlike objects that condensed from the hot, inner portions of the solar nebula.
8. The giant planets formed when solid cores coalesced from planetesimals and then trapped surrounding gas to form their atmospheres. The extensive ring and satellite systems of these planets formed from disks of gas and solid particles in their equatorial planes.
9. In the evolutionary theory, it is expected that many other solar systems should exist around other stars, but so far none have been detected because of the difficulty of observing planets at interstellar distances.

Review Questions

1. We might have mentioned, as another fact that must be explained, the fraction of the total mass of the solar system that is contained in the Sun. Using data from tables in the appendixes, calculate the fraction of the total mass of the solar system that is contained in the Sun.
2. How might the present-day solar system be different if the Sun had been formed with no magnetic field?
3. Summarize the differences between the formation of the terrestrial planets and that of the giant planets in terms of the evolutionary theory described in the text.
4. Why does the evolutionary theory make it doubtful that a major tenth planet exists?
5. The giant planets rotate rather rapidly in comparison to the terrestrial planets. Why is this so? What forces might be acting to slow their rotations?
6. How much dimmer would Jupiter appear from the Earth if it orbited alpha Centauri, the nearest star? The distance to alpha Centauri is about 4.3 light years, or 270,000 AU.

Additional Readings

Beatty, J. K., O'Leary, B., and Chaikin, A., eds. 1981. *The new solar system.* Cambridge, Eng.: Cambridge University Press.

Cameron, A. G. W. 1975. The origin and evolution of the solar system. *Scientific American* 233(3):32.

Falk, S. W., and Schramm, D. N. 1979. Did the solar system start with a bang? *Sky and Telescope* 58(1):18.

Head, J. W., Wood, C. A., and Mutch, T. A. 1977. Geologic evolution of the terrestrial planets. *American Scientist* 65:21.

Reeves, H. 1977. The origin of the solar system. *Mercury* 7(2):7.

Sagan, C. 1975. The solar system. *Scientific American* 233(3):23.

Schramm, D. N., and Clayton, R. N. 1978. Did a supernova trigger the formation of the solar system? *Scientific American* 237(4):98.

Wetherill, G. W. 1981. The formation of the Earth from planetesimals. *Scientific American* 244(6):162.

Chapter 16

The Earth and Its Companion

Earthrise. (*NASA*)

Chapter Preview

The Earth's Atmosphere
The Earth's Interior and Magnetic Field
A Crust in Action
Exploring the Moon
   A Battle-Worn Surface and a Dormant Interior
The Development of the Earth-Moon System
   Formation of the Earth and Its Atmosphere
   Origin of the Moon
   History of the Moon

Of all the bodies in the heavens, the Earth has, of course, been the best studied (Table 16.1), although many mysteries remain. To understand the planet we live on has practical as well as philosophical importance. Furthermore, working at such close quarters with the object of study has numerous advantages, including the ability to observe in great detail and over long periods of time and the possibility of making direct experiments by probing and sampling the surface.

The Earth's satellite, the Moon, has been more thoroughly probed than any other remote body, and it, too, has much to tell us about the past and future of our planet. As we will learn in this chapter, the Earth and Moon probably formed together, but the Moon has been relatively little affected by the forces that have since shaped and altered the earth. Therefore the Moon is a useful laboratory for developing and testing theories of the history and formation of the Earth-Moon system.

We will begin this chapter by discussing the Earth and the Moon as separate entities, but we will draw parallels and contrasts between them, and at the end of the chapter we will treat their formation jointly.

The Earth's Atmosphere

The Earth is a brilliant sight as seen from space (Fig. 16.1). It has an overall blue color, due to scattering of light in its atmosphere and to its oceans. Much of it is white, where there is cloud cover, and here and there brown land masses can be seen. The Earth is so bright because it reflects a large fraction of the incident sunlight. The **albedo** of the Earth, the fraction of incoming light that is reflected back into space, is 0.31. This is a much higher value than is typical for bodies without atmospheres; the Moon, for example, has an albedo of only 0.07, so it is far outshone by the Earth when the two are viewed together from space.

Table 16.1 Earth

Orbital semimajor axis: 1.000 AU (149,600,000 km)
   Perihelion distance: 0.983 AU
   Aphelion distance: 1.017 AU
Orbital period: 365.256 days (1.000 year)
Orbital inclination: 0°0'0''

Rotation period: $23^h\ 56^m\ 4.1^s$
Tilt of axis: 23°27'

Diameter: 12,734 km (1.000 $D_\oplus$)
Mass: 5.974×10^{27} grams (1.000 $M_\oplus$)
Density: 5.517 grams/cm^3
Surface gravity: 980 cm/sec^2 (1.000 g)
Escape velocity: 11.2 km/sec

Surface temperature: 200–300 K
Albedo: 0.31 (average)

Satellites: 1

Figure 16.1 The Earth as Seen from Space. *(NASA)*

We begin by examining the thin layer of gas that surrounds the solid Earth. The Earth's atmosphere is composed of a variety of gases (Table 16.2), as well as a distribution of suspended particles called **aerosols.** The gases, which are evenly mixed up to an altitude of about 80 kilometers, are primarily nitrogen and oxygen, in the form of the molecules N_2 and O_2. It is nearly 80 percent nitrogen and about 20 percent is oxygen, and all the other gases exist only as traces, although in many cases they are important traces. Some of the components of the atmosphere are highly variable in quantity. Among these, water vapor (H_2O) is important because of its direct role in supporting life, and others, such as ozone and carbon dioxide, are also essential, but for less direct reasons. Ozone (O_3) is a form of oxygen in which three oxygen atoms are bound together instead of two. Unlike ordinary oxygen (O_2), ozone in the upper atmosphere acts as a shield by absorbing ultraviolet light from the Sun, which could be harmful to life forms if it penetrated to the ground.

Although there are obvious fluctuations in the temperature of the atmosphere, particularly near the surface, there is a definite, stable temperature structure as a function of height (Fig. 16.2). This structure includes several distinct layers in the atmosphere: the **troposphere,** from the surface to about 10 kilometers altitude, where the temperature decreases slowly with height and where the phenomena that we call **weather** occur; the **stratosphere,** between 10 and 50 kilometers, where the temperature increases with height due to the absorption of sunlight by ozone; the **mesosphere,** extending from 50 to 80 kilometers, where the temperature again decreases; and the **thermosphere,** above 80 kilometers, where the temperature gradually rises with height to a constant value above 200 kilometers. In layers where the temperature decreases with height, there can be substantial vertical motions of the air; in the layers where it gets warmer with height, the air is stable, with mostly horizontal motions.

The primary influences on the global motions of the Earth's atmosphere are heating from the Sun, which is most effective at low latitudes, and the rotation of the Earth. The heating creates regions where the air rises. It must later cool and fall somewhere else, and thus a pattern of overturning motions called **convection** is created (Fig. 16.3). Generally, air rises in the tropics and descends nearer the poles, although the situation is more complicated than that (Fig. 16.4). For example, the continents and the oceans often have different sur-

Table 16.2 Composition of Earth's Atmosphere

Gas	Symbol	Fraction*
Nitrogen	N_2	0.77
Oxygen	O_2	0.21
Water vapor	H_2O	0.001–0.028
Argon	Ar	0.0093
Carbon dioxide	CO_2	3.3×10^{-4}
Neon	Ne	1.8×10^{-5}
Helium	He	5.2×10^{-6}
Methane	CH_4	1.5×10^{-6}
Krypton	Kr	1.1×10^{-6}
Sulfur dioxide	SO_2	1.0×10^{-6}
Ozone	O_3	4×10^{-7}
Nitrous oxide	N_2O	3×10^{-7}
Carbon monoxide	CO	1.2×10^{-7}
Ammonia	NH_3	1×10^{-8}

*The listed values are fractions by number.

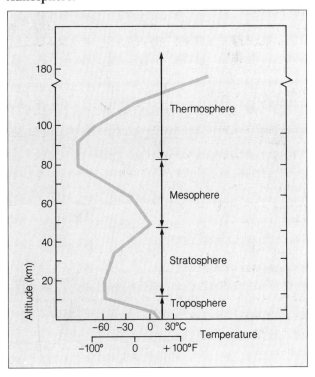

Figure 16.2 The Vertical Structure of the Earth's Atmosphere.

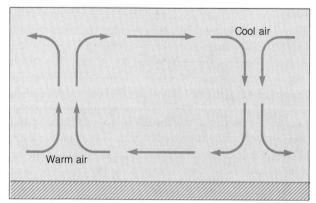

Figure 16.3 Convection in the Earth's Atmosphere. This is a simplified illustration of the principle of convection, in which warm air rises, cools, and descends.

face temperatures, which leads to pressure differences. In summer, when the land masses are warmer than the oceans, cool air tends to descend over the oceans, creating high-pressure regions; low-pressure regions are created where warm air rises over warm land. The situation may be reversed in winter, when the continents are cooler than the oceans. The Earth is the only planet that has large bodies of liquid water, but as we will see, temperature contrasts caused by different mechanisms on other planets can create similar air flows.

Figure 16.5 A Storm System on Earth. This is in the Southern Hemisphere, so the circulation about this low-pressure region is clockwise. *(NASA)*

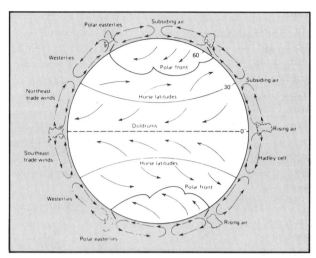

Figure 16.4 The General Circulation of the Earth's Atmosphere. *(From F. K. Lutgens and E. J. Tarbuck 1979, The Atmosphere: An Introduction to Meteorology, Fig 7.3, p. 150 [Englewood Cliffs: Prentice Hall], reprinted with permission of the publisher.)*

The rotation of the Earth causes circular flow patterns, particularly around low-pressure regions (Fig. 16.5). Air from the surrounding region tends to flow into a low-pressure area and is forced into a circular pattern known as a **cyclone.** By contrast, air flowing away from a high-pressure region is forced into a rotary pattern called an **anticyclone.** Cyclonic flows rotate counterclockwise in the Northern Hemisphere and clockwise in the Southern Hemisphere. Anticyclones rotate in the opposite direction. Cyclonic flows can become very intense and are responsible for the most severe storms, such as hurricanes. The locations and severity of these flow patterns change with the seasons because of the changing surface temperature pattern.

The Earth's Interior and Magnetic Field

The primary probes of the Earth's interior are **seismic waves** created in the Earth as a result of major shocks, most commonly earthquakes. These waves take three possible forms, called the **P, S,** and **L** waves. Studies of wave transmission in solids and liquids have shown that the P waves, which are the first to arrive at a site remote from the earthquake location, are **compres-**

sional waves, meaning that the oscillating motions occur parallel to the direction of motion of the wave, creating alternating regions of high and low density without any sideways motions (Fig. 16.6). Sound waves are examples of compressional waves. The S waves are **transverse,** or **shear,** waves, the vibrations occurring at right angles to the direction of motion. These waves require that the material they pass through have some rigidity, and unlike the P waves, they cannot be transmitted through a liquid. The L waves travel only along the surface of the Earth and hence do not provide much information on the deep interior.

By measuring both the timing and the intensity of these seismic waves at various locations away from the site of an earthquake, scientists can determine what the Earth's interior is like (Fig. 16.7). The speed of the P waves depends on the density of the material they pass through, and the distribution of the P and S waves reaching remote sites provides data on the location of liquid zones in the interior.

The general picture that has developed from these studies is that of a layered Earth, something like an onion (Fig. 16.8; Table 16.3). At the surface is a crust whose thickness varies from a few kilometers beneath the oceans to perhaps 60 kilometers under the continents. There is a sharp break between the crust and the underlying material, which is called the **mantle.** The mantle transmits S waves, so it must be solid, but on the other hand, it is known to undergo slow, steady flowing motions in its uppermost regions. Perhaps it is best viewed as a plastic material, one that has some rigidity but can be deformed given sufficient time.

The uppermost part of the mantle and the crust together form a rigid zone called the **lithosphere** (Fig. 16.9). The part of the mantle where the fluid motions occur, just below the lithosphere, is called the **asthenosphere.** Below the asthenosphere, there is a more rigid portion of the mantle that extends nearly halfway to the center of the Earth. The lower mantle is called

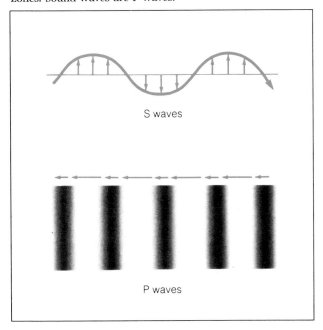

Figure 16.6 S and P Waves. An S, or shear, wave (top) consists of alternating motions transverse (perpendicular) to the direction of propagation. Waves in a tight string or on the surface of water are S waves. Compressional, or P, waves have no transverse motion but consist of alternating dense and rarefied regions created by motions along the direction of propagation. The arrows at the top of the waves represent the relative speeds in and between the dense zones. Sound waves are P waves.

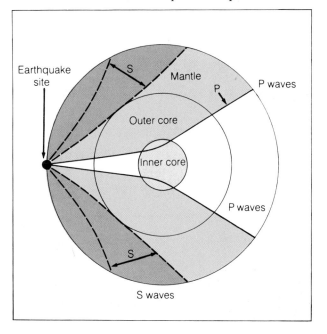

Figure 16.7 Seismic Waves in the Earth. This simplified sketch shows that P waves can pass through the core regions (although their paths may be bent), while S waves cannot. This observation led to the deduction that the outer core is liquid, because S waves, which require an elastic or solid medium, cannot penetrate liquids.

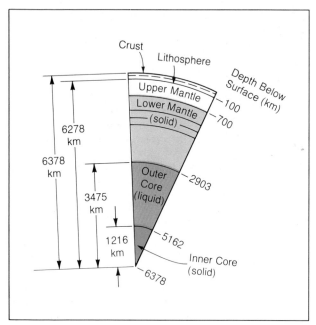

Figure 16.8 The Internal Structure of the Earth.

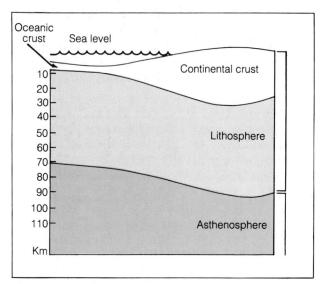

Figure 16.9 The Structure of the Earth's Outer Layers. This illustrates the relative thickness of the crust in continental areas as compared to the seafloor.

the **mesosphere** (not to be confused with the level in the Earth's atmosphere bearing the same name).

Beneath the mantle is the **core,** consisting of an outer core, between 2900 and 5100 kilometers in depth, and an inner core. Because S waves are not transmitted through the outer core, it is thought to be liquid. As a result, the inner core can be directly probed only with P waves. Because these waves travel more rapidly through the inner core, its density is thought to be greater than that of the outer core, and it is surmised that the innermost region is solid. It is known that the

core has great density, so its composition is thought to be primarily iron and nickel, the most abundant heavy elements.

The high density and probable metallic composition of the core are quite significant, for they show that the Earth has undergone **differentiation,** a sorting out of elements according to their weight. This can happen only when a planet is in a molten state, so it shows that the Earth once was largely liquid, probably early in its history, soon after it formed. The Earth's molten state was due to heating caused by its formation and to

Table 16.3 Earth's Interior

Layer	Depth (km)	Density (g/cm^3)	Temperature (K)	Composition
Crust	0–30	2.69–2.9	300–700	Silicates andoxides
Mantle				
Lithosphere	30–70	2.9–3.3	700–1200	Basalt, silicates, oxides
Asthenosphere	70–1000	3.9–4.6	1200–3000	Basalt, silicates, oxides
Mesosphere	1000–2900	4.6–9.7	3000–4500	Basalt, silicates, oxides
Core				
Outer core	2900–5100	9.7–12.7	4500–6000	Molten iron, nickel, cobalt
Inner core	5100–6378	12.7–13.0	6000–6400	Solid iron, nickel, cobalt

radioactivity. Several naturally occurring elements, such as uranium, thorium, and some forms of potassium, are radioactive, that is, their nuclei spontaneously emit subatomic particles and, over a long period of time, produce substantial quantities of heat. Differentiation has not occurred in all the bodies in the solar system, and this tells us something significant about their histories.

The fluid interior portions of the Earth give rise to its magnetic field. It is convenient to visualize the structure of the field by imagining magnetic lines of force connecting the two poles; these lines correspond to the lines along which iron filings lie when placed near a small bar magnet. In cross-sectional view, the Earth's magnetic field is reminiscent of a cut apple, but one that is lopsided, due to the flow of charged particles that constantly sweep past the Earth from the Sun (Fig. 16.10; also see the discussion of the **solar wind** in Chapter 5). The region enclosed by the field lines is called the **magnetosphere,** and it acts as a shield, preventing the charged particles from reaching the Earth's surface.

The axis of the Earth's magnetic field is aligned closely (within 11½°) with the rotation axis of the planet, but this has not always been the case. The past alignment of the magnetic field can be ascertained from studies of certain rocks that contain iron-bearing minerals whose crystalline structure is aligned with the direction of the magnetic field at the time the rocks solidified from a molten state. Thus traces of the Earth's ancient magnetic field, called **paleomagnetism,** can be deduced from the analysis of the magnetic alignments of rocks. Such studies reveal that the magnetic poles have moved about during the Earth's history and that the north and south magnetic poles have completely and rather suddenly reversed from time to time, so that the north magnetic pole has moved to the south and vice versa. This flip-flop of the magnetic poles seems to have happened at irregular intervals, typically thousands to hundreds of thousands of years apart.

The source of the Earth's magnetic field is a mystery, although it is nearly certain that it involves the iron-nickel core. A magnetic field is produced by flowing electrical charges, as when a current flows through a wire wound around a metal rod, and forms an electromagnet. Convection and the Earth's rotation may combine to create systematic flows in the liquid outer core; this would give rise to the magnetic field if the core material carries an electrical charge. This general type

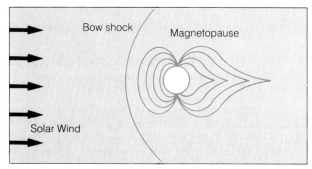

Figure 16.10 The Earth's Magnetic Field Structure. This cross-section shows the Earth's field and how its shape is affected by the stream of charged particles from the Sun known as the solar wind.

of mechanism is called a **magnetic dynamo.** The cause of the reversals of the poles is not well understood but presumably could be related to occasional changes in the direction of the flow of material in the core.

Let us look again at the magnetosphere and its ability to control the motions of charged particles. When the first U.S. satellite was launched in 1958, zones high above the Earth's surface were discovered to contain intense concentrations of charged particles, primarily protons and electrons. There are several distinct zones containing these particles, and they are now called the **Van Allen belts,** after the physicist who first recognized their existence and deduced their properties.

The charged particles, or ions, in the Van Allen belts were captured from space (primarily from the solar wind) and are forced by the magnetic field to spiral around the lines of force. The same situation occurs in the uppermost portion of the outer atmosphere, called the **ionosphere,** which extends upward from a height of about 60 kilometers. While a reversal of the Earth's magnetic poles is taking place, there is, for a short period of time, a much-weakened magnetic field and consequently a major disruption of the Van Allen belts and the ionosphere. The magnetosphere is greatly diminished, and charged particles from space are more likely to penetrate to the ground. These particles, particularly the very rapidly moving ones called **cosmic rays,** can cause important effects on life forms, including genetic mutations. The sporadic reversals of the Earth's magnetic field may have played a major role in shaping the evolution of life on the surface of our planet.

The ionosphere has important effects for us on the

surface, including enhanced radio communications and the beautiful light displays known as **aurora borealis** (northern lights) and **aurora australis** (southern lights). The aurorae are caused by charged particles entering the atmosphere and occur most commonly near the poles, where the magnetic field lines allow particles to penetrate closest to the ground. The ionosphere's reflects radio communications signals in the short-wave band, so that they can travel around the Earth. When there are fluctuations in the solar wind, particularly following solar flares, enhanced fluxes of charged particles entering the ionosphere from space can disrupt radio communications.

A Crust in Action

Nearly three-fourths of the Earth's surface is covered by water, the rest taking the form of several major continents. The basic substance of the Earth's crust is rock, and rocks are classified into three basic groups according to their origin: **igneous, sedimentary,** and **metamorphic.** Igneous rocks were formed from volcanic

Figure 16.11 Continental Drift. These maps show the distribution of the continents today and as it was some 200 million years ago.

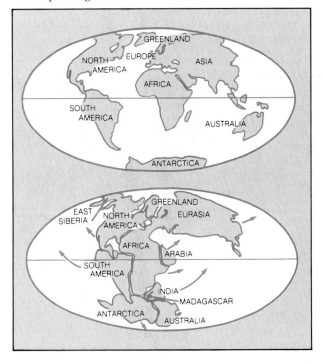

activity and consist of cooled and solidified **magma,** the molten material that flows to the surface during volcanic eruptions. Sedimentary rocks are formed from deposits of gravel and soil that have hardened, usually in layers where old seabeds or coastlines lay. Metamorphic rocks are those that have been altered in structure by heat and pressure created by movements in the Earth's crust. All three forms can be changed from one to another, in a continual recycling process. Rocks are also classified, according to their chemical compositions, as **minerals** of various types. The most common of all minerals are the silicates, which comprise some 90 percent of all rock on the Earth's surface.

Because of various evolutionary processes, the surface of the Earth is continually being renewed. While the age of the Earth is thought to be some 4.5 billion years, the ages of most surface rocks can be measured in the millions or hundreds of millions of years. The distinction between the age of a planet and the age of its surface is an important one, and it will reappear as we discuss the surfaces of other bodies.

The crust of the Earth is not static but is in constant motion. The continents themselves move about, and the world map is variable on geological timescales (Fig. 16.11). The evidence favoring continental drift has several forms, ranging from the obvious fit between land masses on opposite sides of the Atlantic to similarities of mineral types and fossils and the alignment of vestigial magnetic fields in rocks that once were together in the same place. In modern times more direct evidence, such as the detection of seafloor spreading away from undersea ridges and the actual measurement of continental motions using sophisticated laser-ranging techniques, has removed any trace of doubt that pieces of the Earth's crust are in motion.

From all of this evidence has arisen a theory of **plate tectonics,** which postulates that the Earth's crust (the lithosphere) is made of a few large, thin pieces that float on top of the asthenosphere (Fig. 16.12). Due to flowing motions in the asthenosphere, these plates constantly move about, occasionally crashing into each other. The rate of motion is only a few centimeters per year at the most, and the major rearrangements of the continents have taken many millions of years to occur. (The Americas and the European-African system became separated from each other between 150 and 200 million years ago.)

The driving force for the shifting of the plates is not well understood. The most widely suspected cause is convection currents in the mantle (Fig. 16.13). In the

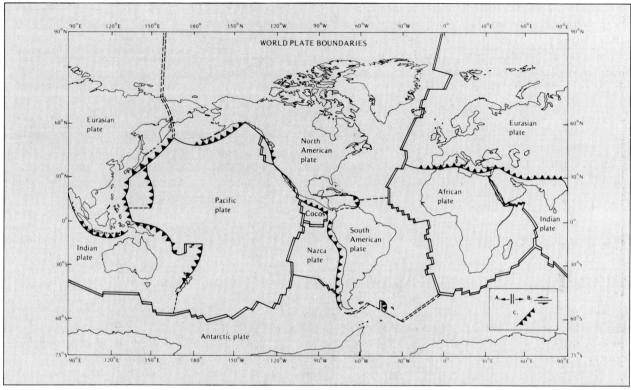

Figure 16.12 The Earth's Crustal Plates. This map shows the plate boundaries and the directions in which the plates are moving. *(From F. S. Sawkins, C. G. Chase, D. C. Darby, and G. Rupp 1978,* The Evolving Earth: A Text in Physical Geology, *Fig. 6.2, p. 163 [New York: Macmillan], © Macmillan. Reprinted with permission of the publisher.)*

Figure 16.13 Schematic of the Mechanisms of Continental Drift. This sketch shows how continental drift is responsible for uplifted mountain ranges and parallel undersea trenches, where one crustal plate sinks below another (subduction zone), and how a midocean ridge is built up where two plates move away from each other.

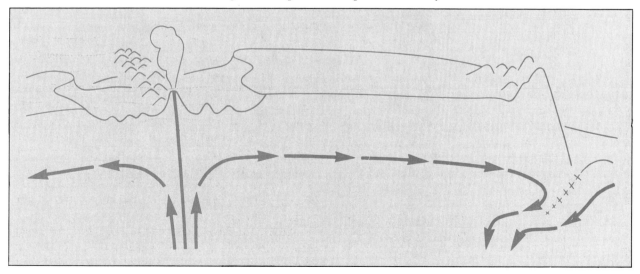

Astronomical Insight 16.1

The Ages of Rock

We have spoken of the ages of rocks and of the Earth itself but have said little about how these ages are measured. The most direct technique, **radioactive dating,** involves measuring the relative abundances of closely related atomic elements in rocks.

Recall from our earlier discussions of atomic structure (Chapter 4) that the nucleus of an atom consists of protons and neutrons and that the number of protons (the **atomic number**) determines the identity of the element. Different **isotopes** of an element have the same number of protons but differing numbers of neutrons. In most elements the number of protons is not very different from the number of neutrons, but there are exceptions. If the imbalance between protons and neutrons is large enough, the element is unstable, that is, it has a natural tendency to correct the imbalance by undergoing a spontaneous nuclear reaction. The reactions that occur in this way usually involve the emission of a subatomic particle, and the element is said to be **radioactive.** The emitted particle may be an **alpha particle** (that is, a helium nucleus, consisting of 2 protons and 2 neutrons) or, as is more often the case, an electron or a **positron,** a tiny particle with the mass of an electron but a positive electrical charge. When a positron or an electron is emitted, the reaction is called a **beta-decay;** at the same time, a proton in the nucleus is converted into a neutron (if a positron is emitted) or a neutron is changed into a proton (if an electron is released). If the number of protons in the nucleus is altered, the identity of the element is changed. A typical example of beta decay is the conversion of potassium 40 (^{40}K, with 19 protons and 21 neutrons in its nucleus) into argon 40 (^{40}Ar, with 18 protons and 22 neutrons).

Several elements are known to be radioactive, naturally changing their identity by emitting particles. The rate of change, expressed in terms of the **half-life,** is known in most cases. The half-life is the time it takes for half of the original element to be converted, and it may be as short as fractions of a second or as long as billions of years. The very slow reactions are useful in measuring the ages of rocks.

If we know the relative abundances of the elements that were present in a rock when it formed, measurement of the ratio of the elements in that rock at the present time can tell us how long the radioactive decay has been at work, that is, the age of the rock. In a very old rock, for example, there might be almost no ^{40}K, but a lot of ^{40}Ar. The decay of ^{40}K to ^{40}Ar has a half-life of 1.3 billion years, so from the exact ratio of these two elements in a rock, we can infer how many periods of 1.3 billion years have passed since the rock formed.

Other processes that are useful in dating rocks include the decay of rubidium 87 (^{87}Rb) into strontium 87 (^{87}Sr), which has a half-life of 47 billion years, and the decays of two different isotopes of uranium to isotopes of lead, ^{235}U to ^{207}Pb and ^{238}U to ^{206}Pb, with half-lives of 700 million years and 4.5 billion years, respectively. (These decays each involve several intermediate steps; the half-lives given represent the total time for all of those steps.)

These dating techniques have shown that the oldest rocks on the Earth's surface are about 3.5 billion years old and that ages of a few hundred million years are more common. The age of the solar system, and therefore of the Earth itself, is estimated (from isotope ratios in meteorites) to be about 4.5 billion years.

mantle, just as in the Earth's atmosphere, temperature differences between levels could cause an overturning motion. If a fluid is hot at the bottom and much cooler at the top, warm material will rise, cool, and descend again, creating a constant churning. The speed of the overturning motions depends largely on the **viscosity** of the fluid, that is, the degree to which it resists flowing freely. The Earth's mantle, as we have already seen, is sufficiently rigid to transmit S waves and must therefore have a high viscosity. Hence if convection is occurring in the mantle, it is reasonable to assume that the motion is very slow.

Whether convection is the cause or not, the effects of tectonic activity are becoming well known. Where plates collide, one may submerge below the other in a process called **subduction**, creating an undersea trench and volcanoes along the adjacent shoreline, or, particularly if both plates carry continents, the collision may force the uplifting of high mountain ranges. The lofty Himalayas are thought to have been created when the Indian plate collided with the Eurasian plate some 50 million years ago. The boundaries where plates either collide or separate are marked by a wide variety of geological activity. Earthquakes are attributed to the sporadic shifting of adjacent plates along fault lines, and volcanic activity is common where material from the mantle can reach the surface, most often along plate boundaries. It has been long recognized, for example, that a great deal of earthquake and volcanic activity is concentrated around the shores of the Pacific, defining the so-called Ring of Fire. Now it is understood that this region represents the boundaries of the Pacific plate.

Chains of volcanoes, such as the Hawaiian Islands, are also attributed to the action of plate tectonics. In several locations around the world, volcanic hot spots that lie deep and are fixed in place bring molten rock to the surface. As the crust passes over one of these hot spots, volcanoes are formed and then carried away, creating a chain of mountains as new material keeps coming to the surface over the hot spot. This process is still active, and the Hawaiian Islands, for example, are still growing. There is volcanic activity on the southeast part of the largest (and youngest) island, Hawaii. Furthermore, a new island, already given the name Loihi, is rising from the sea floor about 20 kilometers south of Hawaii. Its peak has already risen 80 percent of the way to the surface and only has about 1 kilometer to go, but it will take an estimated 50,000 years to break through.

Plate tectonics, then, accounts for many of the most prominent features of the Earth's surface. Of course, other processes, such as running water, wind erosion, and glaciation, are also important in modifying the face of the Earth.

Exploring the Moon

Seen from afar, the Moon (Table 16.4) is a very impressive sight, especially when full (Fig. 16.14). Its 0.5° diameter is pockmarked with a variety of surface features, the most prominent of which are the light and dark areas. The latter, thought by Renaissance astronomers to be bodies of water, are called **maria** (singular: **mare**), from the Latin word for seas. Closer examination of the Moon with even primitive telescopes also revealed numerous circular features called **craters** after similar structures found on volcanoes. The lunar craters are not volcanic, however, but were formed by the impacts of bodies that crashed onto the surface from space.

The pattern of surface markings on the face of the Moon is quite distinctive, and because the pattern never changes, it was recognized long ago that the Moon always keeps one face toward the Earth. The far side cannot be observed from the Earth, but it has now been thoroughly mapped by spacecraft (Fig. 16.15).

The lunar surface in general can be divided into lowlands, primarily the maria, and highlands which

---

Table 16.4 Moon

Mean distance from Earth: 384,401 km (60.4 $R_\oplus$)
  Closest approach: 363,297 km
  Greatest distance: 405,505 km
Orbital sidereal period: $27^d\ 7^h\ 43^m\ 12^s$
Synodic period (lunar month): $29^d\ 12^h\ 44^m\ 3^s$
Orbital inclination: 5°8′43″

Rotation period: 27.32 days
Tilt of axis: 6°41′ (with respect to orbital plane)

Diameter: 3476 km (0.273 $D_\oplus$)
Mass: 7.35×10^{25} grams (0.0123 $M_\oplus$)
Density: 3.34 grams/cm^3
Surface gravity: 0.165 Earth gravity
Escape velocity: 2.4 km/sec

Surface temperature: 400 K (day side); 100 K (dark side)
Albedo: 0.07

Figure 16.14 The Full Moon. This is a high-quality photograph taken through a telescope on the earth. *(Mt. Wilson and Las Campanas Observatories, Carnegie Institution of Washington)*

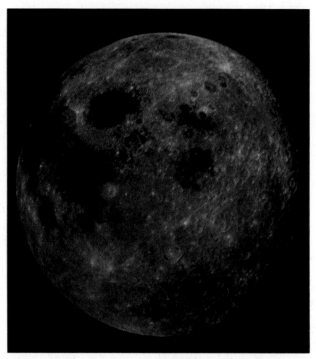

Figure 16.15 The Moon as Seen from Space. Here is a view obtained by one of the *Apollo* missions, showing portions of the near (left) and far (right) sides. On the left horizon is Mare Crisium, and in left center are Mare Marginis (upper) and Mare Smythii (lower). No maria are seen in the right-hand half of this image; there are almost none on the entire lunar far side. *(NASA)*

are vast mountainous regions, that are not as well organized into chains or ranges as the mountains on the Earth. With no trace of an atmosphere (because the escape velocity is so low that all the volatile gases were able to escape long ago), erosion due to weathering does not occur on the Moon, so large-scale features such as craters and mountain ranges retain a jagged appearance when seen from afar. The surface consists everywhere of loosely piled rocks ejected and redistributed by impacts. Other types of features seen on the Moon include **rays**, light-colored streaks emanating from some of the large craters, and **rilles**, winding valleys that resemble earthly canyons.

The Moon's mass is only about 1.2 percent of the mass of the Earth, and its radius is a little more than one-fourth the Earth's radius. The Moon's density, 3.34 grams per cubic centimeter, is below that of the Earth as a whole, more closely resembling the density of ordinary surface rock. This indicates that the Moon probably does not have a large, dense core and that it therefore most likely did not undergo strong differentiation. Infrared measurements showed, even before the space program allowed direct measurements, that the lunar surface is subject to hostile temperatures, ranging from 100 K ($-279°$F) during the two-week night to 400 K ($260°$F) during the lunar day.

Of course, all of the long-range studies of the Moon were almost instantly antiquated by the development of the space program, which has featured extensive exploration of the Moon, first by unmanned robots and then by manned landings (Figs. 16.16 and 16.17). The historic *Apollo 11* mission made the first manned landing on July 20, 1969. It was followed by five more manned landings, of which the last were *Apollo 17*, which took place in late 1972. Each mission incorporated a number of scientific experiments, some involving observations of the Sun and other celestial bodies from the airless Moon, but most devoted to the study of the Moon itself.

Astronomical Insight 16.2

Lunar Geography*

All the major topographic features on the Moon and many minor ones have names. These names are used in maps of the Moon and are recognized by the international community of astronomers. It is interesting to examine the history of how they were assigned.

Galileo was the first to name lunar features, because he was the first to look at the Moon through a telescope that enabled him to see them. He is responsible for the terms *maria* (the dark areas that he thought were seas) and *terrae* (the highlands). Following Galileo's lead, as telescope technology improved and more and more people examined the Moon, a number of early lunar cartographers chose names for prominent craters, mountain ranges, and other regions. Some of the names that were assigned, such as those devised by the court astronomer to the king of Spain (who named features after Spanish nobility), were doomed to be forgotten. Others, like those based on the suggestion of a German astronomer that lunar mountain ranges be named after terrestrial features, have persisted to the present time.

The most influential of the lunar cartographers of medieval times was the Italian priest Joannes Riccioli. With his pupil, Francesco Grimaldi, Riccioli named maria after human moods or experiences and craters after famous scientists. Thus, we have Mare Imbrium (Sea of Showers), Mare Tranquilitatis (Sea of Tranquility, where the first manned moon landing occurred), Mare Serenitatis (Sea of Serenity), and craters named Tycho, Hipparchus, and Archimedes. There were political overtones to the nomenclature of Riccioli and Grimaldi, who published their map in 1651, when the heliocentric hypothesis was not something that people publicly embraced. With this in mind, Riccioli and Grimaldi gave geocentrists like Ptolemy and Tycho large, prominent craters but assigned Galileo only a small one and placed the name of Copernicus on a crater in the Mare Nubium (Sea of Storms or Sea of Clouds).

Other names were added during the eighteenth and nineteenth centuries, but until 1921 there was no formal international agreement on a specific, uniform naming system. In that year, the newly formed International Astronomical Union designated a committee to oversee and standardize lunar nomenclature. There was no further controversy until 1959, when the Soviets obtained the first photographs of the far side of the Moon and promptly named numerous features after prominent Soviets. Many of these names (but not all) were later approved by the International Astronomical Union. Perhaps the most controversial was the Sea of Moscow, which reportedly was finally approved (amid laughter) when the Soviet representative declared—in response to criticism that it was inconsistent with tradition to name a mare after a city—that Moscow is a state of mind just as tranquility and serenity are.

Numerous additional features have been mapped and named as the sophistication of lunar probes has improved and as the manned exploration of the Moon has proceeded. The *Apollo* astronauts assigned rather colloquial names to features at each landing site, and because of all the attendant publicity these names became well known before there was time for the International Astronomical Union even to consider them for approval. Most of them were based on characteristics of the features themselves, such as a terraced crater called Bench and one with a bright rim dubbed Halo. Groups of craters were named for their pattern, such as a pair called Doublet and a group named Snowman, which included individual craters with names such as Head. Some of the features of the terrain mapped by the *Apollo* astronauts

(continued on next page)

Astronomical Insight 16.2

Lunar Geography*

(continued from previous page) received names of prominent scientists, in keeping with the tradition established by Riccioli and Grimaldi over three hundred years earlier. Many of the names invented by the *Apollo* astronauts were eventually approved by the International Astronomical Union, as were those of a dozen craters named for American and Soviet astronauts and cosmonauts.

About the time that lunar cartography was being resolved (after years of confusion brought on by the lunar exploration program), unmanned probes began to obtain images of Mercury, Venus, and Mars, and the whole problem of naming extraterrestrial landmarks arose again. Except for Mars, whose large-scale light and dark regions had been named on the basis of telescopic observations from Earth, no tradition existed for naming features on other planets. After considerable discussion, the International Astronomical Union adopted rules for doing so. New features on Mars will be named after towns and villages on Earth; Venus will bear the names of prominent women and radio and radar scientists; Mercury will have features named after people famous for pursuits other than science; and the outer planets and their satellites will have names assigned from mythology. With this forethought, perhaps it will be possible to map the rest of the solar system with a minimum of controversy.

*Based on information from El-Baz, F. 1979. "Naming the Moon's features created oceans of storms" *Smithsonian* 9(10):96.

Figure 16.16 Man on the Moon. The *Apollo* missions, six of which included successful manned landings on the Moon, represent humankind's only attempt so far to visit another world. *(NASA)*

Figure 16.17 The Lunar Rover. The later *Apollo* missions used these vehicles to travel over the Moon's surface, allowing the astronauts to explore widely in the vicinity of the landing sites. *(NASA)*

A Battle-Worn Surface and a Dormant Interior

Viewed on any size scale, from the largest to the smallest, the Moon's surface is irregular, marked throughout by a variety of features. We have already mentioned the maria, the large, relatively smooth, dark areas (Fig. 16.18). The maria appear darker than their surroundings because they have a relatively low albedo (that is, they reflect less sunlight). Despite their smooth appearance relative to the more chaotic terrain seen elsewhere on the Moon, the maria are marked here and there by craters.

Outside of the maria, much of the lunar surface is covered by rough, mountainous terrain. Even though the maria dominate the near side of the Moon, there are almost none on the far side, and the highland regions actually cover most of the lunar surface.

Figure 16.18 The Lunar "Seas." Here is a broad vista encompassing portions of three maria: Mare Crisium (foreground); Mare Tranquilitatis (beyond Mare Crisium); and Mare Serenitatis (on the horizon at upper right). These relatively smooth areas are younger than most of the lunar surface, having been formed by lava flows after much of the cratering had already occurred. *(NASA)*

Craters are seen everywhere (Figs. 16.19 and 16.20). They range in diameter from hundreds of kilometers to microscopic pits that can be seen only under intense magnification. In some regions the craters are so densely packed together that they overlap. The lack of erosion on the Moon allows craters to survive for billions of years, providing plenty of time for younger craters to form within the older ones. The fact that relatively few craters are seen in the maria indicates that the surface in these regions has been transformed in recent times, after most of the cratering had already occurred.

All of the craters on the Moon are **impact craters**, formed by collisions of interplanetary rocks and debris with the lunar surface, not by volcanic eruptions. This conclusion is based on the shapes of the craters, the central peaks in some of them, and the trails left by **ejecta**, material cast-off by the impacts (Fig. 16.21). The rays seen stretching away from craters are strings of smaller craters formed by the ejecta from the large, central crater.

Figure 16.19 Lunar Craters. This is a view of the far side of the moon, where there is little to interrupt the nearly total coverage by impact craters. The prominent crater in the upper center is Kohlschutter, named by Soviet scientists who first charted the far side of the moon. *(NASA)*

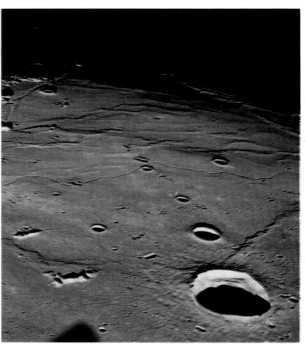

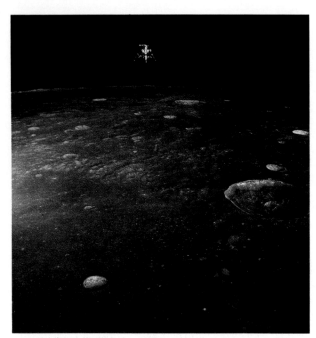

Figure 16.20 *Apollo 12* **Descending near the Crater Herschel.** *(NASA)*

Some of the lunar mountain ranges reach heights greater than any on Earth. They are more jagged than earthly mountains because there is no erosion, and they lack the prominent drainage features usually found in terrestrial ranges.

The **rilles** are rather interesting features that resemble dry riverbeds (Fig. 16.22). Apparently they were formed by flowing lava rather than water. In some cases there are even lava tubes that have partially collapsed, leaving trails of sinkholes. The rilles and the maria indicate that the Moon has undergone stages when large portions of its surface were molten.

The *Apollo* astronauts found a surface strewn in many areas with loose rock, ranging in size from pebbles to boulders as big as a house (Figs. 16.23 and 16.24). The rocks generally show sharp edges, due to the lack of erosion, and occasional cracks and fractures. In most cases the large boulders appear to have been ejected from nearby craters, and are therefore thought to represent material originally beneath the surface.

Figure 16.21 A Crater with Ejecta. This crater on the lunar far side is a good example of a case where material ejected by the impact has created rays of light-colored ejecta. Close examination of such features often reveals secondary craters formed by the impacts of debris blasted out of the lunar surface by the primary impact. *(NASA)*

Figure 16.22 A Rille. The sinuous feature meandering through this image is Hadley Rille, one of many such features seen in the maria. One of the *Apollo* missions landed at the edge of this rille, so that it could be examined at close range. The adjacent mountains are the Apennines. *(NASA)*

Figure 16.23 A Field of Boulders. This jumbled region is inside a relatively young crater; the rocks lying around were disrupted by the impact. *(NASA)*

Figure 16.24 A Large Boulder. Rocks on the lunar surface range in size from tiny pebbles to massive objects like this. *(NASA)*

Over eight hundred pounds of smaller lunar rocks were returned to Earth by the *Apollo* astronauts, giving scientists an opportunity to study the lunar surface characteristics in detail comparable to that possible for Earth rocks and soils (Fig. 16.25). Thus extensive chemical analysis and close-up observation of rocks in place on the Moon were both possible. The rock samples from the Moon are presently housed in numerous scientific laboratories around the world, where analysis continues. Specimens have also found their way into museums, and in at least one case (the Smithsonian's Air and Space Museum in Washington, D.C.) a lunar rock can be touched by visitors.

The lunar soil, called the **regolith**, consists of loosely packed rock fragments and small glassy mineral deposits probably created by the heat of meteor impacts. In addition to the loose soil, a few distinct types of surface specimens were recognized on the basis of morphology. The most common of these are the **breccias** (Fig. 16.25), which consist of small rock fragments cemented together and resembling chunks of concrete. There are similar kinds of rock on Earth, except that those on Earth are formed in stream beds where water plays a role in shaping them. The lunar breccias contain jagged, sharp rock fragments, and they are probably fused together by pressure created by meteor impacts.

All lunar rocks are pittted, with tiny craters called **micrometeorite craters** on the side that is exposed to space. These craters are formed by the impact of tiny bits of interplanetary material no bigger than grains of dust.

Radioactive dating techniques showed the lunar rocks to be very old by earthly standards: as old as 3.5 to 4.5 billion years. Rocks from the maria are not quite so old but still date back 3 billion years or more.

Some of the experiments carried out on the Moon by the *Apollo* astronauts were aimed at revealing its interior conditions by monitoring seismic waves caused by "earthquakes" on the Moon. The astronauts carried with them devices for sensing vibrations in the lunar crust and because it was not known whether natural moonquakes occurred frequently, they also brought along devices for thumping the surface to make it vibrate. It turned out that natural moonquakes do occur, although not with great violence. The seismic measure-

Figure 16.25 Moon Rocks. One of the thousands of lunar samples brought back to Earth by the *Apollo* astronauts, the rock at left is an example of a breccia. At right is a thin slice of a Moon rock, illuminated by polarized light, which causes different crystal structures to appear as different colors. *(NASA)*

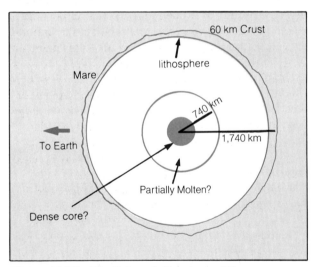

Figure 16.26 The Internal Structure of the Moon. This cross-section illustrates the interior zones inferred from seismological studies. The existence of a dense core is not certain. Note that the maria lie almost exclusively on the side facing the Earth, where the lunar crust is relatively thin.

ments continued after the *Apollo* landings, with data radioed to Earth by instruments left in place on the lunar surface.

The measurements showed that the regolith is typically about 10 meters thick and is supported by a thicker layer of loose rubble. The crust is 50 to 100 kilometers thick at the *Apollo* sites and may be somewhat thicker than this on the far side, where there are no maria (Fig. 16.26). Beneath the crust is a mantle consisting of a well-defined lithosphere, which is rigid, and beneath that an asthenosphere, which is semiliquid. The innermost 500 kilometers consist of a relatively dense core, but not as dense as that of the Earth.

The seismic measurements indicate no truly molten zones in the Moon at the present time, although temperature sensors on the surface discovered a substantial heat flow from the interior, probably caused by radioactive minerals below the surface.

The Moon has no detectable overall magnetic field, further evidence against a molten core. The chief distinction among the internal zones in the lunar interior is density, with the densest material closest to the center. Thus moderate differentiation has occurred in the Moon, implying that it was once at least partially molten.

There is no trace of present-day lunar tectonic activity, perhaps the greatest single departure from the geology of the Earth. The force that has had the greatest influence in shaping the face of the Earth has no role today on the Moon. Instead, the lunar surface is entirely the result of the way the Moon was formed (which may have involved some tectonic activity in the early stages) and the manner in which it has been altered by lava flows and by the incessant bombardment of debris from space.

The Development of the Earth-Moon System

While the formation of the Earth (and of planets in general) is reasonably well understood, there is a great deal of uncertainty in our picture of the Moon's formation. We will discuss the Earth's formation first, then delve into the complexities of the Moon's origin.

Formation of the Earth and Its Atmosphere

The age of the Earth-Moon system, estimated from geological evidence, is about 4.5 billion years. The Earth is thought to have formed from the coalescence of a number of **planetesimals,** small bodies that were the first to condense in the earliest days of the solar system. Most or all of the Earth was molten at some point during the first billion years. The Earth became highly differentiated during its molten period, as the heavy elements tended to sink toward the planetary core. At the same time, volatile gases were emitted by the newly formed rocks at the surface. These gases included hydrogen (H_2), ammonia (NH_3), methane (CH_4), and water vapor (H_2O) and formed the earliest atmosphere of the Earth. The water vapor collected on the surface, and thus from its earliest days our planet had seas. Carbon dioxide (CO_2) was probably abundant for a time but was eventually removed from the atmosphere by being absorbed into rocks. This absorption process, of critical importance to the further evolution of the Earth's atmosphere, depended on the presence of liquid water; if the early Earth had not had oceans, the carbon dioxide might never have left the atmosphere. This would have had fateful consequences for the Earth, which will become clear in our discussion of the evolution of Venus (Chapter 17).

Nearly all of the hydrogen escaped into space by the time the Earth was about a billion years old (about when the first simple life forms apparently appeared). The individual molecules in a gas move around randomly at speeds that depend on both the temperature of the gas and the mass of the molecules. At a given temperature the lightweight molecules move fastest and are therefore most likely to escape the planet's gravitational attraction. Thus the hydrogen in the early atmosphere of the Earth escaped while the heavier gases that now dominate the atmosphere were too massive to escape. Another gas that escaped early in the Earth's history is helium, the second most abundant element in the universe, but one that is so rare on earth that it wasn't discovered until spectroscopic measurements revealed its presence in the sun.

The critical reactions that led to the development of life on Earth must have taken place before all the hydrogen escaped, because the types of reactions that were probably responsible involve this element. The earliest fossil evidence for primitive life dates back at least 3 billion years, when some hydrogen was still left.

Essentially no free oxygen was present in the atmosphere until the development of life forms that released the element as a by-product of their metabolic activities. Most plants release oxygen into the atmosphere during photosynthesis; as a supply of this element built up, the opportunity arose for complex animal forms to evolve. Besides providing the oxygen necessary for the metabolisms of living animals, the buildup of oxygen created a reservoir of ozone (O_3) in the upper atmosphere, which in turn began to screen out the harmful ultraviolet rays from the Sun. Once life forms gained a toehold on the continental land masses, the process of converting the atmosphere to its present state began to accelerate. Soon nitrogen from the decomposition of organic matter began to be released into the air in large quantities (some nitrogen was already present due to volcanic activity), and by the time the Earth was perhaps 2 billion years old the atmosphere had reached approximately the composition it has today.

By this time also the mantle had solidified and the crust had hardened. (The oldest known surface rocks are nearly 3.5 billion years old.) The interior has remained warmer than the surface, because heat can only escape slowly through the crust and because radioactive heating of the interior took place over a long period of time and is probably still effective today.

Origin of the Moon

The Moon's beginnings are much more obscure than the Earth's, despite the close-up examination that has been afforded by the *Apollo* program. There are several different theories regarding the origin of the Moon, including the hypotheses that the Moon and the Earth formed together out of the same material, that the Moon consists of material that split off from the Earth some time after its formation, or that the moon formed elsewhere in the solar system and was later captured by the Earth's gravitational field. For a long time, the simultaneous formation theory was most widely accepted because of difficulties with the other two theories, but careful analysis of lunar samples has recently swung the opinions of many lunar researchers in other directions.

The Moon has major chemical similarities with the Earth but also some important differences. Certain forms of oxygen and other elements, for example, are very similar in the two bodies, arguing for a simultaneous formation, but the Moon has a lower overall iron content and a higher abundance of the so-called **refractory** elements, which are not easily vaporized. This low abundance of easily vaporized **volatile** elements together with evidence found in the radioactive isotopes and apparent melting history of lunar surface rocks, indicate that the entire surface layer of the Moon, down to a depth of some 300 kilometers, was molten for a while during the first few million years after the Moon's formation. These facts argue that the Moon was somehow subjected to much more heating early in its history than the Earth.

There is little doubt that the Moon formed from a coalescence of planetesimals in a manner similar to the Earth's formation. The major uncertainties concern *where* the moon formed and how it came to be orbiting the earth. Until recently, the most widely accepted idea was that the Moon and the Earth formed together, that is, that debris surrounding the infant Earth merged to form the Moon (Fig. 16.27). Because of chemical contrasts, some have argued that the lunar material was actually once part of the young Earth's outer layers, so that differentiation, during which much of the Earth's iron sank to the center, could explain the relative lack of this element in the Moon. According to this view, the material was ejected from the Earth following differentiation and then heated, reducing the abundances of volatile elements. (This heating must have occurred in any case, for the Moon does have a lower abundance of volatile elements.) The major problem with this idea is explaining *how* the Moon was then ejected from the Earth; there is no known mechanism that would do this.

The capture hypothesis has long been subjected to a similar criticism, because if two bodies simply come near each other as they travel their separate orbits about the Sun, they cannot be trapped into mutual orbit unless somehow excess kinetic energy is lost. This could happen in a collision, however, and the modern capture theories invoke such an event. If the Moon, having formed elsewhere in the solar system (probably near the center, where high temperatures could account for its chemical differences with the Earth), encountered the Earth and collided with debris surrounding it, it could have been slowed enough to be trapped into orbit. While such a collision may seem unlikely, it cannot be ruled out, and this scenario has the advantage of explaining most of the chemical differences between the earth and the moon. It is perhaps noteworthy that among the inner planets only the Earth has any significant satellite, so it may be that such planets ordinarily form without moons. (Mars has two tiny moonlets that are almost certainly captured asteroids.)

A very recent suggestion, beginning to be viewed as a real possibility, is that the Moon formed as the result of a grazing collision between the young Earth and a very large planetesimal, perhaps the size of Mars. Such

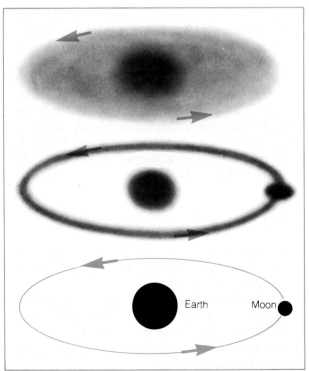

Figure 16.27 Coeval Formation of the Earth and Moon. In this theory, the Moon condensed from a disk of debris that orbited the newly formed Earth.

a collision could have created a disk of debris around the Earth, which could then have coalesced to form the Moon.

History of the Moon

We do know something about the Moon's history since its formation. For example, it is known that the Earth and Moon were once closer together than they are now and that they gradually separated as the Moon's rotation was slowed by tidal forces. Because the total amount of angular momentum in a system must remain constant, the loss of angular momentum as the Moon's spin slowed was compensated by its increased distance from the Earth, which increased the angular momentum due to its orbital motion.

After the Moon's formation, its crust was molten for a time due to heating caused by the heavy bombardment of the surface by objects from space. There undoubtedly was a great deal of gas escaping from the surface during this period, but the Moon's low escape velocity allowed all the gases to dissipate into space, and thus the Moon has no atmosphere. There was little differentiation because the Moon's low mass provided little gravitational force to bring about significant separation of the heavy elements. After a time, the crust cooled and hardened, but gradually the interior became molten due to radioactive heating.

The next major stage in the Moon's life involved extensive volcanic activity. Probably triggered by large impacts from space that cracked the young Moon's crust, molten material from the interior flowed over large areas of the surface, particularly the lowlands, and formed the maria. It appears possible that the extensive maria on the near side of the Moon (the only significant maria the Moon has) were created as the result of a single tremendous impact. This also caused the near side of the Moon to have a somewhat higher surface density than the far side, which helped the Earth's tidal forces alter the Moon's spin. The Moon's original rotation rate was apparently much faster than it is today, but the Earth's tidal force has slowed the rotation, and now the heavy side of the Moon is pointed permanently toward the Earth. The internal friction that had been acting to slow the Moon's rotation has been eliminated. Synchronous rotation such as this is the rule for all satellites that are reasonably close to their parent planets.

Throughout the early history of the Moon, it was continually bombarded, as was the Earth, by the rocky material still floating around the young solar system. Erosion and tectonic processes have erased most evidence of cratering on the Earth, but once the Moon's crust had cooled, craters formed by the impacts of material hitting the surface were able to survive. The heavily cratered regions of the Moon that we see today were shaped at that time, within the first billion years after the Moon had formed. When the lava flows that created the maria occurred, they obliterated craters in the lowlands, and the few craters in the maria were formed by meteorites that hit the Moon after the lava flows had cooled.

Since the creation of the maria some 3 billion years ago, little else has happened to modify the Moon's structure. The rate of cratering decreased as the interplanetary debris either escaped the solar system or got swept up by the planets and other large bodies, so only a moderate amount of cratering has occurred since the maria formed. The Moon's interior gradually cooled, eventually reaching its present nonmolten state.

Perspective

We have itemized the overall properties of the Earth and the Moon, the two best-studied objects in the solar system. In doing so we have discussed many principles of planetary structure and evolution that will be applied to the other planets in the coming chapters. We have seen that the Earth and Moon formed together but have diverged radically in their subsequent evolutions, the Earth remaining a vital, dynamic body while the Moon has been nearly dormant for billions of years. In the next chapter we discuss the other inner planets, Venus, Mars, and Mercury.

Summary

1. The Earth's atmosphere is 80 percent nitrogen and about 20 percent oxygen, with only traces of water vapor and carbon dioxide.
2. The atmosphere is divided vertically into four temperature zones: the troposphere, the stratosphere, the mesosphere, and the thermosphere.
3. The Sun's heating and the Earth's rotation create the global wind patterns.

4. The Earth's interior, explored with seismic waves, consists of a solid inner core, a liquid outer core, the mantle, and the crust.
5. The Earth's magnetic field is probably created by currents in the molten core; it traps charged particles in zones above the atmosphere called radiation belts.
6. The crust is broken into tectonic plates that shift around, accounting for continental drift and for most of the major surface features of the Earth.
7. The lunar surface consists of relatively smooth areas called maria and mountainous regions; it is marked everywhere by impact craters.
8. The lunar soil is called the regolith, and rocks on the surface are all igneous, mostly silicates, with low abundances of volatile gases.
9. Seismic data show that the Moon has a crust 50 to 100 kilometers thick, a mantle, and a core extending about 500 kilometers from the center.
10. The Moon has no present-day tectonic activity and no magnetic field, which indicate that there is probably no liquid core.
11. The Earth's evolution from a largely molten planet with a hydrogen-dominated atmosphere to its present state was caused by the presence of liquid water on its surface, the loss of lightweight gases into space, and the development of life forms on its surface, which helped convert the atmospheric composition to nitrogen and oxygen.
12. The Moon and the Earth have significant chemical contrasts, which argue against a common origin for the two. The true mechanism for the moon's formation is unknown.
13. The Moon's evolution consisted of a molten state, followed by hardening of the crust and subsequent large-scale lava flows that created the maria. Since that time (about 1 billion years after the Moon's formation), the Moon has been geologically quiet.

Review Questions

1. In the first photos from deep space that showed the Earth and the Moon together, the Earth was very bright but the Moon was dark and difficult to see. Both receive about the same intensity of sunlight. Explain why the Moon looked so much darker.
2. Explain why the sun heats the Earth's surface more effectively near the equator than near the poles.
3. Suppose the Earth's interior had a liquid layer just below the crust but above the mantle, so that no S waves could penetrate into the deeper zones. Would it be possible in that case to learn anything at all about the interior from observations of seismic waves?
4. If exploration of a planet revealed no sedimentary or metamorphic rocks, what would you conclude about the geological history of that planet?
5. Would you expect a planet with no molten core to have radiation belts?
6. Suppose you examine a portion of the Moon where craters are so closely packed together that they actually overlap. How could you determine which were formed most recently?
7. Summarize the similarities and contrasts between the chemical composition of rocks on the Earth and on the Moon.
8. Why is the Moon's surface covered with craters, while there are very few impact craters on the Earth?
9. Summarize the differences in the way the surface of the Moon and that of the Earth have been shaped.
10. Explain in your own words, using information from Chapter 3, how the orbit and spin of the Moon have evolved since the formation of the Earth-Moon system.

Additional Readings

Anderson, D. L. 1974. The interior of the Moon. *Physics Today* 27(3):44.
Battan, L. J. 1979. *Fundamentals of meteorology.* Englewood Cliffs, N.J.: Prentice-Hall.
Ben-Avraham, Z. 1981. The movement of continents. *American Scientist* 69:291.
Brownlee, S. 1985. The whacky theory of the Moon's birth. *Discover* 6(3):65 (the Moon's origin attributed to a giant impact).
Cadogan, P. 1983. The Moon's origin. *Mercury* 12(2):34.
Carrigan, C. R., and D. Gubbins. 1979. The source of the Earth's magnetic field. *Scientific American* 240(1):118.
Goldreich, P. 1972. Tides and the Earth-Moon system. *Scientific American* 226(4):42.

Leet, L. D., S. Judson, and M. E. Kauffman. 1978. *Physical geology*. 5th ed. Englewood Cliffs, N.J.: Prentice-Hall.

Siever, R. 1975. The Earth. *Scientific American* 233(3):82.

Toon, O. B., and S. Olson. 1985. The warm Earth. *Science 85* **6**(8):50.

Wahr, J. 1986. The Earth's inconstant rotation. *Sky and Telescope* **71**(6):545.

Wilson, J. T., ed. 1972. Continents adrift. San Francisco: W. H. Freeman. (A collection of readings from *Scientific American*)

Wood, J. A. 1975. The Moon. *Scientific American* 233(3):92.

Chapter 17

The Terrestrial Planets

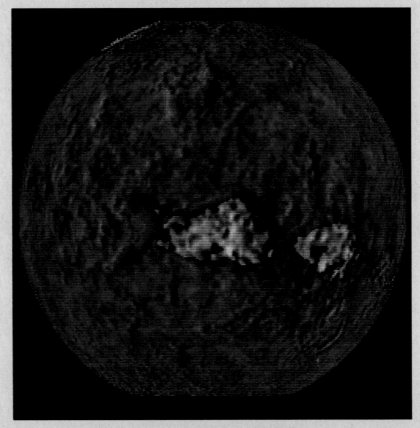

A relief map of Venus. *(NASA/JPL)*

Chapter Preview

Observations of the Terrestrial Planets
General Properties
Atmospheres of the Terrestrial Planets
    Composition of the Terrestrial Atmospheres
    Circulation and Seasonal Effects
Surfaces and Interiors
Evolution of the Terrestrial Planets

We found in our discussion of the solar system as a whole (Chapter 15) that the planets can be divided into two general categories: the terrestrial planets, which are relatively small, dense, rocky bodies (Table 17.1), and the gaseous giant planets, which are large, relatively diffuse bodies. In the preceding chapter (Chapter 16) we discussed in detail the prototype of the terrestrial planets, our home, the Earth. In this chapter we describe the other three terrestrial planets, keeping the Earth in mind for comparison.

This chapter is organized around the general types of processes and phenomena that occur on and in the planets. We begin with a section on observations of the terrestrial planets and then move on to discussions of their properties.

Observations of the Terrestrial Planets

All of the terrestrial planets are visible to the unaided eye and have been known to humans since antiquity. The two inner planets, Mercury and Venus, never are seen far from the Sun in the skies of Earth, and both have been known as the evening star or the morning

Figure 17.1 Mercury as Seen from the Earth. The planet is never far from the Sun and is therefore difficult to observe. *(NASA)*

star, depending on whether they rise just before sunrise or set just after the Sun. Mercury is so close to the Sun (its greatest elongation is only 28°) that many people never see it despite its brilliance. It is easily observed only very near sunset or sunrise. Many of the best telescopic observations are made during daytime, when the planet's image is not so blurred by atmospheric effects as it is when the planet is near the horizon (Fig. 17.1). Venus (Fig. 17.2) wanders farther from the Sun (its greatest elongation is 47°) and is therefore available for easy Earth-based observation more frequently than Mercury. Venus is so bright that it has sometimes been mistaken for a man-made light in the sky. Mars (Fig. 17.3) is quite bright, has a distinctive reddish color, and can be seen in any direction with respect to the Sun, since its orbit lies outside of the Earth's orbit. Mars was the best-studied of the terrestrial planets in ancient and medieval times, and observations of its motions inspired many attempts to unravel the nature of

Table 17.1 The Terrestrial Planets

Planet	Semimajor Axis	Sidereal Period	Mass*	Diameter*	Density	Surface Temperature	Escape Speed
Mercury	0.387 AU	0.241 yr	0.0558	0.382	5.50 g/cm^3	100–700 K	4.3 km/s
Venus	0.723	0.615	0.815	0.951	5.3	730	10.3
Earth	1.000	1.000	1.000	1.000	5.517	200–300	11.2
Mars	1.524	1.881	0.107	0.531	3.96	130–290	5.0

*The masses and diameters are given in units of the Earth's mass and diameter, which are 5.974×10^{27} grams and 12,734 km, respectively.

Figure 17.2 The Crescent Venus. This photograph, taken from Earth, shows Venus when it is near inferior conjunction; thus we see only a sliver of its sunlit side. *(NASA)*

Figure 17.3 The Earth-based View of Mars. *(Lick Observatory Photograph)*

the solar system (of particular importance was Kepler's analysis of the orbit of Mars, which led to his discovery of the laws of planetary motion).

From Earth-based telescopic observations, a wealth of information has been collected on each of the terrestrial planets. Visible-light observations have revealed little about Venus, which is completely cloud-covered, but they have shown that Mercury has a slow rotation rate and has no atmosphere and have provided many important findings on Mars. Besides the red color, which is a clue to the nature of the Martian surface materials, these observations have revealed seasonal variations in surface coloration (Fig. 17.4) and have shown intricate surface markings that were once imagined to be artificial canals (Fig. 17.5). In addition, spectroscopic observations of Venus and Mars have provided some information on the gases in their atmospheres.

Radio data have allowed Earth-bound astronomers to probe the planets in ways not possible with visible-light telescopes. For example, radar measurements (Fig. 17.6) revealed the true rotation period of Mercury, which was found to be exactly equal to two thirds of its orbital period. This is not a coincidence—it is due to tidal forces exerted on Mercury by the Sun. Mercury has a highly noncircular orbit in which the Sun is off-center (Fig. 17.7), so the Sun's tidal force on Mercury is far greater at the closest approach than at other points in Mercury's orbit. Because Mercury apparently is more dense on one side than the other, the tidal forces have acted over time to alter Mercury's spin rate so that the heavy side of the planet is always pointed directly toward or away from the Sun at closest approach. The result is that Mercury spins three times for every two trips around the Sun (Fig. 17.7). This is similar to the synchronous rotation of the Moon, except that in the Moon's case, where the orbit is more circular and the tidal forces more even throughout the orbit, the spin has been slowed until it matches the orbital period. Both the 3:2 spin-to-orbit ratio of Mercury and the 1:1 ratio of the Moon are examples of **spin-orbit coupling.**

Radar observations of Venus provided information on its spin rate as well and also gave astronomers another surprise. The spin of Venus is very slow, with a period of 243 days, and the planet rotates in the retrograde direction, counter to the direction of most of the

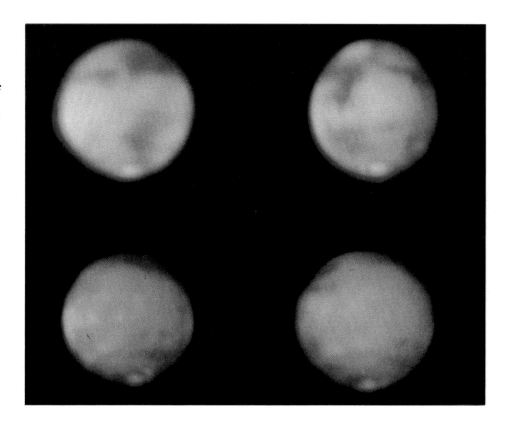

Figure 17.4 Variations in the Martian Surface Markings. This series of photos shows the changes in the polar ice caps and surface markings that occur with the seasons on Mars. *(Mt. Wilson and Las Campanas Observatories, Carnegie Institution of Washington)*

other spin and orbital motions in the solar system. The reason for this slow backward spin is not known, but it is suspected to be the result of an impact by a large planetesimal late in the formation process of the planet. The other surprise that came from radar observations of Venus was the discovery that its surface is very hot, hotter even than the surface of Mercury. The high concentration of carbon dioxide (CO_2) in the atmosphere of Venus creates a **greenhouse effect,** which traps heat near the surface. Carbon dioxide allows sunlight to penetrate to the ground and heat the surface of Venus, but it does not allow the infrared radiation that results from this heating to escape easily (Fig. 17.8). The same effect occurs in the Earth's atmosphere but to a much smaller extent, because we have only a very low concentration of carbon dioxide and other gases

Figure 17.5 The Martian "Canals." These sketches, made by various early observers of Mars, show the linear features eventually interpreted to be artificial channels in which water flowed. *(Historical Pictures Service, Chicago)*

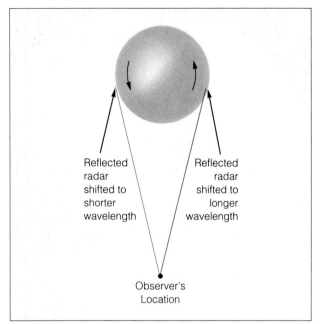

Figure 17.6 Measuring Planetary Rotation. To measure how fast a planet is spinning, scientists use radar to measure the speed of approach of one edge of the planetary disk and the speed of recession of the other edge. This provides a measurement of the rotational speed, and from this speed and the radius of the planet, the rotational period can be derived.

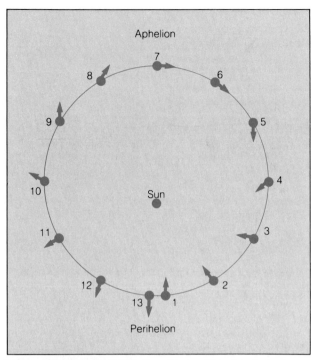

Figure 17.7 Spin-Orbit Coupling. This sketch shows how Mercury spins one-and-a-half times while completing one orbit around the Sun. At perihelion it always has the heavy side facing directly toward or away from the Sun.

that are efficient absorbers of infrared radiation. As we will discuss later in this chapter, the greenhouse effect on Venus gives us a valuable insight into how conditions could change on the Earth if the atmospheric composition is altered significantly.

Since the 1960s the terrestrial planets have been targets for space-based observations and probes (Table 17.2). The U.S. *Mariner* series of spacecraft visited all three of the terrestrial planets, beginning with Venus in 1962. This planet has also been a target for intensive Soviet exploration, culminating in a series of landers that withstood the high temperatures and corrosive atmosphere long enough to send back photos of the surface. Today a U.S. probe called *Pioneer-Venus* still orbits the planet, returning data from ultraviolet observations (this probe also observed Halley's comet in 1986, providing a rather different perspective from the Earth-based view; see Chapter 19).

Mercury has been visited by only one spacecraft, the *Mariner 10* mission, but the path of this probe was adjusted so that it flew past the planet no less than three times, providing extensive maps of its surface (actually, only half of the planet's surface was mapped, because Mercury always had the same face toward the Sun when *Mariner 10* flew past). Mars has been visited by several U.S. probes. One of the most spectacular successes was *Mariner 9,* which went into orbit about the red planet in 1971. When it arrived, astronomers were frustrated by the fact that the entire planet was shrouded in windblown dust so that the surface could not be seen, but within several weeks the dust storm abated, and *Mariner 9* was able to make many important discoveries about surface features on Mars. Some five years later, the two U.S. *Viking* missions reached Mars, each consisting of an orbiter that obtained color photographs of the entire surface (Fig. 17.9) and a lander (Fig. 17.10) that made close-up photos of the surface terrain and carried out experiments on the Martian soil, including searches for microscopic life forms.

Future missions are planned for both Venus and

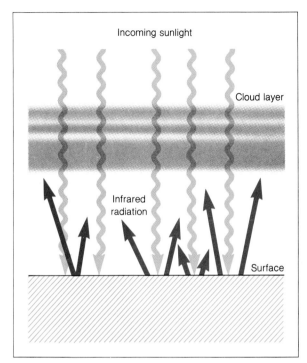

Figure 17.8 The Greenhouse Effect. Visible light from the Sun reaches the surface of Venus and heats it, which causes the ground to emit infrared radiation. The carbon dioxide in the atmosphere efficiently absorbs infrared radiation, so heat is trapped near the surface.

Figure 17.9 Surface of Mars. This is a mosaic of several images obtained by the *Viking* orbiters. The long horizontal streak is Valles Marineris, an enormous canyon. At right are seen some of the huge volcanoes of Mars. *(NASA: image processing by A. S. McEwan, U. S. Geological Survey)*

Mars by the U.S. and the Soviet Union. The U.S. is planning a radar mapper that will obtain finer-quality maps of surface topography than were provided by the radar instrument on *Pioneer-Venus*, and is already developing a *Mars Observer* mission, designed to go into polar orbit about Mars with the specific goal of better understanding the processes affecting water on the planet. Preliminary discussion has even begun on a possible future manned mission to Mars, which may be a joint Soviet-U.S. venture.

As we will see in the coming sections, quite a bit is already known about the terrestrial planets, particularly their surfaces and atmospheres. We will now discuss these three neighbor planets, beginning with an overview of the general processes that mold them, and then proceeding to discussions of their atmospheres, their surfaces and interiors, and what is known of their evolutionary histories.

General Properties

As a class, the terrestrial planets display many similarities (see Table 17.1) Each is basically a dense, rocky object with a thin atmosphere (if any) that contains a miniscule fraction of its mass. These planets are characterized by low abundances of lightweight, volatile gases, due to their formation in the warm inner regions of the solar system (see Chapter 15).

The densities of the terrestrial planets range from roughly 4 g/cm^3 (Mars) to about 5.5 g/cm^3 (Earth and Mercury). The density of the typical rock on the Earth's surface is around 3 g/cm^3, so the terrestrial planets, on average, are more dense than ordinary rock. Their densities thus imply that they must have relatively dense interiors, since their surfaces are known to con-

Table 17.2 Space Probes to the Terrestrial Planets

Spacecraft	Targets	Encounter Date	Type of Encounter
Mariner 2	Venus	Dec. 14, 1962	Flyby, 34,833 km
Mariner 4	Mars	July 14, 1965	Flyby, 10,000 km
Venera 4	Venus	Oct. 18, 1967	Hard landing, dark side
Mariner 5	Venus	Oct. 19, 1967	Flyby, 4100 km
Venera 5	Venus	May 16, 1969	Hard landing, dark side
Venera 6	Venus	May 17, 1969	Hard landing, dark side
Mariner 6	Mars	July 30, 1969	Flyby, 3200 km
Mariner 7	Mars	Aug. 4, 1969	Flyby, 3200 km
Venera 7	Venus	Dec. 15, 1970	Soft landing, dark side
Mariner 9	Mars	Nov. 13, 1971	Orbiter
Venera 8	Venus	July 22, 1972	Soft landing, day side
Mariner 10	Venus, Mercury	Feb., March 1974	Flybys, both planets
Venera 9	Venus	Oct. 22, 1975	Orbiter plus soft lander
Venera 10	Venus	Oct. 25, 1975	Orbiter plus soft lander
Viking 1	Mars	June 16, 1976	Orbiter plus soft lander
Viking 2	Mars	Aug. 7, 1976	Orbiter plus soft lander
Pioneer-Venus 1	Venus	Dec. 4, 1978	Orbiter
Pioneer-Venus 2	Venus	Dec. 9, 1978	4 hard landers
Venera 12	Venus	Dec. 21, 1978	Flyby (25,000 km), lander
Venera 11	Venus	Dec. 25, 1978	Flyby (25,000 km), lander
Venera 13	Venus	Mar. 1, 1982	Flyby (?), lander
Venera 14	Venus	Mar. 5, 1982	Flyby (?), lander
Venera 15/Vega 1	Venus/Halley	June ?, 1985	Flyby (distance unknown)
Venera 16/Vega 2	Venus/Halley	June ?, 1985	Flyby (distance unknown)

sist of materials no denser than ordinary rock. From this we deduce that the other terrestrial planets, like the Earth, must have undergone differentiation to some degree. Each of these planets must have been at least partially molten at some time, so that the heavier elements could sift their way to the center, where they form a dense core.

Among the terrestrial planets, Mars has the lowest density and therefore probably the smallest degree of differentiation. Since differentiation can occur only in molten material, the degree of differentiation a planet has undergone depends largely on whether and how extensively it has been molten in its past. This in turn depends on how much heating has occurred. The two most important sources of heat in a planetary interior are released gravitational potential energy from the formation process and natural radioactivity. The amount of heating due to gravitational potential energy depends largely on the mass of the planet, so Venus and the Earth, the two most massive of the terrestrial planets, probably have undergone the greatest degree of internal heating and hence have probably had the most prolonged molten periods. These two planets also are quite dense, indicating that they have undergone substantial differentiation.

Mercury is just as dense, however, and it is the least massive of the terrestrial planets. In this case, we resort to a different explanation: Mercury formed in the central portion of the solar system, where the temperatures were high enough to prevent the condensation of all but the heaviest elements. Hence Mercury started out with a lower abundance of lightweight elements than the other terrestrial planets. *Mariner 10* found evidence of substantial differentiation in Mercury as well (the planet has a magnetic field, thought to arise in an extended molten core), so evidently it was heated sufficiently to have been molten despite its low mass.

Mars has the lowest density among the terrestrials

and therefore has undergone the least differentiation. Mars has a relatively low mass and is farthest from the Sun, so it is reasonable to expect that this planet was heated relatively little in its interior. Even so, the presence of several massive volcanoes provides evidence that Mars has been at least partially molten in its distant past.

Tectonic activity forms another point of comparison among the terrestrial planets. As we learned in our discussion of the Earth, tectonic activity includes many dynamic processes that can take place on a planet, ranging from volcanic eruptions to earthquakes to continental drift. The degree of tectonic activity is related to internal motions that can take place, which are in turn related to the interior structure of a planet. If convection in the Earth's mantle is responsible for the tectonic activity of our planet, as suggested in the previous chapter, then we might expect other terrestrial planets with similar internal structures also to have tectonic activity. There are apparently many factors that govern this activity, however, and we will find that we cannot easily explain its presence or absence in every case. As we will see, the Earth is the most active tectonically of all the terrestrial planets, while Venus, which is nearly identical in its overall physical properties, apparently has much less activity. Mars had an early period when it was active, but it now seems dormant, and Mercury has an extensive series of cliff systems on its surface that may indicate previous activity.

Finally, the terrestrial planets can be compared in terms of their atmospheres. All but Mercury have thin gaseous envelopes trapped by their gravitational fields. The presence of an atmosphere is governed by the surface gravity of a planet and by the temperature and composition of the gases that are present. Each planet has an **escape speed,** a speed at which a particle moving upward will entirely escape the planet's gravitational attraction and be lost to space. If the molecules of gas in a planet's atmosphere exceed this speed, they will all eventually escape. The average speed of the molecules in a gas is determined by the temperature of the gas and the mass of the molecules: the higher the temperature, the faster molecules of a given mass will move; and for a given temperature, the lower-mass molecules will move faster. Each planet has its own surface gravity and its own temperature, determined by its distance from the Sun and by the processes in its atmosphere that govern heat loss, so the ability of each

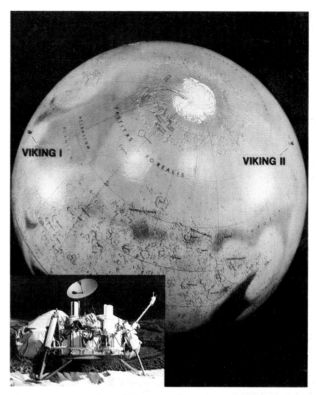

Figure 17.10 The *Viking* Landing Sites. This globe of Mars indicates the locations of the landing sites for the two *Viking* spacecraft. One of the landers is depicted in the inset. *(NASA)*

planet to retain gases in its atmosphere depends on many factors. Mercury, for example, is so hot and has such a small gravitational field that it cannot retain any gases, and thus it has no atmosphere. As we will see in the next chapter, the giant outer planets are so cold and have such strong gravitational fields that they retain all gases, losing none to space. The terrestrial planets other than Mercury represent intermediate cases, with moderate temperatures and gravitational fields, so that some gases are retained and some escape, depending on the masses of the gas particles. On the Earth, as we saw in the previous chapter, hydrogen and helium have escaped, but the more massive oxygen (O_2) and nitrogen (N_2) molecules have been trapped. On Venus, which is much hotter, most of the molecular species that are present are more massive than N_2 and O_2. Mars has retained a thin atmosphere of relatively heavy gases.

Astronomical Insight 17.1

Visiting The Inferior Planets

In Chapter 3 there was some discussion about the mechanics of placing a satellite in the Earth's orbit, but nothing about how to aim a spacecraft that is designed to visit another planet. The principles involved are particularly simple when an inferior planet is to be the target.

The concept of energy was discussed in Chapter 3, where we pointed out that an orbiting body has both kinetic energy owing to its motion, and potential energy owing to the gravitational field of the object that it orbits. For a planet orbiting the Sun, the total energy (the sum of kinetic plus potential) is greater the larger the orbit. Thus, for an object to go from the orbit of the Earth to that of an inner planet, it must lose energy.

A spacecraft sitting on the launchpad is moving with the Earth in its orbit, at a speed of 29 km/sec. To make the rocket fall into a path that intercepts the orbit of an inner planet, we must diminish this speed. Therefore, we launch the rocket backward with respect to the Earth's orbital motion, thereby lowering its velocity with respect to the Sun, and decreasing its orbital energy. If the launch speed is properly chosen the spacecraft will fall into an elliptical orbit that just meets the orbit of the target planet. In order to reach Mercury from the Earth, the rocket must be launched backward with a speed of 7.3 km/sec relative to Earth.

There are, of course, a few additional considerations. For one thing, the rocket has to be launched with enough speed to escape the Earth's gravity. The escape velocity for the Earth is 11.2 km/sec, so in fact we have to launch our Mercury probe with a speed in excess of this value, which is more than the speed needed to attain our trajectory to Mercury. The launch speed must therefore be calculated to take the Earth's gravity into account, so that the rocket escapes the Earth, but in the process is slowed just the right amount to give it the proper course. The gravitational pull of the target planet must also be taken into account, for this

---

The motions of the gases in a planetary atmosphere depend on two general factors: the deposition of heat and the rotation of the planet. Heat deposition for the terrestrial planets comes primarily from the absorption of sunlight, which may occur at the surface or at some level in the atmosphere where the gas is particularly effective at absorbing solar energy, such as in the ozone layer in the Earth's atmosphere. Where heat deposition occurs, there is a general tendency for the warm gas to rise, causing convection to take place. The rotation of a planet can then modify the convective flows into circular patterns. If the heating of the surface is nonuniform, as on the Earth (where the landmasses and oceans have different capacities for absorbing sunlight), or if surface topography (such as mountain ranges) is sufficient to alter atmospheric flow patterns, then a very complex system of atmospheric circulation may arise.

In the following sections we will address several general areas of planetary behavior, and we will see how the general processes described here come into play for each of the terrestrial planets.

Atmospheres of the Terrestrial Planets

The atmosphere of each planet can be characterized by its composition, by its physical properties such as density and pressure, and by the motions that occur in it. Here we discuss these characteristics in a comparative

Astronomical Insight 17.1

Visiting The Inferior Planets

speeds up the spacecraft as it approaches.

Another important consideration is timing; the target planet must be in the right spot in its orbit at just the moment when the spacecraft arrives. It is for this reason that the term *launch window* is used. The Earth and the target planet must be in specific relative positions at the time of the launch. The interval between launch windows for a given planet is simply its synodic period. For a probe to Mercury, the travel time is about 106 days, so Mercury actually makes a little more than one complete orbit while the probe is on its way.

The gravitational pull of the target planet can be used to good advantage in modifying the orbit of a spacecraft that flies by. If the probe is aimed properly, its trajectory may be altered in just the right way to send it on to some other target. This technique was used with *Mariner 10*, first to send it from Venus to Mercury, and then to modify its orbit again so that it returned to Mercury repeatedly thereafter.

Sending spacecraft to the outer planets is a bit more difficult, because the spacecraft has to have more energy than what it gets from the Earth's motion. The launch is therefore made in the forward direction, so that the rocket has the combined speed of the Earth in its orbit plus its own launch speed with respect to the earth.

Ironically, the seemingly simple task of launching a probe to the Sun is one of the most difficult. The most straightforward solution would be to launch the spacecraft backward from the Earth with a speed of 29 km/sec, entirely canceling out the Earth's orbital speed, so that the probe would then fall straight in toward the Sun. This is a prohibitively high launch speed, however, so in practice we will probe the Sun by first sending the spacecraft out around Jupiter. The spacecraft will fly by the massive planet in such a way (backward with respect to Jupiter's orbital motion) that its own orbital energy is reduced; the craft can then fall into the center of the solar system. A mission to explore the outer layers of the Sun in this manner is the planning stages.

---

way, so that the similarities and contrasts among the terrestrial planets can be examined and explanations for them can be sought.

Composition of the Terrestrial Atmospheres

Each of the terrestrial planets has a source of gases that can form atmospheres, and each has processes by which gases are lost. The atmosphere a planet ends up with is determined by a balance between the sources and losses of gas. For Mercury, all gases escape easily due to the high temperature and low surface gravity (hence low escape speed), so no permanent atmosphere is retained. Because a steady stream of particles in the solar wind strikes Mercury, at any given moment there are trace amounts of gas around Mercury, but these do not constitute an atmosphere.

The gases that form a planetary atmosphere can come from at least two different sources: they can be trapped from gases in space around a planet, although this is normally a very insignificant process for the terrestrial planets (it was an important one during the formation of the giant planets), or they can be released from the planet's interior in a process called **outgassing**. A third process, known to be important only on the Earth, is the formation and release of gases by living creatures and plants. The gases present in the atmospheres of the terrestrial planets are almost entirely due to outgassing from the interiors of the planets, except in the case of the Earth, whose atmosphere has been modified by life forms.

The outgassing occurs through general low-level seepage of gases to the surface and sporadic eruptions of volcanoes, which can inject large quantities of gas in brief episodes (there is evidence, for example, that sulfur dioxide is periodically released into the atmosphere of Venus by volcanic eruptions). It is thought that a lot of outgassing occurred in the early history of each terrestrial planet, when the planets were still very hot. Among the most abundant gases released were hydrogen (in molecular form, H_2), helium, hydrogen-bearing molecules such as water vapor (H_2O), methane (CH_4), and ammonia (NH_3), and other species such as carbon dioxide (CO_2).

The present-day atmospheres of the terrestrial planets (Table 17.3) are not necessarily the same as those formed at first. It is thought that each planet ejected large quantities of water vapor, for example, but today only the Earth has any significant quantity of this species left on its surface. On Venus, the surface temperature (due to the greater proximity of the planet to the Sun) was too high for water to persist in liquid form, and on Mars there may have been an early era when water was plentiful, but today the atmospheric pressure is so low that water vaporizes and cannot exist in liquid form.

The presence or absence of liquid water influences the abundances of other gases. On Venus, water remained in the form of vapor, which trapped heat due to the greenhouse effect, a process in which incoming sunlight heats the surface, which in turn emits infrared radiation that is trapped in the atmosphere (Fig. 17.8) by species such as water vapor. The lack of liquid water on Venus meant that carbon dioxide, another abundant gas released from the surface, remained in gaseous form, rather than dissolving in water and being deposited in carbonaceous rocks, as it was on the Earth. Thus the atmosphere of Venus built up a large quantity of carbon dioxide, which has an even stronger greenhouse effect than water vapor and led to additional heating of the atmosphere. The temperature rose to the point where water vapor was dissociated (that is, broken apart into separate atoms of hydrogen and oxygen) and escaped the atmosphere altogether. Today Venus has a very dense atmosphere consisting mostly of carbon dioxide (Table 17.2).

Table 17.3. Atmospheres of the Terrestrial Planets*

Gas	Symbol	Venus	Earth	Mars
Nitrogen	N_2	0.035	0.77	0.027
Oxygen	O_2	—	0.21	0.0013
Water vapor	H_2O	1×10^{-4}	0.001–0.028	3×10^{-4}
Argon	Ar	7×10^{-5}	0.0093	0.016
Carbon dioxide	CO_2	0.96	3.3×10^{-4}	0.95
Neon	Ne	5×10^{-6}	1.8×10^{-5}	2.5×10^{-6}
Helium	He	?	5.2×10^{-6}	?
Methane	CH_4	—	1.5×10^{-6}	—
Krypton	Kr	—	1.1×10^{-6}	3×10^{-7}
Ozone	O_3	—	4×10^{-7}	1×10^{-7}
Nitrous oxide	N_2O	—	3×10^{-7}	—
Carbon monoxide	CO	4×10^{-5}	1.2×10^{-7}	7×10^{-4}
Ammonia	NH_3	—	1×10^{-8}	—
Sulfur dioxide	SO_2	1.5×10^{-4}	—	—
Hydrogen chloride	HCl	4×10^{-7}	—	—
Hydrogen fluoride	HF	1×10^{-8}	—	—
Xenon	Xe	—	?	8×10^{-8}

*The values given are fractional abundances by number; the total for each planet should be 1.0. Question marks indicate gases that are probably present, but data are not available; dashes indicate gases thought not to be present in any significant quantity.

Carbon dioxide has also remained in the atmosphere on Mars, because there has not been enough liquid water to dissolve it into rocks. Surface features on Mars (Fig. 17.11) indicate that there have been times in the past when water flowed, but today there is no liquid water, although traces of water vapor are found in the atmosphere, the polar ice caps consist in part of water ice, and it appears possible that larger quantities of water exist in the form of **permafrost,** ice embedded below the surface in the soil. Mars, like Venus, has an atmosphere dominated by carbon dioxide, but the overall density of the atmosphere is far lower (0.006 times Earth's sea level pressure, compared to the 90 atmospheres of pressure at the surface of Venus).

The Earth's atmosphere, as we have seen, has been altered extensively by the development of life. Even without this unique influence, it would have been somewhat different from the atmospheres of the other terrestrial planets, because it would have had little carbon dioxide due to the presence of oceans of liquid water in which the carbon dioxide can dissolve. Some nitrogen would have been present, because this element is released in volcanic eruptions, and possibly other volcanic products such as sulfur compounds would have made up the rest of the atmosphere. As it happened, however, the development of life forms has given the Earth a very different kind of atmosphere, dominated by nitrogen and oxygen. The nitrogen has been enhanced by the decay of living organisms, and the oxygen is almost solely due to the respiration of plants, which take in carbon dioxide and release oxygen.

Circulation and Seasonal Effects

As mentioned in the preceding section, the circulation of gases in a planet's atmosphere is governed primarily by two factors: the deposition of heat and the rotation of the planet. Each of the terrestrial planets has heat deposited at its surface, where sunlight is absorbed. In

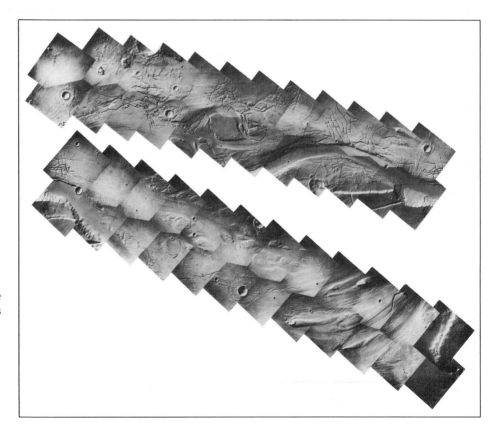

Figure 17.11 A Flood Plain. Here is clear evidence that water once inundated this region of the Martian surface. The surface patterns indicate that the prominent craters in this photo already existed when the water flowed. The two mosaics show parts of Kasei Valles. *(NASA)*

addition, the Earth and Venus have atmospheric layers where heat deposition is significant due to the presence of molecules that are efficient absorbers. In the atmosphere, the ozone layer (at a height of about 60 to 80 km) absorbs sunlight, and in the atmosphere of Venus, sunlight is absorbed by sulfur dioxide (SO_2) at a height of roughly 40 to 60 km.

Where heat deposition occurs, warmed air tends to rise, leading to circulation by convection. Without any other complications, this process might give rise to a very simple circulation pattern in which warm gas rises near the equator and flows toward the polar regions, where it cools and descends (Fig. 17.12). Other factors generally make the circulation pattern more complex, however. Planetary rotation forces convective flows into rotary patterns, whose direction of rotation depends on the pressure differential between the outer and inner regions (see Fig. 16.4 and, generally, Chapter 16). Other complicating factors that modify the flow of the atmosphere include the different heat-absorbing abilities of different surface areas (there is an especially strong contrast between the Earth's oceans and continents, for example) and the influence of surface features such as mountain ranges.

Figure 17.12 Simplified Atmospheric Convection. In the absence of rotation and other complicating factors, the atmospheric circulation of a planet's atmosphere would be driven by convection alone, with the simple flow pattern shown here.

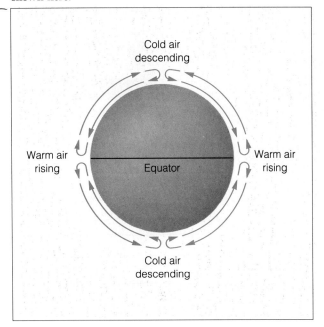

The flow patterns may also be influenced by seasonal variations, which are especially strong on the Earth and Mars, among the terrestrial planets. A planet will have seasonal variations if its spin axis is tilted with respect to the plane of its orbit around the Sun. If this is so, then at one time of the year, the northern hemisphere will be exposed more directly to the Sun, creating summer in the north. Half a year later, the other hemisphere will be more directly exposed to sunlight and will experience summer.

The seasonal variations on Mars are influenced by another effect, in addition to the tilt of the planet's axis. The orbit of Mars is more elongated than those of most other planets (although not as much so as Mercury's orbit); thus the orbit of Mars is not centered very precisely on the Sun, and the distance of Mars from the Sun varies substantially during its year, enough to modify the climate. Mars is only 83 percent as far from the Sun at its closest approach than at its greatest distance, and according to the inverse square law (Chapter 4), sunlight is $(1/0.83)^2 = 1.45$ more intense at that point. It happens that Mars makes its closest approach to the Sun at the same time as summer in the southern hemisphere, so the summer there is especially warm, while the winter in the northern hemisphere is relatively mild. When Mars is at its greatest distance from the Sun, on the other hand, the southern hemisphere is having winter, and therefore winter there is extremely cold while the northern summer is kept from being very warm. Thus, the shape of the Martian orbit causes the seasonal variations in the southern hemisphere to be extreme, while those in the north are relatively moderate.

This unusual seasonal variation is apparently the cause of the major dust storms that sweep Mars every Martian year. These storms arise in the southern spring, when the temperatures are rising rapidly as the season swings from the very cold winter of the south to the warm summer. As the solar heating increases rapidly, temperature contrasts are created with the adjacent south polar cap, and they produce winds that are soon strong enough to pick up fine dust particles and carry them around the planet. These storms can last for weeks or months, sometimes completely covering the surface of Mars, as happened in 1971 when the *Mariner 9* mission arrived (Fig. 17.13). The seasonal dust storms on Mars account for the changes in surface coloration that have already been mentioned; as the dust is alternately picked up by the winds and then redeposited, the surface cover varies and hence there

Chapter 17 Terrestrial Planets 363

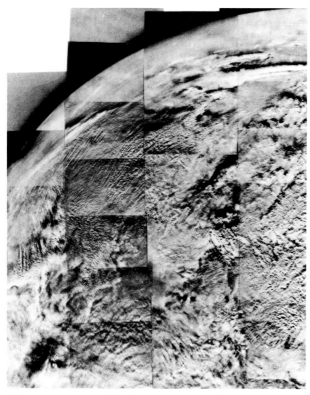

Figure 17.13 A Global Dust Storm. This is the view of Mars that confronted the *Mariner 9* spacecraft for the first several weeks after it went into orbit around the planet. *(NASA)*

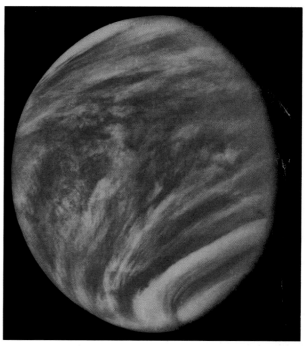

Figure 17.14 A Close-up Portrait of Venus. This is a fulldisk ultraviolet image of Venus obtained by the *Mariner 10* spacecraft; it shows structure in the clouds. *(NASA)*

Figure 17.15 High-Altitude Circulation. Well above the clouds of Venus, which are rotating from right to left in this sketch, gases heated at the subsolar point circulate around the planet toward the dark side, where they cool and descend. *(NASA)*

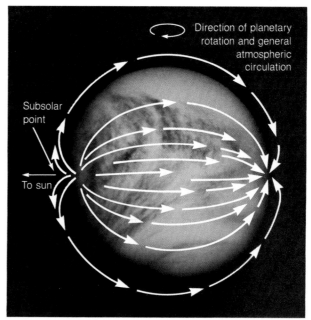

are changes in the appearance of the planet as seen from afar.

Circulation of the atmosphere on Venus is less complex than on the Earth or Mars, partly because the rotation of the planet is so slow and partly because there are few variations in the surface temperature from place to place. The upper clouds flow around the planet at a high rate of speed (Fig. 17.14), taking about four days to make one circuit around the planet in the same direction as the rotation, while there is apparently little or no wind at the surface. Above the clouds there is a convective flow from the side facing the Sun to the dark side (Figs. 17.15 and 17.16). Heat is deposited in the clouds due to absorption of sunlight by sulfur dioxide; the regions where this occurs appear as dark streaks in ultraviolet images of the planet.

The clouds of Venus consist of sulfuric acid (H_2SO_4) droplets that apparently condensed on tiny solid particles. Clouds form where the combination of pressure and temperature is appropriate for droplets to con-

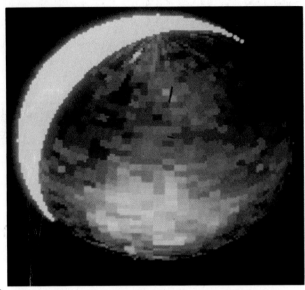

Figure 17.16 Airglow on the Night Side of Venus. Atoms produced by the disruption of molecules on the sunlit side circulate around to the dark side, where they cool and combine once again into molecules. The newly formed molecules have excess energy, which is released in the form of light, creating a glow in the upper atmosphere on the night side. *(LASP, University of Colorado, sponsored by NASA)*

Figure 17.18 Martian Plains on a Frosty Morning. This photo, taken by the *Viking 2* lander, shows a thin coating of water ice on the rocks and soil. *(NASA)*

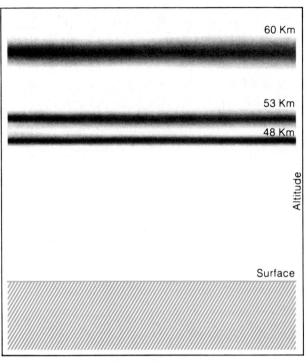

Figure 17.17 The Clouds of Venus. The sulfuric acid clouds are separated into three distinct layers.

dense; in the Earth's atmosphere, water vapor condenses in this way, and in the atmosphere of Venus, conditions are right for the condensation of sulfuric acid. The clouds of Venus occur in three distinct layers (Fig. 17.17). Below the clouds, the air is clear but the pressure is immense, reaching a value of 90 times sea-level pressure on Earth.

Mars also has clouds in its atmosphere, but they are usually small and short-lived. They can be water vapor clouds, particularly evident in the early mornings in some lowland regions, or clouds of carbon dioxide ice particles. Sometimes a frost of water ice forms on the surface during the night (Fig. 17.18).

Surfaces and Interiors

The surfaces of the terrestrial planets are similar in a very broad sense, each one having large areas of lowlands and gently rolling plains, and each one having mountainous regions. Mars is the best observed—its entire surface has been mapped photographically—

Figure 17.19 A Barren Surface. This photo, obtained by one of the *Viking* orbiters, shows that the surface of Mars is dry and barren, with extensive cratering. *(NASA)*

Figure 17.20 Olympus Mons. This is the largest mountain known to exist in the entire solar system. Its base is comparable in size to the state of Colorado, and its height is about three times that of Mount Everest. *(NASA)*

Figure 17.21 A Ground-level Panorama. This is the first photograph made by the *Viking 1* lander. It shows a rock-strewn plain extending in all directions. *(NASA)*

and on Mars we see a crater-pocked surface (Fig. 17.19) mostly covered by plains. A large highland region known as the Tharsis bulge apparently was uplifted some 3 billion years ago, at a time when Mars was somewhat active geologically. Not far from Tharsis are several gigantic volcanoes, the largest of which, Olympus Mons (Fig. 17.20), dwarfs any other mountain in the solar system. Its height above the surrounding plains is 27 km (90,000 feet, more than three times the height of Mt. Everest above sea level!), and its base is some 700 km (400 miles) across. Along one edge of the Tharsis plateau is a very extensive canyon called Valles Marineris, which is about five times larger than the Grand Canyon in every dimension, being long enough to stretch from coast to coast across the United States.

Apart from these large features on the surface of Mars, the plains are covered with rocks and dust (Fig. 17.21). The dust shifts around from season to season with the changing wind patterns. There are some regions where coarser sand covers the surface (Fig. 17.22), and near the edges of the polar caps there are layered regions where coatings of ice and dust are deposited each Martian winter (Fig. 17.23). The reddish color of the Martian surface is attributed to a high iron content; Mars is not highly differentiated, thus its heavy elements such as iron have remained largely on the surface instead of sinking to the core. Iron combines with oxygen to form iron oxide, commonly known as rust, so we can say that the surface of Mars is red because it is rusty.

Venus cannot be mapped photographically, but it has been surveyed by a radar imaging device on the

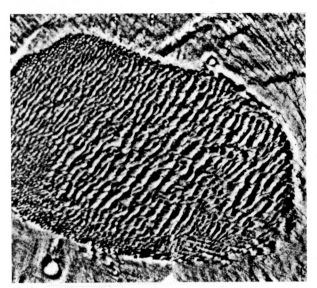

Figure 17.22 Martian Sand Dunes. These dunes are testimonials to the incessant action of winds on Mars. *(NASA)*

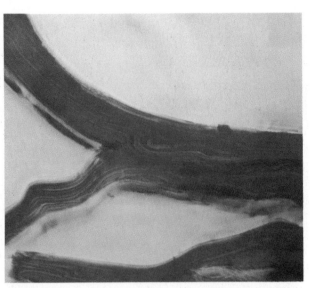

Figure 17.23 Laminated Terrain. This photo shows some of the curious layered surface features seen near the edges of the polar ice caps. *(NASA)*

Figure 17.24 A Relief Map of Venus. This image, constructed on the basis of *Pioneer Venus* radar maps of the surface, shows the three major highland areas: Ishtar Terra at the top, Aphrodite Terra at right center, and Beta Regio at left Center. *(NASA/JPL)*

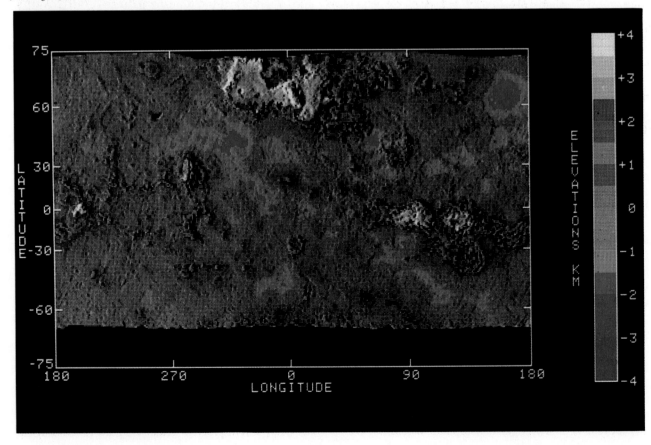

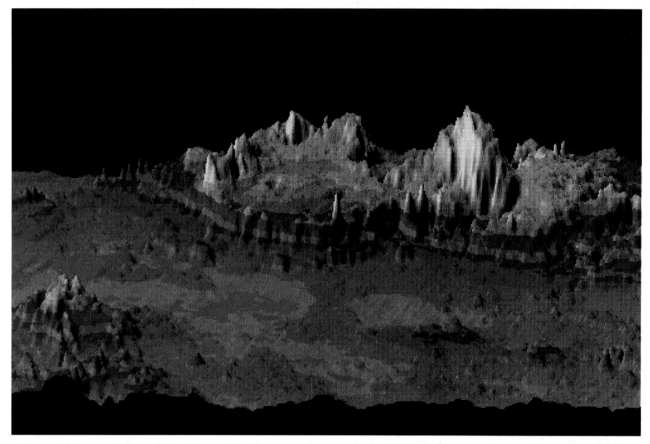

Figure 17.25 A Portion of Ishtar Terra. This computer reconstruction illustrates (on an exaggerated vertical scale) the vertical relief of this highland region, and includes the largest mountain on Venus, Maxwell Montes, more than 11 kilometer above the mean surface level. *(NASA/JPL)*

Pioneer-Venus orbiter. The resulting relief maps (Fig. 17.24) show that Venus has vast regions with relatively few major features, punctuated by three major highland, or continental, regions. There are lowlands, which correspond geologically to the seafloors on Earth, and extensive rolling plains. Some craters are apparently present, but it is possible that they are volcanic in origin, rather than the result of impacts. The largest highland region, Ishtar Terra, has high mountains; one of them (Maxwell Montes, Fig. 17.25) reaches a height 11 km above the mean surface level, exceeding the height of Mt. Everest.

Photographs made of the surface by the Soviet *Venera* landers (Fig. 17.26) show that the surface is littered with rocks. The rocks do not appear to be highly eroded, which was a surprise in view of the tremendous atmospheric density; apparently there is little or no wind at the surface of Venus to scour the rocks. Limited chemical analyses indicate that the rocks are probably basalts, most likely of volcanic origin.

Mercury's surface is almost indistinguishable from that of the Moon (Fig. 17.27). Only half of the surface has been mapped photographically, because the other half was in darkness during the *Mariner 10* encounters, but the portion that is well-studied shows all the major features of the Moon. There are craters everywhere, there are regions similar to the maria, and most of the surface is mountainous. There is a very large impact basin called Caloris Planitia on the side of Mercury that directly faces the Sun on every other closest approach. It is possible that the large body that crashed into Mercury at this point is what made this side of the planet heavier than the other, allowing the Sun's tidal forces to modify the spin and create the spin-orbit coupling described earlier in this chapter.

There are some contrasts between Mercury and the surface of our Moon. Mercury's craters are shallower and the surrounding ejecta less widely distributed (Fig. 17.28), apparently due to the higher surface gravity on Mercury, which prevents crater walls from building up

Figure 17.26 *Venera 13* **Photo of Surface Rocks.** This mission, which took place in the spring of 1982, obtained more extensive and better-quality images of surface rocks than the previous missions, *Venera 9* and *10*. *(TASS from SOVFOTO)*

to such great heights and debris from flying as far. Another contrast is that Mercury is girdled by extensive series of cliffs or **scarps** (Fig. 17.29), which may be the result of a general shrinkage of its crust as the planet cooled and interior gases escaped early in the history of the solar system.

We know relatively little about the interiors of the terrestrial planets, except for what we can guess based on their overall external properties. There is reason to believe, for example, that Venus may have an internal structure very similar to the Earth's, since these two planets have almost identical diameters and masses, and hence very similar overall densities. Therefore, we expect that Venus is similarly differentiated and has internal zones much like those of the Earth. Because the rotation of Venus is so slow, however, there are probably no significant currents in the molten part of the core, if such a molten zone exists. We know that there is no measurable magnetic field around Venus, which is consistent with a lack of currents (or the absence of a molten zone).

Venus has some surface features (similar to the adjacent mountain and trench systems seen at continental plate boundaries on the Earth) that may have resulted from limited continental drift activity early in the planet's history, but it has not undergone extensive tectonic activity comparable to that of the Earth. Apparently the highland masses are not able to slide over the lowlands in the same manner that the continents move over the seafloors on Earth. It has been suggested that the high surface temperature on Venus makes the lowland material so buoyant that it resists being pushed under the highlands, preventing them from moving. Of course, we have no way of knowing whether the same driving mechanism (presumably convection flows in the mantle) operates on Venus, so there may be other explanations for the lack of widespread continental drift. It is suspected that some tectonic activity in the form of volcanic eruptions does occur on Venus. *Pioneer-Venus* data indicate that the sulfur dioxide content of the atmosphere increased dramatically in the late 1970s and has been slowly decreasing since. Some scientists have attributed this to a massive volcanic eruption that injected large quantities of sulfur dioxide into the atmosphere; others say that perhaps there was no such injection, but that high concentrations of the gas were somehow stirred up from lower levels in the atmosphere. The question of current volcanic activity on Venus is still open and may not be fully answered for some time.

Mars, like Venus, shows evidence of limited tectonic activity that occurred long ago. The Tharsis plateau and the gigantic volcanoes date back some 3 billion years, indicating that at that early time in the planet's history there was some geological activity. There is no indication of present-day activity, and it is thought that

Figure 17.27 Mercury as Seen from Space. This *Mariner 10* view is a mosaic of several images. The planet bears a strong resemblance to the Moon, although there are significant contrasts, particularly in internal structure. *(NASA/JPL)*

Figure 17.28 Craters on Mercury. Because Mercury has a greater surface gravity than the Moon, impact craters have lower rims and are shallower, and ejecta do not travel as far. *(NASA/JPL)*

Figure 17.29 A Scarp System on Mercury. The lengthy series of cliffs extending along the upper left is one of many on the planet's surface. *(NASA/JPL)*

the crust of Mars is simply too thick to allow any continental drift. Also, the interior structure of Mars may lack the ingredients that create tectonic activity in the Earth, since there is no very extensive core and probably little or no molten zone. Mars has no measurable magnetic field even though it rotates fairly rapidly (its day is almost equal to that of the Earth), another indication that there is no extensive molten zone.

Mercury has a high overall density and, despite its very slow rotation, a measurable magnetic field. These facts indicate that the planet has a very extensive core (Fig. 17.30) that is at least partially molten. As mentioned earlier in this chapter, the high density and high degree of differentiation may be attributed to the hot environment in which Mercury formed, where volatile, lightweight gases were prevented from condensing.

Evolution of the Terrestrial Planets

In Chapter 15 we discussed the formation of the solar system and described the formation of the terrestrial planets from a buildup and merging of solid material that condensed from the disk of gas and dust that sur-

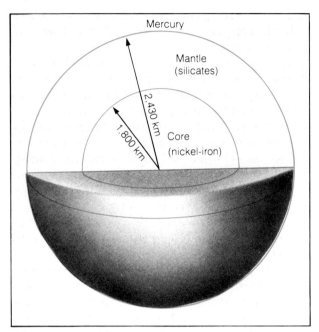

Figure 17.30 Mercury's Internal Structure. The presence of a magnetic field, along with the planet's relatively high density, implies the presence of a large core.

Only the Earth and Mars have satellites, and both may be anomalous in this respect. As we learned in Chapter 16, the Moon may not have formed as a natural result of the Earth's formation but instead may have been captured or formed as the result of a collision. Mars has two tiny satellites (Fig. 17.31), which may be captured asteroids (see Chapter 19). Venus and Mercury have no satellites.

Each terrestrial planet underwent an extensive period of outgassing, as volatile gases from the interior escaped and created the first atmosphere, and each underwent differentiation to some degree. The earliest atmosphere in each case probably consisted of hydrogen and hydrogen-bearing compounds, since hydrogen was by far the most abundant element in the presolar nebula. Mercury probably never retained any atmosphere but simply cooled gradually, its crust shriveling and cracking as interior gases escaped.

Mars apparently released large quantities of water vapor and even had liquid water on its surface for a time. Today there is controversy over whether the water was a sporadic occurrence, as pockets of underground ice occasionally melted, or whether there was a time when water was a standard feature on Mars, forming lakes and rivers. In the latter case it is hypothesized that the planet once had a dense atmosphere, but for unknown reasons the climate changed and this atmosphere escaped, leaving the planet with such low atmospheric pressure that it can no longer maintain liq-

rounded the early Sun. We will now consider the processes that led these planets from their primitive states to the present. Each is thought to have been very hot initially and either fully or partially molten.

Figure 17.31 The Martian Moons Phobos and Deimos. The *Viking* mosaic of Phobos, *left,* shows two interesting features: the parallel grooves that mark the surface, and the major impact crater whose formation may have caused the stresses that created the grooves. Deimos, *right,* the more distant of the two moons of Mars, has a relatively featureless surface. *(NASA)*

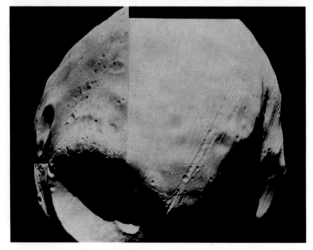
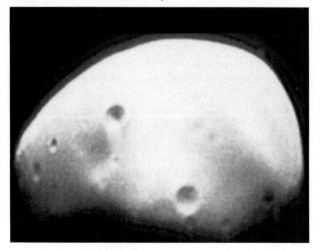

uid water on its surface. Carbon dioxide is the chief constituent of Mars' atmosphere today, the early hydrogen-bearing gases having escaped long ago.

The early atmosphere of Venus, like the early atmosphere of the Earth, probably contained large quantities of water vapor and other hydrogen compounds. On Earth the water condensed to form oceans, while Venus was so hot due to its closer proximity to the Sun that water stayed in vapor form. Carbon dioxide on the Earth was dissolved in seawater and deposited in rocks, whereas it remained in the atmosphere in Venus, enhancing the greenhouse effect. The atmosphere of the Earth remained moderate in temperature, while on Venus it got very hot, eventually breaking apart the water vapor molecules in the atmosphere and allowing the hydrogen and oxygen atoms to escape into space. Venus developed the inhospitable, hellish environment we see today, while the Earth became a comfortable abode for life, all because Venus was fractionally closer to the Sun. At its distance of 0.72 AU from the Sun, Venus receives $(1/0.72)^2 = 1.92$ times greater intensity of sunlight, enough to make all the difference.

It is well for us to keep in mind how close Earth was to being like Venus and how possible it may be to reproduce Venus here, if we allow our atmosphere to become sufficiently polluted with molecules such as carbon dioxide that have a strong greenhouse effect.

Perspective

We have now surveyed the innermost four planets, finding among them many similarities in overall properties but many contrasts in detail. We have some understanding of how these planets formed and how they have evolved to the states we find them in today.

Next we will undertake a similar survey of the outer planets. There are five to be discussed, but less detail is known, and we will find that four of the five are very similar.

Summary

1. The terrestrial planets are all observable with the naked eye, and their existence has been known since ancient times. Modern observations are made from the Earth and from space, and each planet has been visited by space probes.

2. The terrestrial planets have high densities, indicating that they are differentiated and consist mostly of rocky materials.

3. The internal structure of each terrestrial planet is determined in large part by its heating history; sources of internal heat include natural radioactivity and released gravitational potential energy.

4. All the terrestrial planets except Mercury have atmospheres. Venus and Mars are dominated by carbon dioxide, while the Earth's atmosphere is mostly nitrogen and oxygen. The presence of oceans on the Earth is largely responsible for this difference; liquid water absorbs and dissolves carbon dioxide and provides an environment for the formation of life, which in turn has produced much of the nitrogen and oxygen in the atmosphere.

5. The circulation of air in an atmosphere is controlled primarily by heating, which causes convection, and by the planet's rotation, which creates rotary patterns. The Earth and Mars have complex wind systems, whereas the circulation on Venus, which rotates very slowly, is much simpler.

6. Surface features on all four terrestrial planets include mountainous highland regions, lowlands, and extensive plains. Venus has three highland areas similar to continents on the Earth, whereas Mars has a single, very large uplifted region. All are cratered, though erosion has removed the evidence on the Earth, and some of the craters on Venus may be volcanic rather than the result of impacts.

7. The Earth is the most active geologically, with ongoing continental drift, while Mars and Venus both show signs of early tectonic activity, and Venus may still have some volcanic eruptions.

8. Each planet released gases early in its history, primarily hydrogen-bearing compounds, but each has lost its early atmosphere. On Mars and Venus, carbon dioxide has replaced the primitive gases; on Earth, oceans have removed carbon dioxide from the atmosphere, and life forms have replaced it with nitrogen and oxygen.

Review Questions

1. Explain why the density of a planet is an indicator of its internal structure.
2. How is the degree of differentiation a planet has undergone related to its history of heating?

3. Summarize what is known about water on Mars, past and present.
4. How has water on the Earth helped to make its atmosphere and surface conditions different from those on Venus and Mars?
5. Explain why the surface temperature of Venus is so high and why the Earth's temperature is so much cooler.
6. Explain how spin-orbit coupling has affected the rotational period of Mercury and how this compares with the development of synchronous rotation for the Moon.
7. If the large volcanoes on Mars formed over underground hot spots where molten lava comes to the surface, much as the Hawaiian Islands did, can you think of an explanation for the much greater size of the Martian volcanoes?
8. Summarize the evidence concerning the presence or absence of tectonic activity on Mars and Venus, and compare both with the Earth.
9. Using information from Chapter 15 as well as this chapter, summarize the formation of the terrestrial planets.
10. Design a new space mission for exploration of each of the terrestrial planets, saying what you think the most important questions are and how your mission would answer them.

Additional Readings

Arvidson, R. E., A. B. Binder, and K. L. Jones. 1978. The surface of Mars. *Scientific American* 238(3):76.

Beatty, J. K. 1985. A radar tour of Venus. *Sky and Telescope* 69(2):507.

——— 1985. A Soviet space odyssey (*Vega* at Venus). *Sky and Telescope* 70(4):310.

Beatty, J. K., B. O'Leary, and A. Chaikin. eds. 1981. *The New Solar System*. Cambridge, Eng.: Cambridge University Press.

Carr, M. S. 1976. The volcanoes on Mars: *Scientific American* 234(1):32.

——— 1983. The surface of Mars: a post-*Viking* view. *Mercury* 12(1):2.

Cordell, B. M. 1986. Mars, Earth, and ice. *Sky and Telescope* 71(1):17.

Hartmann, W. K. 1976. The significance of the planet Mercury. *Sky and Telescope* 51(5):307.

Head, J. W., C. A. Wood, and T. A. Mutch. 1977. Geologic evolution of the terrestrial planets. *American Scientist* 65:21.

Horowitz, N. H. 1977. The search for life on Mars. *Scientific American* 237(5):52.

Leovy, C. B. 1977. The atmosphere of Mars. *Scientific American* 237(1):34.

Murray, B. C. 1975. Mercury. *Scientific American* 233(3):58.

Pettingill, G. H., D. B. Campbell, and H. Masursky. 1980. The surface of Venus. *Scientific American* 243(2):54.

Schubert, G. and C. Covey. 1981. The atmosphere of Venus. *Scientific American* 245(1):66.

Veverka, J. 1977. Phobos and Deimos. *Scientific American* 236(2):30.

Weaver, K. F. 1975. *Mariner* unveils Venus and Mercury. *National Geographic* 147:848.

Young, A. and L. Young. 1975. Venus. *Scientific American* 233(3):70.

Chapter 18

The Outer Planets

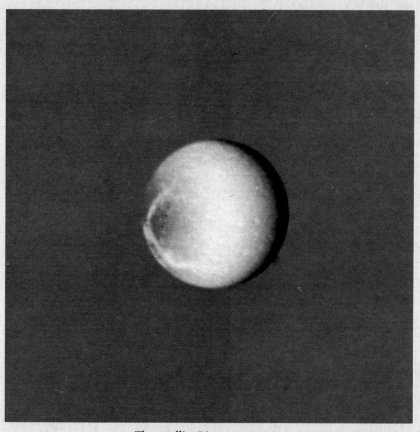

The satellite Dione seen against Saturn's atmosphere. *(NASA/JPL)*

Chapter Preview

Discovery and Observation
General Processes
Atmospheres and Interiors of the Gaseous Giants
   Atmospheric Composition
   Atmospheric Circulation
   Interior Conditions
Magnetic Fields and Particle Belts
Rings and Moons
   Satellites of the Giant Planets
   Ring Systems
Pluto, Planetary Misfit

The outer five planets are very different from the terrestrial planets in almost every respect (Table 18.1). Their compositions are different, all but Pluto are far more massive, the environment is far colder and darker, and the processes by which they formed and have evolved are different.

Four of the outer five planets, Jupiter, Saturn, Uranus, and Neptune, are quite similar in overall properties and are justifiably called gaseous giants. The fifth, Pluto, is anomalous compared to the others and is also very unlike the terrestrial planets. In this chapter we will discuss the first four planets in parallel, as we address their overall properties, their atmospheres and interiors, and their satellite and ring systems. Pluto will be reserved for a brief discussion at the end of the chapter.

Discovery and Observation

Of the four gaseous giant planets, only two, Jupiter and Saturn, are readily visible to the unaided eye and have thus been known since ancient times. Of course, telescopes provide much better views, revealing some detail of the banded atmospheres and color patterns (Fig. 18.1). Jupiter orbits some 5.2 AU from the Sun, while Saturn's orbit lies about 9.6 AU away from the Sun (Table 18.1).

Uranus was discovered in 1781 by the English astronomer William Herschel, who found an unusual object with a bluish color and was able to deduce that it traveled in a solar orbit about 20 AU from the Sun. Even the best telescopic images of Uranus made from Earth reveal little detail of this faraway planet (Fig. 18.2).

Neptune, which is far too faint and distant to be easily noticed as anything but a star, was discovered through astronomical detective work. When Uranus was discovered, its motion was tracked carefully to test how well it adhered to Newton's laws of motion and gravitation, which had been developed only a century earlier. It was soon found that Uranus was not strictly following the expected orbit, and in 1845 two astronomers, John Adams in England and Urbain Leverrier in France, showed that the irregularities could be caused by the gravitational effects of another planet beyond Uranus. Neptune was found at the predicted position, about 30 AU from the Sun, in 1846.

The discovery of Pluto in 1930 was the result of a great deal of hard work and quite a bit of luck. Following the successful prediction of the existence of Neptune based on orbital irregularities of Uranus, several astronomers believed it might be possible to detect the presence of a ninth planet by looking for similar irregularities in the motion of Neptune. One in particular, Percival Lowell, made a prediction based on the motions of Neptune and Uranus and set about the task of finding the expected ninth planet. The search was carried out at Lowell's own observatory (still operating today, in Flagstaff, Arizona), but nothing was found be-

Table 18.1. The Giant Planets

Planet	Semimajor Axis	Sidereal Period	Mass*	Diameter*	Density	Temperature	Escape Speed
Jupiter	5.203 AU	11.86 yr	318.1	10.79	1.33 g/cm^3	130 K	60.0
Saturn	9.555	29.46	95.12	8.91	0.68	95	36.0
Uranus	19.22	84.01	14.54	4.05	1.20	52	21.2
Neptune	30.11	164.79	17.2	3.91	1.58	50	23.5

*The masses and diameters are given in units of the Earth's mass and diameter, which are 5.974×10^{27} grams and 12,734 km, respectively.

Figure 18.1 Jupiter and Saturn as Seen from Earth. These photographs were taken from Earth-based telescopes. In the photograph of Jupiter, *(top)* one moon is visible: The dark spot seen against the planet's disk. The photograph of Saturn *(bottom)* was taken under good conditions, and shows some structure in the ring system. *(top: NASA; bottom: Lick Observatory Photograph)*

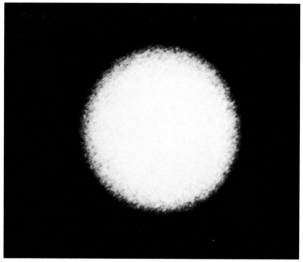

Figure 18.2 Uranus. This exceptionally fine photograph was obtained from a high-altitude balloon, which was able to avoid much of the blurring effect of the Earth's atmosphere. No surface markings are evident, except for some darkening at the edges of the disk. *(Project Stratoscope, Princeton University, supported by the National Science Foundation)*

Figure 18.3 Clyde Tombaugh at the Blink Comparator. This photo shows the discoverer of Pluto peering into the eyepiece of the device he used to find the planet on photographic plates. The blink comparator allows two images of the same field of stars to be examined alternately (with the use of a mirror that flips between two positions) so that moving objects stand out. *(Lowell Observatory photograph)*

fore Lowell's death. Subsequently Clyde Tombaugh (Fig. 18.3), a young staff astronomer at the observatory, found a tiny, dim object that moved with respect to the background stars (Fig. 18.4), and Pluto was discovered. Ironically, it was later demonstrated that Pluto is far too small and distant to have had any observable effect on the motions of Neptune and Uranus and that Lowell's prediction was actually baseless, arising from minor observational errors in the positions of Uranus

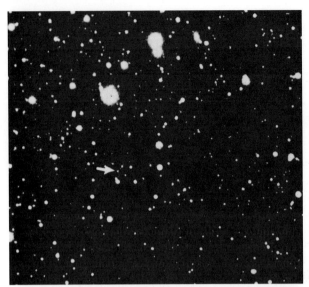

Figure 18.4 The Discovery of Pluto. These are portions of the original photographs on which the ninth planet was discovered in 1930. The position of the planet in each is indicated by an arrow. *(Lowell Observatory photograph)*

and Neptune. Thus the extensive search that resulted in the discovery of Pluto had no sound basis. It is likely that Pluto would have been found in this search, however, even if it had not been near the predicted position, because the search systematically covered the entire ecliptic plane.

Earth-based telescopic observations of the four giant planets provided a wealth of information on them and their satellites (recall that Galileo discovered the four large moons of Jupiter in 1609 and used their motions around Jupiter as evidence that the Earth is not the center of the solar system). Detailed images reveal that the atmospheres of Jupiter and Saturn have banded structures that have been attributed to flows of atmospheric gases around these planets. With more difficulty, similar flows were deduced to be present on Uranus and Neptune, primarily from infrared observations (Fig. 18.5). Spectroscopic measurements revealed that hydrogen and hydrogen-bearing compounds are dominant in the atmospheres of the giants, showing that these planets have retained lightweight gases that have been lost by the terrestrial planets. Both Jupiter and Saturn emit radio waves and excess infrared radiation, indicating that they are hot inside, hotter than they would be if sunlight were their only source of energy. The radio data also indicate that these two planets are surrounded by belts of charged particles trapped by intense magnetic fields.

Each of the four giant planets has numerous satellites; the current count is 16 for Jupiter, 17 (and five more suspected) for Saturn, 15 for Uranus, and 2 for Neptune (but Neptune has yet to be observed at close range by space probe). The brightest few moons in each case are easily seen from Earth (but not with the naked eye) and have been known for some time. In addition to the major moons, however, close-up examination by spacecraft has revealed a number of tiny satellites, usually orbiting very close to the planet.

Saturn has been known since the mid-seventeenth century to be surrounded by a set of concentric rings. Galileo had noticed the unusual shape of Saturn (which depends on the angle of the ring plane with respect to our line of sight) but was unable to deduce the cause. In 1655 the Dutch astronomer Christiaan Huygens realized that the observations indicated that a ring system was present. Much more recently, rings have been detected around Jupiter and Uranus and are suspected to circle Neptune as well, but the rings of these three planets are far dimmer and more difficult to see than the rings of Saturn.

The outer planets have been targets for space probes sent from Earth (Table 18.2), which have revealed an astonishing complexity of detail (Fig. 18.6). The U.S. *Pioneer 10* and *11* spacecraft, launched in 1972 and 1973, and the U.S. *Voyager 1* and *2* missions, launched

Astronomical Insight 18.1

Close Encounters of a Distant Kind

The process of discovery in astronomy is usually a methodical, gradual one; typically an astronomer gathers data for a few nights at an observatory and then spends days or weeks analyzing the data before determining the results. Things are far different, however, on the occasion of a planetary encounter by a space probe such as *Voyager* or its predecessors.

The preparation for a planetary encounter is a combination of literally *years* of design and planning and months or years of waiting for the probe's arrival at the target planet once it has been launched. The full scientific analysis following the encounter is thorough and methodical, and the results go through the normal scientific evaluation before being published months later in technical journals. What makes an encounter unique in astronomy, however, is that brief period of time when the data are arriving from the spacecraft. During that exciting time, the team of scientists responsible for the probe find themselves trying to explain the data and images as they are coming in, with the public and the news media looking over their shoulders.

Probably the best examples of the pressures and excitement of planetary encounters are provided by the *Voyager* spacecraft, which were launched in 1977 and arrived at Jupiter in 1979 and Saturn in 1980 and 1981. Following the Saturn encounters, *Voyager 1* took a course out of the solar system, while *Voyager 2* went on to Uranus, which it reached in early 1986, and Neptune, which it will fly by in 1989.

Like all other NASA planetary exploration missons, the *Voyager* program was managed by the Jet Propulsion Laboratory (called simply JPL), located in Pasadena, California, and operated under NASA funding by a contractual agreement with the California Institute of Technology. The spacecraft itself and the scientific instruments were built by various contractors (primarily groups of university scientists in the case of the instruments) and assembled and tested at JPL. The launches took place at NASA's Kennedy Space Flight Center at Cape Canaveral, Florida, but the in-flight mission operations and the encounter activities were all managed from JPL, through a network of deep-space radio antennae scattered around the world.

During the "cruise" phases of the *Voyager* craft—the lengthy periods between encounters—operations were routine, and most of the scientists spent their time at their home institutions carrying on normal duties. The excitement and the pace increased each time an encounter was imminent, however, with observers spending more and more time at JPL making detailed plans. After all the months of waiting, the encounters themselves took place very rapidly, with the spacecraft passing by the target planet in a matter of hours. Therefore, the sequence of observations had to be carefully planned and programmed into the on-board computers ahead of time. A further complication is the fact that radio commands to the spacecraft take substantial amounts of time to reach them; the light travel time from Earth to Jupiter, for example, was about forty-five minutes when the *Voyager* encounters took place. Thus it was impossible to relay commands to the probes and get an instantaneous response.

During the weeks and days leading up to the encounter, team scientists were busy checking the command sequence and obtaining preliminary data. Among the most important tasks was refining positional data on target objects such as satellites, so that the cameras would succeed in getting the desired images. This was especially crucial for the *Voyager 2* encounter with Uranus because the spacecraft went by the planet very quickly

(continued on next page)

Astronomical Insight 18.1

Close Encounters of a Distant Kind

(continued from previous page) and because the plane of the satellite orbits and of the rings is tipped perpendicular to the ecliptic. *Voyager* travelled in the ecliptic plane, and had to swivel around busily getting all its images in a very short time. The position of each satellite to be observed during the Uranus encounter had to be known very precisely, because even a small error would have caused the cameras to miss the target. Furthermore, the cameras had to be programmed to swivel during exposures, to avoid blurring the images due to the rapid motion of the spacecraft as it flew past. All of this was done very successfully, despite the fact that radio communications during the encounter took nearly 3 hours one way!

During the days or hours of an encounter, most of the team scientists gather at JPL to examine the data as they come in. Much of the time members of the news media watch along with the scientists as the images are displayed on a large screen in an auditorium. There are few other examples in all of science where the public sees the data at the same time as the scientists. Add the fact that every encounter revealed major surprises and mysteries, and imagine the combination of pressure and excitement the scientists experienced as they were called on to explain what they were seeing literally *as* they saw it. This gave the public a rare but distorted opportunity to see how scientists make hypotheses and refine them.

Further close encounters with distant objects such as planets, comets, and asteroids are already being planned. The next major event will be *Voyager 2*'s encounter with Neptune in August of 1989, when again things will happen very quickly and communications will be even slower than they were at Uranus. Other upcoming planetary probes include the *Galileo* mission to Jupiter, which will place a probe in orbit around the planet, allowing for a more leisurely examination, and a mission called *Comet Rendezvous and Asteroid Flyby* or *CRAF*, which will travel along with a comet for several months as it approaches the Sun. The schedule for *Galileo* is currently uncertain; it was to have been launched in the spring of 1986, but major delays have been caused by the shuttle *Challenger* disaster. The *CRAF* mission could be launched in 1991 for a rendezvous with Comet Wild 2 in 1995, encountering an asteroid on the way to the comet. Other plans for future missions include a new Mars orbiter and a Venus radar mapper. We can look forward with anticipation to more exciting times at JPL as each of these missions unfolds.

Table 18.2. Probes to the Outer Planets

Spacecraft	Targets	Flyby dates
Pioneer 10	Jupiter	Dec. 3, 1973
Pioneer 11	Jupiter, Saturn	Dec. 2, 1974; Sept., 1979
Voyager 1	Jupiter, Saturn	Mar. 5, 1979; Nov., 1980
Voyager 2	Jupiter, Saturn, Uranus, Neptune	July 9, 1979; Aug., 1981; Jan., 1986; Aug. 23, 1989
Galileo	Jupiter	1994

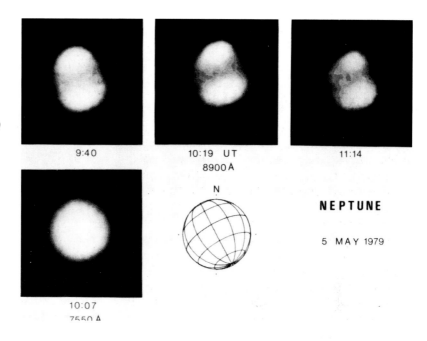

Figure 18.5 Neptune. This series of images was obtained in infrared light. The three upper images were made at a wavelength where methane absorbs light, so the dark areas are regions of methane concentration. The bright areas north and south of the equator are thought to have ice crystals and haze above the methane. *(NASA)*

in 1977, all visited Jupiter, making close-up measurements of magnetic fields, charged particles, radio emissions, and other properties, as well as providing fine images. *Pioneer 10* veered out of the plane of the ecliptic following its encounter with Jupiter, while *Pioneer 11* and both *Voyagers* went on to visit Saturn. *Voyager 2* then followed a path to Uranus and is now on its way to Neptune, where it will arrive in August 1989. It is primarily because of these spacecraft that we know much about the giant planets.

General Processes

Table 18.1 summarizes some of the general properties of the giant planets. From this table alone, we see that these planets must have very different internal structures than the terrestrial planets, because the densities of the giants are very low. The densities range from 0.7 g/cm^3 (less than the density of water) for Saturn to about 1.6 g/cm^3 for Neptune. These densities immediately tell us that these planets are not solid, rocky objects but must instead consist largely of lightweight elements. Indeed, spectroscopic data indicate that the outer layers are dominated by hydrogen, and it is believed that helium is also very abundant, although this element is more difficult to observe. The composition of the giant planets seems to be very similar to that of the Sun (Table 5.3), showing that the giant planets have not been altered significantly since the solar system formed.

Learning about the insides of the giant planets may at first seem like an impossible goal, since we have no opportunity to make direct probes into their interiors. This is true of the terrestrial planets, of course, but in their case at least we have detailed seismic information on the Earth and some confidence that the other terrestrial planets are similar. For the giant planets, we must rely rather heavily on theoretical calculations to tell us what they are like inside. Fortunately, there is good reason to believe that these calculations can tell us a great deal. Studying the structure of a planet is much like studying the structure of a star (discussed in Chapter 6). A set of equations representing the various physical processes is written and solved (usually by a large computer), yielding calculated values for such parameters as density, temperature, interior mass, and energy transport at various levels inside the planet. The solution to the equations provides predicted values for the external properties such as total mass and surface temperature, and these predictions can be compared with

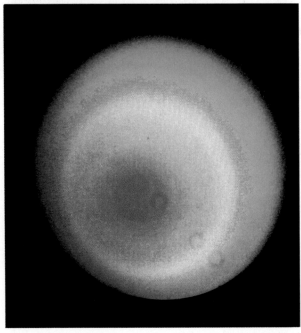

Figure 18.6 *Voyager* **Images of Jupiter, Saturn, and Uranus.** The photograph of Jupiter *(top left)* was obtained by *Voyager 1* when it was still millions of kilometers away. Already a vast amount of detail is evident in the atmospheric structure. The image of Saturn *(top right)* shows atmospheric bands. Color enhancement of a photograph of Uranus *(left)* has been used to increase the contrast and show bands. *(NASA/JPL)*

observations. If they do not agree, adjustments to the internal properties are made until theory and observation match. Because it can be shown that there is only one solution that can produce a given set of surface conditions, scientists have confidence that these model calculations give realistic results for the interior conditions. There is always room for improvement, however, because many simplifications have to be incorporated into the equations in order to make them easy to solve. As more sophisticated methods and better computers become available, it becomes possible to incorporate increasingly complex details in the calculations.

Observations and model calculations suggest that the giant planets do not have solid surfaces at all but instead are fluid throughout most of their interiors. Thus to understand the internal structures of these planets, we must deal with rather different physical processes than we did in discussing the interiors of the terrestrials. One contrast is that the interiors of the giant planets, being fluid, can easily be forced into motion, and we believe that there are complex flows deep within these planets. These flows are governed by the same two general influences that create flows in the atmospheres of the terrestrial planets, that is, heating and planetary rotation.

The major source of heat in the interiors of the giant planets is internal, rather than from the Sun (although the deposition of solar energy in the upper atmospheres does have some effects). The temperature is thought to be very high at the cores of these planets, and thus convection currents are created in the interior. In addition, the giant planets all spin very rapidly (with rotation periods much less than one Earth day), so rotational forces have very significant effects. At the uppermost levels, where the visible clouds exist, the combination of heat from below and rapid planetary rotation creates rapidly moving bands of gas that encircle the entire planet (Fig. 18.7). Details of these motions are described in the next section.

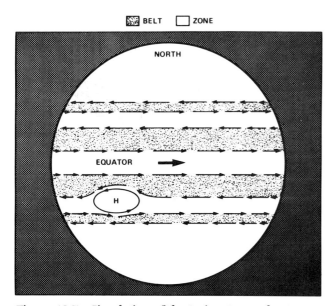

Figure 18.7 Circulation of the Jovian Atmosphere. This sketch illustrates the direction of flow in the belts and zones. *(NASA)*

Model calculations indicate that density and temperature rise to very high levels inside the giant planets, creating conditions unlike any we are familiar with on the Earth's surface and leading to forms of matter that cannot exist under normal conditions on Earth. Below the thin gaseous atmosphere lies a zone where the density is high enough for hydrogen to be compressed into a liquid state, and deeper yet is a region where hydrogen takes on a strange semiliquid, semisolid state known as **liquid metallic hydrogen.** Helium and traces of other elements are present as well in these layers, but hydrogen is by far most abundant and determines the overall physical state.

As we learned in our discussions of the terrestrial planets, differentiation takes place only when a body is fluid inside, and thus the giant planets are all differentiated in the extreme. It is thought that each of these planets has a small, exceedingly dense core that contains virtually all of the heavy elements originally incorporated into that planet when it formed. This core is thought to be solid, so in a sense these planets have solid surfaces buried deep inside.

Magnetic fields are formed by flowing material that carries electrical currents, and it is natural that the fluid interiors of the giant planets would give rise to magnetic fields. As already mentioned, Jupiter and Saturn have long been known from their radio emissions to have charged particle belts, and these belts are thought to be held in place by planetary magnetic fields. Earth-based observations give no strong indication that Uranus and Neptune have similar belts and fields, but now the *Voyager 2* spacecraft has discovered that Uranus does. We will learn whether Neptune has a magnetic field as well when the craft reaches that planet in 1989.

Now that we have reviewed the general processes that govern the properties of the giant planets, we are ready to embark on a more detailed discussion of their characteristics, their similarities, and their contrasts. We begin in the next section with a discussion of the atmospheres of the giant planets.

Atmospheres and Interiors of the Gaseous Giants

As we have already mentioned, the composition and structure of the giant planets are far different from those of the terrestrial planets. The atmospheres of the giants should be viewed as extensions of the interiors of these planets, rather than as distinct zones that are clearly defined in terms of composition and conditions. Here we discuss first the composition of the atmospheres, then the circulation patterns, and then their interiors and magnetic fields.

Atmospheric Composition

The compositions of the giant planets (Table 18.3) are very similar to that of the Sun; they are dominated by hydrogen and helium, with only traces of the heavier elements. The atmospheres of these planets are even more devoid of heavy elements than the Sun, however, because differentiation has occurred. Thus the atmospheres contain hydrogen, helium, carbon, nitrogen, and oxygen and very little else. These elements are mostly in molecular form, for example, hydrogen molecules (H_2), methane (CH_4), and ammonia (NH_3). Conditions are too cold for water vapor (H_2O) to exist in gaseous form, and helium tends to remain in atomic form. In addition to these simple molecular species composed of the most common elements, there are some more complex species present as well, varying in quantity from one planet to the next. Some unknown species creates the reddish colors in the atmosphere of Jupiter, for example, and in the atmosphere of Saturn such molecules as ethane (C_2H_6) have been detected spectroscopically. Apparently the lower temperature of Saturn (compared to Jupiter) has allowed larger and more complex molecules to form, a trend that we might expect to continue in the colder planets beyond Saturn. Ammonia, which is very abundant in the atmosphere of Jupiter, is less common on Saturn, apparently because the low temperature has allowed it to crystallize and precipitate out, like snow on the Earth. Thus, while the overall compositions of the four giant planets are similar, variations in local conditions such as temperature have caused some minor differences in the compositions of their atmospheres.

Atmospheric Circulation

The circulation patterns in the atmospheres of the four giant planets are quite similar, although again minor differences appear. Jupiter's atmosphere displays a vivid pattern of **belts** and **zones,** alternating dark and light bands that stretch around the planet parallel to the equator (Fig. 18.8). The belts are regions where the atmospheric gas is sinking, and the zones are places where it it rising (Fig. 18.9). Each may be viewed as a cyclonic (belts) or anticyclonic (zones) flow, stretched by the rapid rotation of the planet into an elongated rotary pattern. The coloration of the belts is apparently created by some as yet unidentified molecular species that forms under slightly higher pressure conditions than exist in the zones.

The many spots in Jupiter's atmosphere are localized rotary flows; some are dark like belts, and some are light like zones. The Great Red Spot (Fig. 18.10) is dark yet consists of gas that is rising above the average level of the clouds, contrary to normal belt behavior. Apparently the Great Red Spot is a long-lived storm system that has been present in the planet's southern hemisphere for at least 300 years, since the time of the first telescopic observations. There are many regions in the atmosphere of Jupiter that appear to be highly turbulent (Fig. 18.11), and they are often in the vicinity of spots.

Saturn's atmosphere has a similarly banded appearance (Fig. 18.12), but there is less contrast between the belts and zones. At the surface, Saturn has a lower gravitational field strength than Jupiter, thus Saturn's atmosphere is less compressed and extends to greater altitude (Fig. 18.13). As a result, there is an extensive region of gas above the level where the belts and zones form, and this upper region mutes the appearance of the belts and zones to an outside observer. Saturn also has spots and turbulent regions (Fig. 18.14).

One contrast between the circulation patterns on Jupiter and Saturn is that the belts and zones extend closer to the poles on Saturn than on Jupiter, and on

Table 18.3. Atmospheres of Jupiter and Saturn*

Gas	Symbol	Jupiter	Saturn
Hydrogen	H_2	0.89	0.94
Helium	He	0.11	0.06
Methane	CH_4	1.8×10^{-3}	2×10^{-3}
Ammonia	NH_3	1.8×10^{-4}	2×10^{-5}
Ethane	C_2H_2	7×10^{-7}	9×10^{-8}
	C_2H_6	4×10^{-5}	8×10^{-6}
	PH_3	4×10^{-7}	3×10^{-6}

*The values given are relative abundances by number.

Chapter 18 The Outer Planets 383

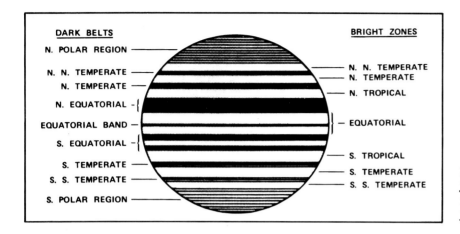

Figure 18.8 Belts and Zones on Jupiter. This schematic drawing names the major features that give Jupiter its banded appearance. *(NASA)*

Saturn the belts and zones are wider in the north than in the south. The latter is likely to be a seasonal effect; Jupiter's axis is very nearly perpendicular to its orbital plane, so Jupiter effectively has no seasons, whereas Saturn's axis is tilted by 27°, so that some seasonal variations are expected. It is difficult to determine just what the seasonal effects are, since close-up observations have been made over such a short time compared to a year on Saturn (almost 30 Earth-years). Another contrast between Jupiter and Saturn is that the wind speeds in the belts and zones are much higher on Sat-

urn (as fast as 1800 km/hr or 1200 mph on Saturn, compared to about half of that on Jupiter).

The *Voyager 2* observations of Uranus have revealed some information on atmospheric flows on the seventh planet. Before the *Voyager* mission, there was some evidence for the presence of belts and zones (see Fig. 18.15) but also some uncertainty. The uncertainty was due in part to the unknown effects of the planet's large tilt—its rotation axis is tipped 98° from perpendicular to the orbital plane (Fig. 18.16). This should surely cause unusual seasonal effects and could possibly cre-

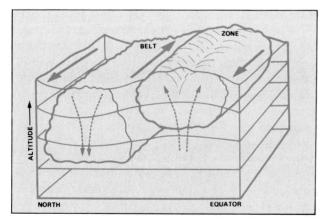

Figure 18.9 Vertical Motions in the Jovian Atmosphere. This cross-section indicates how the horizontal circulation pattern is related to vertical convective motions in the outer layers of the atmosphere. This is analogous to circulation in the Earth's atmosphere, except that the rapid rotation of Jupiter stretches rotary patterns into belts and zones. *(NASA)*

Figure 18.10 The Great Red Spot. This is a *Voyager* image of the rotating column of rising gas that has been present on Jupiter for at least 300 years. *(NASA)*

Figure 18.11 Turbulent Motions in the Jovian Atmosphere. In this *Voyager* image, colors are altered to enhance contrast and reveal complex details. *(NASA)*

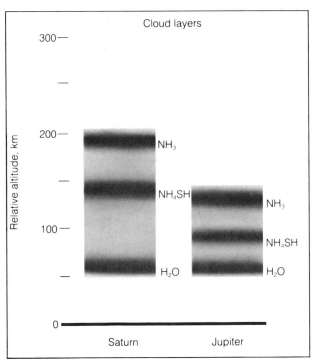

Figure 18.13 The Contrasting Cloud Thicknesses on Jupiter and Saturn. The deeper cloud zone on Saturn is largely responsible for the relative lack of contrast in its visible surface features.

Figure 18.12 A *Voyager* Portrait. This image of Saturn was obtained from several million kilometers away, and it shows much more detail than can be seen in the best Earth-based photos. *(NASA)*

ate atmospheric flow patterns very different from those of the other giant planets. One prediction made before the *Voyager* encounter was that the flows would be primarily in the direction opposite to the planet's rotation, yet this turned out not to be the case. The north pole of Uranus is currently pointed almost directly toward the Sun, and this led to the expectation that the temperature would be higher near the pole than at the equator. Theoretical calculations show that the winds near the equator should move opposite to the rotation direction in this case, yet the opposite situation was found. Surprisingly, it was discovered that the atmospheric temperature was quite uniform over the entire planet, so that the expected temperature difference between pole and equator does not exist.

The belts and zones on Uranus have almost no contrast in appearance and could be detected only when very subtle brightness variations were enhanced by computer image-processing techniques (Fig. 18.15). White clouds seen near the equator on Uranus were quickly spread out by the planetary rotation (Fig.

Figure 18.14 Circulation in the Atmosphere of Saturn. Even though the contrast is not as vivid on Saturn as on Jupiter, the ringed planet does have complex atmospheric motions, driven by a combination of convection and rapid planetary rotation. *(NASA)*

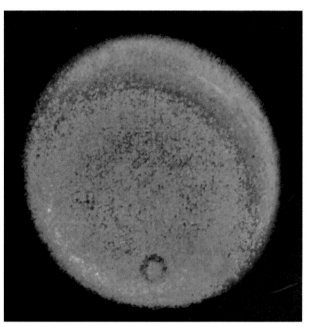

Figure 18.15 Zonal Flows on Uranus. This computer-enhanced image shows the belts and zones on Uranus, which are the equivalent of those on Jupiter and Saturn except that the contrast is very low. Despite the lack of internal heating on Uranus, the flow patterns in its atmosphere are very similar to those of the larger giant planets. *(NASA/JPL)*

18.17). These clouds are thought to be formed by plumes of methane ice crystals rising from below.

The atmosphere of Neptune is thought to have circulation patterns very much like those on Uranus, except for much less pronounced seasonal effects. Neptune has a 28° axial tilt, so we would expect some seasonal variations, but not as severe as those on Uranus. There is very little direct information on atmospheric circulation on Neptune so far, however, except for some infrared observations of methane flows (Fig. 18.5).

Interior Conditions

Theoretical models for the interior structure of two of the four giant planets are illustrated in Fig. 18.18. Each model calculation was guided by what is known of the surface conditions, and each is subject to a great deal of uncertainty. The models for Jupiter and Saturn are based on more complete information than those for

Figure 18.16 Seasons on Uranus. The unusal tilt of Uranus creates bizarre seasonal effects. The solid arrow in each case indicates the direction of the planet's north pole, and the curved arrow, the direction of its rotation.

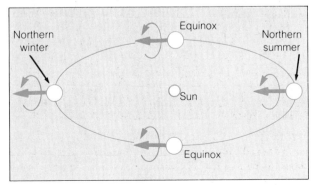

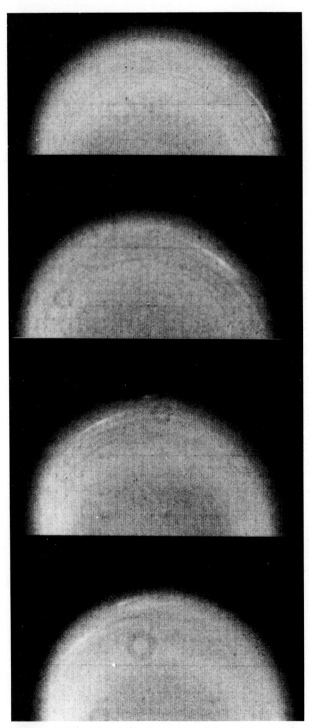

Figure 18.17 Clouds in the Atmosphere of Uranus. The white streak seen here is a cloud of methane ice crystals. Such clouds occasionally rise through the atmosphere near the equator and are then carried around the planet by the winds, as seen in this sequence. *(NASA/JPL)*

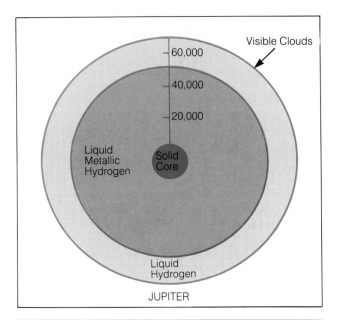

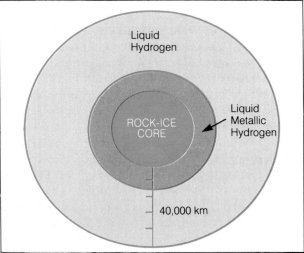

Figure 18.18 The Internal Structures of Jupiter and Saturn. These cross-sectional diagrams of Jupiter *(top)* and Saturn *(bottom)* illustrate the rough model of the planetary interiors.

Uranus and Neptune, but even so, there remains substantial room for improvement as more detailed information becomes available.

The internal structures of Jupiter and Saturn are thought to be similar in overall properties but different in detail. Jupiter has a higher overall density, indicating a greater degree of compression inside, and this is thought to have resulted in some contrasts in physical conditions. One difference is that the liquid metallic hydrogen zone in Jupiter is thought to be much more extensive than in Saturn, and the internal temperature of Jupiter is expected to be much higher. The core of Jupiter, thought to contain 10 to 20 times the mass of the Earth, is probably more dense than the core of Saturn, which may contain some ice mixed in with rocky materials.

Both Jupiter and Saturn emit excess radiation; that is, each emits more energy than it receives from the Sun (by a factor of 1.68 for Jupiter and 1.78 for Saturn), indicating the presence of an internal heat source. Because of the contrasts in interior structure already cited, however, the sources of excess internal energy are thought to be different for the two planets. Jupiter is thought to be warm inside simply as a result of heat trapped there from the time of the planet's formation some 4.5 billion years ago. The outer layers of Jupiter act as a very effective blanket, preventing internal heat energy from easily escaping. Saturn, with its lower internal density, is thought to be unable to retain heat for so long, and a different mechanism is invoked to explain its excess radiation. It is hypothesized that differentiation is still occurring in Saturn and that as heavy elements sink, gravitational potential energy is released in the form of heat. More specifically, it has been suggested that helium is settling toward the core of Saturn, drifting downward through the liquid and liquid metallic hydrogen zones.

The internal structures of Uranus and Neptune have been less thoroughly studied than those of Jupiter and Saturn, although more attention is now being focused on the outer two in view of the *Voyager* encounters. Uranus in particular has been modeled recently, and the results suggest that the internal pressure is not sufficient to create a liquid metallic hydrogen zone. Instead there is thought to be an extended liquid hydrogen zone below the visible clouds and inside of that a small solid core containing about the equivalent of one Earth-mass. Because of the great similarities between Neptune and Uranus in external properties, it is expected that the internal structure of Neptune is very much like that of Uranus.

Magnetic Fields and Particle Belts

Radio observations showed long ago that both Jupiter and Saturn have strong magnetic fields that trap charged particles in belts surrounding them. The radiation is produced by the synchrotron process (discussed in Chapter 8), in which radiation is emitted by rapidly moving electrons in a magnetic field. The magnetic fields appear to arise in internal layers where liquid hydrogen flows carry electrical currents. The field of Jupiter is not aligned with the planet's rotation axis but is tilted by 11° with respect to it, creating some interesting effects in the motions of the charged particles (discussed below). Saturn's field is closely aligned with the rotation axis.

No direct evidence for the presence of magnetic fields was found for either Uranus or Neptune before the *Voyager 2* encounter with Uranus, although some reports had been made of upper atmospheric emission from Uranus that might have been due to charged particles moving in a magnetic field. *Voyager* confirmed that Uranus has a field and revealed some surprising information about it. It remains to be seen whether Neptune also has a magnetic field, but the similarities of its external and, presumably, internal properties to those of Uranus suggest that it probably has.

The field of Uranus is quite unusual. The axis of the field is not aligned even closely with the rotation axis but is tipped by about 60°, meaning that its orientation is not so far from perpendicular to the planet's orbital plane (Fig. 18.19). Furthermore, the magnetic axis is offset from the center of the planet, showing that the region where it forms is not symmetrically distributed about the core. The magnetosphere of Uranus trails away from the Sun due to the solar wind, as do the magnetospheres of the other planets with magnetic fields, but in this case the tail of the magnetosphere wobbles as the planet rotates because of the gross misalignment of the field with the rotation axis of the planet.

The charged particle belts of Jupiter and Saturn are closely confined to the equatorial planes of the planets because their rapid rotation forces the magnetospheres to be highly flattened (Fig. 18.20). The magnetosphere of Jupiter wobbles because of the 11° misalignment between the magnetic field and the rotation axis. This effect is especially marked because of a trail of ionized gas in the orbit of Io, the innermost of the four major moons of Jupiter (Fig. 18.21). This trail of gas, known as the **Io torus,** is formed from gas that escapes the surface of Io and goes into orbit around Jupiter. The torus wobbles as Jupiter rotates, because the magnetic field of Jupiter governs the motions of the charged particles in the torus, and the field is misaligned with the rotation axis of the planet.

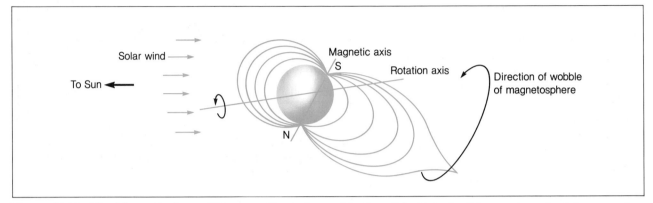

Figure 18.19 The Magnetosphere of Uranus. Due to the combination of the high tilt of the rotation axis of Uranus and the misalignment between its rotation and magnetic axes, the magnetosphere of Uranus resembles that of one of the planets whose rotation axis is more normal. The tail of the Uranian magnetosphere rotates as the planet spins around its highly tilted axis.

Figure 18.20 The Jovian Magnetosphere and Radiation Belts. The shape of the magnetosphere is influenced by the solar wind and by the rapid rotation of Jupiter. The result is a sheet of ionized gas that is closely confined to the equatorial plane but wobbles as the planet's off-axis magnetic field rotates. *(NASA/JPL)*

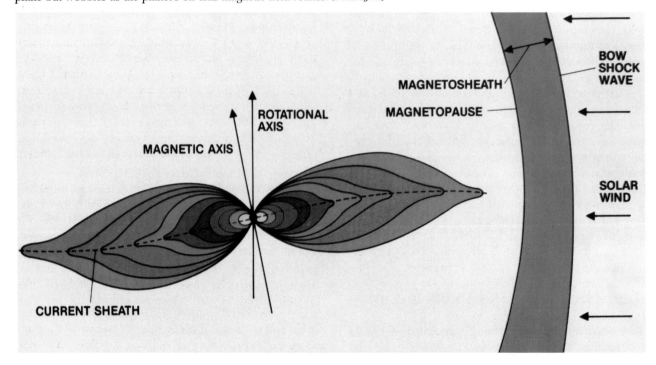

Rings and Moons

All four of the giant planets appear to have ring systems, and at least three of the four have large numbers of satellites. From what we learned in Chapter 15 about the formation of these planets, we believe that the creation of rings and numerous satellites is a natural by-product of the formation process. As each giant planet coalesced from planetesimals, its gravitational field apparently trapped large quantities of gas and solid matter from surrounding space. (This did not happen in the inner solar system, largely because of tidal forces from the Sun and higher temperatures, which prevented the terrestrial planets from trapping volatile gases.) The material gathered in by each giant planet was forced into a disklike shape by the rapid rotation of the planet. In the outer portions of each disk, solid particles were able to coalesce to form satellites, while tidal forces prevented this in the inner portions.

Recall that tidal forces acting on a body in a gravitational field have the net effect of stretching the body (Chapter 3). On the Earth the tidal force causes tides in the oceans; on the Moon, it creates a permanent deformation (and caused the Moon to slow its rotation until it was synchronous). If a swarm of small particles is to coalesce to form a satellite, it must come together under its own gravitational force. If there are tidal forces present that exceed this gravitational attraction, then a satellite can never form. The closer the loose particles are to the parent planet, the stronger the tidal force and the greater the difficulty in forming a moon. For each planet, there is a point close to the planet, within which the tidal force acting to keep particles apart is stronger than the gravitational force acting to pull them together. This point, called the **Roche limit**, determines where the boundary between rings and satellites occurs for each giant planet. Some minor moons can exist within the Roche limit, but major bodies cannot form there.

Both the ring particles and the satellites of the giant planets are composed largely of ice, mixed with rocky material to varying degrees. Thus the densities of the major moons tend to lie near 1 g/cm^3, exceeding that only for satellites massive enough (or exposed to enough tidal stresses) to have undergone some geological processing. In the following sections, we discuss first the satellites and then the ring systems of the giant planets.

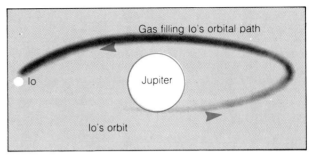

Figure 18.21 The Io Torus. This sketch illustrates the geometry of the ring of gas that fills the orbit of Io.

Satellites of the Giant Planets

Jupiter has 16 known satellites; Saturn, 17 (with more suspected); Uranus, 15; and Neptune, 2 (but perhaps more will be found when *Voyager 2* visits in 1989). The names and properties of all these satellites are listed in Appendix 7. The major satellites of each planet were discovered telescopically from Earth, while the minor ones are far too small to be seen in this way and were found primarily by close-up imaging by space probes. The most famous of the satellites are the Galilean moons of Jupiter, named in honor of Galileo, who discovered them in 1609. Titan, the giant moon of Saturn, was found by Huygens in 1655, and four more moons of Saturn were discovered by the French astronomer G. D. Cassini in the late 1600s. Additional intermediate-sized satellites of Saturn have been discovered periodically since then. Five moons of Uranus were known before the *Voyager* encounter, two of them having been discovered by Herschel himself in 1787, some six years following his discovery of the planet. Triton, the giant moon of Neptune, was discovered in 1846 (the same year that Neptune itself was discovered), but Nereid, the lesser moon, was not found until 1949.

The relatively large satellites of the giant planets have undergone some degree of geological processing. Most of the moons that are larger than about 100 km in diameter are more or less spherical in shape, indicating that they have been at least partially molten and probably have differentiated to some extent (Fig. 18.22). Their compositions are largely ice, with varying quantities of rocky material included. The surfaces of many show linear grooves and other features indicative of crustal shifting, and it is clear that some of the surfaces are younger than others, that is, some have been

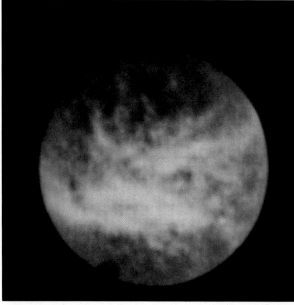

Figure 18.22 Large Satellites of the Giant Planets. This shows examples of some of the large, icy moons of the giant planets. At upper left is Jupiter's moon Callisto; at upper right is Dione and lower left is Rhea, both moons of Saturn; and at lower right is Titania, a satellite of Uranus. All photos were obtained by the *Voyager* probes. (NASA)

subjected to more reprocessing by outflows from the interior, obliterating ancient craters. The degree of tectonic activity depends in part on tidal stresses; the satellites subjected to the greatest tidal force due to their parent planet or to close encounters with neighboring satellites tend to have the youngest surfaces and the strongest evidence of geological processing such as shifting crustal plates. All of the satellites of the giant planets are in synchronous rotation, forced into that state by tidal forces due to their parent planets.

The four Galilean satellites of Jupiter (Fig. 18.23) provide an excellent example of the dependence of in-

Figure 18.23 The Galilean Satellites of Jupiter. These are the four moons discovered by Galileo, shown in correct relative size. They are Io *(upper left)*, Europa *(upper right)*, Ganymede *(lower left)*, and Callisto *(lower right)*. Ganymede is the largest satellite in the solar system. From Callisto, the farthest from Jupiter, to Io, the innermost of the four, there is a wide range of geological and physical properties, which is discussed in the text. *(NASA)*

ternal activity on tidal stresses. Callisto, the outermost, has a very ancient surface densely covered with old craters. Ganymede, next in, shows extensive cratering but also has some linear features due to past tectonic activity that apparently involved the upwelling of water from below, which then froze. Next in is Europa, which is covered by an extensive system of dark grooves thought to be made of water ice formed when water oozed up through cracks in the crust (it is thought that liquid water may still exist below the surface of Europa). Finally, the closest to Jupiter and therefore the one subject to the greatest tidal force is Io, a moon with a high degree of current geological activity. Its surface is very young (no impact craters persist) and is covered with brightly colored volcanic deposits. *Voyager 1* discovered active volcanic eruptions (Fig. 18.24), and later it was determined that as many as 8 to 10 volcanoes were erupting at a single time. The surface shows volcanic vents here and there (Fig. 18.25). The density of Io is larger than those of the other Galilean moons, indicating that a larger fraction of its lightweight, volatile gases have been cooked out of the interior and allowed to escape (forming the Io torus, described in the previous section).

Not only is Io subject to strong tidal forces due to Jupiter itself, but it is also in **orbital resonance** with Europa, that is, the orbital periods of the two moons are simple fractions of each other, so that they are frequently aligned on the same side of Jupiter. Io's orbital period is almost exactly half that of Europa, so that the two are often aligned, adding to the tidal stress on Io (and on Europa, accounting for its relatively young surface and warm interior).

Most of the moons of Saturn are smaller than the four Galilean satellites, and none appear to be as geologically active as Io. The one very large moon, Titan

Figure 18.24 An Eruptin on Io. This image of an eruption from a volcanic vent with a plume of ejected gas dramatically illustrates Io's present state of dynamic activity. *(NASA)*

Figure 18.25 Volcanic Vents on Io. The darkest material seen here indicates the locations of recently active volcanoes. *(NASA)*

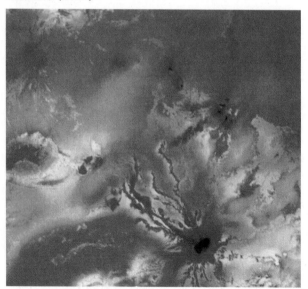

Figure 18.26 Titan. The largest of Saturn's moons, Titan is almost unique among all the satellites in the solar system in having an atmosphere. *(NASA)*

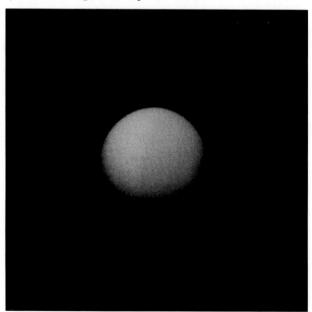

(Fig. 18.26), is unusual in that it has a thick atmosphere, the source of which is unknown. The surface atmospheric pressure is 50 percent greater than sea-level pressure on the Earth, and the atmosphere of Titan extends some five times higher (Fig. 18.27). The composition is dominated by nitrogen, although methane is also present. Calculations show that the surface conditions (pressure and temperature) are near the point where methane can exist in solid, liquid, or gaseous form, so it is speculated that there may be methane lakes there, with methane ice as well. It may also be that nitrogen exists in liquid form (the temperature of liquid nitrogen at atmospheric pressure on Earth is 77 K, and the surface temperature on Titan is about 90 K, but the higher pressure could allow liquid nitrogen to exist there).

The seven intermediate moons of Saturn (Fig. 18.28) are similar in their gross properties such as diameter and density, but they vary considerably in surface appearance. Most have craters, but some, such as Dione, have wispy patterns of frost that were probably deposited as gases escaped from the interior, and some (Dione, Tethys, and Rhea) have linear trenches that are probably the result of crustal movements or fractures caused by massive impacts. Mimas has an impact crater that is about one-fourth as large as the diameter of the satellite itself, representing a blow that may have nearly destroyed it. Enceladus is the shiniest object in the solar system, reflecting virtually 100 percent of the light that strikes it, while Iapetus has a curious mixture of light and dark regions on its surface, perhaps due to the deposition of sootlike material from space. The remaining intermediate moon, Hyperion, has a strangely asymmetric shape.

The five large moons of Uranus (Fig. 18.29) are similar in size and overall properties to the intermediate moons of Saturn. Again, as in the case of the Galilean satellites of Jupiter, there are variations in the degree of geological activity, but here the variations are not so systematically related to the distance from the parent planet. The outermost is Oberon, which has a heavily cratered surface and is probably dormant, although its vast mountain and cliff systems indicate past activity. Next is Titania, which shows signs of more recent pro-

Figure 18.27 Titan's Atmosphere. This diagram compares conditions in the atmospheres of Titan and the Earth. Note that the vertical scales are not the same; Titan's atmosphere is much deeper. *(Data from Beatty, J. K., B. O'Leary, and A. Chaikin, (eds.) 1981, The New Solar System (Cambridge, England: Cambridge University Press)*

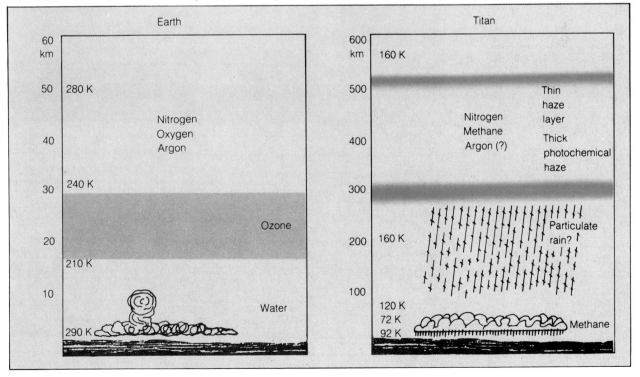

394 Section V The Solar System

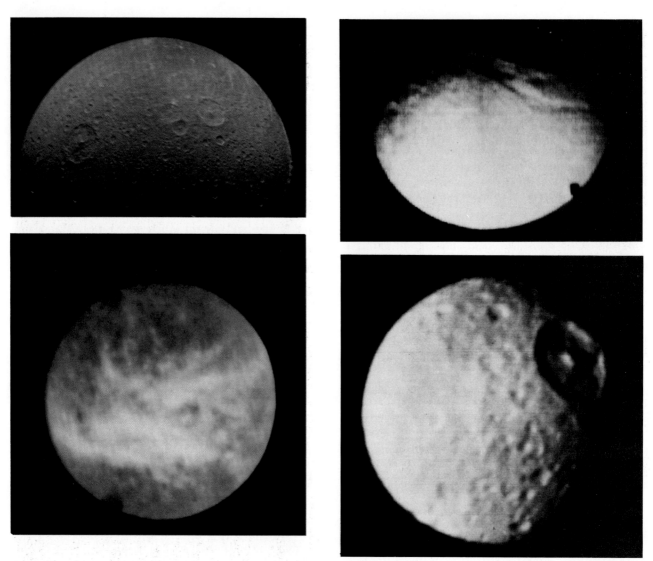

Figure 18.28 The Seven Intermediate Moons of Saturn. *Clockwise from top left:* Dione, Tethys, Mimas, and Rhea. *(Continued on next page.)*

cessing, while Umbriel, the third one in, has an old surface and little evidence of geological activity. Umbriel has a very dark surface, perhaps due to a coating of debris from space. Ariel, the brightest of the five, apparently is coated by ice that has escaped the interior. The innermost of the five is Miranda, whose surface indicates that this moon has undergone major geological turmoil. There are large mountains and huge uplifted regions with rectangular shapes and layered edges (Fig. 18.30).

No explanation for these structures on Miranda has been widely accepted, but one interesting possibility has been proposed. A satellite close to its parent planet is unusually prone to impacts from space, because the planet's gravitational field attracts nearby bodies. It has been suggested that Miranda has been broken into fragments by a large impact, only to re-form due to its own gravitation. When it came together again, according to this hypothesis, some of the lightweight icy material may have ended up buried in the interior, while

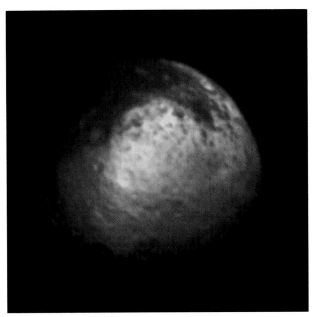
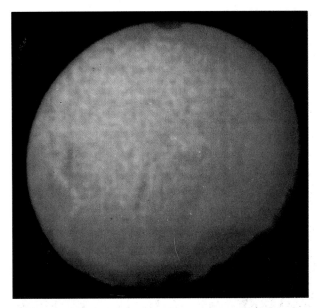
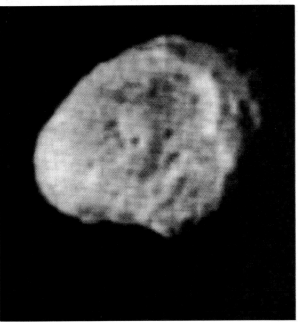

Figure 18.28 *(continued). Clockwise from top left: Iapetus, Enceladus, and Hyperion. (NASA)*

heavier rocky matter ended up in the outer layers. The buoyancy of the ice would then have forced it upward, creating the curious highlands that are seen in the *Voyager* images. This explanation is obviously very speculative, but it is supported by the observation that other major moons of the giant planets have huge impact craters.

The two known moons of Neptune are mysterious objects to us, although we can hope to learn more in 1989, when *Voyager 2* is scheduled to fly by. These two moons, Triton and Nereid, have very unusual orbits (Fig. 18.31). Triton lies close to Neptune and follows a nearly circular orbit, but it goes around Neptune in the retrograde direction. Nereid lies much farther out and goes around in the normal (prograde) direction, but it has a very eccentric orbit, with its distance from Neptune varying by a factor of five from closest approach to farthest retreat. No widely accepted explanation for these strange orbits have been offered, but some have suggested that they may be the result of a near collision between the satellites and a third body.

The minor moons of Jupiter, Saturn, and Uranus tend to lie close in to their parent planets, within the Roche limit for larger bodies. They are irregular in shape and seem to be composed primarily of ice, although detailed analyses are difficult. There are three examples of **co-orbital satellites** orbiting Saturn. In these situations, one or more minor moons share the same orbit with a large one, occupying positions either 60° ahead of or 60° behind the large one (Fig. 18.32). The combined gravitational forces of Saturn and the large moon cause the small moons to be effectively

Figure 18.29 The Five Large Moons of Uranus.
Bottom left: Umbriel; *top left:* Oberon; *top right:* Titania. *(Continued on next page.)*

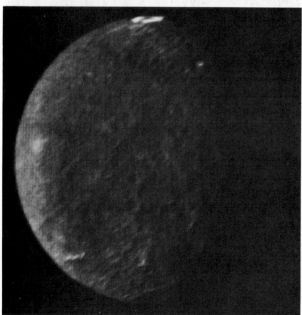

trapped in these positions. The large moon Tethys has two co-orbital satellites, and Janus, itself not very large, has one. The small satellites of Saturn and Uranus will be discussed further in the next section on the ring systems of those planets, because these moons have major effects on the rings.

Ring Systems

The rings of Saturn were observed by Galileo when he first turned his telescope toward that planet, although he was unaware of the cause of its strange appearance. More than four decades passed before Huygens deduced the correct explanation. Subsequent observations showed that since the planet can be seen through the rings, they must be made of myriads of individual particles. Spectroscopic observations in which the particle speeds were deduced from the Doppler shift (discussed in Chapter 4) later demonstrated that the particles follow their own individual orbits around Saturn, in accordance with Kepler's laws. The sizes of the particles could also be estimated from the manner in which they reflect sunlight; it was found that they range in size from very small (a fraction of a centimeter) to large (many centimeters or even meters). More recently, radar observations have also been used to analyze the particle sizes; these observations are particularly important for studying the larger particles.

Uranus was the second planet known to have a ring system, which was first observed in 1977 when Uranus

 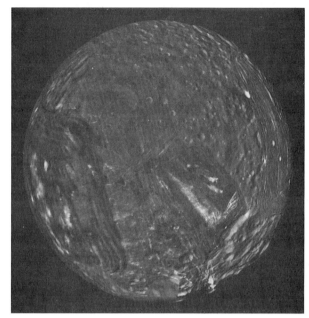

Figure 18.29 *(continued).* *Left:* Ariel; *right:* Miranda. *(NASA/JPL)*

passed directly in front of a distant background star. At this time a **stellar occulation** observation was being carried out to measure the diameter of Uranus accurately by timing how long it took for the planet, moving at its known speed, to move a distance equal to its own diameter as it passed in front of the star. To the surprise of the astronomers making the measurement, the star blinked off and on several times just before and just after the planet itself moved in front of it. The astronomers quickly deduced that Uranus had a set of nine distinct rings, each thinner and dimmer than the rings of Saturn. These rings have since been detected with Earth-based telescopes, particularly infrared ones (Fig. 18.33).

More recently, similar evidence has been found that Neptune may also have a ring system, although there are contradictory reports and the question is still unsettled. Meanwhile, the *Voyager* observations of Jupiter revealed a single, thin ring circling that planet (Fig. 18.34), so we now know that at least three and possibly all four of the giant planets have rings.

Close-up examinations of the ring systems of Jupiter, Saturn, and Uranus have now been possible, thanks to the *Voyager* spacecraft. The rings of Saturn were observed extensively by both *Voyager* probes, which revealed an astonishing array of complex structures. Be-

Figure 18.30 The "Circus Maximus." This unusual feature (right) on Miranda consists of an elevated inner portion, and sharply sloping terraces around it, resembling a Roman racetrack. The cause of this feature is unknown, but it would require some source of uplift in the interior. At lower left is a similar, chevron-shaped feature. *(NASA/JPL)*

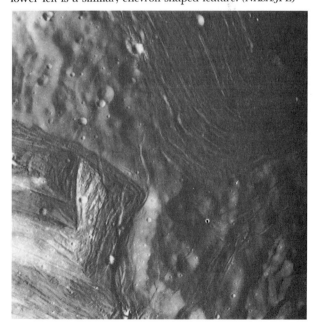

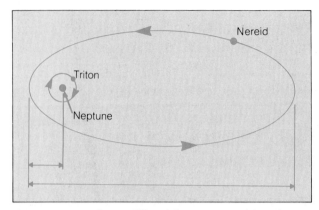

Figure 18.31 The Satellites of Neptune. This scale drawing illustrates the unusal orbital characteristics of the two moons, Triton and Nereid.

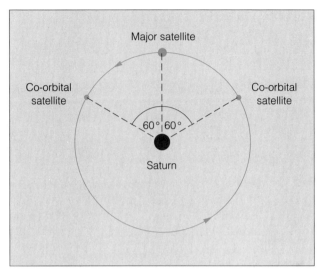

Figure 18.32 Co-Orbital Satellites, The combined gravitational effects of a planet and a major satellite (or any pair of bodies) create positions 60° ahead of and behind the major satellite where other bodies tend to be trapped. Several such co-orbital satellites have been discovered in the system of moons orbiting Saturn.

Figure 18.33 The Rings of Uranus. This infrared image of Uranus uses color to indicate relative temperatures. The blue disk of the planet is surrounded by red for the rings, which are colder. *(NASA)*

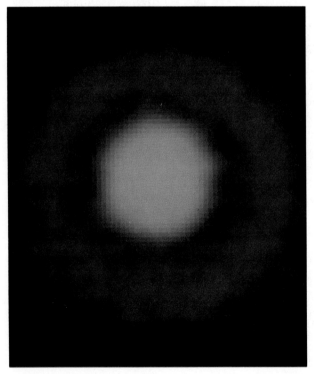

Figure 18.34 The Jovian Ring. This *Voyager 1* image revealed the first evidence of a ring girdling Jupiter. Here a segment of the ring is seen at the right. *(NASA)*

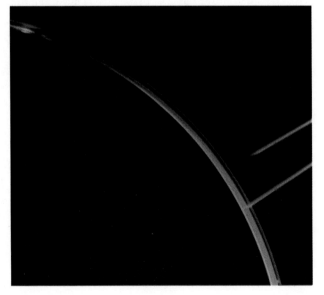

Astronomical Insight 18.2

Resolving The Rings

The *Voyager* missions to Saturn have produced a wealth of new information on the rings, primarily because of the high resolution of the images that they have sent back to earth. Recall, from our discussions in Chapter 4, that resolution is the ability to make out details. In our past discussion, we emphasized that resolution can be enhanced by the use of large telescopes, by placing observatories above the Earth's turbulent atmosphere, or by the use of special interferometry techniques. There is another way to observe fine details, however: get close to the subject of the observations. This is what the *Voyager* spacecraft did.

The cameras aboard *Voyager* 1 and 2 yielded images of Saturn's rings with a maximum resolution of abut 10 kilometers; that is, details as small as 10 kilometers in size could be seen. This was quite an improvement over the best photographs taken from Earth, which have a resolution of about 400 kilometers. The *Space Telescope*, when it is placed in Earth orbit in the late 1980s, will resolve details in the ring system about ten times smaller than this, roughly 400 kilometers, still far inferior to *Voyager* images.

There was an instrument on the *Voyager* spacecraft that provided much better resolution than even the cameras. This was the photopolarimeter, a device for measuring the intensity of light, as well as its polarization. The photopolarimeter on *Voyager 1* failed before that spacecraft reached Saturn, but the one on *Voyager 2* was able to carry out an observation that has provided by far the most detailed information yet on the ring system.

The observation was rather simple: As *Voyager 2* approached Saturn, the photopolarimeter measured the brightness of a star (Delta Scorpii) that happened to lie behind the rings, from the point of view of the spacecraft. As *Voyager 2* swept by the planet with the photopolarimeter trained on the star, the spacecraft motion caused the ring system to move across in front of the star. As it did so, the ringlets and gaps caused the brightness of the star to fluctuate. When a dense ringlet passed in front of the star, its brightness temporarily dimmed; when a gap passed in front, the star appeared brighter. This kind of observation is known as a stellar occultation.

The photopolarimeter was able to record the changes in the star's brightness very rapidly. This, combined with the rate at which the rings appeared to move across in front of the star, determined how fine the details were that could be detected in the ring structure. As it turned out, the resolution of this experiment was about 100 meters, a factor of 100 better than the best images obtained by the *Voyager* cameras! This meant that ringlets or gaps as small as the length of a football field could be detected.

Of course, the photopolarimeter could only measure the ring structure along a single cross-section; it could not provide a full picture of the entire system. Its image of the rings was one-dimensional, but it nevertheless led to profound new understanding, particularly of the spiral density waves described in the text.

The data produced by the photopolarimeter were originally in the form of plots showing the variations in the star's intensity as a function of time. Dips in the intensity corresponded to the passage of ringlets in front of the star. This information was then translated into a plot of ring density as a function of distance from Saturn. The data were processed a step further so that things were easier to visualize; synthetic images of the rings, simulating those taken by a camera, were produced (it was assumed that the rings were circular, and different colors were used to represent varying degrees of ring density; see Fig. 18.38). In this way, false-color images of the rings were produced, somewhat similar to the actual photographs obtained by the cameras, but showing details a factor of 100 smaller.

sides the major gaps known from Earth-based observations, many small gaps and tiny rings were discovered (Fig. 18.35). Rather than consisting of three or four major, broad rings, the system was now seen to consist of countless tiny features, dubbed "ringlets." Explanations for most of this structure were soon found, although some mysteries remain.

At least one major gap (the so-called Cassini division; see Fig. 18.1) is probably created in part by an orbital resonance. Recall that orbital resonance refers to a situation in which two orbiting bodies have orbital periods that are simple multiples of one another. It happens that a ring particle orbiting Saturn at the position of the Cassini division would have exactly one half the period of the moon Mimas. Thus a particle at this distance from the planet would find itself frequently lined up between Mimas and Saturn and subjected to the combined gravitational effects of the two. Apparently these frequent alignments are sufficient to alter the orbit of any particles at that distance from Saturn and thus keep that position clear and maintain the gap there. The details of how this happens are very complex and not yet worked out in full mathematical detail, but the coincidence of the orbital resonance with Mimas suggests strongly that such a mechanism is indeed responsible for creating and maintaining this gap. Apparently another factor is at work, however; it has been suggested that a small moon orbits in the Cassini division, helping to alter the orbits of any particles there and therefore helping to keep the gap clear. This conclusion is based on details of the structure of the gap, despite the fact that the proposed moon has not yet been detected.

At least two other gravitational effects, both involving moons of Saturn, appear to play roles in shaping the ring structure. The thin, wispy F ring (Fig. 18.36) lies between the orbits of two small satellites, and their combined gravitational effects seem to be responsible for keeping the F ring particles in the ring (Fig. 18.37). The small gravitational tugs exerted on the ring particles by these moons keep the particles confined to an orbit lying between the moons. Since this is something like the action of sheepdogs keeping a flock of sheep in line, these moons have been dubbed **shepherd satellites.** The unusual twisted appearance of the F ring is probably caused by the gravitational effects of these moons.

The other gravitational effect on the ring structure was confirmed by a stellar occultation observation made by an instrument on *Voyager 2* that measures brightnesses very accurately. As the spacecraft flew past the ring system, this instrument measured changes in the brightness of a background star seen through the rings, recording fluctuations caused by the rings as

Figure 18.36 The F Ring. Here we see the unusual braided appearance of this thin, outlying ring. The cause of the asymmetric shape is not well understood. *(NASA)*

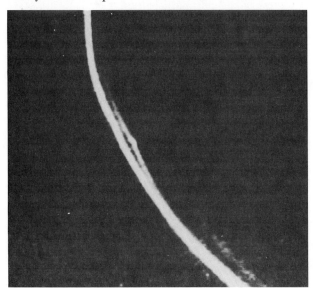

Figure 18.35 The Rings of Saturn. This *Voyager* image reveals some of the fantastically complex structure of the ring system. *(NASA)*

Figure 18.37 The Action of Shepherd Satellites. This illustrates how a pair of small moons can keep particles trapped in orbit between them. The inner shepherd satellite overtakes the particles, accelerating them if they fall inward, thus moving them out. Similarly, the outer shepherd slows particles that move too far out, so that they drift back in (*NASA/JPL*)

they appeared to move in front of the star. These fluctuations were then translated into measurements of the thickness of the rings, revealing details ten times smaller (that is, about 100 meters across) than the finest details seen in the images obtained by *Voyager*'s camera. (Fig. 18.38).

These fine observations showed that in portions of the ring system, the tiny ringlets form a spiral pattern; that is, rather than consisting of separate circular rings, the rings are actually a single trail of particles arranged in a spiral pattern, like the groove on a phonograph record. Such a pattern is expected to form in a rotating, fluid disk that is subjected to gravitational disturbances by outside objects such as satellites orbiting Saturn beyond the ring system. The disk oscillates in a standing wave that has a spiral shape. This **spiral density wave** is exactly analogous to the mechanism that is thought to be responsible for the spiral arms in our galaxy and others like it (Chapter 10). The possibility that the rings of Saturn were subject to spiral density waves had been discussed before the *Voyager* encounters and was confirmed by the observations just described.

There remain some structural features of the rings of Saturn that are more difficult to explain. For example, some of the rings are quite asymmetric, that is, noncircular (Fig. 18.39), and it is still not clear why.

Figure 18.38 Very Fine Structure in the Rings. This is a synthetic image of a small section of the ring system, reconstructed from stellar occultation data obtained by the photopolarimeter experiment on *Voyager 2*. The finest details visible here are only about 100 meters in size, whereas the best *Voyager* camera images have a resolution of about 10 kilometers. Data like these have established that spiral density waves play a role in shaping the rings. (*NASA*)

In addition, there are some dark, radial features called spokes (Fig. 18.40), which appear not to obey gravitational forces at all. The spokes are suspended above and below the plane of the rings, and they are now thought to be due to very fine particles that carry enough electrical charge to have their motions governed by Saturn's magnetic field.

The ring system of Uranus proved to be nearly as complex as that of Saturn, even though it is not as extensive or as bright. The *Voyager 2* spacecraft obtained good images of the rings (Fig. 18.41) and was also able to carry out stellar occultation measurements revealing fine detail (Fig. 18.42). In addition to the nine major rings already known from Earth-based observations, many dimmer features were found, some with surprising characteristics (Fig. 18.41). Some of the rings are noncircular and may even be incomplete, that is, they may be arcs rather than full rings around the planet. It is not known how the noncircular shapes of these rings

Figure 18.39 Asymmetries in the Rings. This composite shows that the rings are not perfectly circular; here the thin ring within the dark gap is thinner on one side of the planet than on the other and slightly displaced. *(NASA)*

Figure 18.40 The Dark Spokes. At left the spokes appear bright because they scatter light in the forward direction, and they were photographed looking toward the Sun. At right they are viewed looking away from the Sun; so they appear dark. *(NASA)*

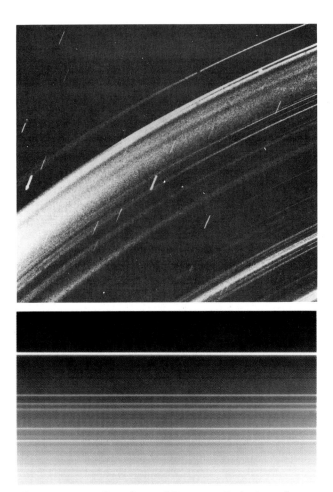

Figure 18.41 The Rings of Uranus. Both views show the epsilon ring at the top, and include all the other major rings inward toward the planet. The color-enhanced image brings out greater detail. *(NASA/JPL)*

Figure 18.42 Detailed Structure in the Rings of Uranus. This is a computer reconstruction of small-scale structure in the rings of Uranus, obtained by one of the instruments aboard *Voyager 2* during a stellar occultation observation. The instrument observed the variations in brightness of a background star as the rings passed in front of the star. The brightness variations were then interpreted in terms of density variations in the rings, leading to this reconstruction. *(LASP, University of Colorado, sponsored by NASA)*

are maintained, but it is suggested that gravitational forces due to the many minor moons may play a role. Another surprise was the absence of small ring particles; the size range of particles in the rings of Uranus is between 10 cm (about 4 in) and 10 meters (40 feet). These large particles are very dark compared to those in the rings of Saturn. Their low reflectivity is thought to be due either to a carbonaceous composition or to processing of the particle surfaces by ions in the planet's charged particle belts.

There is a substantial quantity of fine dust particles in parts of Uranus' ring system, and this presents another mystery: It is calculated that these dust grains should collide frequently with gas particles, causing the dust to lose orbital energy and spiral into Uranus. Thus the presence of dust suggests that there is a source to replenish what is being lost or that the dust is only a temporary feature of the ring system.

There are suggestions that the entire ring system of Uranus may be a changing, unstable entity. As already mentioned, it is not understood how asymmetric rings can stay in their observed shapes. It is also difficult to see how the major rings, which have rather distinct edges, can remain so well defined and sharp, rather than gradually diffusing and merging into a single, broad disk. It is possible that the sharp ring boundaries are maintained by shepherd satellites as yet undetected, or perhaps the rings are simply young and have not yet had time to dissipate. One of the rings definitely is governed by shepherd satellites (Fig. 18.43), but so far similar shepherds for the other rings have not been found.

If the ring system is young, this does not mean that Uranus only recently acquired rings; it means that the rings we see today have been modified or replenished somehow. The most likely mechanism for doing this is thought to be the destruction of a satellite, presumably by impact. Impacts due to passing bodies such as asteroids are more probable near a massive planet, because the planet's gravitational field will attract such bodies. It has even been suggested that collisions are so common that satellites of giant planets cannot survive for the 4.5 billion year age of the solar system without being broken apart at least once. This would imply that both the ring system of Uranus and its present day moons are young, having been re-formed following impacts that destroyed them (recall that this explanation has been suggested for the unusual geological features of Miranda, the innermost major satellite of Uranus). The same suggestion can be extended to the moons of the other giant planets, and it would greatly revise our understanding of their formation and evolution, but it is still a speculative idea, and it has not yet been widely accepted.

Figure 18.43 Shepherd Satellites. Here we see two small moons on either side of a thin ring around Uranus. The gravitational effects of these moons keep the ring confined. *(NASA/JPL)*

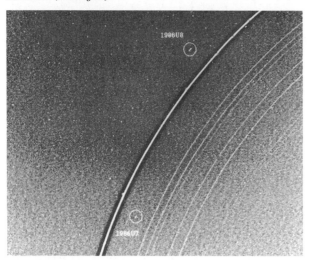

Many of the ideas regarding the rings of Uranus and their asymmetries can be tested by close-up observations of the probable rings of Neptune. Therefore, we can hope to gain new insights into both planets and into rings in general when *Voyager 2* encounters Neptune in 1989.

Pluto, Planetary Misfit

We have now thoroughly discussed the first eight planets from the Sun. In doing so, we were able to make many useful comparisons that added to our overall understanding of the solar system and its formation and evolution. Now we are ready to describe the ninth planet, Pluto, and here we will find almost no connection with the systematic trends we have discussed previously. It is apparent that Pluto is different from the other outer planets, yet is also very unlike the terrestrial planets.

The first irregularity we notice is that the orbit of Pluto is highly unusual, compared to those of the other planets. It is both noncircular and tilted with respect to the ecliptic. At its greatest distance from the Sun, Pluto is nearly 70 percent farther away than at its closest, when for a time it actually moves closer to the Sun than Neptune (in fact, Pluto is temporarily the eighth planet from the Sun right now; it moved inside of Neptune's orbit in 1978 and will not re-emerge until 1998). Pluto's orbit is tilted by 17° to the ecliptic, whereas the tilts of all the other planetary orbits are 7° or less.

In physical characteristics, Pluto also seems not to fit in with the other planets. The other outer members of the solar system are giant planets with thick atmospheres; while it has been difficult to pinpoint the mass of Pluto, it has long been clear that this planet does not resemble the other outer planets. For a long time the only information on the mass of Pluto came from the fact that it had no significant effect on the motion of Neptune, which implied that Pluto's mass had to be less than about 10 percent of that of the Earth. The diameter has been particularly difficult to determine; recent values range from 2200 km (based on infrared data from the *IRAS* satellite) to 3000 km (derived from interferometry; see Chapter 4). In either case the planet is certainly very small, comparable to any of several satellites of the giant planets.

Spectroscopy has shown that Pluto has methane ice and possibly methane gas as well, so there is a thin atmosphere and perhaps also a frost of methane ice on the surface. Most molecular compounds formed of common elements freeze out into solid form at the low temperature (perhaps 40 K) thought to characterize Pluto.

Brightness variations seem to indicate that Pluto has some surface markings and rotates with a period of 6.39 days. The orientation of the rotation axis has been difficult to determine, but new information indicates a large tilt, probably about 112° (with respect to the planet's orbital plane, which is itself tilted 17° with respect to the ecliptic).

Pluto was found in 1977 to have a satellite (Fig. 18.44). This was a bit of a surprise, because some astronomers had suggested that Pluto itself was an escaped moon of one of the giant planets (most likely Neptune). Pluto's moon, named Charon, is nearly as large as Pluto itself (the *IRAS* infrared data suggests a diameter of 1500 km, compared to 2200 km for Pluto). Charon's orbital period is also 6.39 days, and Pluto and Charon are *both* in synchronous rotation, each keeping the same face permanently toward the other.

The orbital plane of Charon is tilted about 68° with respect to the plane of Pluto's orbit about the Sun, and Charon orbits in the retrograde direction (Fig. 18.45). This leads to the conclusion that Pluto's rotation axis is tipped 112° with respect to the perpendicular to its orbital plane, based on the assumption that Charon orbits in the plane of Pluto's equator. (In other words, the rotation axis points 22° below the orbital plane; see Fig. 18.45.)

The discovery of Charon gave astronomers a chance to determine the mass of the planet using Kepler's third law. It was found that Pluto's mass is only about 0.0019 Earth mass, and this leads in turn to a rather low average density, around 2 g/cm^3 if the *IRAS* value for the diameter (2200 km) is correct and much lower (perhaps 0.5 g/cm^3) if the larger suggested values for the diameter (around 3000 km) are adopted. It is very difficult to see how the density could be less than 1 g/cm^3, since this is the density of pure ice. Even a value of 2 g/cm^3 implies a large ice content, however, much like the large moons of Jupiter and Saturn.

Perspective

We have thoroughly digested all of the available information on the outer planets. We have seen the many similarities among them, and to a large extent we have come to understand their overall properties in the con-

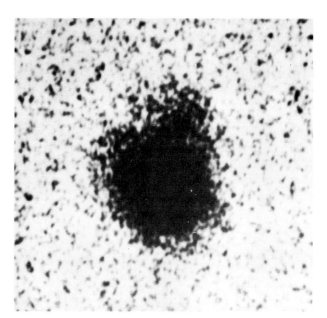

Figure 18.44 the Discovery of Charon. This is the photograph on which Pluto's satellite, the bulge at upper right, was discovered. *(U.S. Naval Observatory)*

Figure 18.45 Charon's Orbit. This shows the orientation of Pluto's rotation axis and Charon's orbit, with respect to the plane of Pluto's orbit about the Sun (which is tilted 17° with respect to the plane of the Earth's orbit). At the present time, the plane of Charon's orbit is aligned closely with the Earth-Pluto direction, so that Pluto and Charon are alternately eclipsing each other. The duration of the eclipses provides new data on the relative sizes of the two bodies.

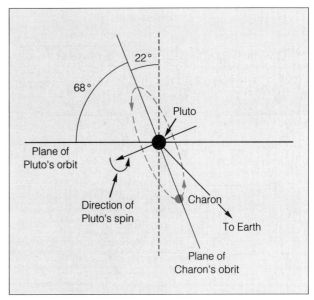

text of the formation of the solar system itself. Only Pluto defies any accepted explanation of its general properties, and we must assume that it formed as a runaway satellite or through some other singular event.

We have now nearly completed our examination of the solar system. We need only examine the nonplanetary bodies roving through space. Therefore we now turn our attention to the minor bodies, the asteroids, comets, and meteroids.

Summary

1. The four giant planets are very different from the terrestrials. They have much lower overall densities and are composed largely of lightweight elements. These planets do not have solid surfaces but are gaseous or liquid throughout most of their interiors.
2. The internal structure of the giant planets is deduced from theoretical calculations whose validity is confirmed by requiring a match between observations and predictions for external properties of the planets.
3. The model calculations indicate that below the gaseous upper levels, the interiors of the giant planets have liquid hydrogen zones and possibly liquid metallic hydrogen zones outside of small, rocky cores. The pressure and temperature are very high in the deep interior.
4. Jupiter, Saturn, and Uranus have charged particle belts held in place by strong magnetic fields, and it is likely that Neptune will also be found to have a magnetic field.
5. The atmospheric composition of the giant planets is dominated by hydrogen, helium, and hydrogen compounds such as methane and ammonia.
6. Circulation in the atmospheres of the giant planets is governed by heating from below and the rapid rotation of the planet, which create belts and zones that encircle the planet.
7. Jupiter, Saturn, and Uranus have rings and many satellites, and there is evidence that Neptune has rings. Rings and satellites are thought to be a natural by-product of the giant planets' formation process, in which each was originally surrounded by a disk of icy debris. In the outer portion of each disk, moons formed from coalescence of the debris; but within the Roche limit, tidal forces prevented this, and a ring system resulted.
8. The large satellites of the giant planets are composed of ice and rock and generally have densities between 1 and 2 g/cm^3. The interior and surface properties depend largely on the degree of geological processing each moon has undergone, and this is related in many cases to internal heating by tidal stresses. Io, the innermost of Jupiter's four major moons, is an example of this; it undergoes continual volcanic eruptions due to tidal forces exerted on it by Jupiter and its neighbor moon, Europa. There are suggestions that many of the moons of the giant planets are relatively young, having been broken apart by collisions and then re-forming, perhaps several times in some cases.
9. Detailed structure in the ring systems is created largely by gravitational effects due to satellites, including orbital resonance, shepherding, and spiral density waves.
10. Pluto, the ninth planet, is very different from the other outer planets. It is a small, icy body more closely resembling one of the large moons of the giant planets than any other planet. The orbit of Pluto is also peculiar—it is tilted and highly eccentric, suggesting that Pluto may have originated as an escaped satellite, perhaps from Neptune. Pluto itself has a large satellite.

Review Questions

1. Summarize the discoveries of the five outer planets. How would you go about searching for a possible tenth planet?
2. Describe the contrasts in the formation process for the giant planets and the formation of the terrestrial planets. Your summary should include an explanation of the ring systems and the large numbers of satellites of the giant planets. (Note: you may need to review some material in Chapter 15).
3. Explain how scientists can be confident that they have some understanding of internal conditions in the giant planets, even though it is impossible to make any direct observations.
4. Are the giant planets differentiated? Explain.
5. Describe the contrasts in the atmospheric flow patterns of Jupiter and Saturn. How does Uranus compare with these two?

6. What are the similarities and differences in the internal structures of Jupiter and Saturn, and how does Uranus compare?
7. Summarize the use of stellar occultation measurements to determine properties of the outer planets and their ring systems.
8. What forms of geological activity affect the satellites of the giant planets, and what causes these forms of activity to occur?
9. What is orbital resonance? Explain how it influences the rings and satellites of the giant planets.
10. Summarize the ways in which Pluto is unusual in comparison to both the giant planets and the terrestrial planets.

Additional Readings

Allen, D. A. 1983. Infrared views of the giant planets. *Sky and Telescope* 65(2):110.
Beatty, J. K. 1985. Pluto and Charon: the dance begins. *Sky and Telescope* 69(6):501.
_____ 1986. A place called Uranus. *Sky and Telescope* 71(4):333.
_____ 1987. Pluto and Charon: the dance goes on. *Sky and Telescope* 74(3):248.
Beatty, J. K., B. O'Leary, and A. Chaikin, eds. 1981. *The new solar system*. Cambridge, Eng.: Cambridge University Press.
Chaikin, A. 1986. *Voyager* among the ice worlds. *Sky and Telescope* 71(4):338.
Elliott, J. L., E. Dunham, and R. L. Mills. 1977. The discovery of the rings of Uranus. *Sky and Telescope* 53(6):412.
Harrington, R. S. and B. J. Harrington. 1980. Pluto: still an enigma after 50 years. *Sky and Telescope* 59(6):452.
Ingersoll, A. P. 1981. Jupiter and Saturn. *Scientific American* 145(6):90.
_____ 1981. The meteorology of Jupiter. *Scientific American* 245(6):90.
Kaufmann, W. 1984. Jupiter: lord of the planets. *Mercury* 13(6):168.
Morrison, D. 1981. The new Saturn system. *Mercury* 10(6):162.
_____ 1982. Voyages to Saturn. NASA Special Publication SP-451. Washington, D.C.:NASA.
_____ 1985. The enigma called Io. *Sky and Telescope* 69(3):198.
Morrison, D. and J. Samz. 1980. Voyage to Jupiter. NASA Special Publication SP-439. Washington, D.C.:NASA.
Overbye, D. 1986. *Voyager* was on target again (at Uranus). *Discover* 7(4):70.
Pollack, J. B. and J. N. Cuzzi. 1981. Rings in the solar system. *Scientific American* 145(5):104.
Soderblom, L. A. 1980. The galilean moons of Jupiter. *Scientific American* 242(1):68.
Soderblom, L. A. and T. V. Johnson. 1982. The moons of Saturn. *Scientific American* 246(1):100.
Squyres, S. W. 1983. Ganymede and Callisto. *American Scientist* 71(1):56.
Tombaugh, C. W. 1979. The search for the ninth planet. *Mercury* 8(1):4.
_____ 1986. The discovery of Pluto. *Mercury* 15(3):66.

Chapter 19

Space Debris

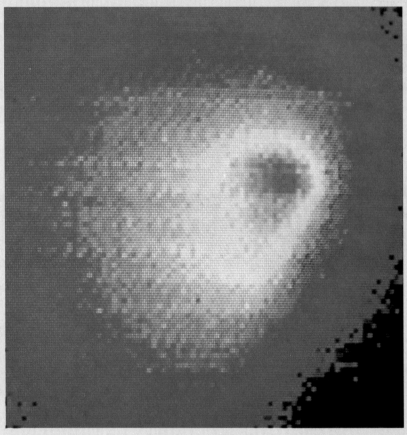

A close-up of Comet Halley,
from the Soviet *Vega* spacecraft.
(*TASS from SOVFOTO*)

Chapter Preview

The Minor Planets
    Kirkwood's Gaps: Orbital Resonances Revisited
    The Origin of Asteroids
Comets: Messengers from the Past
    Halley, Oort, and Cometary Orbits
    The Anatomy of a Comet
Meteors and Meteorites
    Primordial Leftovers
    Dead Comets and Fractured Asteroids
Microscopic Particles: Interplanetary Dust and the
    Interstellar Wind

The planets are the dominant objects among the inhabitants of the solar system (except, of course, for the Sun), but they are not entirely alone as they follow their clockwork paths through space. The abundant craters on planetary and satellite surfaces show us that there must have been a time when interplanetary rocks and gravel were very abundant, raining down continually on any exposed surface. Today the rate of cratering is much lower than it once was, but vestiges of the space debris that caused it remain, orbiting the Sun and occasionally becoming obvious to us as they pass near the Earth or enter its atmosphere.

There are at least four distinct forms of interplanetary matter, some of which are closely related. In this chapter we will discuss **asteroids,** or **minor planets,** myriad rocky chunks up to several hundred kilometers in diameter that orbit the Sun between Mars and Jupiter; **comets,** whose ephemeral and striking appearances have caused humankind to pause and marvel since ancient times; **meteors,** the objects that create the bright streaks often visible in our nighttime skies; and **interplanetary dust,** a collection of very fine particles that inhabit the void between the planets.

The Minor Planets

The discovery of the first minor planet came accidentally when an Italian astronomer named Piazzi noted a new object on the night of January 1, 1801, and within weeks found from its motion that it was probably a solar system body. When the orbit of the new object was calculated, its semimajor axis turned out to be 2.77 AU, which placed the object in solar orbit between Mars and Jupiter, where it had been predicted that a new planet might be found. The new fifth planet was named Ceres.

A little over a year after the discovery of Ceres, a second object was found orbiting the Sun at approximately the same distance; it was named Pallas. From their faintness, it was obvious that both Ceres and Pallas were very small bodies, not respectable planets. By 1807 two more of these **asteroids,** as they were then being called, had been found and designated Juno and Vesta. A fifth, Astrea, was discovered in 1845, and in the next decades vast numbers of these objects began to turn up. The efficiency of finding them was improved greatly when astronomers began to use photographic techniques. A minor planet or asteroid leaves a trail on a long-exposure photograph because of its orbital motion (Fig. 19.1).

Today the number of known asteroids is in the thousands, with about two thousand of them sufficiently well observed to have had their orbits calculated and cataloged. The total number is probably much more, perhaps one hundred thousand.

It is possible to deduce some of the properties of the asteroids from a variety of observational evidence. Measurements of their brightness lead to size estimates; spectroscopy of light reflected from their surfaces provides information on their chemical makeup; and direct analysis of meteorites that may be remnants of asteroids adds data on their internal properties.

The largest asteroids were the first to have their diameters estimated by simply calculating what size they had to be to reflect the amount of light observed. Much more recently, infrared brightness measurements of both large and small asteroids have provided information on their sizes, since at temperatures of a few hundred degrees they glow at infrared wavelengths. The Stefan-Boltzmann law (Chapter 4) can be used to determine the total surface area from the intensity of the emission.

In some cases asteroids have passed sufficiently close to Earth that their angular sizes could be directly determined. In recent times, the angular diameters of some asteroids have been successfully measured by the technique of interferometry.

The result of using these assorted techniques is that asteroids are known to have a wide range of diameters. A few (Table 19.1) are as large as several hundred ki-

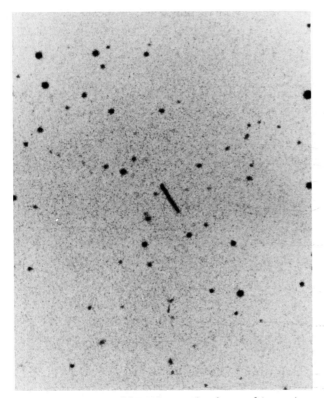

Figure 19.1 Asteroid Motion. The elongated image in this photo is the trail made by an asteroid that moved relative to the fixed stars during the 20-minute exposure. Most new asteroids are discovered on photos of this type, although the first few were spotted visually. *(Eleanor F. Helin, Palomar Observatory)*

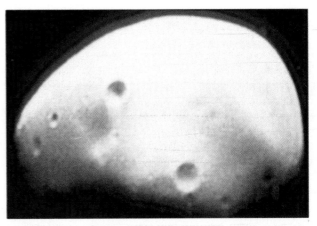

Figure 19.2 A Portrait of an Asteroid? Many asteroids probably bear a general resemblence to this irregular shaped object, which is Deimos, one of the tiny Martian moons. *(NASA)*

lometers in diameter (Ceres, the largest, is about one thousand kilometers), and the majority are rather small, with diameters of one or two hundred kilometers or less. The largest asteroids are apparently spherical in shape, while the smaller ones are often jagged and irregular (Fig. 19.2), varying in brightness as they tumble through space and reflect light with different efficiencies on different sides. A few asteroids are binary, consisting of two chunks that orbit each other as they circle the Sun. The total mass of all the asteroids together is small by planetary standards, amounting to only about 0.04 percent of the mass of the Earth.

Table 19.1 Selected Asteroids

No.	Name	Discovery	Diameter (km)	Mass (G)	Period* (y)	Distance* (AU)
1	Ceres	1801	760	1×10^{24}	4.60	2.766
2	Pallas	1802	480	3×10^{23}	4.61	2.768
3	Juno	1804	200	2×10^{22}	4.36	2.668
4	Vesta	1807	480	2×10^{23}	3.63	2.362
6	Hebe	1847	220	2×10^{22}	3.78	2.426
7	Iris	1847	200	2×10^{22}	3.68	2.386
10	Hygiea	1849	320	6×10^{22}	5.59	3.151
15	Eunomia	1851	280	4×10^{22}	4.30	2.643
16	Psyche	1852	280	4×10^{22}	5.00	2.923
51	Nemausa	1858	80	9×10^{20}	3.64	2.366
511	Davida	1903	260	3×10^{22}	5.67	3.190

*Periods are orbital periods; distances are orbital semimajor axes.

Astronomical Insight 19.1

Trojans, Apollos, and Target Earth

Not all of the asteroids are confined to the main belt between the orbits of Mars and Jupiter. There are exceptions, and many of them fall into two special groups.

The existence of one of these groups was actually predicted on the basis of mathematical calculations carried out in the late 1700s by the French scientist J. L. Lagrange. Using Newton's laws of motion, Lagrange showed that there are certain points in a system consisting of two orbiting bodies where additional objects may orbit in a stable fashion. (Ordinarily, when more than two objects are involved, the system becomes so complex that it is nearly impossible to describe all the possible motions mathematically.) Lagrange showed that if one body orbits another, there are points in its orbit 60 degrees ahead of and 60 degrees behind it where additional bodies can stay. In Chapter 18 we learned that there is a small satellite occupying the orbit of Dione, one of the larger satellites of Saturn, and that this small satellite stays in position 60 degrees ahead of Dione. There are also two such satellites sharing the orbit of Tethys, another moon of Saturn; one is 60 degrees ahead of Tethys in its orbit, and the other is 60 degrees behind. There are similar points in the orbit of our Moon, and there has even been speculation that these might be convenient places to locate permanent manned space stations. (The L-5 Society, a group advocating such stations, takes its name from the fact that one of these points is called the L-5 point, for Lagrange point number 5.)

The same situation exists for the orbit of Jupiter, and after a few asteroids were found in the early decades of this century, concerted searches revealed a large number of asteroids in the Lagrange points, 60 degrees in either direction from Jupiter. There may be hundreds of objects at these two locations, held in place by the combined gravitational pulls of Jupiter and the Sun. These asteroids are called the **Trojan asteroids,** and they are individually named for classical Trojan and Greek heros, such as Achilles and Hector.

Another distinct group of asteroids has been identified and named after the first to be discovered, Apollo. Their distinction is that their orbits bring them within 1 AU of the Sun, so that their paths may cross the orbit of the Earth. There are about thirty Apollo asteroids known, and a couple of them, Icarus and Eros, have provided useful information for scientists on Earth. In 1968 Icarus, which comes closest to the Sun (0.19 AU) of all the Apollo asteroids, came within 16 million kilometers of Mercury, close enough for Mercury's gravitational pull to alter its direction noticeably, and this allowed astronomers to deduce the mass of Mercury. Eros passed the Earth at a distance of about 23 million kilometers in 1931, giving astronomers a clear view of it through telescopes, so that its size and shape could be directly measured. It was seen to be an irregular chunk of rock, tumbling end over end with a period of about 5.3 hours. This same asteroid passed in front of a bright star in 1975, allowing astronomers to measure its diameter precisely. It has an oblong shape, with dimensions of $7 \times 19 \times 30$ kilometers.

It is likely that some of the Apollo asteroids will eventually collide with the Earth. They are so few in number, and the volume of space that they occupy is so large, however, that the average time between such collisions is likely to be very long (millions of years). We can hope, therefore, to escape any truly major impacts for the foreseeable future.

The compositions derived from spectroscopic analyses are highly varied—including metallic compounds, nearly metal-free compounds, carbon-dominated minerals, and silicon-bearing minerals—and there are several classes whose composition is not known. The majority (about three-quarters) of the biggest asteroids in the main belt between Mars and Jupiter are carbonaceous, meaning that they contain carbon in complex molecular forms. Most of the rest are composed chiefly of silicon-bearing compounds, and about 5 percent are rich in metals, primarily nickel-iron mixtures.

Kirkwood's Gaps: Orbital Resonances Revisited

As increasing numbers of asteroids were discovered and cataloged throughout the nineteenth century, calculations of their orbits showed remarkable gaps at certain distances from the Sun. One such gap is at 3.28 AU, and another is at 2.50 AU.

The explanation for these gaps was offered by Daniel Kirkwood in 1866, when he realized that these distances correspond to orbital periods that are simple fractions of the period of Jupiter (Fig. 19.3). The giant planet, at its distance of 5.2 AU from the Sun, takes 11.86 years to make a trip around it, whereas an asteroid at 3.28 AU would have a period exactly half as long, 5.93 years. Thus this asteroid and Jupiter would be lined up in the same way every time Jupiter made one orbit. The asteroid would therefore be subjected to regular tugs by Jupiter's gravity and would gradually alter its orbit, vacating the zone 3.28 AU from the Sun. In this manner Jupiter has cleared out several gaps in the asteroid belt, at this and other distances where orbital periods would result in regular alignments. The gap at 2.50 AU corresponds to orbits with a period exactly one third that of Jupiter's. These orbital resonances are exactly like the ones created by the moons of Saturn, which are responsible for some of the gaps in the ring system of that planet (see Chapter 18). Since Jupiter is the most massive and nearest of the outer planets to the asteroid belt, it has by far the greatest effect in producing gaps, but in principle the other planets could do the same thing.

The Origin of Asteroids

The most natural explanation of the asteroids seemed for a while to be that the breakup of a former planet created a swarm of fragments that continued to orbit

Figure 19.3 Kirkwood's Gaps. This graph shows the distribution of asteroids. Gaps appear at the distances from the Sun where asteroids would have exactly one half, one third, or other simple fractions of the orbital period of Jupiter.

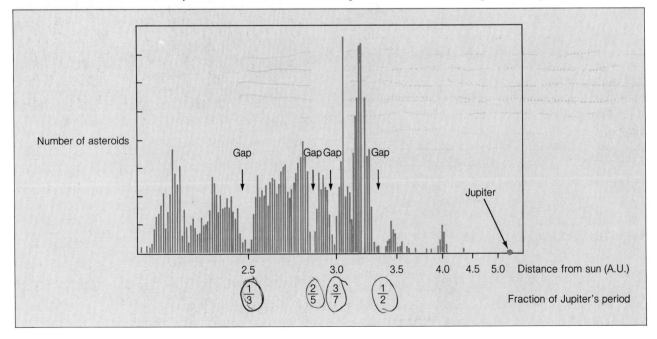

the Sun. This argument was weakened when the total mass of the asteroids was estimated and found to be much less than that of any ordinary planet. There is another rather strong argument against the planetary remnant hypothesis: There is no known reasonable way to break a planet apart once it has formed. However, it is easy to understand why material orbiting the Sun at the position of the asteroid belt could never have combined into a planet in the first place. It is simpler to accept the idea that the debris there was never part of a planet, than to try to find a way to have it first form a planet and then break apart.

The invisible hand of Jupiter's gravity is invoked again. If, as is suspected, the planets formed from a swarm of debris orbiting the young Sun in a disk, no planet could have formed where the pieces of debris could not stick together. Calculations show that once Jupiter had formed, its immense gravitational force would have stirred up the material near its orbit so that collisions between particles would occur at speeds too great to allow them to stick together. It is as though Jupiter wielded a giant spatula, stirring up the debris near it and keeping it spread out as a loose collection of rocky fragments. Even today Jupiter is still at work keeping the asteroids mixed up and occasionally causing collisions between them that break them up further.

The study of meteorites, some of which probably originated as pieces of asteroids that broke apart in collisions, gives us a chance to examine material from the early solar system. It is interesting to note that some of the asteroids apparently have undergone differentiation, developing nickel-iron cores; some meteorites are almost pure chunks of this material. There must have been sufficient heat inside some of the asteroids to create a molten state, allowing the heavy materials to separate from the rest. Unlike the nickel-iron asteroids, some of the other types, particularly the carbonaceous ones, have apparently undergone almost no heating, since they contain large quantities of volatile elements that would have been easily cooked out.

Comets: Messengers from the Past

Among the most spectacular of all the celestial sights are the comets. With their brightly glowing heads, long streaming tails, and infrequent and often unpredictable appearances, these objects have sparked the imagination (and often the fears) of humankind through the ages.

In antiquity, when astrological omens were taken very seriously, great import was attached to a cometary appearance (Fig. 19.4). Ancient descriptions of comets are numerous, and in many cases these objects were thought to be associated with catastrophe and suffering. In contrast, the modern view is that comets are samples of primordial material from which the solar system formed, and it is hoped that the study of comets will lead to better understanding of the origins of our planetary system.

Included among the teachings of Aristotle was the notion that comets were phenomena in the Earth's atmosphere. There was no good evidence for this idea;

Figure 19.4 Calamity on Earth Associated with the Passage of Comets. This drawing is from a seventeenth century book describing the universe. *(The Granger Collection)*

apparently Aristotle adopted it because he believed that any phenomena that were changeable were associated with the Earth, whereas the heavens were perfect and immutable. In any case, this idea was accepted for centuries to come. In 1577, however, Tycho Brahe proved that comets were too distant to be associated with the Earth's atmosphere because they do not exhibit any parallax when viewed from different positions on the Earth. If a comet were really located only a few kilometers or even a few hundred kilometers above the surface of the Earth, its position as seen from the Earth would change from one location to another. Tycho was able to show that this was not the case and that therefore comets belonged to the realm of space.

Halley, Oort, and Cometary Orbits

A major advance in the understanding of comets was made by a contemporary and friend of Newton, Edmund Halley. Aware of the power of Newton's laws of motion and gravitation, Halley reviewed the records of cometary appearances and noted one outstanding regularity. Particularly bright comets seen in 1531, 1607, and 1682 seemed to have similar properties (especially their retrograde motion), and Halley suggested that in fact all three were appearances of the same comet, orbiting the Sun with a seventy-six year period.

Calculations using Kepler's third law showed that for a period of seventy-six years, this object must have a semimajor axis of nearly 18 AU. Halley realized that in order to appear as dominant in our skies as the comet does, it must have a highly elongated orbit (Fig. 19.5), so that it comes close to the Sun at times, even though its average distance is well beyond the orbit of Saturn, nearly as far out as Uranus. Such an eccentric orbit, as a very elongated ellipse is called, had not previously been observed, even though Newton's laws clearly allowed the possibility.

Since Halley's time, searches of ancient reports of comets have revealed that Halley's comet has been making regular appearances for many centuries. The earliest records are provided by the ancient Chinese astronomers, who apparently observed each of its ap-

Figure 19.5 A Cometary Orbit. This is a rough scale drawing of the orbit of Halley's comet. Most comets actually have much more highly elongated orbits than this, and correspondingly longer periods.

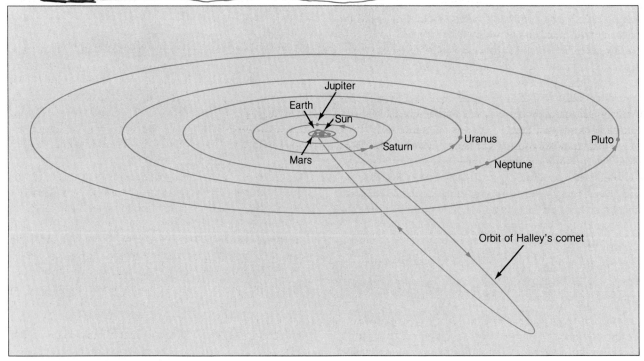

pearances for well over one thousand years, possibly beginning as early as the fifth century B.C.

The most recent visit of Halley's comet was in 1985 and 1986. It did not pass very close to the Earth (Fig. 19.6), and the best views were from the Southern Hemisphere (Fig. 19.7). A much more spectacular appearance occurred on the comet's previous visit in 1910 (Fig. 19.8), when the Earth actually passed through the rarefied gases of its tail.

Other spectacular comets have been seen (Fig. 19.9), and a number have rivaled Halley's comet in brightness. Traditionally a comet is named after its discoverer, and there are astronomers (mostly amateurs) around the world who spend long hours peering at the nighttime sky through telescopes looking for a piece of immortality.

When a new comet is discovered, a few observations of its position are sufficient to allow computation of its orbit. The results of many years of comet watching have shown that there are many comets whose orbits are so incredibly stretched out that their periods are measured in the thousands or even the millions of years. For all practical purposes these comets are seen only once. They return thereafter to spend millennia in the void of space well beyond the orbit of Pluto before visiting the inner solar system again.

The orbits of comets, particularly these so-called long-period ones, are randomly oriented. In strong contrast with the planets, these comets do not show any preference for orbits in the plane of the ecliptic, and about half go around the Sun in the retrograde direction.

Consideration of these orbital characteristics, especially the large orbital sizes, led the Dutch astronomer Jan Oort to suggest that all comets originate in a cloud of objects that surrounds the solar system. The Oort cloud, as it is now called, is envisioned to be a spherical shell with a radius of 50,000 to 150,000 AU, extending therefore a significant fraction of the distance to the nearest star, which is almost 300,000 AU from the Sun.

Occasionally a piece of debris from the Oort cloud is disturbed from its normal path, either by a collision with another object or perhaps by the gravitational tug of a nearby star, and it begins to fall inward toward the Sun. Left to its own devices, a comet falling in from the Oort cloud would follow a highly elongated orbit with a period of millions of years, appearing to us as one of the long-period comets when it made its brief incandescent passage near the Sun. In many cases,

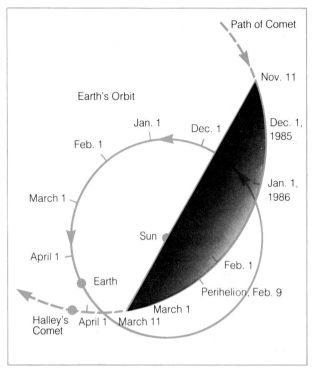

Figure 19.6 The Path of Halley's Comet in 1986.
This graph shows the path of Halley's comet and the relative position of the Earth during the recent appearance. Note that the comet was on the far side of the Sun when it passed closest to it.

however, a comet is not left to its own devices. Instead, it runs afoul of the gravitational pull of one of the giant planets. When this happens, the comet may be so speeded up that it will escape the solar system entirely after it loops around the Sun, or it may be so slowed down (dropping into a smaller orbit with a shorter period) that it becomes one of the numerous comets that reappear frequently (Table 19.2).

The Oort cloud hypothesis accounts for all the observed cometary orbits, and this concept is widely accepted today. But we must still consider how the Oort cloud itself came into being. The best guess is that it consists of material left over from the formation of the solar system and that comets are therefore made of very old matter, representing the primordial stuff from which the planets and the sun were created. It is thought that the Oort cloud consists of solid debris not condensed in the early solar system and was then gradually forced outward by the gravitational effects of the planets.

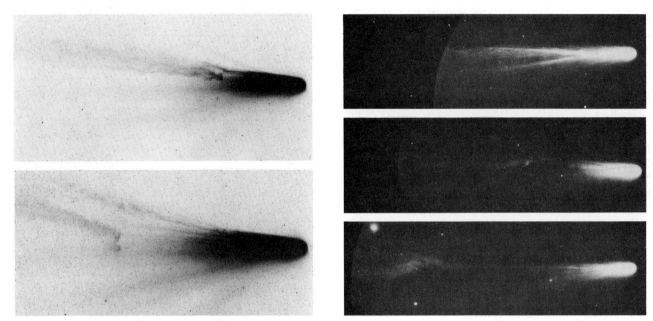

Figure 19.7 Halley's Comet in 1986. The excellent photographs at left were obtained from an observatory in Australia, and those at right were obtained in the United States. Note the complex structure of the tail, and the changes that occurred between photographs taken only days apart, as the ion tail disconnected and reformed. *(left: ©1986 Royal Observatory, Edinburgh; right: Palomar Observatory, California Institute of Technology)*

Table 19.2   Selected Periodic Comets

Comet	Period (y)	Semimajor Axis (AU)	Next Appearance
Encke	3.30	2.21	1987.6
Temple (2)	5.26	3.0	1988.6
Schwassmann-Wasserman (1)	6.52	3.50	1987.8
Wirtanen	6.65	3.55	1987.9
Reinmuth (2)	6.72	3.6	1987.8
Finlay	6.88	3.6	1988.2
Borrelly	7.00	3.67	1988.5
Whipple	7.44	3.80	1993.1
Oterma	7.89	3.96	1990.0
Schaumasse	8.18	4.05	1993.0
Wolf	8.42	4.15	1992.9
Comas Sola	8.58	4.19	1987.0
Vaisala	10.5	4.79	1991.9
Schwassmann-Wasserman (2)	16.1	6.4	1998.6
Neujmin (1)	17.9	6.8	2002.7
Crommelin	27.9	9.2	2012.6
Olbers	69	16.8	2025
Pons-Brooks	71	17.2	2025
Halley	76.1	17.2	2062.5

Figure 19.8 Halley's Comet as it Appeared in 1910.
(Palomar Observatory, California Institute of Technology)

The Anatomy of a Comet

A remarkably successful concept of the nature of a comet was developed many years ago by the American astronomer Fred Whipple. Whipple's theory, sometimes called the "dirty snowball," envisions that for most of its life, a comet is just a frozen chunk of icy material, probably consisting of small particles like gravel or larger boulders embedded in frozen gases. This picture was based on observations of many comets over the years, and its basic features were verified by the close-up examination of Halley's comet in early 1986, when several space probes were able to make observations from close range. Of course, a great many new details were discovered as well.

As a comet passes through the outer reaches of its orbit, far from the Sun, it does not glow, has no tail, and is not visible from Earth. As it approaches the Sun, however, it begins to warm up as it absorbs sunlight, and the added heat causes volatile gases to escape. A spherical cloud of glowing gas called the **coma** develops around the solid **nucleus** (Fig. 19.10). The principal gases found in the coma of Halley's comet are water vapor (H_2O; about 80 percent by number), carbon dioxide (CO_2; about 3.5 percent), and carbon monoxide (CO; most of the remaining 16.5 percent). Other slightly more complex elements such as ammonia (NH_3) and methane (CH_4) are probably present also, along with hydrogen molecules (H_2). The gases glow by a process called **fluorescence**. The molecules absorb ultraviolet and visible light from the Sun, causing them to be excited to high energy levels, and then

Figure 19.9 Another Bright Comet. This is Comet West, which appeared in 1970 and was one of the brightest comets in recent years. *(Fr. R. E. Royer)*

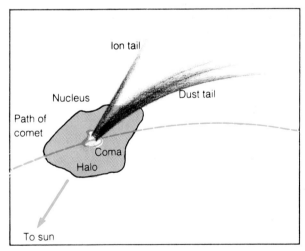

Figure 19.10 Anatomy of a Comet. This sketch illustrates the principal features of a comet, though not necessarily to correct scale.

they re-emit the light, sometimes at different wavelengths, as they return to low energy states. A cloud of hydrogen atoms, resulting from the breakup of water molecules in the coma, extends great distances from the nucleus. The visible coma may be as large as 100,000 kilometers in diameter, while the halo of hydrogen atoms may extend as much as one hundred

Astronomical Insight 19.2

Encounters with Halley's Comet

Even though the 1986 visit of Halley's comet to the inner solar system did not provide very good views from Earth (particularly from the Northern Hemisphere), scientists had an extraordinary opportunity to observe the comet from close range using space probes. No less than five spacecraft were sent from Earth to rendezvous with the comet, all within a few days of each other in March 1986, when Halley crossed the plane of the Earth's orbit after passing behind the Sun. In addition to these five missions, two other spacecraft originally launched for other purposes observed the comet from locations in space far from the earth.

The five missions designed to encounter Halley's comet were the European Space Agency's *Giotto* mission, two Soviet *Vega* spacecraft, and two Japanese probes called *Sakigake* and *Suisei*. This collection of spacecraft carried an assortment of instruments ranging from magnetic field detectors to spectrographs to particle detectors to cameras for obtaining images of the comet. *Giotto* passed closest to the nucleus of Halley, zipping by only about 600 kilometers from it, while the *Vega*s and the Japanese pair passed by at greater distances. Both the *Vega*s and *Giotto* suffered damage from impacts of the comet's dust particles: *Giotto* was sent spinning by an impact just seconds before its closest approach to the nucleus, and communications were lost temporarily. Nevertheless, it succeeded in obtaining unprecedented images of the nucleus.

The two existing spacecraft that observed Halley were the *Pioneer Venus* probe (renamed *Pioneer Halley* for the occasion), which has been orbiting Venus since 1978, and *International Cometary Explorer (ICE)*, a probe originally sent into space to measure interplanetary particles and magnetic fields. The *ICE* craft actually carried out the first space rendezvous with a comet, passing through the tail of Comet Giaccobini-Zinner in September 1985 before making more remote observations of Halley in early April 1986. The *Pioneer Venus* spacecraft, able to view the comet from a rather different vantage point than Earth, observed ultraviolet emission from the extended halo of hydrogen gas that surrounds the comet. Since this hydrogen is a by-product of water, the major constituent of the nucleus, measurements of the hydrogen led directly to estimates of the rate at which the nucleus was losing mass as the Sun's heat caused water ice to vaporize from the nucleus.

As expected, the combination of all these space encounters and observations greatly extended our knowledge of comets. In general, the results supported the "dirty snowball" theory, in which a comet is viewed as a loose conglomerate of rock and gravel held together by ice, which vaporizes as the comet passes through the inner solar system. There were, however, some surprises, along with a significant refinement of this picture; the major discoveries are described in the text.

Besides all their scientific yield, the encounters with Hally's comet also prompted a remarkable degree of international cooperation. The Soviet *Vega* spacecraft, which reached the comet before *Giotto*, provided essential information on the exact position of the nucleus so that *Giotto* could be steered precisely to its close encounter. On the ground, a massive international comet-observing program known as *International Halley Watch* was carried out over several months and involved free and rapid exchange of information among scientists from many countries.

A remarkable future comet encounter mission is now being planned in the United States. A spacecraft called the *Comet Rendezvous and Asteroid Flyby* (or simply *CRAF*) will, if the plan is funded, fly by an asteroid, then match orbits with Comet Wild 2 in 1995 and stay with it in its solar orbit for at least three years thereafter, observing the nucleus from very close range (30 km) as it approaches the Sun and develops its coma and tail. This mission will provide even more detailed information on the life and times of a comet than the recent Halley's comet spectacular.

times farther from the nucleus, as demonstrated by ultraviolet observations of Halley's comet in 1986 (Fig. 19.11). The solid nucleus is relatively tiny, having a diameter of a few kilometers. The nucleus of Halley's comet was found to be irregular in shape, with a long axis of about 15 kilometers and a width ranging to about 10 kilometers (Fig. 19.12). The nucleus was found to have a thin coating of very dark material (albedo only 0.04), probably a matrix of dust left behind by the evacuation of ices that evaporated previously. Hence the dirty snowball idea has been modified to incorporate a partial outer coating of dust, along with the traditional picture of mixed dust and ice in the interior of the nucleus.

Apparently the gas escaping from the nucleus of a comet is often released in jets that can be powerful enough to alter the comet's orbital motion. It has been known for years that comets can be slowed in this way as they approach the Sun, but the close encounters with Halley's comet provided dramatic confirmation, with images revealing spectacular fountains of gas and dust erupting from active regions of the surface of the nucleus (Fig. 19.12). Apparently these gas jets are created by heat that is absorbed by the dark crust and conducted into the icy interior, where the heat causes ice to evaporate in a process called **sublimation,** in which solid ice is converted directly into gas. The released gases expand outward and escape, carrying fine dust with them. The observed jets are thought to occur at places where the dusty crust is thin or nonexistent. As the nucleus rotates, jets are active only on the sunlit side, so each individual jet turns on and off as it rotates toward and then away from the Sun. Surprisingly, examination of photos of Halley's comet in 1910 indicate that the jets may have occurred in the same locations on the nucleus then as they did in 1986. Apparently the jets create craters that persist over the years between encounters with the Sun's warming radiation.

As a comet nears the Sun, the solar radiation and the solar wind force some of the gas from the coma to flow away from the Sun and form the tail, which in some cases is as long as 1 AU. There are often two

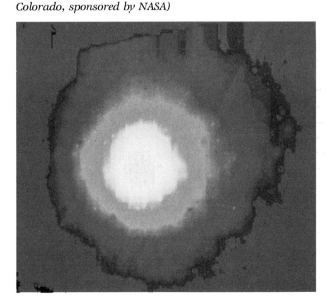

Figure 19.11 The Halo of Halley's Comet. This is an ultraviolet image made by the *Pioneer Venus* spacecraft from Venus orbit; it shows the extent of the hydrogen cloud surrounding the comet's nucleus. The hydrogen is a by-product of water vapor that is ejected from the nucleus and then dissociated by the Sun's ultraviolet radiation. The halo was about 500,000 kilometers, or roughly ⅓ AU in diameter, when this photo was taken. *(LASP, University of Colorado, sponsored by NASA)*

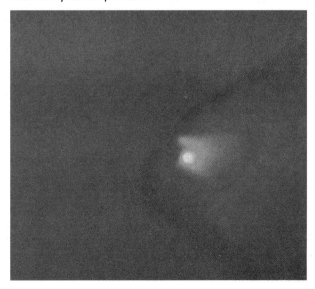

Figure 19.12 A Close-up of Comet Halley. This image was obtained by the European spacecraft *Giotto* from close range. The dark object in the picture is the potato-shaped nucleus of the comet, some 10 kilometers wide at its widest and about 15 kilometers in length. The nucleus has several bright jets extending to the right, toward the Sun. *(Max Planck Institute for Aeronomy, courtesy of A. Delamere, Ball Aerospace Corporation)*

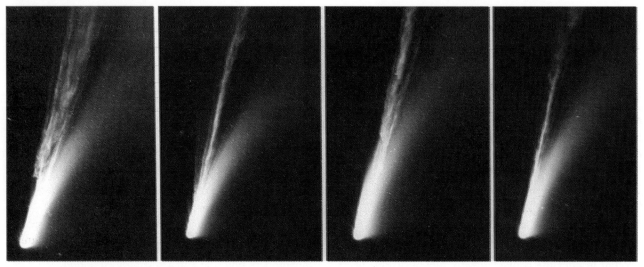

Figure 19.13 Two Tails. These photos of Comet Mrkos (1957) show distinctly the two types of tails that characterize many comets. The ion tail points straight up in these views, while the dust tail curves over to the right. *(Palomar Observatory, California Institute of Techology)*

distinct tails (Fig. 19.13), one formed of ionized gases from the coma (such as CO^+, N_2^+, CO_2^+, and CH^+) and another formed of tiny solid particles released from the ice of the nucleus. The **ion tail**, the one formed of ionized gases, is shaped by the solar wind and therefore points almost exactly straight away from the Sun at all times. It has been observed many times, for Halley and other comets, that all or part of the ion tail can detach itself and dissipate and then be replaced by a new tail (see Fig. 19.7). This apparently is caused by reversals in the ambient magnetic field direction as the comet passes through different regions in the solar wind.

The other tail, called the **dust tail**, usually takes on a curving shape, as the dust particles are pushed away from the Sun by the force of the light they absorb. This **radiation pressure** is not strong enough to force the dust particles into perfectly straight paths away from the Sun, so they follow curved trajectories that are caused by a combination of their orbital motion and the outward push due to sunlight.

One of the major discoveries made by the probes that encountered Halley's comet and Comet Giaccobini-Zinner (see Astronomical Insight 19.2) was that the ionized gases surrounding a comet create a trapped magnetic field. A "bow shock," analogous to a boundary of a planet's magnetosphere, builds up on the sunward side of a comet, where the solar wind encounters the trapped magnetic field and flows around it. In the direction away from the Sun, the magnetic field that is wrapped around the comet forms a layer of trapped charged particles called a **current sheet.**

The gases that escape from the nucleus of a comet as it approaches the Sun are highly volatile and would not be present in the nucleus if it had ever undergone any significant heating. This tells us that comets must have formed and lived their entire lives in a very cold environment, probably never getting even as warm as 100 K before falling into orbits that bring them close to the Sun. If a comet is so easily vaporized, then its days are numbered, once it begins to follow a path that regularly brings it close to the Sun. It has been estimated that Halley's comet was losing some 50 tons of water ice per second when it passed nearest the Sun in early 1986! Despite this high mass-loss rate, the comet should last for many tens of thousands of years before the nucleus is entirely dissipated. It is possible that the dusty crust will eventually become so thick that the ices underneath it are completely shielded from the Sun's heat. If this happens, Halley's comet may continue to orbit the Sun indefinitely, but it will no longer produce a coma and a tail. Several cases have been noted where a comet failed to reappear on schedule but was replaced by a few pieces or perhaps a swarm of fragments (Fig. 19.14). In time, the remains of a dead comet are dispersed all along its orbital path. Each

Figure 19.14 The Breakup of Comet West. This dramatic sequence shows the nucleus of Comet West fragmenting into four pieces. *(New Mexico State University)*

time the Earth passes through such a region, it encounters a vast number of tiny bits of gravel and dust, and we experience a meteor shower.

Meteors and Meteorites

Occasionally one of the countless pieces of debris floating through the solar system enters the Earth's atmosphere, creating a momentary light display as it evaporates in a flash of heat created by the friction of its passage through the air. The streak that is seen in the sky is called a **meteor** (Fig. 19.15). Most of us are familiar with this phenomenon, commonly called a shooting star, since it is often possible to see one in just a few minutes of sky gazing on a clear night. On rare occasions an especially brilliant meteor is seen, possibly persisting for several seconds; these spectacular events are called **fireballs** or **bolides**.

The piece of solid material that causes a meteor is called a **meteoroid.** Most are very small, nothing more than tiny grains of dust or perhaps fine gravel. A few, however, are larger solid chunks, and these are responsible for the bright fireballs.

Occasionally one of the larger meteoroids survives the arduous trip through the atmosphere and reaches the ground intact. Such an object is called a **meteorite** (Fig. 19.16), and examples can be found in museums around the world. Meteorites have always been the subject of intense scrutiny; until the past fifteen years or so they were the only samples of extraterrestrial material scientists could get their hands on.

Figure 19.16 A Meteorite. This is a stony meteorite; the black coloring is due to heating as the object passed through the atmosphere. The light-colored spots are breaks in the fusion crust where interior material is exposed. *(Griffith Observatory, Ronald A. Oriti Collection)*

Figure 19.15 Two Bright Meteors. The streaks of light in this photo are created by a tiny particle entering the Earth's atmosphere from space. *(Yerkes Observatory)*

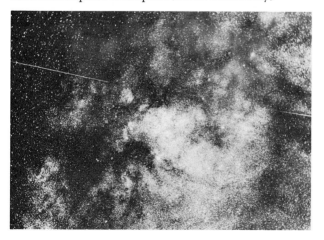

Scientists scoffed at the notion that rocks could fall from the sky until a meteorite was seen to fall near a French village in 1803 and was found and examined just after it fell. Such falls are rare, but they are occasionally observed and have even been known to cause damage (but so far, few injuries).

Primordial Leftovers

Meteorites are much older than most surface rocks on the Earth. This fact enhances the interest of scientists studying them, for in doing so they get a glimpse into the past history of the solar system.

Meteorites can generally be grouped in three classes: the stony meteorites, which comprise about 93 percent of all meteorite falls; the iron meteorites, which account for about 6 percent; and the stony-iron meteorites, which are the rarest. These relative abundances were determined indirectly because the different types of meteorites are not equally easy to find on the ground. The majority of all meteorites found are the iron ones, which, as we have just seen, comprise only a small fraction of those that fall. The stony meteorites look so much like ordinary rocks that they are usually difficult to pick out. A particularly productive place to search for meteorites is Antarctica, where a thick layer of ice conceals the native rock. Meteorites that fall there are relatively easy to find, and there is little chance for confusion with Earth rocks.

The stony meteorites are primarily of a type called **chondrites**, so named because they contain small spherical inclusions called **chondrules** (Fig. 19.17). Chondrules are mineral deposits formed by rapid cooling, most likely at an early time in the history of the solar system, when the first solid material was condensing. A few of the stony meteorites are **carbonaceous chondrites** (Fig. 19.18), thought to be almost completely unprocessed since the solar system was formed and therefore representative of the original stuff of which the planets were made. The primordial nature of the carbonaceous chondrites, like the carbonaceous asteroids mentioned in an earlier section, is deduced from their high volatile content, which indicates that they were never exposed to much heat. One particularly fascinating aspect of these meteorites is that in at least one case, complex organic molecules called **amino acids** were found inside a carbonaceous chondrite meteorite, indicating that some of the ingredients for the development of life were apparently available even before the Earth formed.

The iron meteorites have varying nickel contents and sometimes show an internal crystalline structure (Fig.

Figure 19.18 A Carbonaceous Chondrite. This example, not of the type in which amino acids have been found, shows a large chondrule (the light-colored spot, upper center), that is about 5 millimeters in diameter. *(Griffith Observatory, Ronald A. Oriti Collection)*

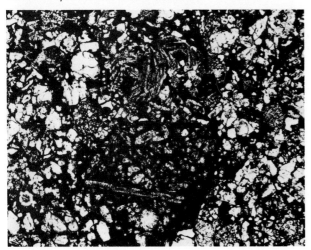

Figure 19.17 Cross-Section of a Chondrite. This shows the many chondrules (light patches) embedded within the structure of this type of stony meteorite. *(Griffith Observatory, Ronald A. Oriti Collection)*

19.19), which indicates a rather slow cooling process in their early histories. This has important implications about their origin, as we shall see.

Dead Comets and Fractured Asteroids

The origins of the meteoroids that enter the Earth's atmosphere can be inferred from what we know of the properties of meteorites and of asteroids and comets. As we have seen, most meteors are caused by relatively tiny particles that do not survive their flaming entry into the Earth's atmosphere. During **meteor showers,** when meteors can be seen as frequently as once per second, all seem to be of this type. As noted earlier, meteor showers are associated with the remains of comets that have disintegrated and left behind a scattering of gravel and dust (Table 19.3). Therefore it is thought that the most common meteors—those created by small, fragile meteoroids—are due to cometary debris.

The larger meteoroids that reach the ground as meteorites may have a different origin. It is likely that the occasional collisions among asteroids are sometimes sufficiently violent to destroy them and disperse the resulting rubble throughout the solar system. Most meteorites are probably fragments of asteroids. The iron meteorites apparently originated in asteroids that had undergone differentiation, while the stony ones came either from the outer portions of differentiated asteroids or from smaller bodies that never underwent differentiation at all. The chondrites probably fall into the latter category, since the chondrules reflect a rapid cooling that would have characterized very small bodies. Thus the chondrites are thought to be the most primitive of the meteorites because they have undergone no processing in the interiors of large bodies.

Figure 19.19 Cross-Section of a Nickel-Iron Meteorite. This example shows the characteristic crystalline structure indicative of a slow cooling process from a previous molten state. Such meteorites are thought to have once been parts of larger bodies that differentiated. *(Griffith Observatory, Ronald A. Oriti Collection)*

From our studies of the other planets and satellites, we know that there was a time long ago when frequent impacts formed most of the craters seen today. Certainly the Earth was not immune and no doubt was

Table 19.3 Major Meteor Showers

Shower	Approximate Date	Associated Comet
Quadrantid	January 3	—
Lyrid	April 21	Comet 1861 I
Eta Aquarid	May 4	Halley's comet
Delta Aquarid	July 30	—
Perseid	August 11	Comet 1862 III
Draconid	October 9	Comet Giaccobini-Zinner
Orionid	October 20	Halley's comet
Taurid	October 31	Comet Encke
Andromedid	November 14	Comet Biela
Leonid	November 16	Comet 1866 I
Geminid	December 13	—

also subjected to heavy bombardment (Table 19.4). The difference, of course, is that the Earth has an atmosphere, flowing water, and glaciation, all of which combine to erase old craters in time. A few traces can still be seen, however; there is a very large basin under the Antarctic ice that is probably an ancient impact crater, and a portion of Hudson Bay in Canada shows a circular shape thought to have a similar origin. A number of other suspected ancient impact craters have been found throughout the world (Fig. 19.20).

Although the frequency of impacts has decreased, there are still rare occasion when major impacts occur. The Barringer crater (Fig. 19.21) near Winslow, Arizona, was formed only about 25,000 years ago, for example, and other large bodies could still hit the Earth. Given a long enough time, it is almost inevitable.

Microscopic Particles: Interplanetary Dust and the Interstellar Wind

The empty space between the planets plays host to some very tiny particles in addition to the larger ones we have just described. There is a general population of small solid particles, perhaps a millionth of a meter in diameter, called interplanetary dust grains. There is also a very tenuous stream of gas particles flowing through the solar system from interstellar space.

The presence of interplanetary dust has been known for some time from two celestial phenomena, both of which can be observed with the unaided eye, although only with difficulty. The dust particles scatter sunlight, so that under the proper conditions a diffuse glow can be seen where the light from the Sun hits the dust. This is analogous to seeing the beam of a searchlight stretching skyward; you see the beam only where there are small particles (either dust or water vapor) that scatter its light so that some of it reaches your eye.

One of the phenomena created by the interplanetary dust is the **zodiacal light** (Fig. 19.22), a faintly illuminated belt of hazy light that can be seen stretching across the sky (along the ecliptic) on clear dark nights just after sunset or before sunrise. The second observable phenomenon caused by interplanetary dust is a small bright spot seen on the ecliptic in the opposite direction from the Sun. This diffuse spot, called the **ge-**

Figure 19.20 Impact Craters on the Earth. This map shows the locations of major craters thought to have been created by impacts of massive objects. *(Griffith Observatory)*

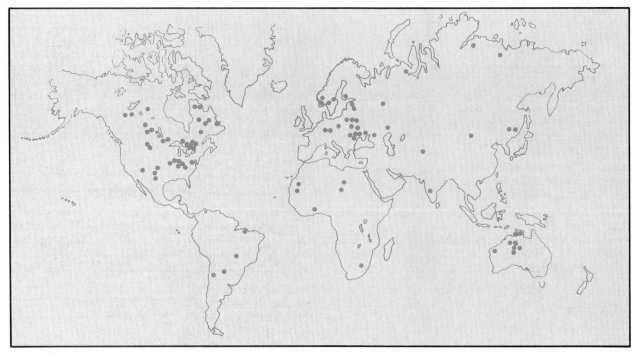

Figure 19.21 Meteor Crater near Winslow, Arizona. The impact that created this crater occurred about 25,000 years ago. *(Meteor Crater Enterprises)*

Table 19.4 Some Known or Suspected Impact Craters on the Earth

Location	Diameter
Amirante Basin, Indian Ocean	300 km
Sudbury, Ontario, Canada	140
Vredefort, Orange Free State, South Africa	100
Manicouagan	100
Sierra Madera, Texas	100
Charlevoix Structure, Quebec, Canada	46
Clearwater Lake West, Quebec, Canada	32
Mistastin Lake, Labrador, Canada	28
Gosses Bluff, Northern Territory, Australia	25
Clearwater Lake East, Quebec, Canada	22
Haughton, Northwest Territories, Canada	20
Wells Creek, Tennessee	14
Deep Bay, Saskatchewan, Canada	13.7
Lake Bosumtwi, Ashanti, Ghana	10.5
Chassenon Structure, Haut-Vienne, France	10
Wolf Creek, Western Australia	8.5
Brent, Ontario, Canada	3.8
Chubb (New Quebec), Quebec, Canada	3.2
Steinem, Swabia, Germany	2.5
Henbury, Northern Territory, Australia	2.2
Boxhole, central Australia	1.8
Barringer Crater, Winslow, Arizona	1.2

Figure 19.22 The Zodiacal Light. This photo shows the diffuse band of light in the plane of the ecliptic that is caused by the scattering of sunlight from tiny interplanetary dust grains. The zodiacal light is most easily visible about an hour before sunrise or after sunset. *(Yerkes Observatory)*

genschein (Fig. 19.23), is created by sunlight reflected straight back by the interplanetary dust concentrated in the plane of the ecliptic. The geganschein is like the bright spot you see on a cloud bank or low-lying mist when you look at it with the Sun directly behind you; the bright spot is just the reflected image of the Sun.

One of the many surprising discoveries made by the *IRAS* satellite was that some of the interplanetary dust in the outer solar system lies well out of the plane of the ecliptic. Gigantic spiraling rings of dust have been found between the orbits of Mars and Jupiter, and they reach substantial distances above and below the ecliptic (Fig. 19.24). These bands of dust might have been created when an asteroid orbiting the Sun in the ecliptic collided with a comet whose orbit passed through the ecliptic at an angle. Dust particles released in the collision would then have streamed out of the ecliptic plane, following spiral paths as they continued to orbit the Sun.

It is possible to collect interplanetary dust particles for direct examination (Fig. 19.25), most commonly by using high-altitude balloons. Recently a surprising new technique has been developed, in which the dust particles are found in sludge scooped off the ocean floor. The Earth is constantly being pelted with dust particles (which add about 8 tons per day to its mass!), and those that fall into the oceans can lie undisturbed on the seabed for a long time. Studies of the grains show that they are probably of cometary origin, having been dispersed through space from the dead nuclei of old comets.

As we learned in Chapter 9, the space between stars in our galaxy is permeated by a rarefied gas medium. In the Sun's vicinity, the average density of this gas is far below that of any man-made vacuum; it amounts to only about 0.1 particle per cubic centimeter (that is, there is one atom, on average, in every volume of 10 cubic centimeters, corresponding to a cube a little less than 1 inch on each side). Because of the motion of the Sun, the interstellar gas streams through the solar system with a velocity of about 20 kilometers per second. This ghostly breeze, consisting mostly of hydrogen and helium atoms and ions, was discovered in the early 1970s, when observations made from satellites revealed very faint ultraviolet emission from the hydrogen and helium atoms in the gas. The tenuous interstellar wind has very little effect on the other components of the solar system, but it is nevertheless studied with some interest for what it may tell us about the interstellar medium.

Perspective

The interplanetary wanderers discussed in this chapter have given us insight into the history of the solar system and have told us much about its present state as well. We have found two primary origins of the various objects: comets and asteroids. The former account for most of the meteors and for the interplanetary dust, and the latter are responsible for the meteorites, including the massive bodies that formed the major impact craters in the solar system.

Next we turn our attention to the formation of life on the Earth and the possibility that it might begin elsewhere.

Summary

1. Thousands of minor planets have been discovered and cataloged. They display a variety of sizes (up

Figure 19.23 The Gegenschein. This photo shows the Milky Way stretching across the upper portion and a diffuse concentration of light at lower center, (the gegenschein). The gegenschein is created by light reflected directly back to Earth from interplanetary dust in the direction opposite the Sun. *(Photo by S. Suyama, courtesy of J. Weinberg)*

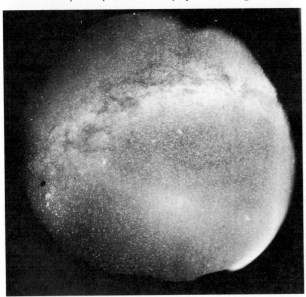

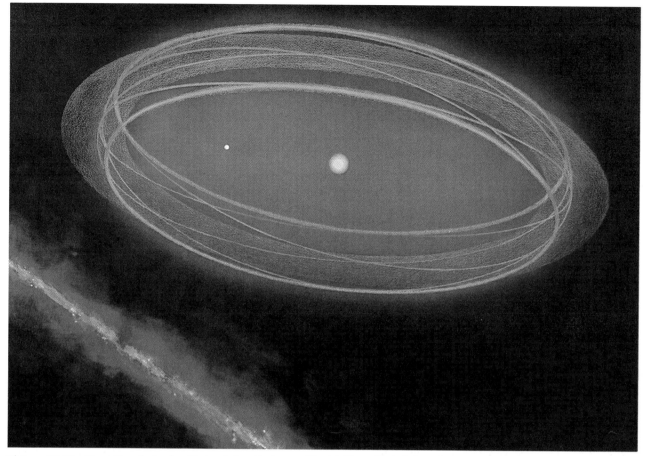

Figure 19.24 Dust Bands in the Outer Solar System. The *IRAS* satellite's infrared sensors detected emission from this spiral-shaped tail of dust, which may have been created by a collision between an asteroid and a comet. At lower left is the Milky Way, as seen in the infrared by *IRAS*. *(NASA)*

to 1000 km in diameter) and compositions (from metals to rocky minerals).

2. Gaps in the asteroid belt are created by orbital resonances with Jupiter, whose gravitational influence was probably also responsible for preventing the asteroids from coalescing into a planet in the first place.
3. Comets are small, icy objects that develop their characteristic comae and tails only when in the inner part of the solar system. The solid part is the nucleus, which consists of dust and ice with a dark dust coating.
4. Comets apparently originate in a cloud of debris very far from the Sun. They occasionally fall inward and either bypass the Sun and return to the distant reaches of the solar system for millennia or are perturbed by the gravitational influence of one of the planets and become periodic comets.

Figure 19.25 An Interplanetary Dust Grain. This is a microscopic view of a tiny particle from interplanetary space. The amorphous structure is highly variable from one grain to another. *(NASA photograph, courtesy of D. E. Brownlee)*

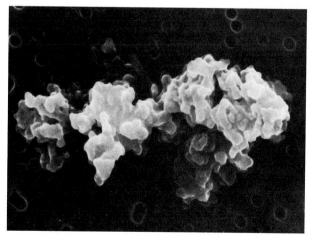

5. When a comet nears the Sun, it ejects gases, mostly in powerful jets, that glow by fluorescence. Water vapor is the most abundant gas. Most periodic comets eventually lose all of their icy substance in this process and disintegrate into swarms of rocky debris.
6. A comet may have two tails, one created by ionized gases and the other made of fine dust particles.
7. A meteor is a flash of light created when a meteoroid enters the Earth's atmosphere from space, and a meteorite is the solid remnant that reaches the ground in some cases.
8. Meteorites are either stony, stony-iron, or iron in composition. They are very old and provide information on the early solar system.
9. Most meteors are created by fine debris from comets, but most meteorites are fragments of asteroids.
10. Interplanetary space is permeated by fine dust particles and by an interstellar wind of hydrogen and helium atoms from the space between the stars.

Review Questions

1. What is the orbital period of a minor planet whose semimajor axis is 2.8 AU?
2. At what distance from the Sun would an asteroid have exactly one tenth the orbital period of Jupiter? Is there a gap in the asteroid belt at that distance?
3. How is the formation of the asteroid belt like the formation of the rings of Saturn?
4. How do the motions of the comets differ from those of planets?
5. Summarize the effects of Jupiter on asteroids, comets, and meteoroids.
6. Summarize the life story of a typical periodic comet.
7. Why are carbonaceous chondrites important clues to the early history of the solar system?
8. Would an observer on the surface of Mercury see meteors in the nighttime sky? Would one find meteorites on the surface of Mercury?
9. What does the concentration of the zodiacal light in the ecliptic tell us about the distribution of interplanetary dust in the solar system?
10. What are the orbital periods of the Trojan asteroids? How do the periods of the Apollo asteroids compare with that of the Earth?

Additional Readings

Beatty, J. K. 1986. An inside look at Halley's comet. *Sky and Telescope* 71(5):438.

Cassidy, W. A., and L. A. Rancitelli. 1982. Antarctic meteorites. *American Scientist* 70(2):156.

Chapman, C. R. 1975. The nature of asteroids. *Scientific American* 232(1):24.

Chapman, R. P., and J. C. Brandt. 1985. An introduction to comets and their origins. *Mercury* XIV(1):2.

Gehrels, T. 1985. Asteroids and comets. *Physics Today* 38(2):32.

Gingerich, O. 1986. Newton, Halley and the comet. *Sky and Telescope* 71(3):230.

Greenstein, G. 1985. Heavenly fire (Tunguska). *Science 85* 6(6):70.

Hartmann, W. K. 1975. The smaller bodies of the solar system. *Scientific American* 233(3):142.

Ronan, C., and M. Mohs. 1986. Scientist of the year (Edmund Halley). *Discover* 7(1):52.

Tatum, J. B. 1982. Halley's comet in 1986. *Mercury* 11(4):126.

Van Allen, J. A. 1975. Interplanetary particles and fields. *Scientific American* 233(3):160.

Whipple, F. L. 1986. Flying sandbanks or dirty snowballs: Discovering the nature of comets. *Mercury* XV(1):2.

———. 1974. The nature of comets. *Scientific American* 230(2):49.

Section VI

Life in the Universe

We finish our survey of the universe with a discussion of living organisms and the prospects for finding life elsewhere in the cosmos. To many, this is the most essential question we can ask. In the sole chapter in this section, we examine the astronomer's attempt to answer it.

In discussing the question of life elsewhere in the universe, we begin by looking into the origins of life here on Earth. In the process we learn of biological evolution and the evidence that has been unearthed to tell us of our own beginnings. To do this we will step outside the arena of astronomy, but we return to it in assessing whether the conditions that led to life on Earth might exist on other planets in our solar system and in the galaxy at large. We discuss the probability that technological civilizations might be thriving here and there in the Milky Way.

We complete our brief treatment of life in the universe with a description of the strategy for searching for other civilizations. It seems most likely that radio communications will be our initial means of contact, and there is an interesting exercise in logic involved with choosing the wavelength at which to seek or send signals.

Chapter 20

The Chances of Companionship

A space station concept. *(NASA)*

Chapter Preview

Life on Earth
Could Life Develop Elsewhere?
The Probability of Detection
The Strategy for Searching

We have attempted to answer all of the fundamental questions about the physical universe that can be treated scientifically. Having done this, we know our place in the cosmos; we know something of its scale and its age; and we realize how insignificant our habitat is.

In this chapter we contemplate whether we as living creatures are unique in the universe or whether we must relinquish even that distinction. Nothing that we have learned so far leads us to rule out the possibility that other life forms, some of them intelligent, exist. We believe that the Earth and the other planets are a natural by-product of the formation of the Sun, and we have evidence that some of the essential ingredients for life were present on the Earth from the time it formed. Similar conditions must have been met countless times in the history of the universe and will occur countless more times in the future.

Science cannot yet tell the full story of how life began, however, and we have not found any evidence that it actually does exist elsewhere. The mystery remains.

Life on Earth

We start by discussing the origins of the only life we know. Besides giving us some insight into the processes thought to have been at work on the Earth, this discussion will help us later when we are speculating whether the same processes have occurred elsewhere.

Before the time of Charles Darwin (Fig. 20.1), in the mid-1800s, the view was widely held that life could arise spontaneously from nonliving matter. Darwin's work in the study of evolution, which showed how species develop gradually as a result of environmental pressures, made such an idea seem improbable.

An alternative to the spontaneous formation of life was proposed in 1907 by the Swedish chemist S. Arrhenius, who suggested that life on the Earth was introduced billions of years ago from space, originally in the form of microscopic spores that float through the cosmos, landing here and there to act as seeds for new biological systems. This idea, called the **panspermia** hypothesis, cannot be ruled out, but several arguments make it seem unlikely. It would take a very long time for such spores to permeate the galaxy, and there would have to be a very high density of them in space in order for one or more of them to reach the Earth by chance. More important, it seems very unlikely that the spores could survive the hazards of space such as ultraviolet light and cosmic rays. Even if the panspermia concept is correct, the question of the ultimate origin of life remains; it is merely transferred to some other location. In view of what is known today about the evolution of life and the early conditions on the Earth, scientists generally agree that life arose through natural processes occurring here and was not introduced from elsewhere.

It is believed that the early atmosphere of the Earth

Figure 20.1 Charles Darwin. Scientific inquiries by Darwin led to an understanding of evolution, one of the most profound concepts of human intellectual development. *(The Granger Collection)*

was composed chiefly of hydrogen and hydrogen-bearing molecules such as ammonia (NH_3) and methane (CH_4), as well as water (H_2O). Therefore the first organisms must have developed in the presence of these ingredients.

In the 1950s experiments began to be carried out in which attempts were made to reproduce the conditions of the early Earth. The starting point of these experiments was to place water in containers filled with the type of atmosphere just described. Water was introduced because it is apparent that the Earth had oceans from very early times and because it is thought that life started in the oceans, where the liquid environment provided a medium in which complex chemical reactions could take place. Reactions occur much more slowly in solids (because the atoms are not free to move about easily and interact) and in gases (because the density of gas is low, and particles are relatively unlikely to encounter each other). Water is the most stable and abundant liquid that can form from the common elements thought to be present when the Earth was young.

The first of these experiments (Fig. 20.2) was carried out in 1953 by the American scientists H. Urey and S. Miller, who concocted a mixture of methane, ammonia, water, and hydrogen and exposed it to electrical discharges, a possible source of energy on the primitive Earth. (Ultraviolet light from the Sun is another, but it was more difficult to work with in the lab.) After a week, the mixture turned a dark brown, and Urey and Miller analyzed its composition. What they found were large quantities of amino acids (complex molecules that form the basis of proteins), which are the fundamental substance of living matter. It was later shown by other experimenters that exposure to ultraviolet light produced the same results. These experiments demonstrated that at least some of the precursors of life probably existed in the primitive oceans almost immediately after the Earth had cooled enough to support liquid water. Other similar experiments have produced more complex molecules, including sugars and larger fragments of proteins.

As noted in Chapter 19 it may be that amino acids were present in the solar system even before the Earth

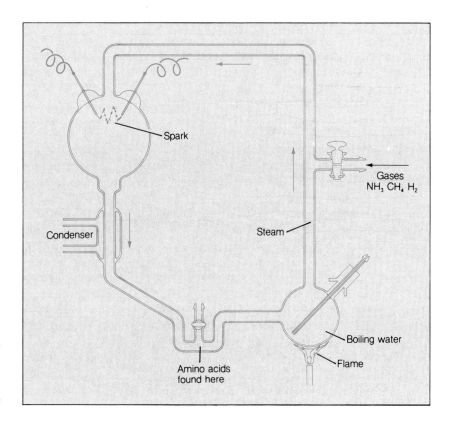

Figure 20.2 Simulating the Early Earth. This apparatus was constructed by Urey and Miller to reproduce conditions on the primitive Earth, in hopes of learning how life forms could have developed. *(© West Publishing Co.)*

Astronomical Insight 20.1

Panspermia Revisited

In the text we discussed the old idea that life originally reached the Earth from space in the form of microscopic spores. This hypothesis has not been considered seriously by most modern scientists, but it has recently been revived by Fred Hoyle, a prominent and highly controversial astrophysicist, along with his colleagues J. V. Narlikar and N. C. Wickramasinghe.

Hoyle's basic argument is that the spontaneous creation of life is simply too unlikely an event to have had even a remote chance of occurring during the lifetime of the Earth. To illustrate his point, Hoyle notes that there are some two thousand enzymes that are critically important for human life and that each, on average, consists of around one hundred amino acids in a very specific sequence and configuration. If we take the point of view that all of these enzymes had to be created initially by random mixing of amino acids, then, as Hoyle calculates, it should take $10^{40,000}$ trial groupings of amino acids to produce the needed two thousand enzymes by chance. This is an impossibly large number, and certainly Hoyle is correct in saying that there has not been sufficient time since the Earth formed for this number of random groupings to have come together. There is a persuasive counterargument, but first let us follow Hoyle to his logical conclusion.

If there has been insufficient time for life to develop on the Earth, the argument goes, life must have reached the Earth from somewhere else. Hoyle and his colleagues choose to assume that life came to Earth (and still does) in the form of tiny spores, very reminiscent of the original panspermia hypothesis. Several reasons are cited for assuming this: For example, spores can be quite durable and long-lived, so that they could survive the rigors of space, which is necessary in Hoyle's hypothesis. As support for this point, Hoyle claims that many epidemics are due to the introduction from space of new spores to which humans have not developed immunity.

Even if we accept Hoyle's assertion that life did not originate on Earth, there remains the question of where it *did* happen. Where and when, in the lifetime of the universe, could the required $10^{40,000}$ random groupings of amino acids have had time to occur? Here Hoyle revives another idea that received notice a decade or two ago: the steady state universe (Hoyle and Narlikar were leading proponents). In the original steady state hypothesis, the universe was postulated to have had no beginning and no end; it simply always existed and always will, in the form seen today. This theory lost all chance for widespread acceptance with the discovery of the 3-degree background radiation, which has been satisfactorily explained only as a remnant of the big bang.

The recent development of the inflationary universe models, in

---

formed, because traces of them have been found in some meteorites (Fig. 20.3) and we know that meteorites are very old, the first solid material in the solar system. We also know that several kinds of complex molecules, including organic (carbon-bearing) molecules, exist inside dense interstellar clouds (see Chapter 9), and we can speculate that perhaps amino acids may also have formed in these regions. In order to have survived on the Earth, these primordial amino acids would have had to reach our planet sometime *after* its molten period.

It is not so clear what direction things took once amino acids and other organic molecules existed. Somehow these building blocks had to combine to form **ribonucleic acid (RNA)** and **deoxyribonucleic acid (DNA;** Fig. 20.4). These very complex molecules carry

Astronomical Insight 20.1

which the universe is assumed to have undergone a very rapid transformation from a highly dense state to a very much lower density, expanded state, has inspired Hoyle to develop a new version of the steady state model. In this version, the universe is assumed to have been in an infinitely long-lived state of moderately high density and temperature, providing the time and conditions necessary for life to develop. At a time some 10 to 20 billion years ago, a "bubble" suddenly expanded as in the inflationary model, and this bubble became our universe, seeded with life forms from the start. Thus Hoyle and his coworkers have proposed a radical picture of both the origin of the universe and the beginning of life, tied together in a single scenario.

Hoyle's ideas are not at all widely accepted, for various reasons. One is that other scientists do not share his pessimism about the chances for the development of life on Earth within a reasonably short time. They point out that it is not necessary, and is in fact probably incorrect, to assume that an enzyme containing 100 amino acids in a specific combination could only have been created by chance—by bringing together a sufficient number of random groupings of 100 or more amino acids at a time so that the required combination would eventually appear. On the contrary, it has been shown more likely that the process occurred sequentially, that amino acids came together first in small combinations that then grew into larger ones. As the small combinations grew larger by the addition of new amino acids, the particular combinations that produced the most viable larger groupings were favored, and other, less successful combinations did not develop further. Thus the process was not at all random but was dictated by evolutionary pressures.

A key point here is that the buildup of the correct sequences of amino acids from random collections of smaller groupings has actually been observed in repeatable laboratory experiments. This shows that the particular sequences found in enzymes are indeed the ones that develop naturally in sequential fashion from smaller groupings of amino acids, and that they do so in very short times.

Another argument against Hoyle's picture may be made as follows: If the universe has been permeated by living spores since its expansion some 10 to 20 billion years ago, then *every* planetary or satellite body in the universe must have been seeded with life forms at a very early time in its history. Therefore, every corner of the galaxy capable of supporting life should have life forms today, yet no evidence for this has been found. Places not capable of supporting life, such as the Moon's surface, should still have undeveloped spores, yet none have been found in the lunar or Martian soils.

---

the genetic codes that allow living creatures to reproduce themselves. Experiments have successfully produced molecules that are fragments of RNA and DNA from conditions like those that prevailed on the early Earth, but not the complete forms required. Maybe it is a simple matter of time; if such experiments could be carried out for years or millennia, perhaps the vital forms of these proteins would appear. This is one of the areas of greatest uncertainty in our present knowledge of how life began.

Fossil records (Fig. 20.5) tell us that the first microorganisms appeared some 3 to 3.5 billion years ago, when the Earth was barely a billion years old. Following their appearance, the evolution of increasingly complex species seems to have followed naturally (Table 20.1). At first the development was very slow, only

Figure 20.3 Primordial Amino Acids. This is a section of the Murchison meteorite, which fell in Australia. Amino acids found in this carbonaceous chondrite were apparently present in it when it fell. *(Photograph by John Fields, the Trustees, the Australian Museum)*

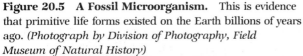

Figure 20.5 A Fossil Microorganism. This is evidence that primitive life forms existed on the Earth billions of years ago. *(Photograph by Division of Photography, Field Museum of Natural History)*

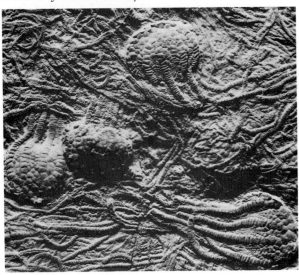

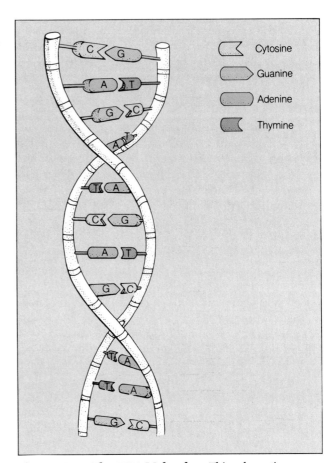

Figure 20.4 The DNA Molecule. This schematic diagram shows a section of a DNA molecule. DNA carries the genetic code that allows organisms to reproduce themselves. A critical question in understanding the development of life on Earth is to learn how DNA arose. *(© West Publishing Co.)*

reaching the level of simple plants such as algae a billion years later. Increasingly elaborate multicellular plant forms followed and gradually altered the Earth's atmosphere by introducing free oxygen. Meanwhile the gases hydrogen and helium, light and fast-moving enough to escape the Earth's gravity, essentially disappeared. Nitrogen, always present from outgassing and volcanic activity, became more predominant through the decay of dead organisms. By about one billion years ago, the Earth's atmosphere had reached its present composition.

The first broad proliferation of animal life occurred about 600 million years ago, and the great reptiles

Table 20.1 Steps in the Evolution of Life on Earth

Era	Period	Age (yrs)	Biological Developments
Archeozoic		$>3.5 \times 10^9$	
	Archean		No life forms
Proteozoic		$1.5\text{--}3.5 \times 10^9$	
	Algonkin		Radiolaria; marine algae
Paleozoic		$0.25\text{--}1.5 \times 10^9$	
	Cambrian		Marine faunas, primitive vegetation
	Ordovician		Fishlike vertebrates
	Silurian		Air-breathing invertebrates, first land plants
	Devonian		Fishes and amphibians; primitive ferns
	Mississippian		Ancient sharks, mosses, ferns
	Pennsylvanian		Amphibians, insects
	Permian		Primitive reptiles, mosses, ferns
Mesozoic		$70\text{--}250 \times 10^6$	
	Triassic		Reptiles, dinosaurs, ferns
	Jurassic		Toothed birds, primitive mammals, palms
	Cretaceous		Decline of dinosaurs; modern insects, birds, snakes
Cenozoic		$1.5\text{--}70 \times 10^6$	
	Paleocene		Large land mammals
	Eocene		Primates, first horse, modern plants
	Oligocene		Larger horse, modern plants
	Miocene		Proliferation of mammals, larger horse, modern plants
	Pliocene		Grassland mammals, earliest hominids
Modern		$1.5 \times 10^6\text{--present}$	
	Pleistocene		Hominids, modern size horse, modern plants
	Holocene		*Homo sapiens*, present-day vegetation

arose some 350 million years later. The dinosaurs died out after about 200 million years, and mammals came to dominance about 65 million years ago. Our primitive ancestors appeared only in the past 3 or 4 million years. Once the development of intelligence had provided the ability to control the environment, the entire world became our ecological home, and our physical evolution essentially stopped. It remains to be seen whether future ecological pressures will create further evolution of the human species.

Could Life Develop Elsewhere?

The scenario just described, if at all accurate, seemingly should occur almost inevitably, given the proper conditions. If this is so, the question of whether life exists elsewhere is equivalent to asking whether the conditions that existed on the primitive Earth could have arisen elsewhere. (For the dissenting view that life was still improbable, see Astronomical Insight 20.2.)

It is clear that no other planet in the solar system could have provided an environment exactly like that on the early Earth. It therefore seems unlikely that Earth-like organisms will be found anywhere else in the solar system. Mars is the only planet so far where life forms have been sought (Fig. 20.6), but even there the conditions are quite unlike those on the Earth. Given the billions of stars in the galaxy, however, and the vast number that are very similar to the Sun, it seems highly probable that the proper conditions must have been reproduced many times in the history of the galaxy.

So far we have worked from the tacit assumption that life on other planets, if it exists, must be similar to life on the Earth, and we have considered only the quesiton of whether other Earth-like environments may exist. We may, however, question the premise that life

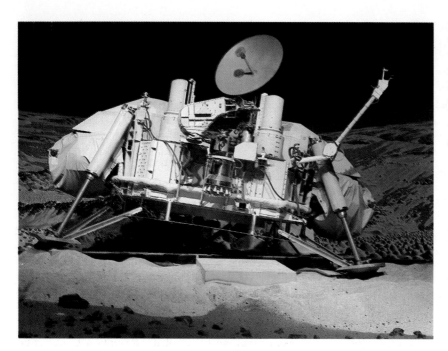

Figure 20.6 Searching for Life in the Solar System. Mars was long thought to be the most likely home in the solar system for extraterrestrial life. Here we see a *Viking* lander in a simulated Martian environment. The *Viking* missions reached Mars in 1976 but found no evidence for life forms. *(NASA)*

could have developed only in the form with which we are familiar. Here, obviously, we must indulge in speculation, since we have no examples of other types of life at hand for examination.

Life as we know it is based on carbon-bearing molecules, and it has been argued that only carbon has the capability of combining chemically in a sufficiently wide variety of ways with other common elements to produce the complexity of molecules thought to be necessary for life. After all, it is clear that the basic elements available to begin with must be the same everywhere, given the homogeneous composition of the universe. This may seem to rule out life forms based on anything other than carbon, but it has been pointed out that another common element, silicon, also has a very complex chemistry and therefore might provide a basis for a radically different type of life. We cannot begin to speculate on the conditions necessary for such life forms to arise.

Another assumption that might be subject to question is that water is a necessary medium. As noted earlier, it is the most abundant liquid that can form under the temperature and pressure conditions of the early Earth, and it is thought that only a liquid medium could support the required level of chemical activity. Other liquids can exist under other conditions, however, and it is interesting to consider whether life forms of a wholly different type might arise in oceans of strange composition. It is interesting to speculate, for example, about what goes on in the lakes of liquid methane that are thought to exist on Titan, the mysterious giant satellite of Saturn.

If we are satisfied that life probably has formed naturally in many places in the universe, we can address a related, and to many a more important, question: Given the existence of life, how likely is it that intelligence will follow? Here we have no means of answering, except to reiterate that as far as we know, the evolution of mankind on Earth was the natural product of environmental pressures.

The Probability of Detection

In view of the limitations that prohibit faster-than-light travel, it is exceedingly unlikely that we will be able to visit other solar systems, seeking out life forms that may live there. We will continue to explore our own system, so there is a reasonable chance that if life exists on any of the other planets of our Sun, we will someday discover it. It seems, however, that our best hope of finding other intelligent races in the galaxy will be to make long-range contact with them, through radio or light signals. Since this requires both a transmitter and a receiver, we can only hope to contact other civilizations as advanced as ours, with the capability of constructing the necessary devices for interstellar communication.

A mathematical exercise in probabilities has been used for some years as a means of assessing, as objectively as possible, the chances for making contact with an extraterrestrial civilization. The aim is to separate the question into several distinct steps, each of which could then be treated independently. The underlying assumption is that the number of technological civilizations in our galaxy today with the capability for interstellar communication is the product of the number of planets that exist with appropriate conditions, the probability that life developed on those planets, the probability that such life developed intelligence that gave rise to a technological civilization, and, finally, the likelihood that the civilization has not killed itself off through evolution or catastrophe.

Mathematically, the so-called **Drake equation** (after Frank Drake, who has been its best-known advocate) is written:

$$N = R_* f_p n_e f_l f_i f_c L,$$

where N is the number of technological civilizations presently in existence, R_* is the number of stars of appropriate spectral type formed per year in the galaxy, f_p is the fraction of these that have planets, n_e is the number of Earth-like planets per star, f_l is the fraction of these on which life arises, f_i is the fraction of those planets on which intelligence has developed, f_c is the fraction of planets with intelligence on which a technological civilization has evolved to the point where interstellar communications would be possible, and L is the average lifetime of such a civilization.

By expressing the number of civilizations in this way, it is possible to isolate the factors about which we can make educated guesses from those about which we are more ignorant. It is an interesting exercise to go through the terms in the equation one by one, to see what conclusions we reach under various assumptions. People who do this have to make sheer guesses for some of the terms, and the result is a variety of answers ranging from very optimistic to very pessimistic. In the following, we will adopt middle-of-the-road numbers for most of the unknown terms.

The first two factors, R_* and f_p, are quantities that in principle can be known with some certainty from observations. The rate of star formation that is relevant is that for stars of a spectral type similar to the Sun's. A much cooler star would not have a temperate zone around it where a planet could have the moderate temperatures needed for life to begin (we are, throughout this exercise, limiting ourselves to life forms similar to our own), and a very hot star would be short-lived, so that there would be insufficient time for life to develop on its planets before the parent star blew up in a supernova explosion. Taking these considerations into account, it is estimated by some that up to ten suitable stars form in our galaxy per year; for the sake of discussion, we will be more cautious and adopt $R_* = 1/\text{year}$. This may even be a little high for the present epoch in galactic history, but the star formation rate was surely much higher early in the lifetime of the galaxy, and low-mass stars of solar type formed that long ago are still in their prime. Thus the adoption of an average formation rate of one Sun-like star per year is probably reasonable.

From what we know of the formation of our solar system, it seems that the formation of planets is almost inevitable, except perhaps in double or multiple star systems. For the sake of discussion, let us assume that $f_p = 1$; that is, that all stars of solar type have planets. Observations made with future instruments such as the *Space Telescope* may soon provide real information on this term.

The number of Earth-like planets, n_e, is highly uncertain, and depends on how wide the zone is where the appropriate temperature conditions could exist. Recent studies show that this zone might be rather small; that is, that the earth would not have been able to support life if its distance from the Sun were only about 5 percent closer or farther than it is. Estimates of n_e vary from 10^{-6} to 1. Let us be moderate and assume $n_e = 0.1$; that is, that in one out of ten planetary systems around solar type stars, there is a planet within the temperate zone where life can arise.

Now we get to the *really* speculative terms in the equation. We have no way of estimating how likely it is that life should begin, given the right conditions. From the seeming naturalness of its development on Earth, it can be argued that life would always begin if given the chance. Let us be optimistic here and agree with this, adopting $f_l = 1$.

The chances of this life developing intelligence and advancing to the point of being capable of interstellar communication are again complete unknowns. All we know is that in the only example that has been observed, both things happened. For the sake of argument, we therefore set both f_i and f_c equal to 1.

At this point it is instructive to put the values adopted so far into the equation. We find:

$$N = (1/\text{year})(1)(0.1)(1)(1)(1)L$$
$$= 0.1L$$

Having taken our chances and guessed at the values for all the other terms, we now face a critical question: How long can a technological civilization last? Ours has been sufficiently advanced to send or receive interstellar radio signals for only about fifty years, and there is sufficient instability in our society to lead some pessimists to think we will not last many more decades. If we take this viewpoint and adopt 50 years as the average lifetime, we find:

$$N = 5,$$

meaning that we should expect the total number of technological civilizations present in the galaxy at one moment to be very small, about five. If this is correct, then the average distance between these outposts of civilization is nearly 20,000 light-years (Fig. 20.7). The time it would take for communications to travel between civilizations would therefore be very much longer than their lifetimes, and there would be no hope of establishing a dialogue with anyone out there. If this estimate is correct, it is no surprise that we haven't heard from anybody yet.

We can be more optimistic, though, and assume that a technological civilization solves its internal problems and lives much longer than fifty years. Extremely optimistic people would argue that a civilization is immortal; that it colonizes other star systems than its own, so that it is immune to any local crises such as planetary wars or suns expanding to become red giants. In that case, allowing a few billion years for the development of such civilizations, we can set $L = 10^{10}$ years (nearly equal to the age of the galaxy), and we find

$$N = 10^9,$$

in which case the average distance between civilizations is only about thirty light-years (Fig. 20.7), coincidentally just a little less than the distance our own radio signals have traveled (Fig. 20.8) since the early

Figure 20.8 Earth's Message to the Cosmos. As our entertainment and communications broadcasts travel out into space, they provide a history of our culture for anyone who may be receiving the signals. At the present time, the growing sphere that is filled with our broadcasts has a radius of over fifty light-years.

Figure 20.7 Possible Values of N. The number of technological civilizations in the galaxy (N) may be very small (upper), in which case the average distance between them is very large; or N may be large, so that the distance between civilizations is relatively small. The chances for communication with alien races are much higher in the latter case, where the distances are only a few tens of light-years.

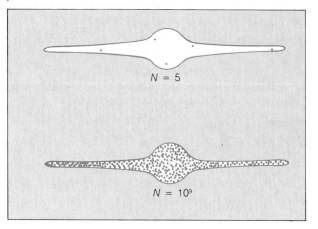

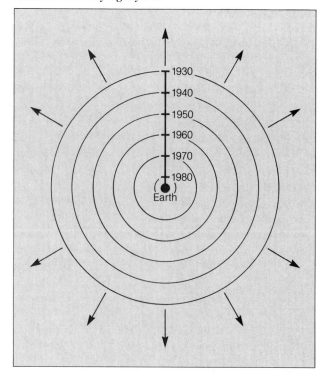

Astronomical Insight 20.2

The Case for a Small Value of N

While it is an interesting exercise to write down an equation for N, the number of technological civilizations in the galaxy, no two astronomers can agree on the values of the various terms.

There is one very pessimistic argument, developed recently, that concludes that we are alone in the galaxy. This argument is based on the likely fact that if there were other civilizations, a significant fraction of them would have arisen long before ours. This is a straightforward consequence of the great age of the galaxy, some 5 or 10 billion years greater than the age of the Earth. This argument postulates further that if technological civilizations live a long time, the galaxy must be inhabited by a number of very advanced races much older and more mature than our own. Up to this point there is little controversy about any of these assertions.

The next logical step in the argument, however, is hotly debated. It says that any advanced civilization that has survived for millions or even billions of years will necessarily have done so by perfecting some form of interstellar travel and colonizing other planets. Once this started, the argument goes, the spread of a given civilization throughout the galaxy would accelerate rapidly, and by this time in galactic history, no habitable planet such as the Earth would remain uncolonized. The first successful galactic civilization, in this view, would quickly rise to complete dominance.

The only logical corollary, if the argument is accepted up to this point, is that no other civilizations exist, because if they did the Earth would have been colonized long ago. The absence of interstellar visitors on the Earth is taken as proof that there are no other civilizations in the galaxy.

Those who adhere to this line of reasoning believe that $N = 1$. Therefore at least one of the terms in Drake's equation must be very much smaller than the more optimistic values discussed in the text. One suggestion is that the temperate zone around a Sun-like star where planets can have moderate conditions is really very much narrower than usually supposed, perhaps due to the more ready development of a Venus-like greenhouse effect than is normally thought to be the case. If the temperate zone is very small, then the term n_e, representing the number of habitable, Earth-like planets per star, would be very much smaller than the value of 0.1 adopted in the text.

Another term that has recently been singled out is f_1, the fraction of Earth-like planets on which life begins. As we learned in the text, somehow the amino acids that were present on the primitive Earth had to arrange themselves into special combinations to produce the long, chain-like protein molecules RNA and DNA. From a purely statistical point of view, the probability that the proper combination would happen to come together by chance is very small. This apparently happened on the Earth, but it may be so unlikely that it has not occurred anywhere else in the galaxy (but see the counterargument in Astronomical Insight 20.1).

Whatever the correct values of the terms in the equation, the debate will rage on until our civilization reaches its life expectancy L and dies, we discover another civilization, or we ourselves expand to colonize the galaxy and find no one else out there.

days of radio and television. If this estimate of N is correct, we should be hearing from somebody very soon.

We have presented two extreme views of the likelihood that other civilizations exist in our part of the galaxy. As we mentioned earlier, opinions among the scientists who seriously study this question vary thoughout this large range. Those who favor the optimistic viewpoint advocate the idea that we should make deliberate attempts to seek out other civilizations.

The Strategy for Searching

The probability arguments outlined in the previous section are amusing and perhaps somewhat instructive but obviously not very accurate. There are entirely too many unknowns in the equation for us to develop a reliable estimate of the chances for galactic companionship. Perhaps we will not know for certain what the answer is until we make contact with another civilization.

The problem of developing an experiment to search for or to send interstellar signals is that we do not know the ground rules. There are infinite possible ways in which a distant civilization might choose to try to communicate, and it is impossible to search for them all. We must try to guess what the most probable technique would be.

In view of the power that is transmitted by radio signals and the relative lack of natural "noise" in the galaxy in that part of the spectrum, it has often been assumed that radio communications are most likely to succeed, although other techniques have been tried (Fig. 20.9). In the early 1960s, the U.S. National Radio Astronomy Observatory (Fig. 20.10) "listened" for transmissions from two nearby Sun-like stars, tau Ceti and epsilon Eridani, without success. More recently, larger surveys of neighboring stars have been carried out in both the United States and the Soviet Union, with similar results. A new search program was started in 1983, using a radio telescope belonging to Harvard University. This telescope is now wholly devoted to the search for extraterrestrial communications and takes advantage of sophisticated computer technology to monitor a wide range of frequencies in many stars.

The wavelength chosen for most of these observa-

Figure 20.9 Another Message from Earth. This recording of a message from Earth is traveling beyond the solar system aboard the *Voyager* spacecraft. Only an advanced race of beings would have the technological skills necessary to learn how to listen to and decode the message of peace that it contains. *(NASA)*

Figure 20.10 A Radio Telescope Used in the Search for Life. The large dish at the left, at the U. S. National Radio Astronomy Observatory, in Green Bank, West Virginia, has been used in efforts to detect signals from alien civilizations. Most of these attempts have been made at the 21-centimeter wavelength of atomic hydrogen emission. *(National Radio Astronomy Observatory, operated by Associated Universities, Inc., under contract with the National Science Foundation)*

tions was 21 centimeters, partly for practical reasons (radio telescopes designed for observing at this wavelength already existed) and partly because it was hypothesized that if a civilization out there wanted deliberately to send signals, it might choose to do so at a wavelength that other astronomers around the galaxy might be observing anyway, even if they weren't trying to find distant civilizations. Another wavelength that has been considered, but tried in only a very limited search, lies in the microwave region between the emission lines of interstellar water vapor (H_2O) and hydroxyl (OH). Dubbed the "water hole," this region has the advantages that it, too, might be a target of extraterrestrial astronomers and that it is in the quietest part of the radio spectrum (various objects, ranging from supernova remnants to radio galaxies, create radio noise that fills the galaxy, and the "water hole" lies in a region of the spectrum where this noise is minimal; see Fig. 20.11).

One pessimistic viewpoint is that we are all listening and no one is sending. In that case, we can only hope to pick up accidental emissions such as entertainment broadcasts on radio or television, and these would be much weaker and more difficult to detect. It has been estimated that at the present level of technology we could deliberately send radio signals that could be received by a similarly advanced civilization up to several thousand parsecs away, while our accidental radio and television signals, broadcast indiscriminately in all directions rather than at a specific target, could be detected only at distances of one or two parsecs. Therefore, it makes a big difference whether someone out there is trying to send a message.

While most thinking about means of communication with other civilizations has been based on the assumption that radio signals are the best choice, there have been other suggestions. One is that a **laser** might be used. A laser is a device that can emit a powerful, narrow beam of visible light, at a single wavelength; lasers today are being used for many purposes ranging from microsurgery to weapons to telephone communications. The lasers developed so far, however, would not have the range of radio transmissions, so at this point most scientists favor a search for and with radio signals. Laser technology is improving, but there is a fundamental limitation on the use of visible light: The universe is far "noisier"—more filled with confusing natural emissions—at visible wavelengths than in the radio portion of the spectrum.

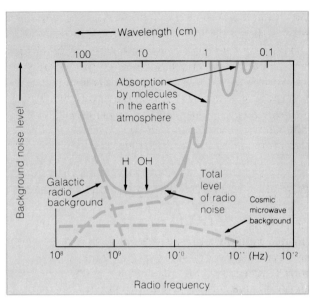

Figure 20.11 The Galactic Noise Spectrum. This diagram shows the relative intensity of the background noise in the galaxy as a function of wavelength. The quietest portion of the spectrum is in the vicinity of the "water hole" near the natural emission wavelengths of H_2O and OH molecules (only OH is specifically shown here). *(Data from D. Goldsmith and T. Owen 1980,* The Search for Life in the Universe *(Menlo Park, Calif.: Benjamin Cumming)*

It is not clear what the future will bring, in terms of deliberate searches for extraterrestrial intelligence. The most ambitious project yet studied seriously was called *Project Cyclops* (Fig. 20.12), but in its full form, it is unlikely ever to be. The plan was funded to build a veritable forest of large radio antennae, along with a very sophisticated data-processing system, so that up to a million stars could be scanned for artificial emissions over a broad wavelength region in a few years' time. By searching over a wide range of wavelengths, the scientists planning this effort hoped to eliminate much of the guesswork about which is the most likely portion of the spectrum to be used by alien civilizations. This method also enhances the chances of detecting accidental emissions, which, if they are anything like our own radio and television signals, may be spread over much of the spectrum. The *Project Cyclops* antennae could also have been used to send powerful radio signals, allowing the Earth to broadcast its own beacon for others to find and providing for the possibility of two-way communications.

Figure 20.12 Project Cyclops. The grandest plan seriously considered for the purpose of communicating with extraterrestrial civilizations, *Project Cyclops* would consist of 1000 to 2500 radio antennae, each 100 feet in diameter. This huge array, some 16 kilometers in extent, would be capable of detecting signals over a large portion of the radio spectrum, from planetary systems 1000 or more parsecs away. *(NASA)*

The principal problem in implementing *Cyclops,* or any other large-scale effort, is the enormous cost. The version of *Project Cyclops* outlined above would cost some 10 billion dollars. Such sums are rarely devoted to astronomy, and if they were, there would still be considerable debate over whether to spend it on this or on some other research programs that might appear more likely to succeed. Presumably, alien astronomers in other solar systems would have to face the same questions, although we can hope that sufficiently advanced civilizations will have overcome the limitations of their planetary resources and will be willing and able to tackle such a massive task.

Fortunately, advanced computer technology is making it possible to achieve most of the goals of *Project Cyclops* at a much lower cost. A less expensive project, called the *Search for Extraterrestrial Intelligence (SETI)* has been funded by NASA. The plan is to use existing radio dishes and specially developed, sophisticated computers to carry out a limited search of a number of stars. Like the more complex *Cyclops* plan, *SETI* will scan a number of wavelengths for possible signals; in fact, modern high-speed computer technology will allow coverage of as broad a spectrum and as great a number of stars as envisioned originally for *Cyclops.* Scientists involved with *SETI* are confident that it will eventually prove successful in detecting alien civilizations if N is at least moderately large.

Perspective

In this chapter we have introduced a bit of speculation, well-founded perhaps in the discussion of life on Earth but less so in the discussions of the possibilities of life elsewhere.

We are prepared now to understand and appreciate new advances in astronomy. Having completed our study of the present knowledge of the universe, we will be able to put into perspective the major new discoveries that are sure to come as technology and theory continue to improve. The *Space Telescope,* the large optical telescopes now being designed, and the comparable advances in radio, infrared, and X-ray techniques will all inevitably lead to novel and unforeseen breakthroughs in our view of the universe. The next few decades will be exciting times for astronomy, and we can anticipate their coming with a sense of excitement and wonder.

Summary

1. While there were early beliefs that life started on Earth spontaneously or by primitive spores from space, most scientists today accept the theory that life began through natural, evolutionary processes.

2. Amino acids, fundamental components of living organisms, are formed readily in experiments designed to simulate early conditions on the Earth.
3. The steps that led to the development of the necessary forms of RNA and DNA are not yet fully understood and probably occurred over a very long period of time.
4. Fossil evidence provides a record of the evolutionary steps leading from the first primitive life forms to modern life.
5. The conditions that prevailed on the early Earth have probably been duplicated on other planets in the galaxy, though probably not on other planets in the solar system.
6. It is often assumed that only Earth-like life could develop, because carbon is nearly unique in the complexity of its chemistry, but there have been suggestions that at least one other element (silicon) may have the necessary properties.
7. Estimates of the number N of technological civilizations now in the galaxy can be made, with great uncertainty, based on what is known of the formation rate of Sun-like stars, what is guessed for the probability of such stars' having planets with the proper conditions, and the probability that life, leading to intelligence and technology, will develop on these planets. Estimates range from $N = 1$ to $N = 10^9$.
8. Some attempts have been made to search for radio signals at 21-centimeter and other wavelengths, with no success.
9. The chances for detecting or being detected by other civilizations depend strongly on whether deliberate attempts are made to send signals.
10. Future projects to search for and send signals have been planned and will be sufficiently powerful to have a high probability of success if a large number of civilizations exist.

Review Questions

1. Suppose life did start on Earth as a result of a chance arrival of microscopic spores from space. If these spores travel through space with velocities typical of interstellar clouds, then their speeds might be about 20 kilometers per second. Calculate how long it would take a spore to travel at this speed from the nearest star (1.4 parsecs away) to Earth. How long would it take for a spore to cross the galactic disk, a distance of 30,000 parsecs?
2. Summarize the evolution of the Earth's atmosphere, as outlined in Chapter 16. Does this provide a possible clue to look for in assessing whether life exists on a planet orbiting a distant star?
3. We have noted here that some of the conditions thought necessary for the formation of life exist in the atmosphere of Jupiter. Some key conditions are probably absent, however. What are they?
4. Based on your answer to question 3, do you think life is likely to have formed inside interstellar clouds? (Recall that rather complex organic molecules have been observed in dark clouds.)
5. Why would a planet only slightly closer to the Sun than the Earth probably not be able to support life? (Hint: Recall what you learned in Chapter 17 about the origin of the contrasting conditions on the Earth and on Venus.)
6. Repeat the calculation of N, the number of technological civilizations in the galaxy, with the same parameters as used in the text, except that the probability of life beginning on an Earth-like planet is only 10^{-4}, rather than 1 (assume that the lifetime L is 10^{10} years).
7. What is N if all the values used in the text are adopted, except that L is 10,000 years?
8. Would you expect technological civilizations in the galaxy to be concentrated along the spiral arms or distributed more or less uniformly throughout the disk? Explain your answer.
9. Would you expect life to be as likely to arise on planets orbiting Population II stars as on those orbiting Population I stars? Explain.
10. Explain why it makes such a big difference whether a deliberate attempt is made by a civilization to send radio signals; that is, explain why deliberate signals can be detected at much greater distances than accidental signals such as radio and television broadcasts.

Additional Readings

Abt, H. A. 1979. The companions of Sun-like stars. *Scientific American* 236(4):96.
Ball, J. A. 1980. Extraterrestrial intelligence: Where is everybody? *American Scientist* 68:656.

Barber, V. 1974. Theories of the chemical origin of life on Earth. *Mercury* 3(5):20.

Goldsmith, D., ed. 1980. *The quest for extraterrestrial life.* Mill Valley, Calif.: University Science Books.

O'Neill, G. K. 1974. The colonization of space. *Physics Today* 27(9):32.

Papagiannis, M. D. 1982. The search for extraterrestrial civilizations—a new approach. *Griffith Observer* 11(1):112.

Pelligrino, C. R. 1979. Organic clues in carbonaceous meteorites. *Sky and Telescope* 57(4):330.

Pollard, W. G. 1979. The prevalence of Earth-like planets. *American Scientist* 67:653.

Rood, R. T., and Trefil, J. S. 1981. *Are we alone?* New York: Scribners.

Sagan, C., and Drake, F. 1975. The search for extraterrestrial intelligence. *Scientific American* 232(5):80.

Sagan, C., and Shklovskii, I. S. 1966. *Intelligent life in the universe.* San Francisco: Holden-Day.

Tipler, F. J. 1982. The most advanced civilization in the galaxy is ours. *Mercury* 11(1):5.

Wetherill, C., and Sullivan, W. T. 1979. Eavesdropping on the Earth. *Mercury* 8(2):23.

Appendices

Appendix 1   Symbols Commonly Used in this Text

Symbol	Meaning	Symbol	Meaning
Å	Angstrom, a unit of length often used to measure wavelength of light; $1\text{ Å} = 10^{-8}$ cm	$\Delta\lambda$	The Greek letters delta lambda, used to designate a shift in wavelength, as in the Doppler effect
c	Standard symbol for the speed of light	γ	The Greek letter gamma, sometimes used to designate a gamma-ray photon
G	Standard symbol for the gravitational constant		
H	Symbol for the constant in the Hubble expansion law	λ	The Greek letter lambda, usually used to designate wavelength
h	Standard symbol for the Planck constant	ν	The Greek letter nu, the standard symbol for frequency, also used to designate a neutrino in nuclear reactions
K	Kelvin, the unit of temperature in the absolute scale		
z	Symbol commonly used to designate the Doppler shift ($z = \Delta\lambda/\lambda = v/c$ for velocities much less than the speed of light)	π	The Greek letter pi, usually used to designate the parallax angle; also used for the ratio of the circumference of a circle to its diameter
α	The Greek letter alpha, sometimes used to designate an alpha particle		

Appendix 2   Physical and Mathematical Constants

Constant	Symbol	Value
Speed of light	c	2.9979250×10^{10} cm/sec
Gravitation constant	G	6.670×10^{-8} dyn cm^2/gm
Planck constant	h	6.62620×10^{-27} erg/sec
Electron mass	m_e	9.10939×10^{-28} gm
Proton mass	m_p	1.672623×10^{-24} gm
Stefan-Boltzmann constant	σ	5.67051×10^{-5} erg/cm^{2}deg^{4}sec
Wien constant	W	0.289789 cm/deg
Boltzmann constant	k	1.38066×10^{-16} erg/deg
Astronomical unit	AU	1.49599×10^8 km
Parsec	pc	3.085678×10^{13} km = 3.261633 light-years
Light-year	ly	9.460530×10^{12} km
Solar mass	$M_\odot$	1.9891×10^{33} gm
Solar radius	$R_\odot$	6.9600×10^{10} cm
Solar luminosity	$L_\odot$	3.827×10^{33} erg/sec
Earth mass	$M_\oplus$	5.9742×10^{27} gm
Earth radius	$R_\oplus$	6378.140 km
Tropical year (equinox to equinox)		365.241219878 days
Sidereal year (with respect to stars)		365.256366 days = 3.155815×10^7 sec

Appendix 3 The Elements and Their Abundances

Element	Symbol	Atomic No.	Atomic Weight*	Abundance†	Element	Symbol	Atomic No.	Atomic Weight*	Abundance†
Hydrogen	H	1	1.0080	1.00	Iodine	I	53	126.9045	4.07×10^{-11}
Helium	He	2	4.0026	0.085	Xenon	Xe	54	131.30	1×10^{-10}
Lithium	Li	3	6.941	1.55×10^{-9}	Cesium	Cs	55	132.905	1.26×10^{-11}
Beryllium	Be	4	9.0122	1.41×10^{-11}	Barium	Ba	56	137.34	6.31×10^{-11}
Boron	B	5	10.811	2.00×10^{-10}	Lanthanum	La	57	138.906	6.46×10^{-11}
Carbon	C	6	12.0111	0.000372	Cerium	Ce	58	140.12	4.37×10^{-11}
Nitrogen	N	7	14.0067	0.000115	Praseodymium	Pr	59	140.908	4.27×10^{-11}
Oxygen	O	8	15.9994	0.000676	Neodymium	Nd	60	144.24	6.61×10^{-11}
Fluorine	F	9	18.9984	3.63×10^{-8}	Promethium	Pm	61	146	—
Neon	Ne	10	20.179	3.72×10^{-5}	Samarium	Sm	62	150.4	4.57×10^{-11}
Sodium	Na	11	22.9898	1.74×10^{-6}	Europium	Eu	63	151.96	3.09×10^{-12}
Magnesium	Mg	12	24.305	3.47×10^{-5}	Gadolinium	Gd	64	157.25	1.32×10^{-11}
Aluminum	Al	13	26.9815	2.51×10^{-6}	Terbium	Tb	65	158.925	2.63×10^{-12}
Silicon	Si	14	28.086	3.55×10^{-5}	Dysprosium	Dy	66	162.50	1.29×10^{-11}
Phosphorus	P	15	30.9738	3.16×10^{-7}	Holmium	Ho	67	164.930	3.1×10^{-12}
Sulfur	S	16	32.06	1.62×10^{-5}	Erbium	Er	68	167.26	5.75×10^{-12}
Chlorine	Cl	17	35.453	2×10^{-7}	Thulium	Tm	69	168.934	2.69×10^{-12}
Argon	Ar	18	39.948	4.47×10^{-6}	Ytterbium	Yb	70	170.04	6.46×10^{-12}
Potassium	K	19	39.102	1.12×10^{-7}	Lutetium	Lu	71	174.97	6.92×10^{-12}
Calcium	Ca	20	40.08	2.14×10^{-6}	Hafnium	Hf	72	178.49	6.3×10^{-12}
Scandium	Sc	21	44.956	1.17×10^{-9}	Tantalum	Ta	73	180.948	2×10^{-12}
Titanium	Ti	22	47.90	5.50×10^{-8}	Tungsten	W	74	183.85	3.72×10^{-10}
Vanadium	V	23	50.9414	1.26×10^{-8}	Rhenium	Re	75	186.2	1.8×10^{-12}
Chromium	Cr	24	51.996	5.01×10^{-7}	Osmium	Os	76	190.2	5.62×10^{-12}
Manganese	Mn	25	54.9380	2.63×10^{-7}	Iridium	Ir	77	192.2	1.62×10^{-10}
Iron	Fe	26	55.847	2.51×10^{-5}	Platinum	Pt	78	195.09	5.62×10^{-11}
Cobalt	Co	27	58.9332	3.16×10^{-8}	Gold	Au	79	196.967	2.09×10^{-12}
Nickel	Ni	28	58.71	1.91×10^{-6}	Mercury	Hg	80	200.59	1×10^{-9}
Copper	Cu	29	63.546	2.82×10^{-8}	Thallium	Tl	81	204.37	1.6×10^{-12}
Zinc	Zn	30	65.37	2.63×10^{-8}	Lead	Pb	82	207.19	7.41×10^{-11}
Gallium	Ga	31	69.72	6.92×10^{-10}	Bismuth	Bi	83	208.981	6.3×10^{-12}
Germanium	Ge	32	72.59	2.09×10^{-9}	Polonium	Po	84	210	—
Arsenic	As	33	74.9216	2×10^{-10}	Astatine	At	85	210	—
Selenium	Se	34	78.96	3.16×10^{-9}	Radon	Rn	86	222	—
Bromine	Br	35	79.904	6.03×10^{-10}	Francium	Fr	87	223	—
Krypton	Kr	36	83.80	1.6×10^{-9}	Radium	Ra	88	226.025	—
Rubidium	Rb	37	85.4678	4.27×10^{-10}	Actinium	Ac	89	227	—
Strontium	Sr	38	87.62	6.61×10^{-10}	Thorium	Th	90	232.038	6.61×10^{-12}
Yttrium	Y	39	88.9059	4.17×10^{-11}	Protactinium	Pa	91	230.040	—
Zirconium	Zr	40	91.22	2.63×10^{-10}	Uranium	U	92	238.029	4.0×10^{-12}
Niobium	Nb	41	92.906	2.0×10^{-10}	Neptunium	Np	93	237.048	—
Molybdenum	Mo	42	95.94	7.94×10^{-11}	Plutonium	Pu	94	242	—
Technetium	Tc	43	98.906	—	Americium	Am	95	242	—
Ruthenium	Ru	44	101.07	3.72×10^{-11}	Curium	Cm	96	245	—
Rhodium	Rh	45	102.905	3.55×10^{-11}	Berkelium	Bk	97	248	—
Palladium	Pd	46	106.4	3.72×10^{-11}	Californium	Cf	98	252	—
Silver	Ag	47	107.868	4.68×10^{-12}	Einsteinium	Es	99	253	—
Cadmium	Cd	48	112.40	9.33×10^{-11}	Fermium	Fm	100	257	—
Indium	In	49	114.82	5.13×10^{-11}	Mendelevium	Md	101	257	—
Tin	Sn	50	118.69	5.13×10^{-11}	Nobelium	No	102	255	—
Antimony	Sb	51	121.75	5.62×10^{-12}	Lawrencium	Lr	103	256	—
Tellurium	Te	52	127.60	1×10^{-10}					

*The atomic weight of an element is its mass in *atomic mass units*. An atomic mass unit is defined as one twelfth of the mass of the most common isotope of carbon and has the value 1.660531×10^{-24} grams. In general, the atomic weight of an element is approximately equal to the total number of protons and neutrons in its nucleus.

†The abundances are given in terms of the number of atoms of each element compared to hydrogen and are based on the composition of the Sun. For very rare elements, particularly those toward the end of the list, the abundances can be quite uncertain, and the values given should not be considered exact.

Appendix 4 Temperature Scales

At the most basic level, temperature can be defined in terms of the motion of particles in a gas (or in a solid or a liquid). We all have an intuitive idea of what heat is, and we are all familiar with at least one scale for measuring temperature.

The most commonly used scales are somewhat arbitrarily defined, with zero points not having any truly fundamental physical basis. The popular *Fahrenheit* scale, for example, has water freezing at a temperature of 32°F and boiling at 212°F. On this scale absolute zero, the lowest possible temperature (where all molecular motions cease), is −459°F.

The centigrade (or Celsius) scale is perhaps better founded, although it is based on the freezing and boiling points of water, rather than the more fundamental absolute zero. In this system, the freezing point is defined as 0°C, and 100°C is the boiling point. This scale has the advantage over the Fahrenheit scale that there are exactly 100° between the freezing and boiling points, rather than 180°, as in the Fahrenheit system. To convert from Fahrenheit to centigrade, we must subtract 32° first, and then multiply the remainder by 100/180, or 5/9. For example, 50°F is equal to 5/9 × (50 − 32) = 10°C. To convert from centigrade to Fahrenheit, first multiply by 9/5 and then add 32°. Thus, −10°C = 9/5 × (−10) + 32 = 14°F. On the centigrade scale, absolute zero occurs at −273°C.

The temperature scale preferred by scientists is a modification of the centigrade system. In this system, named after its founder, the British physicist Lord Kelvin, the same degree is used as in the centigrade scale; that is, one degree is equal to one hundredth of the difference between the freezing and boiling points of water. The zero point is different from the zero point on the centigrade scale, however; it is equal to absolute zero. Hence on this scale water freezes at 273°K and boils at 373°K. Comfortable room temperature is around 300°K. In modern usage the degree symbol (°) is dropped, and we speak simply of temperatures in units of Kelvins (273 K, for example).

Appendix 5 Radiation Laws

Several laws described in Chapter 4 apply to continuous radiation from hot objects such as stars. In the text these laws were discussed in general terms, and a few simple applications were explained. Here the same laws are given in more precise mathematical form, and their use in that form is illustrated.

Wien's Law

In general terms, Wien's law says that the wavelength of maximum emission from a glowing object is inversely proportional to its temperature. Mathematically this can be written as:

$$\lambda_{max} \propto 1/T,$$

where λ_{max} is the wavelength of strongest emission, T is the surface temperature of the object, and $\propto$ is a special mathematical symbol meaning "is proportional to."

Experimentation can determine the *proportionality constant*, specifying the exact relationship between λ_{max} and T, and Wien did this, finding

$$\lambda_{max} = W/T,$$

where W has the value 0.29 if λ_{max} is measured in centimeters and T in Kelvins. With this equation it is possible to calculate λ_{max}, given T, or vice versa. Thus, if we measure the spectrum of a star and find that it emits most strongly at a wavelength of 2000 Å = 2 × 10^{-5} cm, then we can solve for the temperature:

$$T = W/\lambda_{max} = 0.29/2 \times 10^{-5} = 14{,}500 \text{ K}.$$

This is a relatively hot star, and it would appear blue to the eye. Note that it was necessary to measure the spectrum in ultraviolet wavelengths in order to find λ_{max}.

When solving problems using Wien's law, it is always possible to use the equation form, as we have just done in this example. Often, however, it is more convenient to compare the properties of two objects by considering the ratio of their temperatures or of their wavelengths of maximum emission. In effect this is what we did in the text when we compared two objects of different temperatures in order to determine how their λ_{max} values compared. For example, we said that if one object is twice as hot as another, its value of λ_{max} is half that of the other.

We can see how this ratio technique works by writing the equation for Wien's law separately for object 1 and object 2:

$$\lambda_{max_1} = W/T_1, \quad \text{and} \quad \lambda_{max_2} = W/T_2.$$

Now we can divide one equation by the other:

$$\frac{\lambda_{max1}}{\lambda_{max2}} = \frac{W/T_1}{W/T_2}$$

or

$$\frac{\lambda_{max1}}{\lambda_{max2}} = \frac{T_2}{T_1}.$$

The numerical factor W has canceled out, and we are left with a simple expression relating the values of λ_{max} and T for the two objects. Now we see that if $T_1 = 2T_2$ (object 1 is twice as hot as object 2), then

$$\frac{\lambda_{max1}}{\lambda_{max2}} = \frac{T_2}{2T_2} = \frac{1}{2},$$

or λ_{max} for object 1 is one-half that for object 2. In this extremely simple example, it probably would have been easier just to work it out in our heads, but what if we have a case where one object is 3.368 times hotter than the other, for example?

A great deal can be learned from making comparisons in this way. Astronomers often use the Sun as the standard for comparison, expressing various quantities in terms of the solar values. Another reason for using comparisons occurs when the numerical constants (such as Wien's constant in the foregoing examples) are not known. If the trick of comparing is kept in mind, it is often possible to work out answers to astronomical questions simply by carrying around in one's head a few numbers describing the Sun.

The Stefan-Boltzmann Law

As discussed in the text, the energy emitted by a glowing object is proportional to T^4, where T is the surface temperature. This can be written mathematically as

$$E = \sigma T^4,$$

where σ stands for a proportionality constant that has the value 5.7×10^{-5}, if the centimeter-gram-second metric units are used.

If we now consider how much surface area a star has, then the total energy it emits, called its *luminosity* and usually denoted L, is

$$L = \text{surface area} \times E$$
$$= 4\pi R^2 \sigma T^4,$$

where R is the radius of the star. This equation is called the Stefan-Boltzmann law.

As in other cases, the law can be used directly in this form, or we can choose to compare properties of stars by writing the equation separately for two stars and then dividing. If we do this, we find

$$\frac{L_1}{L_2} = \frac{R_1^2 T_1^4}{R_2^2 T_2^4} = \left(\frac{R_1}{R_2}\right)^2 \left(\frac{T_1}{T_2}\right)^4$$

The constant factors 4π and σ have canceled out.

As an example of how to use this expression, suppose we determine that a particular star has twice the temperature of the Sun but only one half the radius, and we wish to know how this star's luminosity compares to that of the Sun. If we designate the star as object 1 and the Sun as object 2, then

$$\frac{L_1}{L_2} = \left(\frac{1}{2}\right)^2 (2)^4 = \frac{1}{4} \times 16 = 4$$

This star has four times the luminosity of the Sun.

The Stefan-Boltzmann law is particularly useful because it relates three of the most important properties of stars to each other.

The Planck Function

The radiation laws described above and in the text are actually specific forms of a much more general law, discovered by the great German physicist Max Planck. Wien's law, the Stefan-Boltzmann law, and some others not mentioned in the text bear the same kind of relation to Planck's law as the laws of planetary motion discovered by Kepler do to Newton's mechanics. Kepler's laws were first discovered by observation, but with Newton's laws of motion it is possible to derive Kepler's laws theoretically. In the same fashion, the radiation laws discussed so far were found experimentally but can be derived mathematically from the much more general and powerful Planck's law.

Planck's law is usually referred to as the Planck function, a mathematical relationship between intensity and wavelength (or frequency) that describes the spectrum of any glowing object at a given temperature. The Planck function specifically applies only to objects that radiate solely because of their temperature and do not reflect any light or have any spectral lines. The popular term for such an object is *blackbody,* and we often refer to radiation from such an object as *blackbody radiation* or, more commonly, *thermal radiation,* the term used in the text. Stars are not perfect radiators, but to a good approximation, they can be treated as such; hence in the text we apply Wien's law and the Stefan-Boltzmann law to stars (and even planets in certain circumstances) without pointing out the fact that to do so is only approximately correct.

The form of the Planck function for the radiation intensity B as a function of wavelength is

$$B = \frac{2hc^2}{\lambda^5} \frac{1}{e^{hc/\lambda kT} - 1},$$

where h is the Planck constant, c is the speed of light, and k is the Boltzmann constant (the values of all three are tabulated in Appendix 2), λ is the wavelength (in cm), and T is the temperature of the object (on the Kelvin scale). The symbol e represents the base of the natural logarithm, something not used elsewhere in this text; for the present purpose, this may be regarded simply as a mathematical constant with the value 2.718.

In terms of frequency ν rather than wavelength, the expression is

$$B = \frac{2h\nu^3}{c^2} \frac{1}{e^{h\nu/kT} - 1}.$$

Either expression may be used to calculate the spectrum of continuous radiation from a glowing object at a specific temperature. In practice the Planck function is used in a wide assortment of theoretical calculations that call for knowledge of the intensity of radiation so that its effects can be assessed on physical conditions such as ionization.

Appendix 6   Major Telescopes of the World (2 Meters or Larger)

Observatory	Location*	Telescope
Keck Observatory (California Institute of Technology and the University of California)	(Mauna Kea, Hawaii)	10.0-m Keck Telescope
Special Astrophysical Observatory	Mount Pastukhov, USSR	6-m Bol'shoi Teleskop Azimutal'nyi
Hale Observatories	Palomar Mountain, California	5.08-m George Ellery Hale Telescope
Smithsonian Observatory	Mount Hopkins, Arizona	4.5-m Multiple Mirror Telescope
Royal Greenwich Observatory	La Palma, Canary Islands	4.2-m William Herschel Telescope
Cerro Tololo Observatory	Cerro Tololo, Chile	4.0-m
Anglo-Australian Observatory	Siding Spring Mountain, Australia	3.9-m Anglo-Australian Telescope
Kitt Peak National Observatory	Kitt Peak, Arizona	3.8-m Nicholas U. Mayall Telescope
Royal Observatory Edinburgh	Mauna Kea, Hawaii	3.8-m United Kingdom Infrared Telescope
Canada-France-Hawaii Observatory	Mauna Kea, Hawaii	3.6-m Canada-France-Hawaii Telescope
European Southern Observatory	Cerro La Silla, Chile	3.57-m
Max Planck Institute (Bonn)	Calar Alto, Spain	3.5-m
Astronomical Research Corporation	(Sacramento Peak, New Mexico)	3.5-m
Lick Observatory	Mount Hamilton, California	3.05-m C. Donald Shane Telescope
Mauna Kea Observatory	Mauna Kea, Hawaii	3.0-m NASA Infrared Telescope
McDonald Observatory	Mount Locke, Texas	2.7-m
Haute Provence Observatory	Saint Michele, France	2.6-m
Crimean Astrophysical Observatory	Simferopol, USSR	2.6-m
Byurakan Astrophysical Observatory	Byurakan, USSR	2.6-m
Mount Wilson and Las Campanas Observatories	Mount Wilson, California	2.5-m Hooker Telescope

(continued on next page)

Appendix 6 Major Telescopes of the World, continued

Mount Wilson and Las Campanas Observatories	Las Campanas, Chile	2.5-m Irenée du Pont Telescope
Royal Greenwich Observatory	Canary Islands	2.5-m Isaac Newton Telescope
Dartmouth College and University of Michigan	Kitt Peak, Arizona	2.4-m McGraw-Hill Telescope
Wyoming Infrared Observatory	Mount Jelm, Wyoming	2.3-m Wyoming Infrared Telescope
Steward Observatory	Kitt Peak, Arizona	2.3-m
Mauna Kea Observatory	Mauna Kea, Hawaii	2.2-m
Max Planck Institute (Bonn)	Calar Alto, Spain	2.2-m
Max Planck Institute	(South West Africa)	2.2-m
University of Mexico Observatory	(Mexico)	2.16-m
La Plata Observatory	La Plata, Argentina	2.15-m
Kitt Peak National Observatory	Kitt Peak, Arizona	2.1-m
McDonald Observatory	Mount Locke, Texas	2.1-m Otto Struve Telescope
Karl Schwarzschild Observatory	Tautenberg, East Germany	2.0-m (largest Schmidt telescope)

*The locations of telescopes under construction are indicated in parentheses.

Appendix 7 Planetary and Satellite Data

Orbital Data for the Planets

Planet	Sidereal Period	Semimajor Axis	Orbital Eccentricity*	Inclination of Orbital Plane	Rotation Period	Tilt of Axis
Mercury	0.241 yr	0.387 AU	0.2056	7°0'15"	$58\overset{d}{.}65$	28°
Venus	0.615	0.723	0.068	3°23'40"	243	3°
Earth	1.000	1.000	0.0167	0°0'0"	23^h56^m	23°27'
Mars	1.881	1.524	0.0934	1°51'0"	24^h37^m	23°59'
Jupiter	11.86	5.203	0.0485	1°18'17"	9^h56^m	3°5'
Saturn	29.46	9.555	0.0556	2°29'33"	10^h39^m	26°44'
Uranus	84.01	19.22	0.0472	0°46'23"	17^h14^m	97°55'
Neptune	164.79	30.11	0.0086	1°46'22"	18^h12^m	28°48'
Pluto	248.5	39.44	0.250	17°10'12"	6^d24^m	125°

*The eccentricity of an orbit is defined as the ratio of the distance between foci to the semimajor axis. In practice it is related to the perihelion distance P and the semimajor axis a by $P = a(1 - e)$, where e is the eccentricity, and to the aphelion distance A by $A = a(1 + e)$.

Appendix 7 Planetary and Satellite Data, continued

Physical Data for the Planets

Planet	Mass*	Diameter*	Density	Surface Gravity	Escape Speed	Temperature	Albedo
Mercury	0.0558	0.382	5.50 g/cc	0.38 g	4.3 km/sec	100–700 K	0.06
Venus	0.815	0.951	5.3	0.90	10.3	730	0.75
Earth	1.000	1.000	5.517	1.00	11.2	200–300	0.31
Mars	0.107	0.531	3.96	0.38	5.0	130–290	0.15
Jupiter	318.1	10.79	1.33	2.64	60.0	130	0.34
Saturn	95.12	8.91	0.68	1.13	36.0	95	0.34
Uranus	14.54	4.05	1.20	0.89	21.0	52	0.34
Neptune	17.2	3.91	1.58	1.13	24.0	50	0.29
Pluto	0.002	0.18	1.7	0.06	1.2	40	0.4

*The masses and diameters are given in units of the Earth's mass and diameter, which are 5.974×10^{27} grams and 12,734 km, respectively.

Satellites

Planet	Satellite	Semimajor Axis	Period	Diameter	Mass	Density
Earth	Moon	3.84×10^5 km	$27^d.322$	3476 km	7.35×10^{25}	3.34 g/cm³
Mars	Phobos	9.38×10^3	0.3189	$27 \times 22 \times 19$	9.6×10^{18}	2:
	Deimos	2.35×10^4	1.2624	$15 \times 12 \times 11$	1.9×10^{18}	2:
Jupiter	Metis	1.280×10^5	0.295	20:	9.5×10^{19}	
	Adrastea	1.290×10^5	0.298	$20 \times 20 \times 15$	1.9×19^{19}	
	Amalthea	1.81×10^5	0.498	$270 \times 170 \times 150$	7.2×10^{21}	
	Thebe	2.22×10^5	0.675	110×90	7.6×10^{20}	
	Io	4.22×10^5	1.769	3630	8.92×10^{25}	3.53
	Europa	6.71×10^5	3.551	3138	4.87×10^{25}	3.03
	Ganymede	1.07×10^6	7.155	5262	1.49×10^{26}	1.93
	Callisto	1.88×10^6	16.689	4800	1.08×10^{26}	1.70
	Leda	1.11×10^7	238.7	16:	5.7×10^{18}	
	Himalia	1.15×10^7	250.6	186:	9.5×10^{21}	
	Lysithea	1.17×10^7	259.2	36:	7.6×10^{19}	
	Elara	1.17×10^7	259.7	76:	7.6×10^{20}	

(continued on next page)

Appendix 7   Planetary and Satellite Data, continued

Planet	Satellite	Semimajor Axis	Period	Diameter	Mass	Density
Jupiter, cont.	Ananke	2.12×10^7	631	30:	3.8×10^{19}	
	Carme	2.26×10^7	692	40:	9.5×10^{19}	
	Pasiphae	2.35×10^7	735	50:	1.9×10^{20}	
	Sinope	2.37×10^7	758	36:	7.6×10^{19}	
Saturn	Atlas	1.377×10^5	0.602	40×20		
	Prometheus	1.394×10^5	0.613	$140 \times 100 \times 80$		
	Pandora	1.417×10^5	0.629	$110 \times 90 \times 80$		
	Epimetheus	1.514×10^5	0.694	$140 \times 120 \times 100$		
	Janus	1.514×10^5	0.695	$220 \times 200 \times 160$		
	Mimas	1.855×10^5	0.942	392	4.5×10^{22}	1.43
	Enceladus	2.381×10^5	1.370	500	7.4×10^{22}	1.13
	Telesto	2.947×10^5	1.888	$34 \times 28 \times 26$		
	Calypso	2.947×10^5	1.888	$34 \times 22 \times 22$		
	Tethys	2.947×10^5	1.888	1060	7.4×10^{23}	1.19
	No Name I	$3.3: \times 10^5$	?	15–20		
	Dione	3.774×10^5	2.737	1120	1.05×10^{24}	1.43
	Helene	3.781×10^5	2.737	$36 \times 32 \times 30$		
	No Name II	$3.8: \times 10^5$	?	15–20		
	No Name III	$4.7: \times 10^5$	?	15–20		
	Rhea	5.271×10^5	4.518	1530	2.50×10^{24}	1.33
	Titan	1.222×10^6	15.945	5150	1.35×10^{26}	1.89
	Hyperion	1.481×10^6	21.277	$410 \times 260 \times 220$	1.71×10^{22}	
	Iapetus	3.561×10^6	79.330	1460	1.88×10^{24}	1.15
	Phoebe	1.295×10^7	550.5	220		
Uranus	1986U7	4.97×10^4	0.33	40		
	1986U8	5.38×10^4	0.38	50		
	1986U9	5.92×10^4	0.43	50		
	1986U3	6.18×10^4	0.46	60		
	1986U6	6.27×10^4	0.48	60		
	1986U2	6.46×10^4	0.49	80		
	1986U10	6.61×10^4	0.51	80		
	1986U4	6.99×10^4	0.56	60		
	1986U5	7.53×10^4	0.62	60		
	1986U1	8.60×10^4	0.76	170		
	Miranda	1.294×10^5	1.413	480	7.5×10^{22}	1.26
	Ariel	1.910×10^5	2.520	1330	1.4×10^{24}	1.14
	Umbriel	2.663×10^5	4.144	1110	1.3×10^{24}	1.82
	Titania	4.359×10^5	8.706	1600	3.5×10^{24}	1.60
	Oberon	5.835×10^5	13.463	1630	2.9×10^{24}	1.28
Neptune	Triton	3.543×10^5	5.877	3800	1.3×10^{26}	
	Nereid	5.512×10^6	360.2	300	2.1×10^{22}	
Pluto	Charon	1.97×10^4	6.387			

Appendix 8 Stellar Data

The Fifty Brightest Stars

Star		Spectral Type	Apparent Magnitude	Distance	Position (1980) Right Ascension	Declination
α Eri	Archernar	B3 V	0.51	36 pc	01^{h}37^m.0	−57°20′
α UMi	Polaris	F8 Ib	1.99	208	02 12.5	+89 11
α Per	Mirfak	F5 Ib	1.80	175	03 22.9	+49 47
α Tau	Aldebaran	K5 III	0.86	21	04 34.8	+16 28
β Ori	Rigel	B8 Ia	0.14	276	05 13.6	−08 13
α Aur	Capella	G8 III	0.05	14	05 15.2	+45 59
γ Ori	Bellatrix	B2 III	1.64	144	05 24.0	+06 20
β Tau	Elnath	B7 III	1.65	92	05 25.0	+28 36
ε Ori	Alnilam	B0 Ia	1.70	490	05 35.2	−01 13
ζ Ori	Alnitak	09.5 Ib	1.79	490	05 39.7	−01 57
α Ori	Betelgeuse	M2 Iab	0.41	159	05 54.0	+07 24
β Aur	Menkalinan	A2 V	1.86	27	05 58.0	+44 57
β CMa		B1 II-III	1.96	230	06 21.8	−17 56
α Car	Canopus	F0 Ib-II	−0.72	30	06 23.5	−52 41
γ Gem	Alhena	A0 IV	1.93	32	06 36.6	+16 25
α CMa	Sirius	A1 V	−1.47	2.7	06 44.2	−16 42
ε CMa	Adhara	B2 II	1.48	209	06 57.8	−28 57
δ CMa		F8 Ia	1.85	644	07 07.6	−26 22
α Gem	Castor	A1 V	1.97	14	07 33.3	+31 56
α CMi	Procyon	F5 IV-V	0.37	3.5	07 38.2	+05 17
β Gem	Pollux	K0 III	1.16	11	07 44.1	+28 05
γ Vel		WC8	1.83	160	08 08.9	−47 18
ε Car	Avior	K3 III?	1.90	104	08 22.1	−59 26
ζ Vel		A2 V	1.95	23	08 44.2	−54 38
β Car	Miaplacidus	A1 III	1.67	26	09 13.0	−69 38
α Hya	Alphard	K4 III	1.98	29	09 26.6	−08 35
α Leo	Regulus	B7 V	1.36	26	10 07.3	+12 04
γ Leo		K0 III	1.99	28	10 18.8	+19 57
α UMa	Dubhe	K0 III	1.81	32	11 02.5	+61 52
α Cru A	Acrux	B0.5 IV	1.39	114	12 25.4	−62 59
α Cru B		B1 V	1.86	114	12 15.4	−62 59
γ Cru	Gacrux	M4 III	1.69	67	12 30.1	−57 00
β Cru		B0.5 III	1.28	150	12 46.6	−59 35
ε UMa	Alioth	A0p	1.79	21	12 53.2	+56 04
α Vir	Spica	B1 V	0.91	67	13 24.1	−11 03
η UMa	Alkaid	B3 V	1.87	64	13 46.8	+49 25
β Cen	Hadar	B1 III	0.63	150	14 02.4	−60 16
α Boo	Arcturus	K2 III	−0.06	11	14 14.8	+19 17
α Cen A	Rigil Kentaurus	G2 V	0.01	1.3	14 38.4	−60 46
α Cen B		K4 V	1.40	1.3	14 38.4	−60 46
α Sco	Antares	M1 Ib	0.92	160	16 28.2	−26 23
α TrA	Atria	K2 Ib	1.93	25	16 46.5	−68 60
λ Sco	Shaula	B1 V	1.60	95	17 32.3	−37 05
θ Sco		F0 Ib	1.86	199	17 35.9	−42 59
ε Sgr	Kaus Australis	B9.5 III	1.81	38	18 22.9	−34 24
α Lyr	Vega	A0 V	0.04	8	18 36.2	+38 46
α Aql	Altair	A7 IV-V	0.77	5	19 49.8	+08 49
α Pav	Peacock	B2.5 V	1.95	95	20 24.1	−56 48
α Cyg	Deneb	A2 Ia	1.26	491	20 40.7	+45 12
α Gru	Al Na'ir	B7 IV	1.76	20	22 06.9	−47 04
α PsA	Fomalhaut	A3 V	1.15	7	22 56.5	−29 44

Appendix 8 Stellar Data, continued

Nearby Stars (within 5 Parsecs of the Sun)*

Star	Spectral Type	Visual Apparent Magnitude	Visual Absolute Magnitude	Parallax	Distance	Proper Motion
α Centarui	G2V	−0.1	4.8	0.753″	1.33 pc	3.68″/yr
Barnard's star	M5V	9.5	13.2	0.544	1.84	10.31
Wolf 359	M8Ve	13.5	16.7	0.432	2.31	4.71
BD + 36°2147	M2V	7.5	10.5	0.400	2.50	4.78
Luyten 726-8	M6Ve	12.5	15.4	0.385	2.60	3.36
Sirius	A1V	−1.5	1.4	0.377	2.65	1.33
Ross 154	M5Ve	10.6	13.3	0.345	2.90	0.72
Ross 248	M6Ve	12.3	14.8	0.319	3.13	1.58
ε Eridani	K2V	3.7	6.1	0.305	3.28	0.98
Ross 128	M5V	11.1	13.5	0.302	3.31	1.37
Luyten 789-6	M6V	12.2	14.6	0.302	3.31	3.26
61 Cygni	K5Ve	5.2	7.5	0.292	3.42	5.22
ε Indi	K5Ve	4.7	7.0	0.291	3.44	4.69
τ Ceti	G8V	3.5	5.9	0.289	3.46	1.92
Procyon	F5V	0.4	2.7	0.285	3.51	1.25
Σ 2398	M4V	8.9	11.2	0.284	3.52	2.28
BD + 43°44	M1Ve	8.1	10.4	0.282	3.55	2.89
CD − 36°15693	M2Ve	7.4	9.6	0.279	3.58	6.90
G51-15		14.8	17.0	0.273	3.66	1.26
L725-32	M5Ve	11.5	13.6	0.264	3.79	1.22
BD + 5°1668	M4V	9.8	12.0	0.264	3.79	3.73
CD − 39°14192	M0Ve	6.7	8.8	0.260	3.85	3.46
Kapteyn's star	M0V	8.8	10.8	0.256	3.91	8.89
Kruger 60	M4V	9.7	11.7	0.254	3.94	0.86
Ross 614	M5Ve	11.3	13.3	0.243	4.12	0.99
BD − 12°4523	M5V	10.0	11.9	0.238	4.20	1.18
Wolf 424	M6Ve	13.2	15.0	0.234	4.27	1.75
van Maanen's star	W.D.	12.4	14.2	0.232	4.31	2.95
CD − 37°15492	M3V	8.6	10.4	0.225	4.44	6.08
Luyten 1159-16	M8V	12.3	14.0	0.221	4.52	2.08
BD + 50°1725	K7V	6.6	8.3	0.217	4.61	1.45
CD − 46°11540	M4V	9.4	11.1	0.216	4.63	1.13
CD − 49°13515	M3V	8.7	10.4	0.214	4.67	0.81
CD − 44°11909	M5V	11.2	12.8	0.213	4.69	1.16
BD + 68°946	M3.5V	9.1	10.8	0.213	4.69	1.33
G158-27		13.7	15.5	0.212	4.72	2.06
G208-44/45		13.4	15.0	0.210	4.76	0.75
BD − 15°6290	M5V	10.2	11.8	0.209	4.78	1.16
40 Eridani	K0V	4.4	6.0	0.207	4.83	4.08
L145-141	W.D.	11.4	12.6	0.206	4.85	2.68
BD + 20°2465	M4.5V	9.4	10.9	0.203	4.93	0.49
70 Ophiuchi	K1V	4.2	5.7	0.203	4.93	1.13
BD + 43°4305	M4.5Ve	10.2	11.7	0.200	5.00	0.83

*Data from Lippencott, L.S., 1978, *Space Science Reviews* 22:153. Many of the stars listed here are multiple systems; in these cases, the data in the table refer only to the brightest member of the system.

Appendix 9 The Constellations

Name	Genitive	Abbreviation	Position Right Ascension	Declination
Andromeda	Andromedae	And	01^h	$+40°$
Antlia	Antliae	Ant	10	-35
Apus	Apodis	Aps	16	-75
Aquarius	Aquarii	Aqr	23	-15
Aquila	Aquilae	Aql	20	$+05$
Ara	Arae	Ara	17	-55
Aries	Arietis	Ari	03	$+20$
Auriga	Aurigae	Aur	06	$+40$
Bootes	Bootis	Boo	15	$+30$
Caelum	Caeli	Cae	05	-40
Camelopardalis	Camelopardalis	Cam	06	-70
Cancer	Cancri	Cnc	09	$+20$
Canes Venatici	Canum Venaticorum	CVn	13	$+40$
Canis Major	Canis Majoris	CMa	07	-20
Canis Minor	Canis Minoris	CMi	08	$+05$
Capricornus	Capricorni	Cap	21	-20
Carina	Carinae	Car	09	-60
Cassiopeia	Cassiopeiae	Cas	01	$+60$
Centaurus	Centauri	Cen	13	-50
Cepheus	Cephei	Cep	22	$+70$
Cetus	Ceti	Cet	02	-10
Chamaeleon	Chamaeleonis	Cha	11	-80
Circinis	Circini	Cir	15	-60
Columba	Columbae	Col	06	-35
Coma Berenices	Comae Berenices	Com	13	$+20$
Corona Australis	Coronae Australis	CrA	19	-40
Coronoa Borealis	Coronae Borealis	CrB	16	$+30$
Corvus	Corvi	Crv	12	-20
Crater	Crateris	Crt	11	-15
Crux	Crucis	Cru	12	-60
Cygnus	Cygni	Cyg	21	$+40$
Delphinus	Delphini	Del	21	$+10$
Dorado	Doradus	Dor	05	-65
Draco	Draconis	Dra	17	$+65$
Equuleus	Equulei	Equ	21	$+10$
Eridanus	Eridani	Eri	03	-20
Fornax	Fornacis	For	03	-30
Gemini	Geminorum	Gem	07	$+20$
Grus	Gruis	Gru	22	-45
Hercules	Herculis	Her	17	$+30$
Horologium	Horologii	Hor	03	-60
Hydra	Hydrae	Hya	10	-20
Hydrus	Hydri	Hyi	02	-75
Indus	Indi	Ind	21	-55
Lacerta	Lacertae	Lac	22	$+45$
Leo	Leonis	Leo	11	$+15$
Leo Minor	Leonis Minoris	LMi	10	$+35$
Lepus	Leporis	Lep	06	-20

(continued on next page)

Appendix 9 The Constellations, continued

Name	Genitive	Abbreviation	Right Ascension	Declination
Libra	Librae	Lib	15	−15
Lupus	Lupi	Lup	15	−45
Lynx	Lincis	Lyn	08	+45
Lyra	Lyrae	Lyr	19	+40
Mensa	Mensae	Men	05	−80
Microscopium	Microscopii	Mic	21	−35
Monoceros	Monocerotis	Mon	07	−05
Musca	Muscae	Mus	12	−70
Norma	Normae	Nor	16	−50
Octans	Octantis	Oct	22	−85
Ophiuchus	Ophiuchi	Oph	17	00
Orion	Orionis	Ori	05	+05
Pavo	Pavonis	Pav	20	−65
Pegasus	Pegasi	Peg	22	+20
Perseus	Persei	Per	03	+45
Phoenix	Phoenicis	Phe	01	−50
Pictor	Pictoris	Pic	06	−55
Pisces	Piscium	Psc	01	+15
Piscis Austrinus	Piscis Austrini	PsA	22	−30
Puppis	Puppis	Pup	08	−40
Pyxis	Pyxidis	Pyx	09	−30
Reticulum	Reticuli	Ret	04	−60
Sagitta	Sagittae	Sge	20	+10
Sagittarius	Sagittarii	Sgr	19	−25
Scorpius	Scorpii	Sco	17	−40
Sculptor	Sculptoris	Scl	00	−30
Scutum	Scuti	Sct	19	−10
Serpens	Serpentis	Ser	17	00
Sextans	Sextantis	Sex	10	00
Taurus	Tauri	Tau	04	+15
Telescopium	Telescopii	Tel	19	−50
Triangulum	Trianguli	Tri	02	+30
Triangulum Australe	Trianguli Australi	TrA	16	−65
Tucana	Tucanae	Tuc	00	−65
Ursa Major	Ursae Majoris	UMa	11	+50
Ursa Minor	Ursae Minoris	UMi	15	+70
Vela	Velorum	Vel	09	−50
Virgo	Virginis	Vir	13	00
Volans	Volantis	Vol	08	−70
Vulpecula	Vulpeculae	Vul	20	+25

Appendix 10   Mathematical Treatment of Stellar Magnitudes

Logarithmic Representation

In the text magnitudes are discussed in terms of the brightness ratios between stars of different magnitudes. We generally avoided discussion of cases where two stars differ by a fraction of a magnitude, because in such cases it is no longer simple to calculate the brightness ratio corresponding to the magnitude difference. If star 1 is 0.5 magnitudes brighter than star 2, for example, what is the brightness ratio? Or if star 1 is a factor of 48.76 fainter than star 2, what is the difference in magnitudes?

Astronomers use an exact mathematical relationship between magnitude differences and brightness ratios, written as

$$m_1 - m_2 = 2.5 \log (b_2/b_1),$$

where m_1 and m_2 are the magnitudes of two stars, and b_2/b_1 is the ratio of their brightnesses. The notation "$\log (b_2/b_1)$" means the **logarithm** of this ratio; a logarithm is the power to which 10 must be raised to give this ratio. Hence, for example, if $b_2/b_1 = 100$, $\log (b_2/b_1) = \log (10^2) = 2$, because 10 must be raised to the second power to give 100. The magnitude difference is $2.5 \log (100) = 2.5 \times 2 = 5$. Similarly, if $b_2/b_1 = 0.001$, then $\log (b_2/b_1) = \log (0.001) = \log (10^{-3}) = -3$, and in this case the magnitude difference is $2.5 \times -3 = -7.5$ (the minus sign indicates that star 1 is brighter than star 2 in this example).

The method works equally well in cases where the power of 10 is not a whole number, as in the example where $b_2/b_1 = 48.76$. Here $\log (b_2/b_1) = \log (48.76) = 1.69$ (this is usually found by consulting tables of logarithms or by using a scientific calculator). In this example, the magnitude difference is $2.5 \log (b_2/b_1) = 2.5 \log(48.76) = 2.5 \times 1.69 = 4.23$, so star 1 is 4.23 magnitudes fainter than star 2.

The equation can be used in other ways as well; solving for b_2/b_1 yields

$$b_2/b_1 = 10^{(m_1-m_2)/2.5}$$
$$= 10^{0.4(m_1-m_2)}.$$

Thus, if we know that the magnitudes of two stars differ by $m_1 - m_2$, we multiply this difference by 0.4 and raise 10 to the power $0.4(m_1 - m_2)$, again using a calculator or tables, to get the brightness ratio b_2/b_1. As a simple example, suppose $m_1 - m_2 = 5$; then $0.4 (m_1 - m_2) = 0.4 \times 5 = 2$, and $10^{0.4(m_1-m_2)} = 10^2 = 100$, as we knew it should. As a more complex example, consider the stars Betelgeuse (magnitude $+0.41$) and Deneb (magnitude $+1.26$). From the equation above, we see that Betelgeuse is $10^{0.4(1.26-0.41)} = 10^{0.34} = 2.19$ times brighter than Deneb.

While in most cases it is possible to follow the discussions of magnitudes and brightness ratios in the text without using this exact mathmetical technique, it is still useful to be familiar with it.

The Distance Modulus

Whenever we know both the apparent and absolute magnitudes of a star, a comparison of the two will give its distance. In the text we have seen how to make this calculation by the following several steps:

1. Convert the difference $m - M$ between apparent and absolute magnitude into a brightness ratio, that is, a numerical factor indicating how much brighter or fainter the star would appear at 10 parsecs distance than at its actual distance.
2. Using the inverse square law, determine the change in distance required to produce this change in brightness.
3. Multiply this distance factor by 10 parsecs to find the distance to the star.

To do calculations mentally in this way can be laborious, especially in cases where the magnitude difference does not correspond neatly to a simple numerical factor, as it did in the examples given in the text. Hence astronomers use a mathematical equation expressing the relationship between distance and the distance modulus $m - M$. This equation, which works equally well for all cases, is

$$d = 10^{1+.2(m-M)},$$

where d is the distance in parsecs to a star whose apparent magnitude is m and whose absolute magnitude is M.

In a simple example, where $m = 9$ and $M = -6$, we have

$$d = 10^{1+.2\,(15)}$$
$$= 10^{1+3}$$
$$= 10^4 \text{ parsecs.}$$

Now let's try a more complex case. Suppose the star is an M2 main sequence star, so that $M = 13$, as found from the H-R diagram. The apparent magnitude is $m = 16$. Our equation tells us that

$$d = 10^{1+.2(16-13)}$$
$$= 10^{1+.6}$$
$$= 10^{1.6}$$
$$= 39.8 \text{ parsecs.}$$

It is necessary to use a slide rule, calculator, or mathematical table to do this of course, but it is still relatively straightforward compared to following the mental steps outlined above.

The Effect of Extinction on the Distance Modulus

When we discussed distance determination techniques (in Chapter 6), we ignored the effects of interstellar extinction. In any method that depends on the apparent brightness of a star, however, extinction can be important, particularly for very distant stars. Because the effect of extinction is to make a star appear fainter than it otherwise would, the tendency is to overestimate distances if no allowance is made for it.

Recall that in the spectroscopic parallax technique, the distance modulus $m - M$ is used to find the distance to a star from the equation given above:

$$d = 10^{1+.2(m-M)},$$

where m is the apparent magnitude, M is the absolute magnitude, and d is the distance to the star in parsecs.

To correct this equation for extinction, we add a term, A_v, which refers to the extinction (in magnitudes) in visual light. Thus if the extinction toward a particular star makes that star appear 2 magnitudes fainter than it otherwise would, $A_v = 2$. If we insert this into the equation, we find:

$$d = 10^{1+.2(m-M-A_v)}.$$

Let us consider a simple example, a star whose apparent magnitude is $m = 12.4$ and whose absolute magnitude is $M = 2.4$. First let us calculate its distance if extinction is ignored:

$$d = 10^{1+.2(12.4-2.4)} = 10^3 = 1000 \text{ parsecs.}$$

Now suppose it is determined that the extinction in the direction of this star amounts to one magnitude. Now the distance is

$$d = 10^{1+.2(12.4-2.4-1)} = 10^{2.8} = 631 \text{ parsecs.}$$

One magnitude is only a modest amount of extinction, yet by neglecting it, we overestimated the distance to this star by almost 60 percent. We can see from this example that extinction can have a drastic effect on distance estimates.

It is worthwhile to add a note about how the extinction A_v is determined. It is possible to determine how much redder, in terms of the $B - V$ color index, a star appears because of extinction. To carry that a step further, astronomers define a *color excess* called $E(B-V)$, which is the difference between the observed and intrinsic values of $B - V$:

$$E(B-V) = (B-V)_{\text{observed}} - (B-V)_{\text{intrinsic}}.$$

Studies of the variation of interstellar extinction with wavelength show that the extinction at the visual wavelength is approximately three times the color excess; that is:

$$A_v = 3E(B-V).$$

Hence determination of excess reddening leads to an estimate of A_v, and this in turn can be used in the modified equation for the distance.

Appendix 11 Nuclear Reactions in Stars

In the text we did not spell out the details of the reactions that occur in stellar cores, although they are quite simple. To do so here, we will use the notation of nuclear physics, which is basically a shorthand in symbols. For example, a helium nucleus, containing two protons and two neutrons, is designated ^4_2He, the subscript indicating the **atomic number** (the number of protons) and the superscript the **atomic weight** (the total number of protons and neutrons). Similarly, a hydrogen nucleus is $^{1}_{1}H$, and deuterium, a form of hydrogen with an extra neutron in the nucleus, is $^{2}_{1}H$. A special symbol (ν) is used for the **neutrino,** a massless subatomic particle emitted in some reactions, and e^{+} indicates a **positron,** which is equivalent to an electron but has a positive electrical charge. The symbol γ indicates a gamma ray, a very short wavelength photon of light emitted in some reactions.

The Proton-Proton Chain

Using this notation system, we can now spell out the proton-proton chain:

$$^{1}_{1}H + {}^{1}_{1}H \rightarrow {}^{2}_{1}H + e^{+} + \nu$$
$$^{2}_{1}H + {}^{1}_{1}H \rightarrow {}^{3}_{2}He + \gamma.$$

The $^{3}_{2}He$ particle is a form of helium, but not the common form. Once we have this particle it will combine with another:

$$^{3}_{2}He + {}^{3}_{2}He \rightarrow {}^{4}_{2}He + {}^{1}_{1}H + {}^{1}_{1}H.$$

We end up with a normal helium nucleus. A total of six hydrogen nuclei went into the reaction (remember, the first two steps had to occur twice, in order to produce two $^{3}_{2}He$ particles for the final reaction), while there were two left at the end, so the net result is the conversion of four hydrogen nuclei into one helium nucleus.

The CNO Cycle

The CNO cycle, which dominates at higher temperatures, is more complex, involving not only carbon but also nitrogen and oxygen. Each of these elements has more than one form, with differing numbers of neutrons. Some of these **isotopes** are unstable and spontaneously emit positrons, decaying into other species in the process. Here is the CNO cycle:

$$^{12}_{6}C + {}^{1}_{1}H \rightarrow {}^{13}_{7}N + \gamma$$
$$^{13}_{7}N \rightarrow {}^{13}_{6}C + e^{+} + \nu$$
$$^{13}_{6}C + {}^{1}_{1}H \rightarrow {}^{14}_{7}N + \gamma$$
$$^{14}_{7}N + {}^{1}_{1}H \rightarrow {}^{15}_{8}O + \gamma$$
$$^{15}_{8}O \rightarrow {}^{15}_{7}N + e^{+} + \nu$$
$$^{15}_{7}N + {}^{1}_{1}H \rightarrow {}^{12}_{6}C + {}^{4}_{2}He.$$

Here we end up with a helium nucleus and a carbon nucleus, while the particles going into the reaction were four hydrogen nuclei and a carbon nucleus. Along the way, three isotopes of nitrogen and one of oxygen were created and then converted into something else, leaving neither element at the end. As in the proton-proton chain, the net result is the conversion of four hydrogen nuclei into one helium nucleus.

The Triple-Alpha Reaction

Stars that have used up all the available hydrogen in their cores may become hot enough to undergo a new reaction in which helium is converted into carbon. Helium nuclei, consisting of two protons and two neutrons, are called alpha particles, and the reaction is called the triple-alpha reaction since three of these particles are involved:

$$^{4}_{2}He + {}^{4}_{2}He \rightarrow {}^{8}_{4}Be$$
$$^{8}_{4}Be + {}^{4}_{2}He \rightarrow {}^{12}_{6}C + \gamma.$$

In this reaction sequence, $^{8}_{4}Be$ is a form of the element beryllium, and the end product, $^{12}_{6}C$, is the most common form of carbon. The second step must follow very quickly after the first because $^{8}_{4}Be$ is unstable and will break apart into two $^{4}_{2}He$ particles in a very short time. Therefore the third $^{4}_{2}He$ particle must react with the $^{8}_{4}Be$ particle almost immediately so that this reaction sequence can be viewed as a three-particle reaction.

Other Reactions

Following helium burning in the triple-alpha reaction, other reactions can take place if the stellar core becomes hot enough. The first of these reactions are **alpha-capture reactions,** in which one form of nucleus adds an alpha particle to become a new form with two more protons and two additional neutrons. One example is the carbon alpha capture:

$$^{12}_{6}C + {}^{4}_{2}He \rightarrow {}^{16}_{8}O + \gamma,$$

in which $^{16}_{8}O$, the most common form of oxygen, is produced. In another sequence of alpha captures, nitrogen is converted into another isotope of oxygen, which can then be converted into neon:

$$^{14}_{7}N + {}^{4}_{2}He \rightarrow {}^{18}_{8}O + e^{+} + \nu,$$

and

$$^{18}_{8}O + {}^{4}_{2}He \rightarrow {}^{22}_{10}Ne + \gamma.$$

Additional alpha-capture reactions can occur, but at high enough temperatures other, more complex, reactions also take place. For example, two carbon nuclei can react, forming a number of different products, including sodium, neon, and magnesium. Two oxygen nuclei can also react, creating such species as sulfur, phosphorus, silicon, and magnesium. As a massive star evolves and its core contracts and heats following each reaction stage, a wide variety of elements is created with generally increasing atomic numbers. As explained in the text, the heaviest and most complex element produced in stable reactions in stellar cores is iron ($^{56}_{26}Fe$). Once a star has an iron core, it cannot undergo any further reaction stages without major disruption by a supernova explosion or collapse to a neutron star or black hole.

Appendix 12 Detected Interstellar Molecules

Number of Atoms	Symbol	Name	Number of Atoms	Symbol	Name
2	H_2	Molecular hydrogen	4	$HCNH^+$	Protonated hydrogen cyanide
	C_2	Diatomic carbon		HNCS	Isothiocyanic acid
	CH	Methylidyne		C_3N	Cyanoethynyl
	CH^+	Methylidyne ion		C_3O	Tricarbon monoxide
	CN	Cyanogen		H_2CS	Thioformaldehyde
	CO	Carbon monoxide	5	C_4H	Butadiynyl
	CS	Carbon monosulfide		C_3H_2	Cyclopropenylidene
	OH	Hydroxyl		HCO_2H	Formic acid
	NO	Nitric oxide		CH_2CO	Ketene
	NS	Nitrogen sulfide		HC_3N	Cyanoacetylene
	SiO	Silicon monoxide		NH_2CN	Cyanamide
	SiS	Silicon sulfide		CH_2NH	Methanimine
	SO	Sulfur monoxide		CH_4	Methane
3	H_2D^+	Protonated hydrogen deuteride		SiH_4	Silane*
	C_2H	Ethynyl	6	C_5H	Pentynylidyne*
	HCN	Hydrogen cyanide		CH_3OH	Methanol
	HNC	Hydrogen isocyanide		CH_3CN	Methyl cyanide
	HCO	Formyl		CH_3SH	Methyl mercaptan
	HCO^+	Formyl ion		NH_2CHO	Formamide
	N_2H^+	Protonated nitrogen	7	CH_2CHCN	Vinyl cyanide
	HNO	Nitroxyl		CH_3C_2H	Methylacetylene
	H_2O	Water		CH_3CHO	Acetaldehyde
	HCS^+	Thioformyl ion		CH_3NH_2	Methylamine
	H_2S	Hydrogen sulfide		HC_5N	Cyanodiacetylene
	OCS	Carbonyl sulfide	8	$HCOOCH_3$	Methyl formate
	SO_2	Sulfur dioxide		CH_3C_3N	Methylcyanoacetylene
	SiC_2	Silicon dicarbide*	9	CH_3C_4H	Methyldiacetylene
4	C_2H_2	Acetylene*		CH_3CH_3O	Dimethyl ether
	C_3H	Propynylidyne		CH_3CH_2CN	Ethyl cyanide
	H_2CO	Formaldehyde		CH_3CH_2OH	Ethanol
	NH_3	Ammonia		HC_7N	Cyano-hexa-tri-yne
	HNCO	Isocyanic acid	11	HC_9N	Cyano-octa-tetra-yne
	$HOCO^+$	Protonated carbon monoxide	13	$HC_{11}N$	Cyano-deca-penta-yne

Note: This list does not include isotopic variations (identical molecules except that one or more atoms are in rare isotopic forms, such as deuterium in place of hydrogen, or ^{13}C instead of the much more common ^{12}C).
*These species have been detected only in the dense circumstellar clouds surrounding red giant or supergiant stars.

Appendix 13 Clusters of Galaxies

Galaxies of the Local Group

Galaxy*	Type‡	Absolute Magnitude	Position Right Ascension	Declination
M31 (Andromeda)	Sb	−21.1	$00^h 40.^m 0$	+41° 00′
Milky Way	Sbc	−20.5	17 42.5	−28 59
M33 = NGC 598	Sc	−18.9	01 31.1	+30 24
Large Magellanic Cloud	Irr	−18.5	05 24	−69 50
IC 10	Irr	−17.6	00 17.6	+59 02
Small Magellanic Cloud	Irr	−16.8	00 51	−73 10
M32 = NGC 221	E2	−16.4	00 40.0	+40 36
NGC 205	E6	−16.4	00 37.6	+41 25
NCG 6822	Irr	−15.7	19 42.1	−14 53
NGC 185	Dwarf E	−15.2	00 36.1	+48 04
NGC 147	Dwarf E	−14.9	00 30.4	+48 14
IC 1613	Irr	−14.8	01 02.3	+01 51
WLM	Irr	−14.7	23 59.4	−15 44
Fornax	Dwarf sph	−13.6	02 37.5	−34 44
Leo A	Irr	−13.6	09 56.5	+30 59
IC 5152	Irr	−13.5	21 59.6	−51 32
Pegasus	Irr	−13.4	23 26.1	+14 28
Sculptor	Dwarf sph	−11.7	00 57.5	−33 58
And I	Dwarf sph	−11	00 42.8	+37 46
And II	Dwarf sph	−11	01 13.6	+33 11
And III	Dwarf sph	−11	00 32.7	+36 14
Aquarius	Irr	−11	20 44.1	−13 02
Leo I	Dwarf sph	−11	10 0.58	+12 33
Sagittarius	Irr	−10	19 27.1	−17 47
Leo II	Dwarf sph	−9.4	11 10.8	+22 26
Ursa Minor	Dwarf sph	−8.8	15 08.2	+67 18
Draco	Dwarf sph	−8.6	17 19.4	+57 58
Carina	Dwarf sph		06 40.4	−50 55
Pisces	Irr	−8.5	00 01.2	+21 37

*Galaxy names are derived from a variety of sources, including several catalogs (such as those designated M, NGC, and IC) and colloquial names bestowed by discoverers. Many in this list are simply named after the constellation where they are found.

‡The galaxy types listed here are described in the text, except for the *Dwarf sph* designation, which stands for "dwarf spheroidal" and refers to dwarf galaxies that do not fit easily into the designation of dwarf ellipticals. Note that the absolute magnitudes of some of these are comparable to those of the brightest individual stars in our galaxy.

Appendix 13 Clusters of Galaxies, continued

Clusters of Galaxies within 1000 Mpc*

Cluster	Distance	Radial Velocity	Diameter	Number of Galaxies	Density of Galaxies
Virgo	19 Mpc	1180 km/sec	4 Mpc	2500	500/Mpc^3
Pegasus I	65	3700	1	100	1100
Pisces	66	250	12	100	250
Cancer	80	4800	4	150	500
Perseus	97	5400	7	500	300
Coma	113	6700	8	800	40
UMa III	132		2	90	200
Hercules	175	10,300	0.3	300	
Cluster A	240	15,800	4	400	200
Centaurus	250		9	300	10
UMa I	270	15,400	3	300	100
Leo	310	19,500	3	300	200
Cluster B	330		4	300	200
Gemini	350	23,300	3	200	100
CrB	350	21,600	3	400	250
Bootes	650	39,400	3	150	100
UMa II	680	41,000	2	200	400
Hydra	1000	60,600			

*Data from Allen C. W., 1973, *Astrophysical quantities*, 3d ed. (London: Athlone Press).

Appendix 14 The Relativistic Doppler Effect

The simple formula given in the text relating the Doppler shift of an object's spectrum to its velocity is accurate only when the velocity is much less than the speed of light. That simple relation is

$$v = (\Delta\lambda/\lambda)c,$$

where v is the object's velocity, $\Delta\lambda$ is the shift in wavelength of a spectral line whose rest wavelength is λ, and c is the speed of light.

When v is a significant fraction of c, the correct equation must be used. The error in determining v caused by using the incorrect formula is 1 percent of v when $\Delta\lambda/\lambda$ is only 0.02; and the error becomes 5 percent when $\Delta\lambda/\lambda$ is 0.1. Thus failure to use the correct, relativistic formula becomes important for speeds of only a few percent of the speed of light.

For simplicity of notation, astronomers usually use the symbol z to represent the Doppler shift; that is,

$$z = \Delta\lambda/\lambda.$$

From Einstein's theory of special relativity, it can be shown that the correct relationship between the shift in wavelength and the speed of the emitting object is

$$z = \Delta\lambda/\lambda = \sqrt{\frac{1 + v/c}{1 - v/c}} - 1,$$

which leads to the following solution for the velocity:

$$v = c\left[\frac{(z+1)^2 - 1}{(z+1)^2 + 1}\right].$$

Let us apply this to the redshifts of quasars. One of the first quasars discovered, 3C273, has $z = 0.16$, which would imply a velocity of $0.16c = 48,000$ km/sec, if we used the simple, nonrelativistic equation. Use of the relativistic formula leads instead to $v = 0.147c = 44,200$ km/sec. In this case use of the wrong formula leads to an error of almost 9 percent. The quasars with the highest known redshifts have z greater than 4.0, which would lead to the nonsensical conclusion that their speeds are more than four times that of light, if the nonrelativistic equation were used. Use of the correct equation for $z = 4.0$ gives $v = 0.923c = 277,000$ km/sec, still an enormous velocity.

Appendix 15 The Messier Catalog

Number	Right Ascension	Declination	Magnitude	Description
M1	$05^h33.^m3$	$+22°01'$	11.3	Crab nebula in Taurus
M2	21 32.4	$-00\ 54$	6.3	Globular cluster in Aquarius
M3	13 41.3	$+28\ 29$	6.2	Globular cluster in Canes Venatici
M4	16 22.4	$-26\ 27$	6.1	Globular cluster in Scorpio
M5	15 17.5	$+02\ 07$	6	Globular cluster in Serpens
M6	17 38.9	$-32\ 11$	6	Open cluster in Scorpio
M7	17 52.6	$-34\ 48$	5	Open cluster in Scorpio
M8	18 02.4	$-24\ 23$		Lagoon nebula in Sagittarius
M9	17 18.1	$-18\ 30$	7.6	Globular cluster in Ophiuchus
M10	16 56.0	$-04\ 05$	6.4	Globular cluster in Ophiuchus
M11	18 50.0	$-06\ 18$	7	Open cluster in Scutum
M12	16 46.1	$-01\ 55$	6.7	Globular cluster in Ophiuchus
M13	16 41.0	$+36\ 30$	5.8	Globular cluster in Hercules
M14	17 36.5	$-03\ 14$	7.8	Globular cluster in Ophiuchus
M15	21 29.1	$+12\ 05$	6.3	Globular cluster in Pegasus
M16	18 17.8	$-13\ 48$	7	Open cluster in Serpens
M17	18 19.7	$-16\ 12$	7	Omega nebula in Sagittarius
M18	18 18.8	$-17\ 09$	7	Open cluster in Sagittarius
M19	17 01.3	$-26\ 14$	6.9	Globular cluster in Ophiuchus
M20	18 01.2	$-23\ 02$		Triffid nebula in Sagittarius
M21	18 03.4	$-22\ 30$	7	Open cluster in Sagittarius
M22	18 35.2	$-23\ 55$	5.2	Globular cluster in Sagittarius
M23	17 55.7	$-19\ 00$	6	Open cluster in Sagittarius
M24	18 17.3	$-18\ 27$	6	Open cluster in Sagittarius
M25	18 30.5	$-19\ 16$	6	Open cluster in Sagittarius
M26	18 44.1	$-09\ 25$	9	Open cluster in Scutum
M27	19 58.8	$+22\ 40$	8.2	Dumbbell nebula; planetary nebula in Vulpecula
M28	18 23.2	$-24\ 52$	7.1	Globular cluster in Sagittarius
M29	20 23.3	$+38\ 27$	8	Open cluster in Cygnus
M30	21 39.2	$-23\ 15$	7.6	Globular cluster in Capricornus
M31	00 41.6	$+41\ 09$	3.7	Andromeda galaxy
M32	00 41.6	$+40\ 45$	8.5	Elliptical galaxy, companion of M31
M33	01 32.8	$+30\ 33$	5.9	Spiral galaxy in Triangulum
M34	02 40.7	$+42\ 43$	6	Open cluster in Perseus
M35	06 07.6	$+24\ 21$	6	Open cluster in Gemini
M36	05 35.0	$+34\ 05$	6	Open cluster in Auriga
M37	05 51.5	$+32\ 33$	6	Open cluster in Auriga
M38	05 27.3	$+35\ 48$	6	Open cluster in Auriga
M39	21 31.5	$+48\ 21$	6	Open cluster in Cygnus
M40	12 20	$+59$		Double star cluster in Ursa Major
M41	06 46.2	$-20\ 43$	6	Open cluster in Canis Major
M42	05 34.4	$-05\ 24$		Orion nebula
M43	05 34.6	$-05\ 18$		Small extension of the Orion nebula
M44	08 38.8	$+20\ 04$	4	Praesepe; open cluster in Cancer
M45	03 46.3	$+24\ 03$	2	The Pleiades; open cluster in Taurus

(continued on next page)

Appendix 15 The Messier Catalog, continued

Number	Right Ascension	Declination	Magnitude	Description
M46	07 40.9	−14 46	7	Open cluster in Puppis
M47	07 35.6	−14 27	5	Open cluster in Puppis
M48	08 12.5	−05 43	6	Open cluster in Hydra
M49	12 28.8	+08 07	8.9	Elliptical galaxy in Virgo
M50	07 02.0	−08 19	7	Open cluster in Monocerotis
M51	13 29.0	+47 18	8.4	Whirlpool galaxy; spiral galaxy in Canes Venatici
M52	23 23.3	+61 29	7	Open cluster in Cassiopeia
M53	13 12.0	+18 17	7.7	Globular cluster in Coma Berenices
M54	18 53.8	−30 30	7.7	Globular cluster in Sagittarius
M55	19 38.7	−31 00	6.1	Globular cluster in Sagittarius
M56	19 15.8	+30 08	8.3	Globular cluster in Lyra
M57	18 52.9	+33 01	9.0	Ring nebula; planetary nebula in Lyra
M58	12 36.7	+11 56	9.9	Spiral galaxy in Virgo
M59	12 41.0	+11 47	10.3	Elliptical galaxy in Virgo
M60	12 42.6	+11 41	9.3	Elliptical galaxy in Virgo
M61	12 20.8	+04 36	9.7	Spiral galaxy in Virgo
M62	16 59.9	−30 05	7.2	Globular cluster in Scorpio
M63	13 14.8	+42 08	8.8	Spiral galaxy in Canes Venatici
M64	12 55.7	+21 48	8.7	Spiral galaxy in Coma Berenices
M65	11 17.8	+13 13	9.6	Spiral galaxy in Leo
M66	11 19.1	+13 07	9.2	Spiral galaxy, companion of M65
M67	08 50.0	+11 54	7	Open cluster in Cancer
M68	12 38.3	−26 38	8	Globular cluster in Hydra
M69	18 30.1	−32 23	7.7	Globular cluster in Sagittarius
M70	18 42.0	−32 18	8.2	Globular cluster in Sagittarius
M71	19 52.8	+18 44	6.9	Globular cluster in Sagitta
M72	20 52.3	−12 39	9.2	Globular cluster in Aquarius
M73	20 57.8	−12 44		Open cluster in Aquarius
M74	01 35.6	+15 41	9.5	Spiral galaxy in Pisces
M75	20 04.9	−21 59	8.3	Globular cluster in Sagittarius
M76	01 40.9	+51 28	11.4	Planetary nebula in Perseus
M77	02 41.6	−00 04	9.1	Spiral galaxy in Cetus
M78	05 45.8	+00 02		Emission nebula in Orion
M79	05 23.3	−24 32	7.3	Globular cluster in Lepus
M80	16 15.8	−22 56	7.2	Globular cluster in Scorpio
M81	09 54.2	+69 09	6.9	Spiral galaxy in Ursa Major
M82	09 54.4	+69 47	8.7	Irregular galaxy in Ursa Major
M83	13 35.9	−29 46	7.5	Spiral galaxy in Hydra
M84	12 24.1	+13 00	9.8	Elliptical galaxy in Virgo
M85	12 24.3	+18 18	9.5	Elliptical galaxy in Coma Berenices

Appendix 15 The Messier Catalog, continued

Number	Right Ascension	Declination	Magnitude	Description
M86	12 25.1	+13 03	9.8	Elliptical galaxy in Virgo
M87	12 29.7	+12 30	9.3	Giant elliptical galaxy in Virgo
M88	12 30.9	+14 32	9.7	Spiral galaxy in Coma Berenices
M89	12 34.6	+12 40	10.3	Elliptical galaxy in Virgo
M90	12 35.8	+13 16	9.7	Spiral galaxy in Virgo
M91				Not identified; possibly M58
M92	17 16.5	+43 10	6.3	Globular cluster in Hercules
M93	07 43.6	−23 49	6	Open cluster in Puppis
M94	12 50.1	+41 14	8.1	Spiral galaxy in Canes Venatici
M95	10 42.8	+11 49	9.9	Barred spiral galaxy in Leo
M96	10 45.6	+11 56	9.4	Spiral galaxy in Leo
M97	11 13.7	+55 08	11.1	Owl nebula; planetary nebula in Ursa Major
M98	12 12.7	+15 01	10.4	Spiral galaxy in Coma Berenices
M99	12 17.8	+14 32	9.9	Spiral galaxy in Coma Berenices
M100	12 21.9	+15 56	9.6	Spiral galaxy in Coma Berenices
M101	14 02.5	+54 27	8.1	Spiral galaxy in Ursa Major
M102				Not identified; possibly M101
M103	01 31.9	+60 35	7	Open cluster in Cassiopeia
M104	12 39.0	−11 35	8	Sombrero galaxy; spiral galaxy in Virgo
M105	10 46.8	+12 51	9.5	Elliptical galaxy in Leo
M106	12 18.0	+47 25	9	Spiral galaxy in Canes Venatici
M107	16 31.8	−13 01	9	Globular cluster in Ophiuchus
M108	11 10.5	+55 47	10.5	Spiral galaxy in Ursa Major
M109	11 56.6	+53 29	10.6	Barred spiral galaxy in Ursa Major

Glossary

Absolute magnitude The magnitude a star would have if it were precisely 10 parsecs away from the Sun.
Absolute zero The temperature where all molecular or atomic motion stops, equal to $-273°$ C or $-459°$ F.
Absorption line A wavelength at which light is absorbed, producing a dark feature in the spectrum.
Acceleration Any change—either of speed or direction—in the state of rest or motion of a body.
Accretion disk A rotating disk of gas surrounding a compact object (such as a neutron star or black hole), formed by material falling in.
Albedo The fraction of incident light that is reflected from a surface, such as that of a planet.
Alpha-capture reaction A nuclear fusion reaction in which an alpha particle merges with an atomic nucleus. A typical example is the formation of ^{16}O by the fusion of an alpha particle with ^{12}C.
Alpha particle A nucleus of ordinary helium containing two protons and two neutrons.
Amino acid A complex organic molecule of the type that forms proteins. Amino acids are fundamental constituents of all living matter.
Andromeda galaxy The large spiral galaxy located some 700,000 parsecs from the Sun; the most distant object visible to the unaided eye.
Angstrom The unit normally used in measuring wavelengths of visible and ultraviolet light; one angstrom is equal to 10^{-8} centimeter.
Angular diameter The diameter of an object as seen on the sky, measured in units of angle.
Angular momentum A measure of the mass, radius, and rotational velocity of a rotating or orbiting body. In the simple case of an object in circular orbit, the angular momentum is equal to the mass of the object times its distance from the center of the orbit times its orbital speed.
Annual motions Motions in the sky caused by the Earth's orbital motion about the Sun. These include the seasonal variations of the Sun's latitude and the Sun's motion through the zodiac.
Annular eclipse A solar eclipse that occurs when the Moon is near its greatest distance from the Earth, so that its angular diameter is slightly smaller than that of the Sun and a ring, or annulus, of the Sun's disk is visible surrounding the disk of the Moon.
Anticyclone A rotating wind system around a high-pressure area. On the Earth, an anticyclone rotates clockwise in the Northern Hemisphere and counterclockwise in the Southern Hemisphere.
Antimatter Matter composed of the antiparticles of ordinary matter. For each subatomic particle, there is an antiparticle that is its opposite in such properties as electrical charge but its equivalent in mass. Matter and antimatter, if combined, annihilate each other, producing energy in the form of gamma rays according to the formula $E = mc^2$.
Aphelion The point in the orbit of a solar system object where it is farthest from the Sun.
Apollo asteroid An asteroid whose orbit brings it closer to the Sun than 1 astronomical unit (AU).
Asteroid Any of the thousands of small, irregular bodies orbiting the Sun, primarily in the asteroid belt between the orbits of Mars and Jupiter.
Asthenosphere The deep portions of the Earth's mantle, below the zone (the lithosphere) where convection currents are thought to operate. The term is also applied to similar zones in the interiors of the Moon and other planets.
Astrology The ancient belief that earthly affairs and human lives are influenced by the positions of the Sun, Moon, and planets with respect to the zodiac.
Astrometric binary A double star recognized as such because the visible star or stars undergo periodic motion that is detected by astrometric measurements.
Astrometry The science of accurately measuring stellar positions.
Astronomical unit (AU) A unit of distance used in astronomy, equal to the average distance between the Sun and the Earth; 1 AU is equal to 1.4959787×10^8 kilometer.
Atomic number The number of protons in the nucleus of an element. The atomic number defines the identity of an element.
Atomic weight The mass of an atomic nucleus in atomic mass units [one atomic mass unit (amu) is defined as the average mass of the protons and neutrons in a nucleus of ordinary carbon, ^{12}C]. For most atoms, the atomic weight is approximately equal to the total number of protons and neutrons in the nucleus.
Autumnal equinox The point where the Sun crosses the celestial equator from north to south, around September 21. See also **equinox** and **vernal equinox**.

Barred spiral galaxy A spiral galaxy whose nucleus has linear extensions on opposing sides, giving it a barlike shape. The spiral arms usually appear to emanate from the ends of the bar.
Basalt An igneous silicate rock common in regions formed by lava flows on the Earth, the Moon, and probably on the other terrestrial planets.
Beta decay A spontaneous nuclear reaction in which a neutron decays into a proton and an electron, with a neutrino being emitted also. The term has been generalized to mean any spontaneous reaction in which an electron and a neutrino (or their antimatter equivalents) are emitted.

Big bang A term referring to any theory of cosmogony in which the universe began at a single point, was very hot initially, and has been expanding from that state since.

Binary star A double star system in which the two stars orbit a common center of mass.

Black hole An object that has collapsed under its own gravitation to such a small radius that its gravitational force traps photons of light.

Bolide An extremely bright meteor that explodes in the upper atmosphere.

Bolometric magnitude A magnitude in which all wavelengths of light are included.

Breccia Lunar rocks consisting of pebbles and soil fused together by meteorite impacts.

Brown Dwarf A dim object larger than an ordinary planet, but not massive enough to have internal nuclear reactions and become a star.

Burster A sporadic source of intense X rays, probably consisting of a neutron star onto which new matter falls at irregular intervals.

Carbonaceous chondrite A meteorite containing chondrules having a high abundance of carbon and other volatile elements. Carbonaceous chondrites, thought to be very old, have apparently been unaltered since the formation of the solar system.

Cassegrain focus A focal arrangement for a reflecting telescope in which a convex secondary mirror reflects the image through a hole in the primary mirror to a focus at the bottom of the telescope tube. This arrangement is commonly used in situations where relatively lightweight instruments for analyzing the light are attached directly to the telescope.

Catastrophic theory Any theory in which observed phenomena are attributed to sudden changes in conditions or to the intervention of an outside force or body.

Celestial equator The imaginary circle formed by the intersection of the Earth's equatorial plane with the celestial sphere. The celestial equator is the reference line for north–south (declination) measurements in the standard equatorial coordinate system.

Celestial pole The point on the celestial sphere directly overhead at either of the Earth's poles.

Celestial sphere The imaginary sphere formed by the sky. It is a convenient device for discussing and measuring the positions of astronomical objects.

Center of mass In a binary star system, or any system consisting of several objects, the point about which the mass is "balanced"; that is, the point that moves with a constant velocity through space while the individual bodies in the system move about it.

Cepheid variable A pulsating variable star, of a class named after the prototype δ Cephei. Cepheid variables obey a period-luminosity relationship and are therefore useful as distance indicators. There are two classes of Cepheid variables, the so-called classical Cepheids, which belong to Population I, and the Population II Cepheids, also known as W Virginis stars.

Chondrite A stony meteorite containing chondrules.

Chondrule A spherical inclusion in certain meteorites, usually composed of silicates and always of very great age.

Chromosphere A thin layer of hot gas just outside the photosphere in the Sun and other cool stars. The temperature in the chromosphere rises from about 4000 K at its inner edge to 10,000 or 20,000 K at its outer boundary. The chromosphere is characterized by the strong red emission line of hydrogen.

Closed universe A possible state of the universe in which the expansion will eventually be reversed and which is characterized by positive curvature, being finite in extent but having no boundaries.

CNO cycle A nuclear fusion reaction sequence in which hydrogen nuclei are combined to form helium nuclei, and other nuclei, such as isotopes of carbon, oxygen, and nitrogen, appear as catalysts or by-products. The CNO cycle is dominant in the cores of stars on the upper main sequence.

Color index The difference $B - V$ between the blue (B) and visual (V) magnitudes of a star. If B is less than V (that is, if the star is brighter in blue than in visual light), the star has a negative color index and is a relatively hot star. If B is greater than V, the color index is positive and the star is relatively cool.

Coma The extended glowing region that surrounds the nucleus of a comet.

Comet An interplanetary body, composed of loosely bound rocky and icy material, which forms a glowing head and extended tail when it enters the inner solar system.

Configuration The position of a planet or the Moon relative to the Sun–Earth line.

Conjunction The alignment of two celestial bodies on the sky. In connection with the planets, a conjunction is the alignment of a planet with the Sun, an inferior conjunction being the occasion when an inferior planet is directly between the Sun and the Earth, and a superior conjunction being the occasion when any planet is directly behind the Sun as seen from the Earth.

Constellation A prominent pattern of bright stars, historically associated with mythological figures. In modern usage each constellation incorporates a precisely defined region of the sky.

Continental drift The slow motion of the continental masses over the surface of the Earth, caused by the motions of the Earth's tectonic plates, which in turn are probably caused by convection in the underlying asthenosphere.

Continuous radiation Electromagnetic radiation that is emitted in a smooth distribution with wavelength, without spectral features such as emission and absorption lines.

Convection The transport of energy by fluid motions, occurring in gases, liquids, or semirigid material such as the Earth's mantle. These motions are usually driven by the buoyancy of heated material, which tends to rise while cooler material descends.

Co-orbital Satellites Two or more satellites following the same orbit around a planet. Normally one or more minor satellites orbit 60° ahead or 60° behind a major moon, because these positions are stabilized by the combined gravity of the planet and the major moon.

Corona The very hot, extended outer atmosphere of the Sun and other cool, main sequence stars. The high temperature in the corona ($1-2 \times 10^6$ K) is probably caused by the dissipation of mechanical energy from the convective zone just below the photosphere.

Cosmic background radiation The primordial radiation field that fills the universe, created in the form of gamma rays at the time of the big bang, but having cooled since so that today its temperature is 3 K and its peak wavelength is near 1.1 millimeters, in the microwave portion of the spectrum. Also known as the 3-degree background radiation.

Cosmic ray A rapidly moving atomic nucleus from space. Some cosmic rays are produced in the Sun, while others come from interstellar space and probably originate in supernova explosions.

Cosmogony The study of the origins of the universe.
Cosmological constant A term added to the field equations by Einstein to allow solutions in which the universe was static (neither expanding nor contracting). Although the need for the term disappeared when it was discovered that the universe is expanding, the cosmological constant is retained in the field equations by modern cosmologists but is usually assigned the value zero.
Cosmological principle The postulate, made by most cosmologists, that the universe is both homogeneous and isotropic. It is sometimes stated as, "The universe looks the same to all observers everywhere."
Cosmological redshift A Doppler shift toward longer wavelengths that is caused by a galaxy's motion of recession due to the expansion of the universe.
Cosmology The study of the universe as a whole.
Coudé focus A focal arrangement for a reflecting telescope in which the image is reflected by a series of mirrors to a remote, fixed location where a massive, immovable instrument can be used to analyze it.
Cyclone A rotating wind system about a low-pressure center, often associated with storms on the earth. On the Earth, cyclones rotate counterclockwise in the Northern Hemisphere and clockwise in the Southern Hemisphere.

Declination The coordinate in the equatorial system that measures positions in the north–south direction, with the celestial equator as the reference line. Declinations are measured in units of degrees, minutes, and seconds of arc.
Deferent The large circle centered on or near the Earth on which the epicycle for a given planet moved, in the geocentric theory of the solar system developed by ancient Greek astronomers such as Hipparchus and Ptolemy.
Degenerate gas A gas in which either free electrons or free neutrons are as densely spaced as allowed by laws of quantum mechanics. Such a gas has extraordinarily high density, and its pressure is not dependent on temperature as it is in an ordinary gas. Degenerate electron gas provides the pressure that supports white dwarfs against collapse, and degenerate neutron gas similarly supports neutron stars.
Deuterium An isotope of hydrogen containing one proton and one neutron in its nucleus.
Differential gravitational force A gravitational force acting on an extended object, so that the portions of the object closer to the source of gravitation feel a stronger force than the portions that are farther away. Such a force, also known as a tidal force, acts to deform or disrupt the object and is responsible for many phenomena, ranging from synchronous rotation of moons or double stars to planetary ring systems and the disruption of galaxies in clusters.
Differentiation The sinking of relatively heavy elements into the core of a planet or other body. Differentiation can only occur in fluid bodies, so any planet that has undergone this process must once have been at least partially molten.
Distance modulus The difference $m - M$ between the apparent and absolute magnitudes for a given star. This difference, which must be corrected for the effects of interstellar extinction, is a direct measure of the distance to the star.
Diurnal motion Any motion related to the rotation of the Earth. Diurnal motions include the daily risings and settings of all celestial objects.

Doppler effect The shift in wavelength of light caused by relative motion between the source of light and the observer. The Doppler shift, $\Delta\lambda$, is defined as the difference between the observed and rest (laboratory) wavelengths for a given spectral line.
Dwarf elliptical galaxy A member of a class of small spheroidal galaxies similar to standard elliptical galaxies except for their small size and low luminosity. Dwarf galaxies are probably the most common in the universe but cannot be detected at distances beyond the Local Group of galaxies.
Dwarf nova A close binary system containing a white dwarf in which material from the companion star falls onto the other at sporadic intervals, creating brief nuclear outbursts.

Eclipse An occurrence in which one object is partially or totally blocked from view by another or passes through the shadow of another.
Eclipsing binary A double star system in which one or both stars are periodically eclipsed by the other as seen from Earth. This situation can occur only when the orbital plane of the binary is viewed edge-on from the Earth.
Ecliptic The plane of the Earth's orbit about the Sun, which is approximately the plane of the solar system as a whole. The apparent path of the Sun across the sky is the projection of the ecliptic onto the celestial sphere.
Electromagnetic force The force created by the interaction of electric and magnetic fields. The electromagnetic force can be either attractive or repulsive and is important in countless situations in astrophysics.
Electromagnetic radiation Waves consisting of alternating electric and magnetic fields. Depending on the wavelength, these waves may be known as gamma rays, X rays, ultraviolet radiation, visible light, infrared radiation, or radio radiation.
Electromagnetic spectrum The entire array of electromagnetic radiation arranged according to wavelength.
Electron A tiny, negatively charged particle that orbits the nucleus of an atom. The charge is equal and opposite to that of a proton in the nucleus, and in a normal atom the number of electrons and protons is equal so that the overall electrical charge is zero. The electrons emit and absorb electromagnetic radiation by making transitions between fixed energy levels.
Ellipse A geometrical shape such that the sum of the distances from any point on it to two fixed points called foci is constant. In any bound system where two objects orbit a common center of mass, their orbits are ellipses with the center of mass at one focus.
Elliptical galaxy One of a class of galaxies characterized by smooth spheroidal forms, few young stars, and little interstellar matter.
Emission line A wavelength at which radiation is emitted, creating a bright line in the spectrum.
Emission nebula A cloud of interstellar gas that glows by the light of emission lines. The source of excitation that causes the gas to emit may be radiation from a nearby star or heating by any of a variety of mechanisms.
Endothermic reaction Any nuclear or chemical reaction that requires more energy to occur than it produces.
Energy The ability to do work. Energy can be in either kinetic form, when it is a measure of the motion of an object, or potential form, when it is stored but capable of being released into kinetic form.

Epicycle A small circle on which a planet revolves, which in turn orbits another, distant body. Epicycles were used in ancient theories of the solar system to devise a cosmology that placed the Earth at the center but accurately accounted for the observed planetary motions.

Equatorial coordinates The astronomical coordinate system in which positions are measured with respect to the celestial equator (in the north–south direction) and a fixed direction (in the east–west dimension). The coordinates used are declination (north–south, in units of angle) and right ascension (east–west, in units of time).

Equinox Either of two points on the sky where the planes of the ecliptic and the Earth's equator intersect. When the Sun is at one of these two points, the lengths of night and day on the Earth are equal. See also **autumnal equinox** and **vernal equinox**.

Erg A unit of energy equal to the kinetic energy of an object of 2 grams mass moving at a speed of 1 centimeter per second, but defined technically as the work required to accelerate at 1 cm/sec^2 a mass of 1 gram through a distance of 1 centimeter.

Escape velocity The velocity required for an object to escape the gravitational field of a body such as a planet. In a more technical sense, the escape velocity is the velocity at which the kinetic energy of the object equals its gravitational potential energy; if the object moves any faster, its kinetic energy exceeds its potential energy and it can escape the gravitational field.

Event horizon The "surface" of a black hole; the boundary of the region from within which no light can escape.

Evolutionary theory Any theory in which observed phenomena are thought to have arisen as a result of natural processes requiring no outside intervention or sudden changes.

Excitation A state in which one or more electrons of an atom or ion are in energy levels above the lowest possible one.

Fluorescence The emission of light at a particular wavelength following excitation of the electron by absorption of light at another, shorter wavelength.

Focus (1) The point at which light collected by a telescope is brought together to form an image; (2) one of two fixed points that define an ellipse (see also **ellipse**).

Force Any agent or phenomenon that produces acceleration of a mass.

Frequency The rate (in units of hertz, or cycles per second) at which electromagnetic waves pass a fixed point. The frequency, usually designated ν, is related to the wavelength λ and the speed of light c by $\nu = c/\lambda$.

Fusion reaction A nuclear reaction in which atomic nuclei combine to form more massive nuclei.

Galactic cluster A loose cluster of stars located in the disk or spiral arms of the galaxy.

Gamma ray A photon of electromagnetic radiation whose wavelength is very short and whose energy is very high. Radiation whose wavelength is less than one angstrom is usually considered to be gamma-ray radiation.

Gegenschein The diffuse glowing spot seen on the ecliptic opposite the Sun's direction, created by sunlight reflected off interplanetary dust.

Giant planets The large, gaseous planets Jupiter, Saturn, Uranus, and Neptune.

Globular cluster A large spherical cluster of stars located in the halo of the galaxy. These clusters, containing up to several hundred thousand members, are thought to be among the oldest objects in the galaxy.

Gram A unit of mass, equal to the quantity of mass contained in 1 cubic centimeter of water.

Granulation The spotty appearance of the solar surface (the photosphere) caused by convection in the layers just below.

Gravitational redshift A Doppler shift toward long wavelengths caused by the effect of a gravitational field on photons of light. Photons escaping a gravitational field lose energy to the field, and the redshift results.

Greatest elongation The greatest angular distance from the Sun that an inferior planet can reach, as seen from Earth.

Greenhouse effect The trapping of heat near the surface of a planet by atmospheric molecules (such as carbon dioxide) that absorb infrared radiation emitted by the surface.

H II region A volume of ionized gas surrounding a hot star (see also **emission nebula**).

H-R diagram See **Hertzsprung-Russell diagram**.

Half-life The time required for half of the nuclei of an unstable (radioactive) isotope to decay.

Halo (1) The extended outer portions of a galaxy (thought to contain a large fraction of the total mass of the galaxy, mostly in the form of dim stars and interstellar gas); (2) the extensive cloud of gas surrounding the head of a comet.

Helium flash A rapid burst of nuclear reactions in the degenerate core of a moderate mass star in the hydrogen shell-burning phase. The flash occurs when the core temperature reaches a sufficiently high temperature to trigger the triple-alpha reaction.

Hertz A unit of frequency used in describing electromagnetic radiation; 1 hertz (1 Hz) is equal to one cycle or wave per second.

Hertzsprung-Russell diagram A diagram on which stars are represented according to their absolute magnitudes (on the vertical axis) and spectral types (on the horizontal axis). Because the physical properties of stars are interrelated, they do not fall randomly on such a diagram but lie in well-defined regions according to their state of evolution. Very similar diagrams can be constructed using luminosity instead of absolute magnitude and temperature or color index in place of spectral type.

High-velocity star A star whose velocity relative to the solar system is large. As a rule, high-velocity stars are Population II objects following orbital paths that are highly inclined to the plane of the galactic disk.

Homogeneous Having the quality of being uniform in properties throughout. In astronomy this term is often applied to the universe as a whole, which is postulated to be homogeneous.

Horizontal branch A sequence of stars in the H-R diagram of a globular cluster, extending horizontally across the diagram to the left from the red giant region. These stars are probably undergoing helium burning in their cores by the triple-alpha reaction.

Hubble constant The numerical factor, usually denoted H, which describes the rate of expansion of the universe. It is the proportionality constant in the Hubble law $\nu = Hd$, which relates the speed of recession of a galaxy (ν) to its distance (d). The present value of H is not well known, with estimates ranging between 55 and 90 kilometers per second per megaparsec.

Hydrostatic equilibrium The state of balance between gravitational and pressure forces that exists at all points inside any stable object such as a star or planet.

Igneous rock A rock that was formed by cooling and hardening from a molten state.

Impact crater A crater formed on the surface of a terrestrial planet or a satellite by the impact of a meteoroid or planetesimal.

Inertia The tendency of an object to remain in its state of rest or of uniform motion. This tendency is directly related to the mass of the object.

Inferior planet One of the planets whose orbits lie closer to the Sun than that of the Earth (Mercury or Venus).

Infrared radiation Electromagnetic radiation in the wavelength region just longer than that of visible light; that is, radiation whose wavelength lies roughly between 7000 Å and 0.01 centimeter.

Interferometry The use of interference phenomena in electromagnetic waves to measure positions precisely or to achieve gains in resolution. Interferometry in radio astronomy entails the use of two or more antennae to overcome the normally very coarse resolution of a single radio telescope; in visible-light observations, the object is to eliminate the distorting effects of the Earth's atmosphere.

Interstellar cloud A region of relatively high density in the interstellar medium. Interstellar clouds have densities ranging between 1 and 10^6 particles per cubic centimeter and, in aggregate, contain most of the mass in interstellar space.

Interstellar extinction The obscuration of starlight by interstellar dust. Light is scattered off dust grains, so that a distant star appears dimmer than it otherwise would. The scattering process is most effective at short (blue) wavelengths, so that stars seen through interstellar dust are reddened and dimmed.

Inverse square law In general, any law describing a force or other phenomenon that decreases in strength as the square of the distance from some central reference point. In particular, the term *inverse square law* is often used by itself to mean the law stating that the intensity of light emitted by a source such as a star diminishes as the square of the distance from the source.

Ion Any subatomic particle with a nonzero electrical charge. In astronomy, the term *ion* is usually applied only to positively charged particles, such as atoms missing one or more electrons.

Ionization Any process by which an electron or electrons are freed from an atom or ion. Ionization generally occurs in two ways: by the absorption of a photon with sufficient energy or by collision with another particle.

Ionosphere The zone of the Earth's upper atmosphere, between 80 and 500 kilometers altitude, where charged subatomic particles (chiefly protons and electrons) are trapped by the Earth's magnetic field. See also **Van Allen belts.**

Io torus The ring of charged particles that follow the orbital path of Jupiter's moon Io.

Isotope Any form of a given chemical element. Different isotopes of the same element have the same number of protons in their nuclei, but different numbers of neutrons.

Isotropic Having the property of appearing the same in all directions. In astronomy this term is often postulated to apply to the universe as a whole.

Jovian planets The giant planets.

Kelvin A unit of temperature, equal to one hundredth of the difference between the freezing and boiling points of water, used in a scale whose zero point is absolute zero. A Kelvin is usually denoted simply by K.

Kiloparsec A unit of distance, equal to 1000 parsecs.

Kinetic energy The energy of motion. The kinetic energy of a moving object is equal to one-half times its mass times the square of its velocity.

Kirkwood's gaps Narrow gaps in the asteroid belt created by orbital resonance with Jupiter.

Light-gathering power The ability of a telescope to collect light from an astronomical source; the light-gathering power is directly related to the area of the primary mirror or lens.

Limb darkening The dark region around the edge of the visible disk of the Sun or of a planet caused by a decrease in temperature with height in the atmosphere.

Liquid metallic hydrogen Hydrogen in a state of semirigidity that can exist only under conditions of extremely high pressure, as in the interiors of Jupiter and Saturn.

Lithosphere The layer in the Earth, Moon, and terrestrial planets that includes the crust and the outer part of the mantle.

Local Group The cluster of about thirty galaxies to which the Milky Way belongs.

Logarithm The logarithm of a number is the power to which 10 must be raised to equal that number. For example, $100 = 10^2$; so the logarithm of 100 is 2.

Luminosity The total energy emitted by an object per second (the power of the object). For stars the luminosity is usually measured in units of ergs per second.

Luminosity class One of several classes to which a star can be assigned on the basis of certain luminosity indicators in its spectrum. The classes range from I for supergiants to V for main sequence stars (also known as dwarfs).

Lunar month The synodic period of the Moon, equal to 27 days, 7 hours, 43 minutes, 11.5 seconds.

L wave A type of seismic wave that travels only over the surface of the Earth.

Magellanic Clouds The two irregular galaxies that are the nearest neighbors to the Milky Way and are visible to the unaided eye in the Southern Hemisphere.

Magma Molten rock from the Earth's interior.

Magnetic braking The slowing of the spin of a young star such as the early Sun by magnetic forces exerted on the surrounding ionized gas.

Magnetic dynamo A rotating internal zone inside the Sun or a planet, thought to carry the electrical currents that create the solar or planetary magnetic field.

Magnetosphere The region surrounding a star or planet that is permeated by the magnetic field of that body.

Magnitude A measure of the brightness of a star, based on a system established by Hipparchus in which stars were ranked according to how bright they appeared to the unaided eye. In the modern system, a difference of five magnitudes corresponds exactly to a brightness ratio of 100, so that a star of a given magnitude has a brightness that is $100^{1/5} = 2.512$ times that of a star one magnitude fainter.

Main sequence The strip running from upper left to lower right in the H-R diagram, where most stars that are converting hydrogen to helium by nuclear reactions in their cores are found.

Main sequence fitting A distance-determination technique in which an H-R diagram for a cluster of stars is compared with a standard H-R diagram to establish the absolute magnitude scale for the cluster H-R diagram.

Main sequence turnoff In an H-R diagram for a cluster of stars, the point where the main sequence turns off toward the upper right. The main sequence turnoff, showing which stars in the cluster have evolved to become red giants, is an indicator of the age of the cluster.

Mantle The semirigid outer portion of the Earth's interior, extending from roughly the midway point nearly to the surface and consisting of the mesosphere (the lower portion) and the asthenosphere.

Mare (*pl.* maria) Any of several extensive, smooth lowland areas on the surface of the Moon or Mercury that were created by extensive lava flows early in the history of the solar system.

Mass-to-light ratio The mass of a galaxy, in units of solar masses, divided by its luminosity, in units of the Sun's luminosity. The mass-to-light ratio is an indicator of the relative quantities of Population I and Population II stars in a galaxy.

Mean solar day The average length of the solar day as measured throughout the year; the mean solar day is precisely 24 hours.

Megaparsec (Mpc) A unit of distance, equal to 10^6 parsecs.

Meridian The great circle on the celestial sphere that passes through both poles and directly overhead; that is, the north–south line directly overhead.

Mesosphere (1) The layer of the Earth's atmosphere between roughly 50 and 80 kilometers in altitude, where the temperature decreases with height; (2) the layer below the asthenosphere in the Earth's mantle.

Metamorphic rock A rock formed by heat and pressure in the Earth's interior.

Meteor A bright streak or flash of light created when a meteoroid enters the Earth's atmosphere from space.

Meteorite The remnant of a meteoroid that survives a fall through the Earth's atmosphere and reaches the ground.

Meteoroid A small interplanetary body.

Meteor shower A period during which meteors are seen with high frequency, occurring when the Earth passes through a swarm of meteoroids.

Micrometeorite A microscopically small meteorite.

Microwave background See **cosmic background radiation**.

Milky Way Historically, the diffuse band of light stretching across the sky; our cross-sectional view of the disk of our galaxy. In modern usage, the term *Milky Way* refers to our galaxy as a whole.

Minor planets Asteroids.

Neutrino A subatomic particle without mass or electrical charge that is emitted in certain nuclear reactions.

Neutron A subatomic particle with no electrical charge and a mass nearly equal to that of the proton. Neutrons and protons are the chief components of the atomic nucleus.

Neutron star A very compact, dense stellar remnant whose interior consists entirely of neutrons and which is supported against collapse by degenerate neutron gas pressure.

Newtonian focus A focal arrangement for reflecting telescopes in which a flat mirror is used to reflect the image through a hole in the side of the telescope tube.

Nonthermal radiation Radiation not due only to the temperature of an object. The term is most often applied to sources of continuous radiation such as synchrotron radiation.

Nova A star that temporarily flares up in brightness, most likely as the result of nuclear reactions caused by the deposition of new nuclear fuel on the surface of a white dwarf in a binary system. See also **recurrent nova**.

Nucleus The central, dense concentration in an atom, a comet, or a galaxy.

OB association A group of young stars whose luminosity is dominated by O and B stars.

Occam's razor The principle that the simplest explanation of any natural phenomenon is most likely the correct one. *Simple* in this case usually means requiring few assumptions or unverifiable postulates.

Oort cloud The cloud of bodies, hypothesized to be orbiting the Sun at a great distance, from which comets originate.

Open universe A possible state of the universe in which its expansion will never stop; it is characterized by negative curvature, being infinite in extent and having no boundaries.

Opposition A planetary configuration in which a superior planet is positioned in exactly the opposite direction from the Sun as seen from Earth.

Optical binary A pair of stars that happen to appear near each other on the sky but are not in orbit; not a true binary.

Orbital resonance A situation in which the periods of two orbiting bodies are simple multiples of each other so that they are frequently aligned and gravitational forces due to the outer body may move the inner body out of its original orbit. This is one mechanism thought responsible for creating the gaps in the rings of Saturn and Kirkwood's gaps in the asteroid belt.

Organic molecule Any of a large class of carbon-bearing molecules found in living matter.

Outgassing The release of gases from a planet's interior, through volcanic processes or gradual evaporation or seepage.

Paleomagnetism Vestigial traces or artifacts of ancient magnetic fields.

Parallax Any apparent shift in position caused by an actual motion or displacement of the observer. See also **stellar parallax**.

Parsec A unit of distance, equal to the distance to a star whose stellar parallax is 1 arcsecond. A parsec is equal to 206, 265 AU, 3.03×10^{13} kilometers, or 3.26 light-years.

Peculiar velocity The deviation in a star's velocity from perfect circular motion about the galactic center.

Penumbra (1) The light, outer portion of a shadow, such as the portion of the Earth's shadow where the Moon is not totally obscured during a lunar eclipse; (2) the light, outer portion of a sunspot.

Perfect cosmological principle The postulate, adopted by advocates of the so-called steady state theory of cosmology, stating that the universe is homogeneous and isotropic with respect to space and time. It is commonly stated as, "The universe looks the same to all observers everywhere, in all directions, and at all times."

Perihelion The point in the orbit of any Sun-orbiting body where it most closely approaches the Sun.

Permafrost A permanent layer of ice just below the surface of certain regions on the Earth and probably on Mars.

Photometer A device, usually using a photoelectric cell, for measuring the brightnesses of astronomical objects.

Photon A particle of light having wave properties but also acting as a discrete unit.

Photosphere The visible surface layer of the Sun and stars; the layer from which continuous radiation escapes and where absorption lines form.

Planck constant The numerical factor h relating the frequency ν of a photon to its energy E in the expression $E = h\nu$. The Planck constant has the value $h = 6.62620 \times 10^{-27}$ erg second.

Planck function (also known as the Planck law) The mathematical expression describing the continuous thermal spectrum of a glowing object. For a given temperature, the Planck function specifies the intensity of radiation as a function of either frequency or wavelength.

Planetary nebula A cloud of glowing, ionized gas, usually taking the form of a hollow sphere or shell, ejected by a star in the late stages of its evolution.

Planetesimal A small (diameter up to several hundred kilometers) solar system body of the type that first condensed from the solar nebula. Planetesimals are thought to have been the principal bodies that combined to form the planets.

Plate tectonics A general term referring to the motions of lithospheric plates over the surface of the Earth or other terrestrial planets. See also **continental drift.**

Polarization The preferential alignment of the magnetic and electric fields in electromagnetic radiation.

Population I The class of stars in a galaxy with relatively high abundances of heavy elements. These stars are generally found in the disk and spiral arms of spiral galaxies and are relatively young. The term *Population I* is also commonly applied to other components of galaxies associated with the star formation, such as the interstellar material.

Population II The class of stars in a galaxy with relatively low abundances of heavy elements. These stars are generally found in a spheroidal distribution about the galactic center and throughout the halo and are relatively old.

Positron A subatomic particle with the same mass as the electron but with a positive electrical charge; the antiparticle of the electron.

Potential energy Energy that is stored and may be converted into kinetic energy under certain circumstances. In astronomy the most common form of potential energy is gravitational potential energy.

Precession The slow shifting of star positions on the celestial sphere caused by the 26,000-year periodic wobble of the Earth's rotation axis.

Primary mirror The principal light-gathering mirror in a reflecting telescope.

Prime focus The focal arrangement in a reflecting telescope in which the image is allowed to form inside the telescope structure at the focal point of the primary mirror, so that no secondary mirror is needed.

Prograde motion Orbital or spin motion in the "normal" direction; in the solar system, this is counterclockwise as viewed from above the North Pole.

Proper motion The motion of a star across the sky, usually measured in units of arcseconds per year.

Proton-proton chain The sequence of nuclear reactions in which four hydrogen nuclei combine, through intermediate steps involving deuterium and ^{3}He, to form one helium nucleus. The proton-proton chain is responsible for energy production in the cores of stars on the lower main sequence.

Protostar A star in the process of formation, specifically, one that has entered the slow gravitational contraction phase.

Pulsar A rapidly rotating neutron star that emits periodic pulses of electromagnetic radiation, probably by the emission of beams of radiation from the magnetic poles. These pulses sweep across the sky as the star rotates.

P wave A seismic wave that is a compressional, or density, wave. P waves can travel through both solid and liquid portions of the Earth and are the first to reach any remote location from an earthquake site.

QSO See **quasi-stellar object.**

Quadrature The configuration where a superior planet or the Moon is 90 degrees away from the Sun, as seen from the Earth.

Quantum The amount of energy associated with a photon, equal to $h\nu$, where h is the Planck constant, and ν is the frequency. The quantum is the smallest amount of energy that can exist at a given frequency.

Quantum mechanics The physics of atomic structure and the behavior of subatomic particles, based on the principle of the quantum.

Quasar See **quasi-stellar object.**

Quasi-stellar object Any of a class of extragalactic objects characterized by emission lines with very large redshifts. The quasi-stellar objects are thought to lie at great distances, in which case they existed only at earlier times in the history of the universe; they may be young galaxies.

Radiation pressure Pressure created by the forces exerted by photons of light when they are absorbed or reflected.

Radiative transport The transport of energy, inside a star or in other situations, by radiation.

Radioactive dating A technique for estimating the age of material such as rock, based on the known initial isotopic composition and the known rate of radioactive decay for unstable isotopes initially present.

Radio galaxy Any of a class of galaxies whose luminosity is greatest in radio wavelengths. Radio galaxies are usually large elliptical galaxies, with synchrotron radiation emitted from one or more pairs of lobes located on opposite sides of the visible galaxy.

Ray A bright streak of ejecta emanating from an impact crater, especially on the Moon or on Mercury.

Recurrent nova A star known to flare up in nova outbursts more than once. A recurrent nova is thought to be a binary system containing a white dwarf and a mass-losing star, in which the white dwarf sporadically flares up when material falls onto it from the companion.

Red giant A star that has completed its core hydrogen-burning stage and has begun hydrogen shell-burning, which causes its outer layers to become very extended and cool.

Reflecting telescope A telescope that brings light to a focus by using mirrors.

Reflection nebula An interstellar cloud containing dust that shines by light reflected from a nearby star.

Refracting telescope A telescope that uses lenses to bring light to a focus.

Refractory The property of being able to exist in solid form under conditions of very high temperature. A refractory element is one that is characterized by a high temperature of vaporization; refractory elements are the first to condense into solid form when a gas cools, as in the solar nebula.

Regolith The layer of debris on the surface of the Moon created by the impact of meteorites; the lunar surface layer.

Resolution In an image, the ability to separate closely spaced features, that is, the clarity or fineness of the image. In a spectrum, the ability to separate features that are close together in wavelength.

Retrograde motion Orbital or spin motion in the opposite direction from prograde motion; in the solar system, retrograde motions are clockwise as seen from above the North Pole.

Right ascension The east–west coordinate in the equatorial coordinate system. The right ascension is measured in units of hours, minutes, and seconds to the east from a fixed direction on the sky, which itself is defined as the line of intersection of the ecliptic and the celestial equator.

Rille A type of winding, sinuous valley commonly found on the Moon.

Roche limit The point near a massive body such as a planet or star inside which the tidal forces acting on an orbiting body exceed the gravitational force holding it together. The location of the Roche limit depends on the size of the orbiting body.

RR Lyrae variable A member of a class of pulsational variable stars named after the prototype star, RR Lyrae. These stars are blue-white giants with pulsational periods of less than one day and are Population II objects found primarily in globular clusters.

Saros cycle An eighteen-year, eleven-day repeating pattern of solar and lunar eclipses caused by a combination of the tilt of the lunar orbit with respect to the ecliptic and the precession of the plane of the Moon's orbit.

Scattering The random reflection of photons by particles such as atoms or ions in a gas or dust particles in interstellar space.

Schwarzschild radius The radius within which an object has collapsed at the point when light can no longer escape the gravitational field as the object becomes a black hole.

Secondary mirror The second mirror in a reflecting telescope (after the primary mirror), usually either convex in shape, to reflect the image out a hole in the bottom of the telescope to the cassegrain focus, or flat, to reflect the image out of the side of the telescope to the Newtonian focus or along the telescope mount axis to the coudé focus.

Sedimentary rock A rock formed by the deposition and hardening of layers of sediment, usually either underwater or in an area subject to flooding.

Seeing The blurring and distortion of point sources of light, such as stars, caused by turbulent motions in the Earth's atmosphere.

Seismic wave A wave created in a planetary or satellite interior, usually caused by an earthquake.

Selection effect The tendency for a conclusion based on observations to be influenced by the method used to select the objects for observation. An example was the early belief that all quasars are radio sources, when the principal method used to discover quasars was to look for radio sources and then see if they had other properties associated with quasars.

Semimajor axis One half of the major, or long, axis of an ellipse.

Seyfert galaxy Any of a class of spiral galaxies, first recognized by Carl Seyfert, with unusually bright blue nuclei.

Shear wave A wave that consists of transverse motions, that is, motions perpendicular to the direction of wave travel.

Shepherd satellites A pair of minor satellites in orbits near each other, whose combined gravitational forces trap particles between them maintaining a ring there.

Sidereal day The rotation period of the Earth with respect to the stars or as seen by a distant observer, equal to 23 hours, 56 minutes, 4.091 seconds.

Sidereal period The orbital or rotational period of any object with respect to the fixed stars or as seen by a distant observer.

Solar day The synodic rotation period of the Earth with respect to the Sun, that is, the length of time from one local noon, when the Sun is on the meridian, to the next local noon.

Solar flare An explosive outburst of ionized gas from the Sun, usually accompanied by X-ray emission and the injection of large quantities of charged particles into the solar wind.

Solar motion The deviation of the Sun's velocity from perfect circular motion about the center of the galaxy; that is, the Sun's peculiar velocity.

Solar nebula The primordial gas and dust cloud from which the Sun and the planets condensed.

Solar wind The stream of charged subatomic particles flowing steadily outward from the Sun.

Solstice The occasion when the Sun, as viewed from the Earth, reaches its farthest northern (summer solstice) or southern (winter solstice) point.

Spectrogram A photograph of a spectrum.

Spectrograph An instrument for recording the spectra of astronomical bodies or other sources of light.

Spectroscope An instrument allowing an observer to view the spectrum of a source of light.

Spectroscopic binary A binary system recognized as a binary because its spectral lines undergo periodic Doppler shifts as the orbital motions of the two stars cause them to move toward and away from the Earth. If lines of only one star are seen, it is a single-lined spectroscopic binary; if lines of both stars are seen, it is a double-lined spectroscopic binary.

Spectroscopic parallax The technique of distance determination for stars in which the absolute magnitude is inferred from the H-R diagram and then compared with the observed apparent magnitude to yield the distance.

Spectroscopy The science of analyzing the spectra of stars or other sources of light.

Spectrum An arrangement of electromagnetic radiation according to wavelength.

Spectrum binary A binary system recognized as a binary because its spectrum contains lines of two stars of different spectral types.

Spin-orbit coupling A simple relationship between the orbital and spin periods of a satellite or planet, caused by tidal forces that have slowed the rate of rotation of the orbiting body. Synchronous rotation is the simplest and most common form of spin-orbit coupling.

Spiral density wave A spiral wave pattern in a rotating, thin disk, such as the rings of Saturn or the plane of a spiral galaxy like the Milky Way.

Spiral galaxy Any of a large class of galaxies exhibiting a disk with spiral arms.

Standard candle A general term for any astronomical object, the absolute magnitude of which can be inferred from its other observed characteristics and which is therefore useful as a distance indicator.

Steady state theory The theory of cosmology in which the universe is thought to have had no beginning and is postulated not to change with time.

Stefan-Boltzmann law The law of continuous radiation stating that for a spherical glowing object such as a star, the luminosity is proportional to the square of the radius and the fourth power of the temperature.

Stefan's law An experimentally derived law of continuous radiation stating that the energy emitted by a glowing body per square centimeter of surface area is proportional to the fourth power of the absolute temperature.

Stellar parallax The apparent annual shifting of position of a nearby star with respect to more distant background stars. The term *stellar parallax* is often assumed to mean the parallax angle, which is one-half of the total angular motion a star undergoes. See also **parallax** and **parsec**.

Stellar wind Any stream of gas flowing outward from a star, including the very rapid winds from hot, luminous stars, the intermediate-velocity, rarefied winds from stars like the Sun, and the slow, dense winds from cool supergiant stars.

Stratosphere The layer of the Earth's atmosphere between 10 and 50 kilometers in altitude, where the temperature increases with height.

Subduction The process in which one tectonic plate is submerged below another along a line where two plates collide. A subduction zone is usually characterized by a deep trench and an adjoining mountain range.

Supercluster A cluster of clusters of galaxies.

Supergiant A star in its late stages of evolution that is undergoing shell burning and is therefore very extended in size and very cool; an extremely large giant.

Supergranulation The pattern of large cells seen in the Sun's chromosphere when viewed in the light of the strong emission line of ionized hydrogen.

Superior planet Any planet whose orbit lies beyond the Earth's orbit around the Sun.

Supernova The explosive destruction of a massive star that occurs when all sources of nuclear fuel have been consumed and the star collapses catastrophically.

S wave A type of seismic wave that is a transverse, or shear, wave and can travel only through rigid materials.

Synchronous rotation A situation in which the rotational and orbital periods of an orbiting body are equal so that the same side is always facing the companion object.

Synchrotron radiation Continuous radiation produced by rapidly moving electrons traveling along lines of magnetic force.

Synodic period The orbital or rotational period of an object as seen by an observer on the Earth; for the Moon or a planet the synodic period is the interval between repetitions of the same phase or configuration.

Tectonic activity Geophysical processes involving motions of tectonic plates and associated volcanic and earthquake activity. See also **plate tectonics** and **continental drift**.

Temperature A measure of the internal energy of a substance. At the atomic level, temperature is related to the motions of atoms and molecules that make up the substance.

Terrestrial planets Relatively small, rocky planets; specifically Mercury, Venus, Earth, and Mars.

Thermal radiation Continuous radiation emitted by any body whose temperature is above absolute zero.

Thermal spectrum The spectrum of continuous radiation from a body that glows because it has a temperature above absolute zero. A thermal spectrum is one that is described mathematically by the Planck function.

Thermosphere The layer of the Earth's atmosphere above the mesosphere, extending upward from a height of 80 kilometers, where the temperature rises with altitude.

3-degree background The radiation field filling the universe, left over from the big bang. See also **cosmic background radiation**.

Tidal force A gravitational force that tends to stretch or distort an extended object. See **differential gravitational force**.

Total eclipse Any eclipse in which the eclipsed body is totally blocked from view or totally immersed in shadow.

Transit telescope A telescope designed to point straight overhead and accurately measure the times at which stars cross the meridian.

Transverse wave See **shear wave**.

Triple-alpha reaction A nuclear fusion reaction in which three helium nuclei (or alpha particles) combine to form a carbon nucleus.

Trojan asteroid Any of several asteroids orbiting the sun at stable positions in the orbit of Jupiter, either 60° ahead of the planet or 60° behind it.

Troposphere The lowest temperature zone in the earth's atmosphere, extending from the surface to a height of about 10 kilometers, in which the temperature decreases with altitude.

T Tauri star A young star still associated with the interstellar material from which it formed, typically exhibiting brightness variations and a stellar wind.

24-hour anisotrophy The daily fluctuation in the observed temperature of the cosmic background radiation caused by a combination of the Earth's motion with respect to the background and its rotation.

21-centimeter line The emission line of atomic hydrogen, whose wavelength is 21.11 centimeters, emitted when the spin of the electron reverses itself with respect to that of the proton. The 21-centimeter line is the most widely used and effective means of tracing the distribution of interstellar gas in the Milky Way and other galaxies.

Ultraviolet The portion of the electromagnetic spectrum between roughly 100 and 3000 angstroms.

Umbra (1) The dark inner portion of a shadow, such as the part of the Earth's shadow where the moon is in total eclipse during a lunar eclipse; (2) the dark central portion of a sunspot.

Van Allen belts Zones in the Earth's magnetosphere where charged particles are confined by the Earth's magnetic field. There are two main belts, one centered at an altitude of roughly 1.5 times the Earth's radius, and the other between 4.5 and 6.0 times the Earth's radius.

Velocity curve A plot showing the orbital velocity of stars in a spiral galaxy versus distance from the galactic center.

Velocity dispersion A measure of the average velocity of particles or stars in a group or cluster with random internal motions. In globular clusters and elliptical galaxies, the velocity dispersion can be used to infer the central mass.

Vernal equinox The occasion when the Sun crosses the celestial equator from south to north, usually occurring around March 21. See also **autumnal equinox** and **equinox**.

Visual binary Any binary system in which both stars can be seen through a telescope or on photographs.
Volatile Easily vaporized; volatile elements stay in gaseous form except at very low temperatures and did not condense into solid form during the formation of the solar system.

Wavelength The distance between wavecrests in any type of wave.
White dwarf The compact remnant of a low-mass star, supported against further gravitational collapse by the pressure of the degenerate electron gas that fills its interior.
Wien's law An experimentally discovered law applicable to thermal continuum radiation; it states that the wavelength of maximum emission intensity is inversely proportional to the absolute temperature.

X ray A photon of electromagnetic radiation in the wavelength interval between about 1 and 100 angstroms.

Zeeman effect The broadening or splitting of spectral lines due to the presence of a magnetic field in the gas where the lines are formed.
Zenith The point directly overhead.
Zero-age main sequence The main sequence in the H-R diagram, formed by stars that have just begun their hydrogen-burning lifetimes and have not yet converted any significant fraction of their core mass into helium. The zero-age main sequence forms the lower left boundary of the broader band representing the general main sequence.
Zodiac A band circling the celestial sphere along the ecliptic, broad enough to encompass the paths of all the planets visible to the naked eye; in some usages the sequence of constellations lying along the ecliptic.
Zodiacal light A diffuse band of light visible along the ecliptic near sunrise and sunset, created by sunlight scattered off interplanetary dust.

Index

absorption lines, *see* spectral lines
acceleration, 49–50
accretion disks, 147–148, 185
accretion theory of solar system formation, 317
active galactic nuclei, *see* galaxies, active nuclei
Adams, J. C., 374
Alfven, H., 317
Algol, 164–165
α Canis Majoris, *see* Sirius
alpha-capture reactions, 145, 160, 459–460
α Orionis *see* Betelgeuse
American Astronomical Society, 6
amino acids, 422, 431
Anaxagoras, 30
Anaximander, 30
Andromeda galaxy, 13–14, 244, 245, 247–248, 262
angular diameter, 28
angular momentum, 55, 316
annual motions, 20–23
antarctic circle, *see* latitude zones
Apollo, 411
Apollo asteroids, 411
Apollo program, 338
Apollanius, 30
arctic circle, *see* latitude zones
Ariel, 394, 397
Aristarchus, 30, 33
Aristotle, 30, 33, 64, 413–414
Arrhenius, S., 430
Ashbrook, J., 165
asteroids, 311, 409–413
asthenosphere, 331
Astrea, 409
astrology, 5, 8
astrometry, 115–116
astronomers, 6–7
astronomical unit (AU), 12, 43
astronomy
 history of, 29–36, 39–48
 nature of, 4–5
astrophysics, 4
astrophysicists, 7
atomic structure, 71–72
aurorae, 334

Babylonian astronomy, 29
Barnard's star, 324
barred spiral galaxies, *see* galaxies, spiral, barred
β Orionis, *see* Rigel
β Persei, *see* Algol
Betelgeuse, 32, 131
big bang, *see* universe, evolution of
binary stars, *see* stars, binary
binary x-ray sources, *see* stars, binary, x-ray
black dwarfs, 177
black holes, 97, 164–165, 188–191, 278, 289
BL Lac objects, 282
blueshift, *see* Doppler effect
Bohr, N., 71, 94
Boss General Catalog, 117
Brahe, T., 40–41, 182, 414
brecchia, *see* Moon, surface of
brown dwarfs, 324
Buffon, G. L., 316
bursters, 187
Bunsen, R., 70
butterfly diagram, 107, 108

Callisto, 391
Cannon, A. J., 120, 121, 123
carbonaceous chondrites, 422
cassegrain focus, 78
Cassini, G. D., 389
Cassini division, *see* Saturn, rings of
Cassiopeia A, 183
cataclysmic variables, 180
catastrophic theory of solar system formation, 316–317
celestial equator, *see* equatorial coordinates
celestial navigation, 19
celestial sphere, 9–10, 11
Centaurus A, 273, 275
center of mass, 55
Cepheid variables, 125, 198–200
Ceres, 409
Chamberlain, T. C., 316
Chandrasekhar, S., 178
Charon, 405
chondrites, 422

chondrules, 422
chromosphere
 of the Sun, *see* Sun, chromosphere of
 of stars, *see* stars, chromospheres of
cluster variables, *see* RR Lyrae stars
CNO cycle, 145, 159, 459
color index, 119, 129
Comet Giaccobini-Zinner, 418
comets, 323, 413–421
 orbits of, 414–415
 origin and evolution of, 415, 419–421
 structure of, 417–420
Comet Mrkos, 420
Comet Rendezvous and Flyby mission, 378
Comet West, 417
configurations
 of the Moon, *see* Moon, configurations of
 of the planets, *see* planets, configurations of
constellations, 32, A11–A12
continental drift, *see* tectonic activity
continuous radiation, 67–70
convection, 147
co-orbital satellites, 395–396, 398
Copernicus, N., 7, 39–40
Copernicus satellite, 206, 302
corona
 of the Sun, *see* Sun, corona of
 of stars, *see* stars, coronae of
coronal holes, 104
Cosmic Background Explorer, 439
cosmic background radiation, 266–269
 discovery of, 266–267
 isotropy of, 268–269
 spectrum of, 267–268
cosmic rays, 333
cosmic strings, 254
cosmogony, 293
cosmological constant, 296
Cosmological Principle, 294
cosmology, *see* universe, evolution of
coude focus, 78–79
Crab nebula, 181, 182
CRAF mission, *see Comet Rendezvous and Flyby* mission
craters, *see* entry for surface of body in question

I1

Cygnus A, 273, 275
Cygnus X-1, 191

Darwin, C., 430
Da Vinci, L., 65
declination, see equatorial coordinates
degenerate electron gas, 156, 177
degenerate neutron gas, 161, 183
Deimos, 370
δ Cephei stars, see Cepheid variables
density wave theory, see spiral density waves
deoxyribonucleic acid, 432–433
De Revolutionibus, 40
Descartes, R., 65, 312–313
de Sitter, W., 296
detectors, 78
deuterium, 302, 317
Dialogue on the Two Chief World Systems, 46–48
Dicke, R. H., 267
differential gravitational force, see tidal force
differentiation (see also internal structure of individual bodies), 332, 356
Dione, 393, 394, 411
disk population, see stellar populations
distance modulus, 131–132, 457
distance pyramid, 265–266
diurnal motions, 17–18
DNA, see deoxyribonucleic acid
Doppler effect, 74–75, 135
  relativistic, 280, 462
Drake, F. D., 437
Drake equation, 437–440
Draper, H., 122

Earth, 328–337
  atmosphere, 328–330
  basic properties, 328
  evolution of, 345
  interior, 330–333
  magnetic field, 333–334
  surface, 334, 336
eclipses, 28–29
  annular, 28
  lunar, 28
  solar, 28, 29
ecliptic, 7, 21
Einstein, A., 56, 94, 188, 280, 294, 296
Einstein Observatory, 283
electromagnetic force, 51
electromagnetic radiation, 65
electromagnetic spectrum, 63–67
elementary particles, 303, 306
elements, 446
elements, formation of, 303–304
ellipse, 43
emission lines, see spectral lines
Enceladus, 393, 395

energy, 54
  kinetic, 54
  potential, 54
epicycles, 33–35, 39
equator, see latitude zones
equatorial coordinates, 20, 21
equinoxes, 23
equivalence principle, 294–295
Eratosthenes, 30
Eros, 411
escape speed, 58, 357
Eudoxus, 30
Europa, 391
event horizon, 189
evolutionary theory of solar system formation, 313, 318–323
Ewen, H. I., 210
excitation, 73–74
extinction, see interstellar extinction

field equations, 294–296
Fleming, W., 122
force, 50–51
Fraunhofer, J., 70
Fraunhofer lines, 70
frequency, 63
Friedmann, A., 296
fusion reactions, see nuclear reactions

galaxies, 233–255, 273–279
  active nuclei, 273–278
  classification of, 233–236
  clusters of, 14, 243–250, 461–462
    Local Group, 244–248, 269
    masses of, 250
    origins of, 252–255
  colors of, 242
  discovery of, 233
  distances of, 237–240
  distribution of, 251–252, 253–254
  dwarf elliptical, 242, 246
  elliptical, 233, 234
  evolution of, 243
  halos of, 243
  irregular, 235
  jets from, 274–278
  masses of, 240–242
  mass-to-light ratios of, 242
  origins of, 243, 252–254
  radio, 273–278
  rotation curves of, 241
  Seyfert, 278–279, 286
  S0, 234
  spectra of, 242
  spiral, 233–234
    barred, 225, 234, 237
    conversion to ellipticals, 248–249
  starburst, 228

  superclusters of, 14, 243, 251–252
Galaxy, the, see Milky Way galaxy
Galen, 64
Galilean satellites of Jupiter, 389, 390–392
Galileo, g., 44–48, 339, 389, 396
Galileo spacecraft, 378
gamma rays, 65, 67
Gamow, G., 266–267
Ganymede, 391
gegenschein, 424–426
general relativity, 56–57, 188–189, 280, 294–296
giant planets, 311, 315, 322–323, 374–404
Ginga, 169, 173
Giotto mission, 418
Global Oscillation Network Group, 100
globular clusters, see stars, clusters of, globular
Goodricke, J., 164
Grand Unified Theory, 51, 303
gravitation, 52–54
gravitational lenses, 287–288
gravitational redshift, see redshift, gravitational
Great Red Spot, 382
Greek astronomy, 29–36
greenhouse effect, 353
Grimaldi, F., 339
GUT, see Grand Unified Theory
gyroscopes, 19

Halley, E., 48, 414
Halley's comet, 7, 414, 415, 416, 418, 419
harmonic law, see Kepler's laws
Hawking, S., 299
helium flash, 157
Henry Draper Catalog, 122–123, 124
Herschel, W., 164, 374, 389
Hertzsprung, E., 122–123, 129
Hertzsprung-Russell diagram, 129–132
  of clusters of stars, 141–142
high-velocity stars, see stars, high-velocity
Hipparchus, 30, 34, 116
Hippocrates, 64
Homer, 30, 31, 32
horizontal branch, 157, 214
Hoyle, F., 317, 432–433
H-R diagram, see Hertzsprung-Russell diagram
HII regions, 207
Hubble, E., 233, 234, 259
Hubble constant, 262–264, 298
Hubble Space Telescope, 82, 84, 86, 169, 215, 324
Huber, D., 164
Huygens, C., 63, 65, 376, 389
hydrostatic equilibrium, 92, 142
Hyperion, 393, 395

Iapetus, 393, 395
ICE mission, *see* International Cometary Explorer
inertia, 49
Infrared Astronomical Satellite, 82, 426, 427
infrared radiation, 64, 67
interferometry, 76, 81–84, 132
International Cometary Explorer, 418
International Halley Watch, 418
International Ultraviolet Explorer, 82, 84, 169, 173
interplanetary dust, 424–426
interstellar dust, 200–201
interstellar extinction, 203
interstellar gas, 203–211
interstellar medium, 203–211
intracluster gas, 249–250
inverse Compton scattering, 277
inverse square law, 69–70
Io, 391, 392
Io torus, 387
ionization, 72–73
ionosphere, 333
IRAS *see* Infrared Astronomical Satellite
IUE, *see* International Ultraviolet Explorer

Jeans, J. H., 316
Jeffreys, H., 316
Jet Propulsion Laboratory, 377
jets, from galaxies, *see* galaxies, jets from
JPL, *see* Jet Propulsion Laboratory
Jupiter
  atmosphere of, 382–383
  basic properties, 374
  internal structure of, 385–387
  magnetic field of, 387, 388
  radiation belts of, 387, 388
  ring of, 397, 398
  satellites of, 389–391

Kant, I., 313
Kapteyn, J. C., 200, 201
Keck Telescope, 81, 84
Kepler, J., 41–44, 65, 182, 352
Kepler's laws, 43–44, 54–58, 133, 202–203, 240
Kirchhoff, G., 70, 72
Kirchhoff's laws, 72
Kirkwood, D., 412
Kirkwood's gaps, 412–413
Kitt Peak National Observatory, 81, 83

Lagrange, J. L., 411
laminated terrain, *see* Mars, surface of
LaPlace, P. S. de, 313
lasers, 95, 441
latitude zones, 23
Leavitt, H., 123

LeMaitre, G., 296
leptons, 303
Leverrier, U., 374
life
  extraterrestrial, 435–442
    communication with, 440–442
    probability of, 436–440
    search for, 354
  origin of, 430–435
limb darkening, 103
Lindblad, B., 201
liquid metallic hydrogen, 381
lithosphere, 331
Local Group of galaxies, *see* galaxies, clusters of, Local Group
long-period variables, 125
Lowell, P., 374
luminosity, *see* stars, luminosity of
luminosity classes, 130–131
lunar month, *see* synodic period, of the Moon
Lyttleton, R. A., 317

Magellanic Clouds, 13, 169, 171, 246–247
magnetic braking, 318
magnetic monopoles, 301
magnitudes, 116–120, 128–129, A13–A14
  absolute, 128–129
  apparent, 116–120
  B and V, 119
  bolometric, 119–120
magnetosphere, 332
main sequence, *see* Hertzsprung-Russell diagram
main sequence fitting, 140
main sequence turn-off, 141
mantle, 331
maria, *see* Moon, surface of
Mariner program, 354, 356
Mars
  atmosphere of, 362
  basic properties, 351, 355–356
  evolution of, 370
  interior of, 368–369
  life on, 354, 436
  satellites of, 370
  seasons of, 362
  surface of, 365
  water on, 361
Mars Observer mission, 355
mass, 49
mass exchange, *see* stars, binary, mass exchange in
mass loss, *see* stellar winds
mass-luminosity relation, 134
Mauna Kea Observatory, 81
Maunder, E. W., 111
Maury, A., 122–123
Maxwell, J. C., 50–51

mean solar day, 18
mechanics, 45, 49–59
M87, *see* Virgo A
Mercury
  basic properties, 351, 356
  interior of, 369
  magnetic field of, 369
  orbit of, 352
  surface of, 367-368
mesosphere, 329, 331–332
Messier Catalog, A19–A21
meteorites, 413, 421, 422–423
meteors, 7, 421–424
meteoroids, 421
meteor showers, 423
microwave background, *see* cosmic background radiation
Milky Way galaxy, 6, 9, 12, 197–215
  center of, *see* nucleus of
  chemical enrichment of, 221, 226
  evolution of, 226
  formation of, 226
  halo of, 197, 214–215
  mass of, 202–203
  motion of, 269
  nucleus of, 197, 211–213, 286
  rotation of, 202
  rotation curve of, 202
  spiral arms of, 209–211, 222–225
  structure of, 197–215
  Sun's location in, 200–201
Miller, S., 431
Mimas, 393, 394
minerals, 334
minor planets, *see* asteroids
Miranda, 394–397
Mira variables, *see* long-period variables
molecules, interstellar, 207, 460
Montanari, G., 164
moon, 9, 337–347
  basic properties, 337
  configurations of, 24–26
  evolution of, 345–347
  exploration of, 337–338
  interior of, 343–345
  motions of, 24–26
  phases of, 24–26
  surface of, 341–343
Moulton, F. R., 316
moving cluster method, 265–266
Multiple-Mirror Telescope, 78, 80, 81

Narlikar, J. V., 432
NASA, *see* National Aeronautics and Space Administration
National Aeronautics and Space Administration, 7
National New Technology Telescope, 77

National Science Foundation, 7
nebulae, 207, 233
Nelson, G. D., 499
Neptune,
  atmosphere of, 385
  basic properties, 374
  discovery of, 374
  rings of, 397
  satellites of, 395
Nereid, 395
neutrinos, 96, 97–98, 171, 172, 299
neutron stars, 161, 182–187
Newton, I., 5, 48–49, 52, 63, 65, 414
Newtonian focus, 76
Newton's laws of motion, 49–59
NNTT, see National New Technology Telescope
novae, 179–180
nuclear reactions, 94–98, 143–146, A14–A16
  endothermic, 162–163
  fission, 94
  fusion, 94–98, 143–146
  in stars, 143–146
  in the Sun, 94–98

OB associations, see stars, clusters of, OB associations
Oberon, 393, 396
Occam's Razor, 191
Olbers, W., 295
Olber's paradox, 295
Oort, J., 201, 211, 415
Oort cloud, 415
orbital resonance, 391, 400
orbits, 54–58, 358–359

paleomagnetism, 333
Pallas, 409
panspermia, 430, 432–433
parallax, 34–35, 41, 116
  spectroscopic, 132
  stellar, 34–35, 41
parsec, 116
pertide physics, 306
peculiar velocity, 202
Penzias, A., 267
period-luminosity relation (for variable stars), 123, 198–200
permafrost, 361
Philolaus, 30
Phobos, 370
photometer, 78, 117–118
photon, 64
photosphere
  of the Sun, see Sun, photosphere of
  of stars, see stars, photospheres of
Piazzi, G., 409
Pickering, E. C., 122

Pioneer program, 106, 355, 356, 376–378
Pioneer-Venus, 355, 356, 418
Planck, M., 69, 94
Planck function, A5
planetary nebulae, 158–159
planetary systems, search for, 324
planets
  configurations of, 26–28
  motions of, 26–28, 43–44, 311
  summary of data on, A6–A7
planetesimals, 322, 345
plate tectonics, see tectonic activity
Plato, 30, 31–33
Pleiades cluster, 8
Pluto, 404–405
  atmosphere, 404–405
  basic properties, 404–405
  discovery of, 374–377
  orbit of, 404
  satellite of, 405
polarization, 66
Population I, see stellar populations
Population II, see stellar populations
Population III, see stellar populations
power, 54
precession, 20, 21
Principia, 48, 49
Project Cyclops, 441–442
proton-proton chain, 96, 143–144, 459
protostars, see stars, formation of
Ptolemy, C., 30, 35–36, 64, 117
pulsars (see also neutron stars), 183–185
Purcell, E. M., 210
Pythagoras, 30, 31

QSO's, see quasars
quantum, 69
quantum mechanics, 71
quarks, 303
quasistellar objects, see quasars
quasars, 279–289
  discovery of, 279–280
  energy source of, 289
  explanation of, 285–289
  properties of, 283–285
  redshifts of, 279–283
  spectra of, 279, 284–285

radioactive dating, 336
radioactivity, 336
radiation pressure, 147, 420
radio astronomy, see telescopes, radio
redshifts (see also Doppler effect)
  cosmological, 280–283
  gravitational, 177, 280
refractory elements, 322
regolith, see Moon, surface of

relativity, see general relativity or special relativity
resolution, 76
retrograde motion, 27–28
Rhea, 393, 394
ribonucleic acid, 432–433
Riccioli J., 339
right ascension, see equatorial coordinates
rilles, see Moon, surface of
RNA, see ribonucleic acid
Roche limit, 389
rotation curves, see Milky Way galaxy, rotation curve of, or galaxies, rotation curves of
RR Lyrae stars, 125, 198–200
Rudolfine Tables, 44
Russell, H. N., 129, 178, 317

Sakigake mission, 418
Saros, 28
satellites, basic data on, 451–452
Saturn
  atmosphere of, 382–383
  basic properties, 374
  interior of, 386–387
  magnetic field, 387
  rings of, 376, 396, 397–403
  satellites of, 391–393
Schmidt, O. Y., 317
Schwarzschild radius, 188–189
Search for Extraterrestrial Intelligence, 442
seasons
  on the Earth, 21–23
  on Mars, 362
Secchi, Fr. A., 120, 122
seismic waves, 330–331
selection effects, 281, 283
SETI, see Search for Extraterrestrial Intelligence
Sewall, W., 164
sextant, 115
Shapley, H., 200, 201, 211
Shelton, I., 169
shepherd satellites, 400, 401, 404
sidereal day, 19
sidereal period
  of the Moon, 23
  of a planet, 27
singularity, 189
Sirius, 178
Sirius B, 178
Slipher, V. M., 259
Smithsonian Astrophysical Observatory Catalog, 279
Solar constant, 91
solar flares, 109–110
solar motion, 202

solar nebula, 319
solar oscillations, see Sun, pulsations of
solar system, origin of, 312–324
solar wind, 105–107, 333
Space Telescope, see Hubble Space Telescope
special relativity, 56–57
spectral classes, 120–124
spectral lines, 67, 70–75
spectrographs, 78
spectroscopic parallax, 132
spectroscopy, 63, 120–125
spin-orbit coupling, 352
spiral arms, see Milky Way galaxy, spiral arms of, or galaxies, spiral
spiral density waves, 223–224, 401
standard candles, 237
stars
    atmospheres of, 142
    binary, 125–128
        astrometric, 126
        eclipsing, 126–127
        mass exchange in, 147–148
        spectroscopic, 127
        spectrum, 127
        visual, 125
        x-ray, 185–186, 190–191
    brightest, A9
    brightnesses of, see magnitudes
    catalogs of, 117
    chromospheres of, 147, 149
    closest, A10
    clusters of, 139–141
        galactic, 140
        globular, 140, 213–214
        OB associations, 140
        open, 140
    composition of, 134–135
    coronae of, 147, 149
    diameters of, 132
    distances of, 116–117
    energy production in, 143–146
    energy transport in, 146–147
    evolution of, 153–167
        in close binary systems, 165–167
        massive stars, 162–165
        moderately massive stars, 159–162
        one solar mass stars, 154–159
    formation of, 149–153
    fundamental properties of, 128–135
    giant, 130
    high-velocity, 202, 221
    lifetimes of, 146
    luminosities of, 54, 128–129
    magnetic fields of, 135
    masses of, 133
    models of, 142–143, 144
    motions of, 201, 202
    observations of, 115–128
    photospheres of, 142
    populations of, see stellar populations
    pulsations of, 124, 125
    remnants of, 176–192
    rotation of, 135
    spectra of, 120–125
        peculiar, 124–125
    structure of, 141–149
    supergiant, 130
    temperatures of, 129
    variable, 124, 125, 197–200
    winds from, see stellar winds
Stefan's law, see Stefan-Boltzmann law
Stefan-Boltzmann law, 68–69, A4
stellar magnitudes, see magnitudes
stellar parallax, see parallax, stellar
stellar populations 219–221, 226
stellar winds, 147
Stonehenge, 5
stratosphere, 329
strong nuclear force, 51
sublimation, 419
Sun, 91–112
    activity cycle of, 107–112
    atmosphere of, 98–105
    basic properties, 91
    chromosphere of, 98, 103
    composition of, 93
    corona of, 98, 103–105
    energy generation in, 92, 93–98
    formation and evolution of, 318–323
    internal structure of, 91–94
    magnetic field in, 107–109
    photosphere of, 98–103
    pulsations of, 99–100
    rotation of, 93–94, 99–100
sunspots, 107–108
supernovae, 161, 162–164, 180
Supernova 1987A, 169–174
supernova remnants, 181–182
Suisei mission, 418
synchronous rotation (*see also* spin-orbit coupling), 347
synchrotron radiation, 182, 275–276
synodic period
    of the Moon, 23–24
    of a planet, 26, 27

tectonic activity (*see also* entries for individual bodies), 334–337, 368
telescopes, 45, 75–84, A5–A6
    gamma-ray, 80, 82
    infrared, 79, 82
    radio, 76, 79–80, 81–82, 87
    reflecting, 77–78
    refracting, 77, 81
    transit, 18
    ultraviolet, 79, 82, 84
    x-ray, 80, 82, 83
temperature scales, 447
terrestrial planets, 311, 314, 322, 351–371
Tethys, 393, 394
Thales, 30, 31
thermal radiation, 67–70
thermosphere, 329
3C48, 279–280
3C273, 279–280, 283
3-degree background, see cosmic background radiation
tidal forces, 58–59
tides, 58–59
timekeeping, 18
titan, 391–393
Titania, 393, 396
Tombaugh, C., 375
triple-alpha reaction, 145, 459
Triton, 395
Trojan asteroids, 411
tropics, see latitude zones
troposphere, 329
T Tauri stars, 149–150, 323
T Tauri wind, 323
Tully-Fisher method, 239–240
tuning fork diagram, see galaxies, classification of
24-hour anisotropy, see cosmic background radiation, isotropy of
21-cm emission line, 209–211
Tycho, see Brahe, T.

UFO's, see unidentified flying objects
ultraviolet radiation, 64, 67
Umbriel, 394, 396
unified field theory, see Grand Unified Theory
universe
    age of, 262–264
    big-bang theory of, 296–306
    evolution of, 293–306
        early history of, 303–304, 306
        future of, 304–305
    expansion of, 259–266
        deceleration of, 300–303
    density of, 298–299
    homogeneity of, 293–294
    inflationary theory of, 301, 303
    isotropy of, 293–294
Uranus
    atmosphere of, 383–385
    basic properties, 374
    inclination of, 312, 383–385
    interior of, 387
    magnetic field of, 387
    rings of, 396–397
    satellites of, 393–394
Urey, H., 431

Van Allen belts, 333
Vega missions, 418
velocity dispersion, 241, 250
Venera program, 356
Venus
    atmosphere of 353, 360, 363–364
    basic properties, 351, 355–356
    evolution of, 371
    rotation of, 352–353
    surface of, 365–367
Very Large Array, 76, 81, 212
Very Long Baseline Interferometry, 77
Vesta, 409
Viking program, 354, 356
Virgo A, 274

volatile elements, 322
von Weiszacker, C. F., 317–318
Voyager program, 376–378

wavelength, 63, 67
weakly-interacting massive particles, 97–98
weak nuclear force, 51
weather, 329
weird terrain, *see* Mercury, surface of
Whipple, F., 417
white dwarfs, 131, 159, 161, 176–180
Wickramasinghe, N. C., 432
Wien, W., 68
Wien's law, 68, A3

Wilson, R., 267
WIMP's *see* weakly-interacting massive particles
Wolf-Rayet stars, 162

x-ray binaries, *see* stars, binary, x-ray
x rays, 65, 67

Yale Bright Star Catalog, 117

Zeeman effect, 107, 135
zero-age main sequence, 155
zodiac, 21, 22
zodiacal light, 424

THE NIGHT SKY IN SEPTEMBER

Latitude of chart is 34°N, but it is practical throughout the continental United States.

To use: Hold chart vertically and turn it so the direction you are facing shows at the bottom.

Chart time (Local Standard):
  10 p.m. First of month
  9 p.m. Middle of month
  8 p.m. Last of month

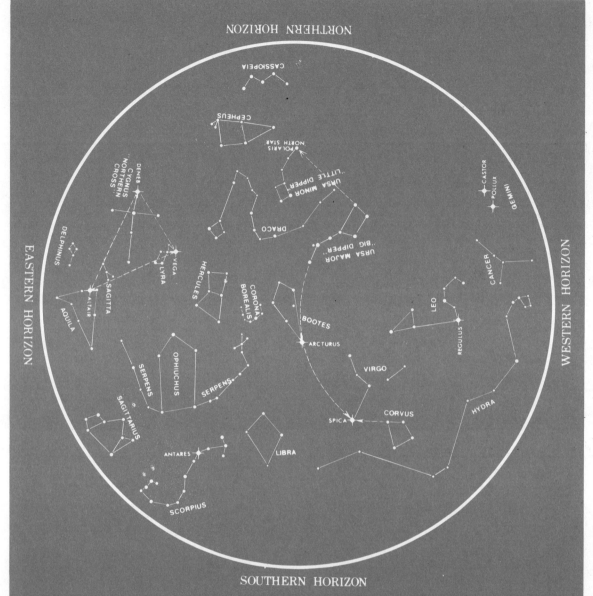

THE NIGHT SKY IN JUNE

Latitude of chart is 34°N, but it is practical throughout the continental United States.

To use: Hold chart vertically and turn it so the direction you are facing shows at the bottom.

Chart time (Local Standard):
  10 p.m. First of month
  9 p.m. Middle of month
  8 p.m. Last of month